OXFORD STUDIES IN PROBABILITY

OXFORD STUDIES IN PROBABILITY

---

1. F. B. Knight: *Foundations of the prediction process*
2. A. D. Barbour, L. Holst, and S. Janson: *Poisson approximation*
3. J. F. C. Kingman: *Poisson processes*
4. V. V. Petrov: *Limit theorems of probability theory*
5. M. Penrose: *Random geometric graphs*

Random Geometric Graphs

MATHEW PENROSE

University of Durham

OXFORD
UNIVERSITY PRESS

OXFORD
UNIVERSITY PRESS

Great Clarendon Street, Oxford OX2 6DP

Oxford University Press is a department of the University of Oxford.
It furthers the University's objective of excellence in research, scholarship,
and education by publishing worldwide in

Oxford New York

Auckland Bangkok Buenos Aires Cape Town Chennai
Dar es Salaam Delhi Hong Kong Istanbul Karachi Kolkata
Kuala Lumpur Madrid Melbourne Mexico City Mumbai Nairobi
São Paulo Shanghai Taipei Tokyo Toronto

Oxford is a registered trade mark of Oxford University Press
in the UK and in certain other countries

Published in the United States
by Oxford University Press Inc., New York

First published 2003

A catalogue record for this title is available from the British Library

Library of Congress Cataloging in Publication Data
(Data available)

ISBN 0 19 850626 0

10 9 8 7 6 5 4 3 2 1

Typeset by the author using LaTeX

Printed in Great Britain
on acid-free paper by
T. J. International Ltd, Padstow

PREFACE

Random geometric graphs are easily described. A set of points is randomly scattered over a region of space according to some probability distribution, and any two points separated by a distance less than a certain specified value are connected by an edge. This book is an attempt to describe the mathematical theory of the resulting graphs and to give a flavour of some of the applications.

I started to contemplate writing this book in the summer of 1998, when it occurred to me, firstly, that random geometric graphs are a natural alternative to the classical Erdös–Rényi random graph schemes, and secondly, that an account of them in monograph form could provide a useful collection of techniques in geometric probability. Although the project has taken longer than expected, I hope these assumptions retain their force, and that resulting book can be useful both to mathematicians with an interest in geometrical probability, and to practitioners in various subjects including communications engineering, classification, and computer science, wishing to see how far the mathematical theory has progressed.

This monograph is self-contained, and could be used as the basis of a graduate-level course (or courses). An overview of the topics covered appears in Section 1.4. The reader will find proofs in the text, and will find prior knowledge of the probabilistic concepts briefly reviewed in Section 1.6 to be useful. Other preliminaries are minimal; a small number of results in adjacent subjects such as measure theory, topology, and graph theory are used and are stated without proof in the text.

With regard to citations, I have tried to provide the most useful references for the reader, without always giving full historical details. Thus, there may be some results for which the reader is referred to some standard text, rather than to the original work containing those results. Likewise, any claims made regarding the novelty of work in this book are necessarily subject to the limits of my own knowledge, and I apologize in advance to the authors of any relevant works which I have failed to mention through ignorance. References to related work are generally given in the notes the end of each chapter, along with relevant open problems.

It is a pleasure to thank the following people and institutions for their assistance. The Fields Institute in Toronto provided hospitality for ten weeks in the spring of 1999. Jordi Petit provided the software used to produce diagrams of random geometric graphs in this book. Pauline Coolen-Schrijner, Joseph Yukich, and Andrew Wade read and commented on earlier drafts of some of the chapters;

however, I wish to take full credit myself for any remaining errors, which I intend to monitor on a web page if and when they come to light.

Durham, UK M.P.
September 2002

CONTENTS

Notation xi

1 Introduction 1
1.1 Motivation and history 1
1.2 Statistical background 4
1.3 Computer science background 7
1.4 Outline of results 9
1.5 Some basic definitions 11
1.6 Elements of probability 14
1.7 Poissonization 18
1.8 Notes and open problems 21

2 Probabilistic ingredients 22
2.1 Dependency graphs and Poisson approximation 22
2.2 Multivariate Poisson approximation 25
2.3 Normal approximation 27
2.4 Martingale theory 33
2.5 De-Poissonization 37
2.6 Notes 46

3 Subgraph and component counts 47
3.1 Expectations 48
3.2 Poisson approximation 52
3.3 Second moments in a Poisson process 55
3.4 Normal approximation for Poisson processes 60
3.5 Normal approximation: de-Poissonization 65
3.6 Strong laws of large numbers 69
3.7 Notes 73

4 Typical vertex degrees 74
4.1 The setup 75
4.2 Laws of large numbers 76
4.3 Asymptotic covariances 78
4.4 Moments for de-Poissonization 82
4.5 Finite-dimensional central limit theorems 87
4.6 Convergence in Skorohod space 91
4.7 Notes and open problems 93

5 Geometrical ingredients 95
5.1 Consequences of the Lebesgue density theorem 95
5.2 Covering, packing, and slicing 97

5.3 The Brunn–Minkowski inequality 102
5.4 Expanding sets in the orthant 104

6 Maximum degree, cliques, and colourings 109
6.1 Focusing 110
6.2 Subconnective laws of large numbers 118
6.3 More laws of large numbers for maximum degree 120
6.4 Laws of large numbers for clique number 126
6.5 The chromatic number 130
6.6 Notes and open problems 134

7 Minimum degree: laws of large numbers 136
7.1 Thresholds in smoothly bounded regions 136
7.2 Strong laws for thresholds in the cube 145
7.3 Strong laws for the minimum degree 151
7.4 Notes 154

8 Minimum degree: convergence in distribution 155
8.1 Uniformly distributed points I 156
8.2 Uniformly distributed points II 160
8.3 Normally distributed points I 167
8.4 Normally distributed points II 173
8.5 Notes and open problems 176

9 Percolative ingredients 177
9.1 Unicoherence 177
9.2 Connectivity and Peierls arguments 177
9.3 Bernoulli percolation 180
9.4 k-Dependent percolation 186
9.5 Ergodic theory 187
9.6 Continuum percolation: fundamentals 188

10 Percolation and the largest component 194
10.1 The subcritical regime 195
10.2 Existence of a crossing component 200
10.3 Uniqueness of the giant component 205
10.4 Sub-exponential decay for supercritical percolation 210
10.5 The second-largest component 216
10.6 Large deviations in the supercritical regime 220
10.7 Fluctuations of the giant component 224
10.8 Notes and open problems 230

11 The largest component for a binomial process 231
11.1 The subcritical case 231
11.2 The supercritical case on the cube 234
11.3 Fractional consistency of single-linkage clustering 240
11.4 Consistency of the RUNT test for unimodality 247

11.5 Fluctuations of the giant component 252
11.6 Notes and open problems 257

12 Ordering and partitioning problems 259
12.1 Background on layout problems 259
12.2 The subcritical case 262
12.3 The supercritical case 268
12.4 The superconnectivity regime 275
12.5 Notes and open problems 279

13 Connectivity and the number of components 281
13.1 Multiple connectivity 282
13.2 Strong laws for points in the cube or torus 283
13.3 SLLN in smoothly bounded regions 289
13.4 Convergence in distribution 295
13.5 Further results on points in the cube 302
13.6 Normally distributed points 306
13.7 The component count in the thermodynamic limit 309
13.8 Notes and open problems 316

References 318

Index 328

NOTATION

In this list, section numbers refer to the places where the notation is defined. If only a chapter number is given, the notation is introduced at the start of that chapter. Some items of notation whose use is localized are omitted from this list.

Symbol	Usage	Section
$\mathbf{0}$	The origin of $\mathbb{R}^d$	1.5
$\mathbf{1}_A$	Indicator random variable or indicator function of A	1.6
$B(x;r)$	Ball of radius r centred at x	1.5
$B^*(x;r,\eta,e)$	Segment of the ball of radius r centred at x	5.2
$B_\delta(x;r)$	Segment of ball of radius r centred at x	8.3
$B(s)$	The box $[-s/2,s/2]^d$	9.6
$B_{\mathbb{Z}}(m)$	Lattice box of side m	9.2
$B'_{\mathbb{Z}}(n)$	Lattice box of side n centred at the origin	10.5
$\mathcal{B}_p$	Bernoulli process (random subset of $\mathbb{Z}^d$) induced by Z^p	9.3
$\mathrm{Bi}(n,p)$	Binomial random variable	1.6
C	The unit cube $[-\frac{1}{2},\frac{1}{2}]^d$	13.4
$\mathcal{C}(G),\mathcal{C}_n,\mathcal{C}'_n$	Clique numbers of $G, G(\mathcal{X}_n;r_n)$, and $G(\mathcal{P}_n;r_n)$, respectively	6
c.c.	Complete convergence	1.6
$\mathrm{card}(\mathcal{X})$	Number of elements of a point set $\mathcal{X}$	1.5
$D(x;r,e)$	Cylinder centred at x, radius r, orientation e	5.2
$D^*(x;r,\eta,e)$	A part of the cylinder $D(x;r,e)$	5.2
diam	Diameter based on the norm of choice	1.5
diam_∞	Diameter based on the l_∞ norm	1.5
d_{TV}	Total variation distance between probability distributions	1.6
∂A	Topological boundary of A	3, 5.2
F	Probability distribution on $\mathbb{R}^d$ with density function f	1.5
f	Underlying probability density function on $\mathbb{R}^d$	1.1, 1.5
$f_{\max}$	Essential supremum of f	1.5
f_{U}	Uniform density function on unit cube in $\mathbb{R}^d$	1.5
f_0	Essential infimum of the restriction of f to its support	5.1
f_1	Essential infimum of the restriction of f to the boundary of its support	5.2, 7
$G(n,p)$	Erdös–Rényi random graph (independent edges)	1.1
$G(\mathcal{X};r)$	Geometric graph on point set $\mathcal{X}$ with distance parameter r	1.1, 1.5

Symbol	Usage	Section
$G_n(\Gamma)$	Number of induced Γ-subgraphs in $G(\mathcal{X}_n; r_n)$	3
$G'_n(\Gamma)$	Number of induced Γ-subgraphs in $G(\mathcal{P}_n; r_n)$	3
$G_{\mathbb{Z}}(\cdot; r)$	Geometric graph with vertices in the integer lattice	9.3
$H(\cdot)$	The function $H(a) = 1 - a + a \log a,\ a > 0\ \ (H(0) = 1)$	1.6
H_+^{-1}, H_-^{-1}	Inverses of the function $H(\cdot)$	6.3
$\mathcal{H}_\lambda$	Homogeneous Poisson process on $\mathbb{R}^d$ of intensity λ	1.7, 3.1, 9.6
$\mathcal{H}_{\lambda,s}$	Homogeneous Poisson process on $B(s)$ of intensity λ	9.6
$h_\Gamma(\cdot)$	An indicator function associated with the graph Γ	3.1
$J_n(\Gamma)$	Number of Γ-components in $G(\mathcal{X}_n; r_n)$	3
$J'_n(\Gamma)$	Number of Γ-components in $G(\mathcal{P}_n; r_n)$	3
$K(G)$	Number of components of graph G	13.7
$K(\mathcal{X})$	Number of components of graph $G(\mathcal{X}; 1)$	13.7
K_n	Number of components of $G(\mathcal{X}_n; r_n)$	13
K'_n	Number of components of $G(\mathcal{P}_n; r_n)$	13
L_s	A certain level set of the density f	4.1
$L_j(G)$	Order of the jth largest component of graph G	9.3, 10
Leb	Lebesgue measure	3
LMP	Left-most point	3
$M_k(\mathcal{X})$	Largest k-nearest-neighbour link in $\mathcal{X}$	7
MBIS	Minimum bisection cost	1.3, 12.1
MBW	Minimum bandwidth cost	1.3, 12.1
MLA	Minimum linear arrangement cost	1.3, 12.1
N_λ	Total number of points of $\mathcal{P}_\lambda$	1.7
$\mathcal{N}(0, \sigma^2)$	Normal random variable	1.6
$p_\infty(\lambda)$	Continuum percolation probability	9.6
$p_c, p_c(r)$	Critical probabilities for lattice percolation	9.3
$p_k(\lambda)$	Probability mass function for the component containing the origin in continuum percolation	9.6
$\mathcal{P}_\lambda$	Poisson process with underlying density $\lambda f(\cdot)$ coupled to $\mathcal{X}_n$	1.7
$\mathrm{Po}(\lambda)$	Poisson variable with parameter λ	1.6
$S_k(\mathcal{X})$	Smallest k-nearest-neighbour link in $\mathcal{X}$	6
$T_k(\mathcal{X})$	k-connectivity threshold	13.1
$W_{k,n}$	Number of vertices of degree k in $G(\mathcal{X}_n; r_n)$	6.1, 8
$W'_{k,\lambda}$	Number of vertices of degree k in $G(\mathcal{P}_\lambda; r_n)$	6.1, 8
$X_1, X_2, \ldots$	Independent random d-vectors with common density f	1.1, 1.5
$\mathcal{X}_n$	The binomial point process $\{X_1, \ldots, X_n\}$	1.1, 1.5
$Z'_\infty(t)$	Weak limit of $Z'_n(t)$, scaled and centred	4.5
$Z_n(t)$	Number of vertices of degree at least k_n in $G(\mathcal{X}_n; r_n(t))$	4.1
$Z'_n(t)$	Number of vertices of degree at least k_n in $G(\mathcal{P}_n; r_n(t))$	4.1
Z^p	Lattice-indexed family of independent Bernoulli(p) variables	9.3

Symbol	Usage	Section
$\Delta(\mathcal{X})$	Add one cost	2.5
Δ_n, Δ'_n	Maximum degree of $G(\mathcal{X}_n; r_n)$, respectively $G(\mathcal{P}_n; r_n)$	6
δ_n	Minimum degree of $G(\mathcal{X}_n; r_n)$	7
$\zeta(\lambda)$	Rate of exponential decay for the component containing the origin	10.1
θ	Volume of the unit ball in the norm of choice	1.5
θ_{d-1}	Volume of the $(d-1)$-dimensional unit ball	8.2
$\theta_{\mathbb{Z}}(p), \theta_{\mathbb{Z}}(p;r)$	Percolation probabilities for lattice percolation	9.3
λ_{c}	Critical value (continuum percolation threshold) for λ	9.6
μ_Γ	An integral associated with the graph Γ	3.1
$\rho(\mathcal{X}; Q)$	Threshold distance for property Q	1.4
$\varphi(\cdot)$	Ordering on a graph	12.1, 4.1
$\Phi(\cdot)$	Standard normal distribution function	2.3, 4.1
$\phi(\cdot)$	Standard normal density function	2.3, 4.1
$\phi(B)$	Packing density of a set B	6.5
$\phi_L(B)$	Lattice packing density of a set B	6.5
$\chi(G), \chi_n$	Chromatic number of graph G, respectively $G(\mathcal{X}_n; r_n)$	6.5
Ω	The support of f	5.1
$\Vert \cdot \Vert$	Norm of choice on $\mathbb{R}^d$ used in defining geometric graphs	1.1, 1.5
$\Vert \cdot \Vert_p$	l_p-norm on $\mathbb{R}^d$	1.5
$\oplus$	Minkowski addition of sets	2.5, 5.3
$\geq_{\mathrm{st}}$	Stochastically dominates	9.4
$\cong$	Is isomorphic to	1.5
$\xrightarrow{\mathcal{D}}$	Converges in distribution to	1.6
$\xrightarrow{P}$	Converges in probability to	1.6
$\xrightarrow{L^p}$	Converges in pth moment to	1.6

1

INTRODUCTION

1.1 Motivation and history

A collection of trees is scattered in a forest, and a disease is passed between nearby trees. A set of nests of animals or birds is scattered in a region, and there is communication of some kind between nearby nests. A set of communications stations is distributed across a country or continent, and one is interested in communication properties between these stations. A brain cortex is viewed as a sheet of nerve cells with connections between nearby cells. A neural network consists of a collection of computational units with connections between nearby units. An astronomer wishes to group stars into constellations according to their positions in the sky. A statistician wishes to classify individuals; based on numerical measurements of d attributes for each individual, the statistician assesses two individuals as similar if the measurements are close together.

In each of these cases, and many others, one may be interested in properties of a *graph* consisting of nodes placed in d-dimensional space $\mathbb{R}^d$, with edges added to connect pairs of points which are close to each other. A mathematical model for the above situations goes as follows. Let $\|\cdot\|$ be some norm on $\mathbb{R}^d$, for example the Euclidean norm (for a formal definition see Section 1.5), and let r be some positive parameter. Given a finite set $\mathcal{X} \subset \mathbb{R}^d$, we denote by $G(\mathcal{X}; r)$ the undirected graph with vertex set $\mathcal{X}$ and with undirected edges connecting all those pairs $\{X, Y\}$ with $\|Y - X\| \leq r$. We shall call this a *geometric graph*; other terms which have been used for these graphs include *interval graphs* (when $d = 1$), *disk graphs* (when $d = 2$), and *proximity graphs*. One may be interested in many properties of geometric graphs, such as connectedness, distribution of degrees, component sizes, clique number, to name but a few.

Rather than any specific geometric graph, this monograph is concerned with an ensemble of geometric graphs. In other words, we consider geometric graphs on *random* point configurations. There are several reasons for doing so. The precise configuration of points may not be known, although one may be in a position to control the spatial density of trees (radio transmitters, etc.). Some properties of graphs are unfeasible to compute for large graphs, and understanding their average case behaviour may be a useful alternative to exact computation (see Section 1.3). Various statistical tests are based on aspects of these graphs, and understanding the probability theory of such graphs aids the construction of significance tests, confidence intervals, and so on (see Section 1.2).

The probabilistic model underlying this monograph is as follows. Let f be some specified probability density function on $\mathbb{R}^d$, and let $X_1, X_2, \ldots$ be

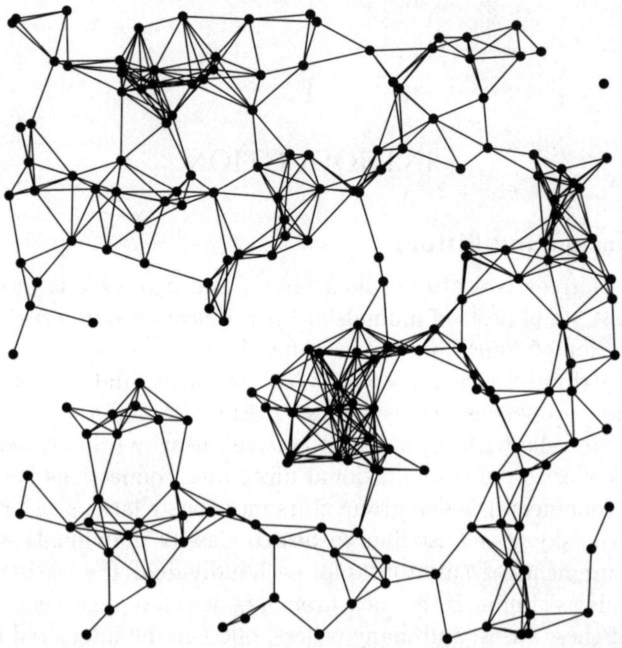

FIG. 1.1. An example of a random geometric graph.

independent and identically distributed d-dimensional variables with common density f. Let $\mathcal{X}_n = \{X_1, X_2, \ldots, X_n\}$. Our main subject is the graph $G(\mathcal{X}_n; r)$, which we shall call a *random geometric graph* (we shall also consider geometric graphs on Poisson point processes). See Fig. 1.1 for an example with $d = 2$, $n = 200$, $r = 0.11$ and f the density of the uniform distribution on $[0,1]^2$.

A more familiar random graph model, initiated by P. Erdös and A. Rényi in the late 1950s, consists of a graph on vertex set $\{1, \ldots, n\}$, either selected uniformly at random from all such graphs with a specified number of edges, or obtained by including some of the edges of the complete graph on $\{1, 2, \ldots, n\}$, each edge being included independently with probability p. The graph derived by the latter scheme will be denoted $G(n, p)$. Erdös–Rényi random graphs have been intensively studied, and many of their properties are by now well understood; see, for example, Bollobás (1985), Alon *et al.* (1992), and Janson *et al.* (2000).

Erdös–Rényi random graphs have the property of independence or near-independence between the status of different edges. This is *not* the case for geometric graphs; in the geometric setting, if X_i is close to X_j, and X_j is close to X_k, then X_i will be fairly close to X_k. In the context of examining statistical tests, this triangle property is often more realistic than the independence of edges in the Erdös–Rényi model; again, in the various modelling settings described above, the geometric random graph is more realistic than the Erdös–Rényi random graph.

It is interesting to compare results for random geometric graphs with their counterparts in the Erdös–Rényi models. Proofs of results tend to be very different in the two settings; combinatorial methods are more powerful for Erdös–Rényi random graphs. Proofs in random geometric graph theory often involve a pleasant blend of stochastic geometry and combinatorics. One motivating factor behind the study of random graphs has been their use to prove the existence of graphs having certain properties. This motivation seems to be more important in the Erdös–Rényi setting (Alon *et al.* 1992), but also has some relevance in the geometric setting (see, e.g., Pach and Agarwal (1995, Chapter 7) and also Solomon (1967)).

The study of *infinite* random geometric graphs begins with Gilbert (1961). In the infinite-space case where the underlying point process is a stationary Poisson (or other) point process, the topic is known as *continuum percolation.* Motivated largely by interest in the statistical physics of inhomogeneous materials (Torquato 2002), percolation is an important branch of modern probability theory (Grimmett 1999); *continuum* percolation is the subject of a monograph by Meester and Roy (1996).

The focus here is on asymptotics for large finite graphs. Precise computation of probabilities for properties of $G(\mathcal{X}_n; r)$ is usually unfeasible except for small values of n, and this motivates our interest in asymptotic theory; we take some sequence (r_n) and consider properties of $G(\mathcal{X}_n; r_n)$. Particularly in the later chapters, our results complement those in existing texts on percolation such as Meester and Roy (1996), Grimmett (1999).

Early work on finite random geometric graphs was done by Hafner (1972). More recently, several groups of researchers worked independently on these graphs in the 1990s. Following the book of Godehardt (1990) on applications of graph theory to statistics, probabilistic and statistical aspects of these graphs (mainly in one dimension) were further investigated by Godehardt and Jaworski (1996), Harris and Godehardt (1998), and Godehardt *et al.* (1998). In higher dimensions, mathematical contributions come from Appel and Russo (1997*a*, *b*, 2002) and McDiarmid (2003), and applications, particularly to wireless communications networks, are discussed by Clark *et al.* (1990), and by Gupta and Kumar (1998) (the former is concerned only with the non-random setting). The author became interested in this subject from the direction of certain percolation and minimal spanning tree problems, and much of the work presented here refines ideas in Penrose (1995, 1997, 1998, 1999*a*–*c*, 2000*a*, *b*), Penrose and Pisztora (1996), and Penrose and Yukich (2001). At the same time, many of the results given here are new.

Questions concerning connected components of $G(\mathcal{X}; r)$ can be rephrased in terms of components of the coverage process, consisting of balls of equal radius $r/2$ centred at the points of $\mathcal{X}$. Such coverage processes have been much studied, mainly on a Poisson point process; see, for example, Hall (1988). Also, understanding questions concerning the minimum degree of $G(\mathcal{X};r)$ is a basic problem in computational geometry; see Steele and Tierney (1986) and references therein.

Related literature includes the following books. General work on theoretical and statistical aspects of stochastic geometry includes books by Santalo (1976), Hall (1988), Ambartzumian (1990), Stoyan *et al.* (1995), Meester and Roy (1996), and Molchanov (1997). One difference in the current work is the focus on specifically graph-theoretic aspects. The books by Steele (1997) and by Yukich (1998) are on properties of the complete graph on $\mathcal{X}_n$ with edges weighted by length. There is a small overlap with the topics discussed here, but in general the methods are very different. In the non-random setting, McKee and McMorris (1999) consider *intersection graphs*, which include geometric graphs as a special case.

1.2 Statistical background

An important motivating factor for the study of random geometric graphs is multivariate statistics. The points X_i, $1 \leq i \leq n$, might represent spatial data (e.g. deposits of some mineral), or spatial–temporal data (e.g. incidences of some disease). More generally, they can represent multivariate data, the measurements of d attributes on the ith individual in a group of n individuals.

One use of random geometric graphs is as a basis for various hypothesis tests. One example arises in the simple *goodness-of-fit* problem where the null hypothesis is that the underlying density f of points is some specified distribution g. For example, one may wish to test a null hypothesis of a uniform distribution. Various test statistics in this setting have been proposed, and some of these are based on the geometric graph. These include: a simple edge count of $G(\mathcal{X};r)$, or more generally a count of the number of complete subgraphs of $G(\mathcal{X};r)$ of specified order k (see, e.g., Silverman and Brown (1978)); the *scan statistic*, which is essentially the clique number of the graph $G(\mathcal{X};r)$ (see Glaz *et al.* (2001)); the empirical distribution of nearest-neighbour distances amongst the points (see Bickel and Breiman (1983)); and the largest nearest-neighbour distance amongst the points (see Henze (1982)). In the last two cases, one can use kth-nearest neighbours with $k > 1$ an integer.

Other problems which have been addressed using tests based on geometric graphs or related concepts include the compound goodness-of-fit problem (testing, e.g., a null hypothesis of normality or of unimodality), and the question of existence of *outliers*.

Perhaps the most natural statistical setting in which geometric graphs arise is *cluster analysis*, also known as *classification* or *taxonomy*. This is the science of dividing a large collection of individuals into groups, based on measurements made for each individual. Typically the number of groups is not known *a priori* and needs to be decided on by the researcher. For example, based on medical data on individuals' symptoms, it may be desired to classify them by illness. Given measurements on fossils, it may be desired to classify by species.

Many classification techniques are based on the structure of the graph $G(\mathcal{X}_n; r)$, and to understand their power we need to understand the probability theory of this graph. We briefly discuss some of those issues of cluster analysis which are relevant to this work. For a much fuller discussion of these issues, and a more exhaustive set of references, see Godehardt (1990). See also the extensive surveys of Bock (1996*a*, *b*). Older books on various mathematical and statistical aspects of cluster analysis include Jardine and Sibson (1971), Sneath and Sokal (1973), and Hartigan (1975). For further references on clustering by graph-theoretic methods, see Brito *et al.* (1997).

Suppose that the number of attributes measured for each individual is denoted d, and each attribute measured is a continuous variable (this latter condition will not always be satisfied in practice, but only continuous variables are within the scope of this study). Based on the measurements, one can make an assessment of the 'dissimilarity' between two individuals, and then construct clusters based on these dissimilarities.

In the above scheme, each individual is represented by an element in $\mathbb{R}^d$. One possible choice of measurement of dissimilarity between two individuals is the Euclidean distance between the corresponding two points in $\mathbb{R}^d$; another is l_∞ distance, that is, the maximum of the absolute values of the differences in the different components. A problem with all this is that the choice of units for measurements of the different variables can affect the relative levels of dissimilarity within the group. In essence, the choice of units reflects the researcher's assessment of the relative importance of different variables. One possibility is to measure each variable on a scale such that measurements on that scale have unit variance. Another possibility is to measure dissimilarity by Mahalanobis distance (see Hartigan (1975) and Mardia *et al.* (1979)). Steering clear of the deeper waters of multivariate analysis, let us just say that it is reasonable here to measure dissimilarity according to some norm on $\mathbb{R}^d$.

Once established, the numerical dissimilarities between individuals can be used as the basis for a *similarity relation* between individuals; choose some threshold r and deem two individuals to be *similar* if their dissimilarity is at most r. Representing this relationship in the obvious way as a graph gives us precisely the graph $G(\mathcal{X}; r)$, where $\mathcal{X}$ denotes the set of points in $\mathbb{R}^d$ representing the individuals.

Given measurements in $\mathbb{R}^d$, many methods of constructing clusters based on the measurements have been proposed. Without attempting to describe them all, we concentrate on those which are based on the distances between them, and in particular on the graph $G(\mathcal{X}; r)$.

One of the oldest and most studied methods of constructing clusters is *single linkage*. The single-linkage clusters at level r are simply the connected components of $G(\mathcal{X}; r)$. Clearly, for each r the single-linkage clusters at a level r form a partition of the data, generally a desirable property for a clustering algorithm.

One is required to specify the parameter r; thus, there is a whole hierarchy of partitions according to the parameter r. Single-linkage clustering is *hierarchical*, meaning that for any two single-linkage clusters (not necessarily at the same level), either they are disjoint or one is contained in the other.

A related concept is the *minimal spanning tree* (MST) on the vertex set $\mathcal{X}$. This is the connected graph on vertex set $\mathcal{X}$ whose total edge length is minimal. There are efficient algorithms for constructing the minimal spanning tree, which can be used to describe the single-linkage clusters; see Gower and Ross (1969). In particular, if all edges of the MST of length greater than h are removed, the components of the resulting graph are precisely the single-linkage clusters at level h. Applications of the MST in statistical testing are various (see, e.g., Rohlf (1975) and Friedman and Rafsky (1979)).

The simplest probabilistic model for the positions of the points of the set $\mathcal{X} = \{X_1, \ldots, X_n\}$ is to suppose the X_i are independent identically distributed d-dimensional random variables. In the context of this work, the assumption is that they have a common joint density function f. If the purpose of cluster analysis is to divide the data points into clusters, a corresponding goal in terms of inference is to identify distinct disjoint regions of $\mathbb{R}^d$ in which f is high, or *population clusters*. Formally, a population cluster at level h is a connected component of $\{x : f(x) \geq h\}$. Then one has a hierarchy of population clusters. The distribution can be said to be *unimodal* if there are no two disjoint population clusters.

Given clusters of points, identified by some clustering algorithm, one desirable property is that they should correspond to population clusters. It is quite natural to construct formal test statistics for unimodality based on the clusters. For single-linkage clusters, such tests have indeed been proposed by Hartigan (1981), Hartigan and Mohanty (1992) and Tabakis (1996), and properties of their test statistics are included in the results below.

Tabakis (1996) suggested a graphical test for unimodality based on the *connectivity threshold* for $\mathcal{X}$, that is, the threshold value of r above which $G(\mathcal{X}; r)$ is connected, which is also the length of the longest edge of the minimal spanning tree. This also has been proposed as a test for *outliers* in multivariate data by Rohlf (1975). What Tabakis (1996) actually considered was the connectivity threshold for $\mathcal{X}^\delta$, where $\mathcal{X}^\delta$ denotes the set of points of $\mathcal{X}$ for which the estimated density f exceeds some small specified $\delta > 0$.

A much-discussed feature of single-linkage clusters is *chaining*. This occurs when two groups of data points in $\mathbb{R}^d$ are well separated, apart from a narrow chain of points linking one group to the other. The single-linkage clustering method may not be able to distinguish between the two groups. One attempt to deal with the worst chaining effects is by taking *strong k-linkage clusters*, where $k > 0$ is an integer parameter. Using terminology from Godehardt (1990) (see also 'integer link linkage' in Sneath and Sokal (1973)), we say that two vertices are in the same strong k-linkage cluster at level r if there are k or more edge-disjoint paths connecting them in $G(\mathcal{X}; r)$. Equivalently, they are in the same strong k-linkage cluster at level r if there is no way to disconnect them by removal of k or fewer edges (the equivalence comes from Menger's theorem which will be described later on). Strong k-linkage clustering can be shown to partition the vertex set. Two groups connected by a single chain will lie in distinct strong k-linkage clusters if $k > 1$, but if they are connected by k chains they will not be distinguished by this method.

An intermediate clustering method is *weak k-linkage*, proposed by Ling (1973) and also described in Godehardt (1990). Weak k-linkage clusters are obtained by first removing all vertices of degree strictly less than k, then taking components of the resulting graph. To get a partition one can also include each of the removed vertices as a weak k-linkage cluster consisting of a single point. This method is intermediate in the sense that for any graph, the strong k-linkage clusters will form a refinement of the weak k-linkage clusters, which in turn will form a refinement of the single-linkage clusters. Some of the results given here, particularly in Chapter 13, have relevance to strong k-linkage clusters and weak k-linkage clusters.

Some other types of clustering in the literature do not partition the data points. For example, a ***k-overlap cluster*** is a maximal collection of $k + 1$ or more vertices with the property that every pair of vertices in the collection is connected by at least $k + 1$ vertex-disjoint paths. A *complete-linkage* cluster is a *clique* of the graph. Again, our results will have something to say about these.

1.3 Computer science background

The NP-complete problems form a large and important class of computational optimization problems for which there is no known algorithm guaranteed to produce a solution in polynomial time; see, for example, Garey and Johnson (1979). One of the best known of these is the travelling salesman problem (TSP) of finding a tour through a given set of points in Euclidean space, of minimal total length.

For such problems, it may be sufficient in practice to use an approximate algorithm or *heuristic*, that is, a procedure which is intended to generate a nearly optimal solution most of the time; computer scientists are interested in examining the performance of such heuristics. Such a description begs the question of the meaning of the phrase 'most of the time'; one interpretation is that one has in mind some probability distribution on the space of instances of a given optimization problem, and a heuristic is effective 'most of the time' if the probability of instances of the problem where it fails to deliver a near-optimal solution is small; this notion of an effective polynomial-time heuristic was introduced by Karp (1976, 1977). Moreover, two or more heuristics can be (and often are) compared empirically by repeated Monte-Carlo simulation of random instances of the problem in question; again one requires some probability measure on the space of instances of the problem.

If the chosen probability measure for the simulated graph is the random geometric graph scheme, then the mathematical theory of random geometric graphs provides a complementary theoretical underpinning for the assessment of the heuristic(s) by simulations.

Considering, for example, the TSP, a natural probability measure is obtained by dropping n points at random into Euclidean space (e.g. uniformly over the unit square). This idea leads to the study of the TSP and related problems (such as finding the minimal matching (MM) or the minimal spanning tree (MST)) on randomly distributed points, which began with a celebrated paper by Beardwood *et al.* (1959), was taken further by Karp (1976, 1977), and has subsequently led to a beautiful and extensive mathematical theory which is described in Steele (1997) and Yukich (1998). Problems such as the TSP, MM or MST on Euclidean points can be viewed as problems where the input is the complete graph on those points, with weights on edges given by inter-point distances.

Many NP-complete problems are defined on unweighted graphs. These include a variety of *layout problems*, where the aim is to order the vertices so that adjacent vertices are close together in the ordering. A (one-dimensional) layout of a finite input graph G is a bijection φ between its vertex set and a set of integers. Given a layout, the weight $\sigma(e)$ of an edge e is the absolute value of the difference between the integers associated with the two end-points.

A layout problem involves choosing φ so as to minimize some cost functional determined by the edge weights. For example, for the *minimum bandwidth* (MBW) problem the cost functional is $\max_e \sigma(e)$, while for the *minimum linear arrangement* (MLA) problem the cost functional is $\sum_e \sigma(e)$. Moreover, the *minimum bisection* (MBIS) problem of partitioning the vertices into two equal-sized sets so as to minimize the number of edges between them can also be formulated as a layout problem. Some other related problems of similar type are described in Chapter 12. Areas of application of such problems include integrated circuit design, parallel processing, numerical analysis, computational biology, brain cortex modelling and even archaeological dating. These applications will be discussed further in Chapter 12.

For each of these problems, finding an optimal layout is known to be NP-complete for general graphs; see the references in Díaz *et al.* (2001*a*). Moreover, a number of other problems that will concern us, involving graph colouring and finding independent sets, are also NP-complete for geometric graphs; see Clark *et al.* (1990). The chromatic number of a geometric graph is of particular interest in the context of the frequency assignment problem for radio transmitters (when different frequencies are required for transmitters with overlapping ranges); for further discussion, see Chapter 6.

For these NP-complete problems, there is an interest in comparing the performance of heuristics for these algorithms on randomly generated graphs. One method of randomly generating graphs that might be used when comparing algorithms for layout problems is the Erdös–Rényi random graph model. These have indeed been studied in this context, but it turns out that for many of the layout problems described above, Erdös–Rényi random graphs fail to differentiate good from bad heuristics, in the sense that with high probability *all* orderings on such graphs have approximately the same behaviour (see Turner (1986), Bui *et al.* (1987), and Díaz *et al.* (2001*b*)).

This leads us to the study of these problems on random *geometric* graphs; moreover, as already mentioned, geometric graphs are often a reasonable model for graphs that occur in practice, such as finite element graphs, integrated circuits, and communication graphs. Empirical studies of layout and partitioning problems have often used random geometric graphs, typically by the experimental comparison of different heuristics for layout problems, by trying them out on repeatedly simulated random geometric graphs. For these reasons, a mathematical theory for layout problems on random geometric graphs is useful in providing a benchmark for assessing particular heuristics. This theory is described in Chapter 12.

1.4 Outline of results

Except in the most trivial cases, exact formulae for properties of $G(\mathcal{X}_n; r)$ tend to be very unwieldy, if available at all, especially in more than one dimension. In this book we concentrate on *asymptotic* properties of the graph $G(\mathcal{X}_n; r_n)$ for some sequence of parameters r_n, usually tending to zero. For the most part, we shall ignore results which are specific to one dimension, concentrating instead on results which hold for all $d \geq 2$ or, in many cases, for all $d \geq 1$.

We are often interested in *monotone increasing* properties of graphs. Given a finite set $\mathcal{X}$, a property Q of graphs with vertex set $\mathcal{X}$ is said to be monotone increasing if, whenever G is a subgraph of H and G has property Q, so does H. Given a monotone increasing property Q, and a set of points $\mathcal{X} \subset \mathbb{R}^d$, we define the *threshold distance* $\rho(\mathcal{X}; Q)$ to be the infimum of all r such that $G(\mathcal{X}; r)$ has that property. For example, the threshold distance above which $G(\mathcal{X}; r)$ has at least one edge is the smallest inter-point distance in $\mathcal{X}$. We are interested in a variety of threshold distances for $\mathcal{X}_n$.

Two limiting regimes for (r_n) are of special interest. One of these is the *thermodynamic limit* in which $r_n \sim \text{const.} \times n^{-1/d}$, so that the expected degree of a typical vertex tends to a constant. The terminology 'thermodynamic limit' is borrowed from the statistical physics literature; this limiting regime is equivalent to observing n points in a large region of volume proportional to n, letting n grow with a fixed range r of inter-point interaction. As we shall see, if the limiting constant in the thermodynamic limit is taken above a certain critical value, there is likely to be a giant component of $G(\mathcal{X}_n; r_n)$ containing a strictly positive fraction of the points, a phenomenon known as *percolation.* When the limiting constant is above the critical value we refer to this as the *supercritical thermodynamic limit*, and when the constant is below the critical value we refer to this as the *subcritical thermodynamic limit.* We refer the cases $nr_n^d \to 0$ and $nr_n^d \to \infty$ as *sparse* and *dense* limiting regimes, respectively, referring to the fact that the points are sparsely (respectively, densely) scattered if viewed on the scale at which connections are made. This is slightly in conflict with the more usual graph-theoretic terminology whereby any graph with n vertices and $o(n^2)$ edges would be regarded as 'sparse'.

The second limiting regime of special interest is the *connectivity regime*, which is the special case of the dense limit regime in which $r_n \sim \alpha((\log n)/n)^{-1/d}$, with α a constant, so that the typical vertex degree grows logarithmically in n. The terminology is motivated as follows. If the expected degree of a point is asymptotic to $c \log n$, then (by Poisson approximation) the probability that it is isolated can be expected to obey an n^{-c} power law, so that the mean number of isolated points is of order n^{1-c} and tends to infinity or zero according to whether $c < 1$ or $c > 1$. Clearly, a necessary condition for connectivity is that there be no isolated points, and this turns out to be sufficient with high probability as $n \to \infty$. Thus we can expect the connectivity regime to exhibit a phase transition in α, with respect to the property of connectivity of the geometric graph. When $nr_n^d/\log n$ tends to infinity we shall refer to the limiting regime as the *superconnectivity* regime, and limiting regimes with $nr_n^d/\log n \to 0$ will sometimes be referred to as *subconnective*.

We are interested both in *convergence in distribution* (also known as *weak convergence*) and *laws of large numbers*. For convergence in distribution results, given a sequence $(r_n)_{n\geq 1}$ one seeks convergence of the (possibly scaled and centred) distribution of some graph invariant evaluated on $G(\mathcal{X}_n; r_n)$ to a non-trivial limiting distribution as $n \to \infty$; alternatively, in the case of a monotone increasing graph property, one may seek the limiting distribution for the threshold distance for a given property, suitably scaled and centred. When available, weak convergence results can be used to estimate p-values in statistical tests.

In the case of laws of large numbers, we usually give *strong laws* with almost sure convergence, as $n \to \infty$, of some scaled version of a threshold distance or (if $(r_n)_{n\geq 1}$ is given) a graph invariant, to a non-zero limit. These results give an idea of orders of magnitude for threshold distances or for graph invariants, without providing precise limiting probabilities.

The remaining three sections of the present chapter contain essential preliminaries in the form of notation and technical background information at a fairly elementary level, which will be used throughout the book. In particular, notions of probability, at the level of an advanced undergraduate or first year postgraduate course, are reviewed in Section 1.6.

After this chapter, the remainder of the book divides roughly into three parts. Each begins with a chapter whose title contains the word 'ingredients', containing results that are included mainly for application later on rather than for their own sake. While Part II is not entirely free of dependence on material in Part I, and Part III is not entirely free of dependence on material from Parts I and II, the parts are sufficiently self-contained that it should be possible to use any individual part separately as part of a graduate course or reading programme.

Part I starts with Chapter 2 and is concerned with sums of quantities which are locally determined in some sense, such as the number of edges or the number of vertices of a given degree. Generalizing both of these quantities, we consider the number of copies of some arbitrary specified connected finite graph embedded in the graph $G(\mathcal{X}_n; r)$. We give Poisson and normal limits, according to the limiting regime; we also consider (by similar methods) the number of *components* isomorphic to a specified graph.

Next we consider is the empirical distribution of the vertex degrees. Given $k \in \mathbb{N}$, how many of the vertices have degree at least k? Questions of this sort are naturally expressed in terms of threshold functions; for example, the threshold function for the property of having all vertices of degree at least k is the *largest* k*-nearest-neighbour link*. In the limit it is possible to have either k fixed or $k = k_n$ increasing with n; we consider both strong laws and weak convergence in these limiting regimes.

Part II starts with Chapter 5, and is concerned with *extremes* of locally determined quantities, including the maximum degree, the minimum degree, the clique number and the chromatic number. Both strong laws of large numbers, and (in some cases) weak convergence results are obtained for these quantities.

Part III starts at Chapter 9, and is concerned with *globally determined* properties of a graph. For example, to know whether the graph is connected one needs to look at the whole graph, not just at neighbourhoods of individual vertices. It is at this stage that technical material related to percolation theory becomes important.

Most of Part III is devoted to the *giant component* and related topics. In addition to laws of large numbers, central limit theorems and large deviations for the order of the largest component, it contains a significant amount of material on continuum percolation that is of interest in its own right. Applications considered here include consistency results for statistical tests for contours of the underlying density and for unimodality, which were suggested by Hartigan (1981) and Hartigan and Mohanty (1992). In Chapter 12 the theory of the giant component is applied to layout problems of the type described in Section 1.3 above.

Chapter 13 is concerned with the *connectivity* of a random geometric graph, and with the number of components. Results include limit laws for the threshold at which the graph $G(\mathcal{X}_n; \cdot)$ becomes connected, and also results on multiple connectivity. Both laws of large numbers and (in some cases) weak convergence results are given. This final chapter makes the greatest use in Part III of material appearing earlier, not just in Part III but in Parts I and II as well.

1.5 Some basic definitions

We use the following standard notation. The symbol $:=$ denotes definition but simply $=$ can also denote definition when the context is clear. Also c, c', and so on stand for strictly positive, finite constants whose exact values are unimportant, and are allowed to change from line to line. The set of real numbers is denoted $\mathbb{R}$ and the set of natural numbers $\{1, 2, 3, \ldots\}$ is denoted $\mathbb{N}$; the set of integers is denoted $\mathbb{Z}$ and the set of non-negative integers is denoted $\mathbb{Z}_+$ (or $\mathbb{N} \cup \{0\}$). Given $t \in \mathbb{R}$, we write $\lfloor t \rfloor$ for the value of t rounded down to the nearest integer, and $\lceil t \rceil$ for the value of t rounded up to the nearest integer. All logarithms are to base e.

Suppose $(a_n)_{n\geq 1}$ and $(b_n)_{n\geq 1}$ are sequences of real numbers with $b_n > 0$ for all n. We write $a_n = O(b_n)$ if $\limsup_{n\to\infty}(|a_n|/b_n) < \infty$, and write $a_n = o(b_n)$ if $\lim_{n\to\infty}(|a_n|/b_n) = 0$. If also $a_n > 0$ for all n, we write $a_n = \Theta(b_n)$ if both $a_n = O(b_n)$ and $b_n = O(a_n)$, and we shall say the sequence $(a_n)_{n\geq 1}$ *decays exponentially in* b_n if $\lim_{n\to\infty} b_n = \infty$ and

$$\limsup_{n\to\infty} ((\log a_n)/b_n) < 0.$$

If the sequence $(a_n)_{n\geq 1}$ decays exponentially in n^r for some $r \in (0,1)$, then we say that it decays *sub-exponentially* in n.

Throughout this monograph, it is assumed that the points $X_1, X_2, X_3, \ldots$ are independent random d-vectors having common probability density function $f : \mathbb{R}^d \to [0, \infty)$. The point process $\mathcal{X}_n$ is the union of the first n points $\mathcal{X}_n := \cup_{i=1}^n \{X_i\}$. In all theorems concerning $\mathcal{X}_n$, the density function f is assumed fixed but arbitrary unless stated otherwise, subject only to the conditions that f is measurable and satisfies $\int_{\mathbb{R}^d} f(x)\mathrm{d}x = 1$ (i.e. f really is a probability density function), and that f is *bounded.* Also, F denotes the common probability distribution of each point X_i, that is, for Borel $A \subseteq \mathbb{R}^d$, we set

$$F(A) = P[X_1 \in A] = \int_A f(x)\mathrm{d}x.$$

Let $f_{\max}$ denote the essential supremum of f, that is, the infimum of all h such that $P[f(X_1) \leq h] = 1$. Since we assume throughout this monograph that f is bounded, $f_{\max} < \infty$.

An important special case is the *uniform case* in which f is the density f_{U} of the uniform distribution on the d-dimensional unit cube $[-\frac{1}{2}, \frac{1}{2}]^d$, defined by

$$f_{\mathrm{U}}(x) := 1 \text{ if } x \in \left[-\tfrac{1}{2}, \tfrac{1}{2}\right]^d, \quad f_{\mathrm{U}}(x) := 0 \text{ otherwise.} \tag{1.1}$$

We write $\mathbf{0}$ for the zero vector $(0, 0, \ldots, 0) \in \mathbb{R}^d$. A *norm* is a real-valued function $\|\cdot\|$ on $\mathbb{R}^d$ with the property that $\|x\| \geq 0$ for all $x \in \mathbb{R}^d$ with equality only if $x = \mathbf{0}$, and $\|ax\| = |a|\|x\|$ for all $a \in \mathbb{R}, x \in \mathbb{R}^d$, and $\|x + y\| \leq \|x\| + \|y\|$ for all $x, y \in \mathbb{R}^d$. The so-called l_p norms $\|\cdot\|_p$ are defined for $1 \leq p < \infty$ by

$$\|(x_1, \ldots, x_d)\|_p := \left(\sum_{i=1}^d |x_i|^p\right)^{1/p}$$

and for $p = \infty$ by $\|(x_1, \ldots, x_d)\|_\infty := \max_{1\leq i\leq d} |x_i|$. The l_2 norm is also denoted the *Euclidean* norm.

A basic fact about norms is the *equivalence of all norms* on the finite-dimensional space $\mathbb{R}^d$. This says that for any two norms $\|\cdot\|$ and $\|\cdot\|'$ on

$\mathbb{R}^d$, there exist constants $0 < c < C < \infty$ such that $c\|x\| \leq \|x\|' \leq C\|x\|$ for all $x \in \mathbb{R}^d$ (see, e.g., Hoffman (1975, Section 6.2)).

With one exception, all our results on geometric graphs refer to some norm $\|\cdot\|$ on $\mathbb{R}^d$ which is fixed, but arbitrary unless otherwise stated. Given the norm $\|\cdot\|$, and given $\mathcal{X} \subset \mathbb{R}^d$, the geometric graph $G(\mathcal{X}; r)$ has vertex set $\mathcal{X}$ and includes as edges all pairs $\{x, y\}$ with $\|x - y\| \leq r$. Also, we use the same norm to define the diameter of subsets of $\mathbb{R}^d$, that is, for $A \subseteq \mathbb{R}^d$, we set

$$\mathrm{diam}(A) := \sup\{\|x - y\| : x \in A, y \in A\}. \tag{1.2}$$

The sole exception to the assumption that our geometric graph $G(\mathcal{X}; r)$ is given by a norm occurs in cases when we assume that the points X_i are uniformly distributed on the unit *torus.* In this case the underlying density f is the uniform density f_{U}, and we specify a norm $\|\cdot\|$ as usual, but for $X, Y \in [-\frac{1}{2}, \frac{1}{2})^d$ the distance between them is defined by

$$\mathrm{dist}(X, Y) = \min\{\|X + z - Y\| : z \in \mathbb{Z}^d\}.$$

For points on the torus, the graph $G(\mathcal{X}; r)$ has vertex set $\mathcal{X}$ and has an edge connecting each pair of points $X, Y \in \mathcal{X}$ with $\mathrm{dist}(X, Y) \leq r$.

Given $x \in \mathbb{R}^d$ and $r \geq 0$, $B(x; r)$ denotes the ball $\{y \in \mathbb{R}^d : \|y - x\| \leq r\}$. The *volume* (Lebesgue measure) *of the unit ball* $B(\mathbf{0}; 1)$ *is denoted* θ (with apologies to percolation theorists who may be used to a different use of this letter). Given any finite set $\mathcal{X}$ we write either $|\mathcal{X}|$ or $\mathrm{card}(\mathcal{X})$ for the cardinality (number of elements) of $\mathcal{X}$. If $\mathcal{X}$ is a locally finite subset of $\mathbb{R}^d$ (i.e. one which has finite intersection with any bounded subset of $\mathbb{R}^d$), and if A is a subset of $\mathbb{R}^d$, we write $\mathcal{X}(A)$ for the number of elements of the set $\mathcal{X} \cap A$. If also $a \geq 0$ then we write $a\mathcal{X}$ for $\{ax : x \in \mathcal{X}\}$.

This section concludes with some basic terminology from graph theory. For a general reference on graphs, see, for example, Bollobás (1979). A *graph* is a pair $G = (V, E)$, where V is a set and E is a set, each of whose elements is an unordered pair $\{x, y\}$ of distinct elements of V. Elements of V are called *vertices* and elements of E are called *edges.* If $\{x, y\} \in E$ we say vertices x and y are *adjacent.* The *order* of such a graph is the number of elements in V. A *path* in G from vertex $u \in V$ to vertex $v \in V$ is a sequence $x_0 = u, x_1, \ldots, x_n = v$ of distinct elements of V such that $\{x_{i-1}, x_i\}$ lies in E for each $i = 1, 2, \ldots, n$. Two paths $(x_0, x_1, \ldots, x_m)$ and $(y_0, y_1, \ldots, y_n)$ from u to v are *independent* if they have no vertices in common except for their end-points, that is, if $\{x_0, \ldots, x_m\} \cap \{y_0, \ldots, y_n\} = \{x_0, x_m\}$. The graph G is *connected* if for any two vertices $u, v \in V$ there is a path from u to v.

A *subgraph* of G is a graph $G' = (V', E')$ for which $V' \subseteq V$ and $E' \subseteq E$. If V' is a subset of V, then the *subgraph induced by* V' is the subgraph (V', E') with E' consisting of *all* edges of G having both end-points in V'.

Two graphs G_1, G_2 are *isomorphic* if there is a one-to-one correspondence between their vertex sets, which preserves adjacency. We shall write $G_1 \cong G_2$

when this is the case. A graph (V, E) is *connected* if there is a path from u to v for all $u, v \in V$. A *component* of G is a maximal connected subgraph of G, that is, a connected subgraph of G that is not a proper subgraph of any other connected subgraph of G.

Menger's theorem tells us that for any two non-adjacent vertices $u, v \in V$, the minimal number of vertices whose removal leaves u and v in distinct components equals the maximal number of independent paths from u to v. The *edge version* of Menger's theorem states that the minimal number of edges whose removal leaves u and v in distinct components equals the maximal number of edge-disjoint paths from u to v (and does not require u, v to be non-adjacent).

1.6 Elements of probability

In this monograph it is assumed that the reader has some familiarity with basic notions of probability. Useful texts include Billingsley (1979), Shiryayev (1984), Durrett (1991), Williams (1991), and Grimmett and Stirzaker (2001). This section contains a brief review of relevant probabilistic concepts.

If A is an event then let $P[A]$ denote its probability, and let $\mathbf{1}_A$ be the indicator random variable taking the value 1 if A occurs and 0 if not; likewise, for $A \subseteq \mathbb{R}^m$ let $\mathbf{1}_A : \mathbb{R}^m \to \{0, 1\}$ be the indicator function with $\mathbf{1}_A(x) = 1$ for $x \in A$, $\mathbf{1}_A(x) = 0$ for $x \notin A$. If ξ is a (real-valued) random variable, then $E[\xi]$ (or just $E\xi$) denotes its expected value and $\mathrm{Var}(\xi)$ denotes its variance. If ξ takes only non-negative values, the *integration by parts* formula for expectation tells us that $E[\xi] = \int_0^\infty P[\xi \geq t]\mathrm{d}t$; see Feller (1971).

If ξ' is another random variable on the same probability space, $\mathrm{Cov}(\xi, \xi')$ denotes the covariance of ξ and ξ'.

Boole's inequality says that if $A_1, A_2, \ldots$ is a (finite or infinite) sequence of events on the same sample space, then $P[\cup_i A_i] \leq \sum_i P[A_i]$. *Markov's inequality* says that if ξ is a random variable with $P[\xi \geq 0] = 1$, and λ is a positive constant, then $P[\xi \geq \lambda] \leq \lambda^{-1} E[\xi]$. *Chebyshev's inequality* states that if $\mathrm{Var}(\xi) < \infty$ and ν is a positive constant, then $P[\,|\xi - E\xi| > \nu] \leq \nu^{-2}\mathrm{Var}(\xi)$. The *Cauchy–Schwarz inequality* tells us that if ξ_1, ξ_2 are random variables on the same sample space with $E[\xi_i^2] < \infty$ for $i = 1, 2$, then $(E[\xi_1\xi_2])^2 \leq E[\xi_1^2]E[\xi_2^2]$. An event occurs *almost surely* (a.s.) if it has probability 1. The (first) *Borel–Cantelli lemma* says that if A_n is a sequence of events on the same sample space, and $\sum_{n=1}^\infty P[A_n] < \infty$, then with probability 1, A_n occurs for only finitely many n.

Suppose $\xi, \xi_1, \xi_2, \ldots$ are random variables all defined on the same sample space $(\Omega, \mathcal{F}, P)$. Then we say ξ_n converges to ξ almost surely, and write $\xi_n \to \xi$ a.s., if the event $\{\omega \in \Omega : \lim_{n\to\infty} \xi_n(\omega) = \xi(\omega)\}$ has probability 1. We say ξ_n converges to ξ in probability, and write $\xi_n \xrightarrow{P} \xi$, if for any $\varepsilon > 0$, $P[|\xi_n - \xi| > \varepsilon] \to 0$ as $n \to \infty$. Alternatively, we say ξ_n converges to ξ in probability if any subsequence of $\{1, 2, 3, \ldots\}$ has a sub-subsequence such that $\xi_n \to \xi$ a.s. as $n \to \infty$ along the sub-subsequence. These two definitions of convergence in probability are equivalent; see, for example, Williams (1991, A13.2). Given

$p \geq 1$, we write $\xi_n \xrightarrow{L^p} \xi$ if $E[|\xi_n - \xi|^p] \to 0$ as $n \to \infty$. We say the variables ξ_n are *uniformly integrable* if $\sup_{n \geq 1} E[|\xi_n| \mathbf{1}_{\{|\xi_n| > K\}}]$ tends to 0 as $K \to \infty$. A sufficient condition for uniform integrability is that for some $q > 1$ we have $\sup_n E[|\xi_n|^q] < \infty$.

A stronger version of almost sure convergence is *complete convergence*; variables ξ_n converge to a constant b with complete convergence (written $\xi_n \to b$ c.c.), if for all $\varepsilon > 0$ we have $\sum_n P[|\xi_n - b| > \varepsilon] < \infty$. By the Borel–Cantelli lemma, complete convergence $\xi_n \to b$ c.c. implies almost sure convergence $\xi_n \to b$ a.s. For a discussion of complete convergence, see Yukich (1998).

If $(\xi_n)_{n \geq 1}$ is a uniformly integrable sequence of random variables converging in probability to ξ, then $E[\xi]$ exists and $\lim_{n \to \infty} E[\xi_n] = E[\xi]$. For example, suppose that $\xi_n \to \xi$ a.s., and also that there exists ξ_0 with $E[|\xi_0|] < \infty$, such that $|\xi_n(\omega)| \leq |\xi_0(\omega)|$ for almost all $\omega \in \Omega$; then $\lim_{n \to \infty} E[\xi_n] = E[\xi]$. This is a special case of the preceding result, and is known as the *dominated convergence theorem.* A related result is *Fatou's lemma*, which says that if $(\xi_n)_{n \geq 1}$ is a sequence of non-negative random variables then $E \liminf_{n \to \infty} \xi_n \leq \liminf_{n \to \infty} E\xi_n$.

Now recall some notions concerned with conditional expectation. If X is an integrable random variable on a probability space $(\Omega, \mathcal{F}, P)$, and $\mathcal{G}$ is a sub-σ-field of $\mathcal{F}$, then the random variable $E[X|\mathcal{G}]$ is the conditional expectation of X with respect to $\mathcal{G}$. If also $g : \mathbb{R} \to \mathbb{R}$ is convex, and $E[|g(X)|] < \infty$, then the conditional version of *Jensen's inequality* says that $g(E[X|\mathcal{G}]) \leq E[g(X)|\mathcal{G}]$, almost surely (the unconditional version says that $g(E[X]) \leq E[g(X)]$). Given a filtration $(\mathcal{F}_1, \mathcal{F}_2, \ldots, \mathcal{F}_n)$ (i.e. an increasing sequence of sub-σ-fields of $\mathcal{F}$), a *martingale* with respect to the filtration is a sequence of integrable random variables $(M_1, \ldots, M_n)$ satisfying $E[M_{k+1}|\mathcal{F}_k] = M_k$, almost surely, for $k = 1, 2, \ldots, n-1$.

A *random d-vector* on a probability space $(\Omega, \mathcal{F}, P)$ is a measurable function $\xi : \Omega \to \mathbb{R}^d$. Suppose $\xi, \xi_1, \xi_2, \ldots$ are random d-vectors, not necessarily defined on the same sample space. We say ξ_n *converges to* ξ *in distribution*, and write $\xi_n \xrightarrow{\mathcal{D}} \xi$, if $E[h(\xi_n)] \to E[h(\xi)]$ as $n \to \infty$ for any bounded continuous $h : \mathbb{R}^d \to \mathbb{R}$. The *Cramér–Wold device* (see, e.g., Durrett (1991)) says that a sufficient condition for $\xi_n \xrightarrow{\mathcal{D}} \xi$ is that for all $\boldsymbol{a} \in \mathbb{R}^d$ we have $\boldsymbol{a} \cdot \xi_n \xrightarrow{\mathcal{D}} \boldsymbol{a} \cdot \xi$ as $n \to \infty$, where $\cdot$ is the Euclidean inner product. If $d = 1$, and $\xi_n \xrightarrow{\mathcal{D}} \xi$ with $\{\xi_n, n \geq 1\}$ uniformly integrable, then $E[\xi_n] \to E[\xi]$ as $n \to \infty$ (see Billingsley (1979, Theorem 25.12)). If $d = 1$, and $\xi_n \xrightarrow{\mathcal{D}} \xi$ and $\zeta_n \xrightarrow{P} 0$, then $(\xi_n + \zeta_n) \xrightarrow{\mathcal{D}} \xi$, a fact which is sometimes known as *Slutsky's theorem* (see, e.g., Durrett (1991)).

The *total variation distance* between two integer-valued random variables ξ, ζ (more correctly, between their distributions) is given by

$$d_{\mathrm{TV}}(\xi, \zeta) = \sup_{A \subseteq \mathbb{Z}} |P[\xi \in A] - P[\zeta \in A]|. \tag{1.3}$$

Recall from Section 1.5 that $B(x; r)$ denotes the r-ball centred at $x \in \mathbb{R}^d$, and that f denotes the common probability density function of the random d-vectors

X_i underlying the random geometric graph model. A *Lebesgue point* of f is a point $x \in \mathbb{R}^d$ with the property that

$$\lim_{r \downarrow 0} \left(r^{-d} \int_{B(x;r)} |f(y) - f(x)| \mathrm{d}y \right) = 0,$$

and the *Lebesgue density theorem* tells us that almost every $x \in \mathbb{R}^d$ is a Lebesgue point of f. By using this theorem we can often prove results that might otherwise be apparent only in the case where f is almost everywhere continuous (see, for example, Rudin (1987) for a proof of the Lebesgue density theorem).

For $\sigma \geq 0$, we denote by $\mathcal{N}(0, \sigma^2)$ the random variable σZ, where Z is a continuous random variable with density function $(2\pi)^{-1/2} \exp(-x^2/2)$, $x \in \mathbb{R}$. Note that $\sigma = 0$ is allowed in this definition of a normal variable. A random k-vector $\xi = (\xi_1, \ldots, \xi_k)$ is *centred multivariate normal* with covariance matrix $\Sigma = (\sigma_{ij}, 1 \leq i, j \leq k)$ if, for all $(a_1, \ldots, a_k) \in \mathbb{R}^k$, the distribution of $\sum_{i=1}^k a_i \xi_i$ is that of $\mathcal{N}(0, \sum_{i=1}^k \sum_{j=1}^k a_i a_j \sigma_{ij})$.

For $0 \leq p \leq 1$, a Bernoulli(p) random variable is one which takes the value 1 with probability p and takes the value 0 with probability $1 - p$. We write Bi(n, p) for any binomial random variable with the distribution of the sum of n independent Bernoulli(p) random variables, and we write Po(λ) for any Poisson random variable with parameter λ.

The next two results give uniform upper bounds on the probability that the value of a binomial variable Bi(n, p) or a Poisson variable Po(λ) is larger or smaller than expected. We define the function $H : [0, \infty) \to [0, \infty)$ (which will recur many times through the monograph) by $H(0) = 1$ and

$$H(a) = 1 - a + a \log a, \qquad a > 0. \tag{1.4}$$

Note that $H(1) = 0$, and that the unique turning point of H is the minimum at 1.

Lemma 1.1 *Suppose $n \in \mathbb{N}$, $p \in (0, 1)$, and $0 < k < n$. Let $\mu = np$. If $k \geq \mu$ then*

$$P[\mathrm{Bi}(n, p) \geq k] \leq \exp\left(-\mu H\left(\frac{k}{\mu} \right) \right) \tag{1.5}$$

and if $k \leq \mu$ then

$$P[\mathrm{Bi}(n, p) \leq k] \leq \exp\left(-\mu H\left(\frac{k}{\mu} \right) \right). \tag{1.6}$$

Finally, if $k \geq \mathrm{e}^2 \mu$ then

$$P[\mathrm{Bi}(n, p) \geq k] \leq \exp\left(-\left(\frac{k}{2} \right) \log\left(\frac{k}{\mu} \right) \right). \tag{1.7}$$

Proof Let $X = \mathrm{Bi}(n,p)$, and set $q := 1-p$. By Markov's inequality, for $z \geq 1$ the probability $P[X \geq k]$ is bounded above by

$$z^{-k}E[z^X] = z^{-k}(pz+q)^n, \tag{1.8}$$

while if $z \leq 1$ the probability $P[X \leq k]$ is bounded above by the same expression (1.8).

Set $z := kq/((n-k)p)$, which is at least 1 for $k \geq \mu$ and at most 1 for $k \leq \mu$. Then $pz + q = (nq)/(n-k)$ so the bound (1.8) becomes

$$\left(\frac{(n-k)p}{kq}\right)^k \left(\frac{nq}{n-k}\right)^n = \left(\frac{np}{k}\right)^k \left(\frac{nq}{n-k}\right)^{n-k}. \tag{1.9}$$

Apply the inequality $x \leq \mathrm{e}^{x-1}$, true for all $x > 0$, to $x = nq/(n-k)$. For this choice of x we have $x - 1 = (k-np)/(n-k)$ so that the bound (1.9) is in turn bounded by the expression

$$\begin{aligned}\left(\frac{np}{k}\right)^k \mathrm{e}^{k-np} &= \left(\frac{\mathrm{e}\mu}{k}\right)^k \mathrm{e}^{-\mu} \\ &= \exp\left(-\mu - k\log\left(\frac{k}{\mu}\right) + k\right) = \exp\left(-\mu H\left(\frac{k}{\mu}\right)\right),\end{aligned}$$

completing the proof of (1.5) and (1.6). If $k \geq \mathrm{e}^2\mu$, the fact that for $a \geq \mathrm{e}^2$ we have $H(a) \geq a(\log a - 1) \geq \frac{1}{2}a\log a$, applied to (1.5), yields (1.7). □

Lemma 1.2 *Suppose $k > 0$, $\lambda > 0$. If $k \geq \lambda$ then*

$$P[\mathrm{Po}(\lambda) \geq k] \leq \exp\left(-\lambda H\left(\frac{k}{\lambda}\right)\right), \tag{1.10}$$

and if $k \leq \lambda$ then

$$P[\mathrm{Po}(\lambda) \leq k] \leq \exp\left(-\lambda H\left(\frac{k}{\lambda}\right)\right). \tag{1.11}$$

Finally, if $k \geq \mathrm{e}^2\lambda$ then

$$P[\mathrm{Po}(\lambda) \geq k] \leq \exp\left(-\left(\frac{k}{2}\right)\log\left(\frac{k}{\lambda}\right)\right). \tag{1.12}$$

Proof Let $X = \mathrm{Po}(\lambda)$. By Markov's inequality, for $z \geq 1$ the probability $P[X \geq k]$ is bounded above by

$$z^{-k}E[z^X] = z^{-k}\mathrm{e}^{\lambda(z-1)} \tag{1.13}$$

and the same expression bounds $P[X \leq k]$ if $z \leq 1$. Putting $z = k/\lambda$, the expression (1.13) becomes

$$\left(\frac{\lambda}{k}\right)^k \mathrm{e}^{k-\lambda} = \exp\left(-\lambda H\left(\frac{k}{\lambda}\right)\right),$$

completing the proof of (1.10) and (1.11). If $k \geq \mathrm{e}^2\lambda$, the fact that for $a \geq \mathrm{e}^2$ we have $H(a) \geq a(\log a - 1) \geq \frac{1}{2}a\log a$, applied to (1.10), yields (1.12). □

Next we give a lower bound on Poisson probabilities, which will show that the preceding upper bounds on the tails are close to being sharp.

Lemma 1.3 *Let* $\mu \geq 0$ *and* $k \in \mathbb{N}$. *Then*

$$P[\mathrm{Po}(\mu) = k] \geq \left(\frac{\mathrm{e}^{-1/(12k)}}{\sqrt{2\pi k}}\right) \exp\left(-\mu H\left(\frac{k}{\mu}\right)\right). \tag{1.14}$$

If $k \geq \mu$, *then*

$$P[\mathrm{Po}(\mu) = k] \geq \left(\frac{\mathrm{e}^{-1/(12k)}}{\sqrt{2\pi k}}\right) \exp\left(-k \log\left(\frac{k}{\mu}\right)\right). \tag{1.15}$$

Proof Robbins' refinement of Stirling's formula (Feller 1968, Section II.9) says that

$$\left(\frac{k}{\mathrm{e}}\right)^k \left(\sqrt{2\pi k}\right) \mathrm{e}^{1/(12k+1)} \leq k! \leq \left(\frac{k}{\mathrm{e}}\right)^k \left(\sqrt{2\pi k}\right) \mathrm{e}^{1/(12k)}$$

and the second inequality yields

$$\begin{aligned} P[\mathrm{Po}(\mu) = k] = \frac{\mathrm{e}^{-\mu}\mu^k}{k!} &\geq \left(\frac{\mathrm{e}^{-1/(12k)}}{\sqrt{2\pi k}}\right) \mathrm{e}^{-\mu}\mu^k \left(\frac{\mathrm{e}}{k}\right)^k \\ &= \left(\frac{\mathrm{e}^{-1/(12k)}}{\sqrt{2\pi k}}\right) \exp\left(-\mu + k - k\log\left(\frac{k}{\mu}\right)\right) \end{aligned}$$

which is the same as the bound in (1.14), and which also implies (1.15) when $k \geq \mu$. □

In some of the proofs it is useful to Poissonize, that is, to first consider instead of $\mathcal{X}_n$ a coupled Poisson process $\mathcal{P}_\lambda$ with λ close to n (see the next section). The following result helps us deduce results about $\mathcal{X}_n$ from results about $\mathcal{P}_\lambda$.

Lemma 1.4 *Let* $\gamma > \frac{1}{2}$. *Then there exists a constant* $\lambda_1 = \lambda_1(\gamma) > 0$ *such that for all* $\lambda > \lambda_1$,

$$\begin{aligned} P[\mathrm{Po}(\lambda) > \lambda + \lambda^\gamma/2] &\leq \exp(-\lambda^{2\gamma-1}/9), \\ P[\mathrm{Po}(\lambda) < \lambda - \lambda^\gamma/2] &\leq \exp(-\lambda^{2\gamma-1}/9). \end{aligned}$$

Proof Since $H''(1) = 1$, Taylor's theorem yields $H(1 + x/2) \geq x^2/9$ for small x. Apply Lemma 1.2 to obtain the result. □

1.7 Poissonization

Poissonization is a key technique in geometric probability. Given $\lambda > 0$, let N_λ be a Poisson random variable, independent of $\{X_1, X_2, X_3, \ldots\}$, and let

$$\mathcal{P}_\lambda := \{X_1, \ldots, X_{N_\lambda}\}. \tag{1.16}$$

As we shall see below, $\mathcal{P}_\lambda$ is a Poisson point process. It is to be assumed throughout the book that for any n, λ, the binomial process $\mathcal{X}_n$ and the Poisson process

$\mathcal{P}_\lambda$ are coupled in this manner. We shall often start by proving limit theorems about $\mathcal{P}_\lambda$ as $\lambda \to \infty$, and then deduce results about $\mathcal{X}_n$ from these.

The next result shows that the point process $\mathcal{P}_\lambda$ has a spatial independence property. Because of this, it is often easier to work with geometric graphs of the form $G(\mathcal{P}_\lambda; r)$ rather than $G(\mathcal{X}_n; r)$. This is somewhat reminiscent of the technique in the Erdös–Rényi setting of proving results first for the case in which edges have independent status, and then deducing similar results for the case where the number of edges included is fixed (see Bollobás (1985, p. 34) and Janson *et al.* (2000, p. 14)).

Suppose $g : \mathbb{R}^d \to [0, \infty)$ is a bounded measurable function. A *Poisson process* with intensity function g is a point process $\mathcal{P}$ in $\mathbb{R}^d$ with the property that for Borel $A \subseteq \mathbb{R}^d$ the random variable $\mathcal{P}(A)$ is Poisson with parameter $\int_A g(x)\mathrm{d}x$ whenever this integral is finite, and if $A_1, \ldots, A_k$ are disjoint Borel subsets of $\mathbb{R}^d$, then the variables $\mathcal{P}(A_i), 1 \leq i \leq k$, are mutually independent. See Kingman (1993) for general information about Poisson processes.

Proposition 1.5 *The point process $\mathcal{P}_\lambda$ is a Poisson process on $\mathbb{R}^d$ with intensity $\lambda f(\cdot)$.*

Proof Suppose $A_1, \ldots, A_k$ are Borel sets forming a partition of $\mathbb{R}^d$. Then for integers $n_1, \ldots, n_k$, if we set $n = \sum_{i=1}^k n_i$,

$$
\begin{aligned}
P\left[\bigcap_{i=1}^k \{\mathcal{P}_\lambda(A_i) = n_i\}\right] &= P[N_\lambda = n] P\left[\bigcap_{i=1}^k \{\mathcal{P}_\lambda(A_i) = n_i\} | N_\lambda = n\right] \\
&= \left(\frac{\mathrm{e}^{-\lambda}\lambda^n}{n!}\right) \frac{n!}{n_1! n_2! \cdots n_k!} \prod_{i=1}^k \left(\int_{A_i} f(x)\mathrm{d}x\right)^{n_i} \\
&\times \prod_{i=1}^k \left(\frac{\exp(-\lambda \int_{A_i} f(x)\mathrm{d}x)(\lambda \int_{A_i} f(x)\mathrm{d}x)^{n_i}}{n_i!}\right).
\end{aligned}
$$

Thus $\mathcal{P}_\lambda(A_i), 1 \leq i \leq k$, are independent Poisson variables with $E\mathcal{P}_\lambda(A_i) = \int_{A_i} \lambda f(x)\mathrm{d}x$ for each i, which proves the result. □

We shall also have occasion to consider a *homogeneous* Poisson point process of intensity λ, denoted $\mathcal{H}_\lambda$. This is a Poisson process on $\mathbb{R}^d$ with constant intensity function $g(x) = \lambda, x \in \mathbb{R}^d$. To reiterate the distinction, throughout this monograph, $\mathcal{P}_\lambda$ is a non-homogeneous Poisson process whose total number of points has mean λ, while $\mathcal{H}_\lambda$ is a homogeneous Poisson process whose total number of points is almost surely infinite.

One aspect of the spatial independence of the Poisson process is the next result, which says that a Poisson process is its own Palm point process; loosely speaking, if it is conditioned to have points at particular locations, the distribution of Poisson points elsewhere is unchanged (see (4.4.3) of Stoyan *et al.* (1995)).

Theorem 1.6 (Palm theory for Poisson processes) *Let $\lambda > 0$. Suppose $j \in \mathbb{N}$ and suppose $h(\mathcal{Y}, \mathcal{X})$ is a bounded measurable function defined on all pairs of the form $(\mathcal{Y}, \mathcal{X})$ with $\mathcal{X}$ a finite subset of $\mathbb{R}^d$ and $\mathcal{Y}$ a subset of $\mathcal{X}$, satisfying $h(\mathcal{Y}, \mathcal{X}) = 0$ except when $\mathcal{Y}$ has j elements. Then*

$$E \sum_{\mathcal{Y} \subseteq \mathcal{P}_\lambda} h(\mathcal{Y}, \mathcal{P}_\lambda) = \frac{\lambda^j}{j!} E h(\mathcal{X}'_j, \mathcal{X}'_j \cup \mathcal{P}_\lambda), \tag{1.17}$$

where the sum on the left-hand side is over all subsets $\mathcal{Y}$ of the random point set $\mathcal{P}_\lambda$, and on the right-hand side the set $\mathcal{X}'_j$ is an independent copy of $\mathcal{X}_j$, independent of $\mathcal{P}_\lambda$.

Proof Conditional on $N_\lambda = n$, the distribution of $\mathcal{P}_\lambda$ is that of a collection $\mathcal{X}_n$ of n independent points with common density f; there are $\binom{n}{j}$ ways to partition this set of points into an ordered pair of disjoint sets of cardinalities $n - j$ and j respectively. By conditioning on N_λ we obtain

$$\begin{aligned} E \sum_{\mathcal{Y} \subseteq \mathcal{P}_\lambda} h(\mathcal{Y}, \mathcal{P}_\lambda) &= \sum_{n=j}^{\infty} \left(\frac{e^{-\lambda} \lambda^n}{n!} \right) \binom{n}{j} E[h(\mathcal{X}'_j, \mathcal{X}'_j \cup \mathcal{X}_{n-j})] \\ &= \frac{\lambda^j}{j!} \sum_{m=0}^{\infty} \left(\frac{e^{-\lambda} \lambda^m}{m!} \right) E\left[h(\mathcal{X}'_j, \mathcal{X}'_j \cup \mathcal{X}_m) \right], \end{aligned} \tag{1.18}$$

where in the last sum we took $m = n - j$. Since expression (1.18) equals the right-hand side of (1.17), we are done. □

Theorem 1.7 *Let $\lambda > 0$. Suppose that $k \in \mathbb{N}$, that $(j_1, j_2, \ldots j_k) \in \mathbb{N}^k$, and, for $i = 1, 2, \ldots, k$, that $h_i(\mathcal{Y})$ is a bounded measurable function defined on all finite subsets $\mathcal{Y} \subset \mathbb{R}^d$ and satisfying $h_i(\mathcal{Y}) = 0$ except when $\mathcal{Y}$ has j_i elements. Then*

$$\begin{aligned} &E \left[\sum_{\mathcal{Y}^{(1)} \subseteq \mathcal{P}_\lambda} \cdots \sum_{\mathcal{Y}^{(k)} \subseteq \mathcal{P}_\lambda} \left(\prod_{i=1}^k h_i(\mathcal{Y}^{(i)}) \right) \mathbf{1}_{\{\mathcal{Y}^{(i)} \cap \mathcal{Y}^{(j)} = \emptyset \text{ for } 1 \leq i < j \leq k\}} \right] \\ &= \prod_{i=1}^k E \sum_{\mathcal{Y} \subseteq \mathcal{P}_\lambda} h_i(\mathcal{Y}^{(i)}). \end{aligned} \tag{1.19}$$

Proof To ease notation we just consider the case $k = 2$, leaving to the reader the straightforward generalization to higher values of k. If $k = 2$, then the left-hand side of (1.19) is the expectation of the sum over all disjoint ordered pairs of subsets of $\mathcal{P}_\lambda$, one with j_1 elements and one with j_2 elements, of the product of h_1 evaluated on the first set with h_2 evaluated on the second. Define the function h on all finite subsets of $\mathbb{R}^d$ as follows: if $\mathcal{Y} \subset \mathbb{R}^d$ has $j_1 + j_2$ elements then set

$$h(\mathcal{Y}) := \sum h_1(\mathcal{Y}_1) h_2(\mathcal{Y} \setminus \mathcal{Y}_1),$$

where the sum is over all $\mathcal{Y}_1 \subset \mathcal{Y}$ of cardinality j_1. Set $h(\mathcal{Y}) = 0$ if $\mathcal{Y}$ does not have $j_1 + j_2$ elements.

Then the left-hand side of (1.19) is equal to $E \sum_{\mathcal{Y} \subset \mathcal{P}} h(\mathcal{Y})$ and by Theorem 1.6 this is equal to

$$\left(\frac{\mu^{j_1+j_2}}{(j_1+j_2)!} \right) Eh(\mathcal{X}_{j_1+j_2}).$$

Since there are $\binom{j_1+j_2}{j_1}$ ways to choose a subset of $\{X_1, X_2, \ldots, X_{j_1+j_2}\}$ with cardinality j_1, this is equal to

$$\begin{aligned}\left(\frac{\mu^{j_1+j_2}}{j_1! j_2!} \right) Eh_1(\mathcal{X}_{j_1}) Eh_2(\mathcal{X}_{j_2}) &= \left(\frac{\mu^{j_1}}{j_1!} \right) Eh_1(\mathcal{X}_{j_1}) \left(\frac{\mu^{j_2}}{j_2!} \right) Eh_2(X_{j_2}) \\ &= \left(E \sum_{\mathcal{Y} \subset \mathcal{P}} h_1(\mathcal{Y}) \right) \left(E \sum_{\mathcal{Y} \subset \mathcal{P}} h_2(\mathcal{Y}) \right),\end{aligned}$$

where the last line comes from a further application of Theorem 1.6. □

1.8 Notes and open problems

At the end of each chapter, any relevant open problems that occur to the author will be given. In this chapter, we describe some general related graphical systems for which one might envisage carrying out a similar programme of research to that described in the present monograph.

Related graph constructions include those where the decision on whether to connect two nearby points depends not only on the distance between them, but also on the positions of other points. Such constructions include the minimal spanning tree, and also graphs such as the nearest-neighbour graph and the Delaunay graph; in the latter, points lying in neighbouring Voronoi cells are connected. For many of these related graph constructions, some of the asymptotic theory is described in Yukich (1998). For further results see Penrose and Yukich (2001, 2003).

Random connection and Boolean models. One generalization of the current setup is to connect two points with a probability which is a decreasing function (the connection function) of the distance between them; another is to make each point have a random type, and to connect two points with a probability which depends on their types as well as the distance between them. Essentially, these extensions are the random connection model and Boolean model, respectively, as described in Meester and Roy (1996). At least in the case of a finite range connection function, much of the present programme can be expected to carry through to these more general models.

Other point processes. In this monograph we restrict attention to geometric graphs on the simplest types of point processes, namely binomial or Poisson point processes. Other point processes of interest in statistical modelling include *Gibbs* and *Markov* point processes; see for example Stoyan *et al.* (1995) and van Lieshout (2000). These may be taken as an alternative to a null hypothesis of a binomial point process, and hence, extending parts of the present programme to such point processes may be of interest.

2

PROBABILISTIC INGREDIENTS

This chapter is concerned with various probabilistic techniques which turn out to be useful in the study of random geometric graphs (non-probabilistic technical material is given elsewhere in the book). These techniques are largely concerned with Poisson and normal approximations; in the first three sections, we use a method developed first by C. Stein and L. Chen in the early 1970s, with many subsequent refinements by others. Stein's method is by no means restricted to the dependency graph setting considered here, or indeed to approximation by Poisson or normal distributions; however, these are the only contexts we shall consider here. See Stein (1986) and Barbour *et al.* (1992) for many other applications of such methods. Subsequent sections of this chapter are concerned with certain martingale-based techniques, and with *ad hoc* but nevertheless useful methods for 'de-Poissonizing' central limit theorems derived for Poisson point processes.

2.1 Dependency graphs and Poisson approximation

Many generalizations are known for the fundamental fact that the distribution of the sum of many independent Bernoulli random variables is approximately Poisson, if their means are all small, and is approximately normal, if their means are all bounded away from zero and from 1. Of particular interest to us here are cases where most, but not all, of the pairs of variables are independent. In this case the notion of *dependency graphs* gives a useful way to express this near-independence.

Suppose (I, E) is a graph with finite or countable vertex set I. For $i, j \in I$ write $i \sim j$ if $\{i, j\} \in E$. For $i \in I$, let $\mathcal{N}_i$ denote the adjacency neighbourhood of i, that is, the set $\{i\} \cup \{j \in I : j \sim i\}$. We say that the graph $(I, \sim)$ is a *dependency graph* for a collection of random variables $(\xi_i, i \in I)$ if for any two disjoint subsets I_1, I_2 of I such that there are no edges connecting I_1 to I_2, the collection of random variables $(\xi_i, i \in I_1)$ is independent of $(\xi_i, i \in I_2)$.

This section contains Poisson approximation results for sums of Bernoulli variables indexed by the vertices of a dependency graph, proved using the Stein–Chen method. Recall the definition of total variation distance d_{TV} at (1.3).

Theorem 2.1 (Arratia *et al.* 1989) *Suppose $(\xi_i, i \in I)$ is a finite collection of Bernoulli random variables with dependency graph $(I, \sim)$. Set $p_i := E[\xi_i] = P[\xi_i = 1]$, and set $p_{ij} := E[\xi_i \xi_j]$. Let $\lambda := \sum_{i \in I} p_i$, and suppose λ is finite. Let $W := \sum_{i \in I} \xi_i$. Then*

$$d_{\rm TV}(W, {\rm Po}(\lambda)) \leq \min(3, \lambda^{-1}) \left(\sum_{i \in I} \sum_{j \in \mathcal{N}_i \setminus \{i\}} p_{ij} + \sum_{i \in I} \sum_{j \in \mathcal{N}_i} p_i p_j \right). \tag{2.1}$$

The Stein–Chen idea goes roughly as follows. Suppose W is a variable with mean $\lambda > 0$, which we suspect to be approximately Poisson. Let $Z := {\rm Po}(\lambda)$. Let $A \subseteq \mathbb{Z}_+$; we need to show that $P[W \in A]$ is close to $P[Z \in A]$. To do this, let $h : \mathbb{Z}_+ \to [0,1]$ be the indicator function $\mathbf{1}_A$, and look for bounded $f = f_A : \mathbb{Z}_+ \to \mathbb{R}$, with $f(0) = 0$, such that for all $w \in \mathbb{Z}_+$,

$$\lambda f(w+1) - w f(w) = h(w) - Eh(Z). \tag{2.2}$$

Once such a function f is found, our objective will be achieved by showing that $E[\lambda f(W+1) - W f(W)]$ is small.

Lemma 2.2 *The solution f to* (2.2)*, with $f(0) = 0$, is bounded and satisfies $|f(k)| \leq 1.25$, for all $k \in \mathbb{Z}_+$, and*

$$|f(k+1) - f(k)| \leq \min(3, \lambda^{-1}), \qquad \forall k \in \mathbb{Z}_+. \tag{2.3}$$

Remark For many purposes, the bound $|f(k+1) - f(k)| \leq 3$ is all we need from (2.3). The full bound (2.3) requires some extra work, and is useful when λ is large.

Proof of Lemma 2.2 To solve the difference equation (2.2), first set $w = 0$ in (2.2) to obtain

$$\lambda f(1) = h(0) - Eh(Z). \tag{2.4}$$

Next, multiply (2.2) by $\lambda^w / w!$ to obtain

$$\frac{\lambda^{w+1} f(w+1)}{w!} - \frac{\lambda^w f(w)}{(w-1)!} = \frac{\lambda^w}{w!}(h(w) - Eh(Z)),$$

and sum from $w = 1$ to $w = k-1$, using also (2.4), to obtain

$$f(k) = \frac{(k-1)!}{\lambda^k} \sum_{w=0}^{k-1} \frac{\lambda^w}{w!}(h(w) - Eh(Z)). \tag{2.5}$$

Since $\sum_{w \geq 0} (\lambda^w / w!)(h(w) - Eh(Z)) = 0$, eqn (2.5) implies that

$$f(k) = -\frac{(k-1)!}{\lambda^k} \sum_{w=k}^{\infty} \frac{\lambda^w}{w!}(h(w) - Eh(Z)). \tag{2.6}$$

Since $|h(w) - Eh(Z)| \leq 1$, putting $m = k - 1 - w$ in (2.5), for $k - 1 < \lambda$ we obtain

$$\begin{aligned} |f(k)| &\leq \frac{1}{\lambda} \sum_{m=0}^{k-1} \frac{(k-1)!}{\lambda^m (k-1-m)!} \\ &\leq \frac{1}{\lambda} \sum_{m=0}^{\infty} \left(\frac{k-1}{\lambda} \right)^m = (\lambda - (k-1))^{-1}. \end{aligned} \tag{2.7}$$

Similarly, putting $m = w - k$ in (2.6), for $k + 1 > \lambda$ we obtain

$$|f(k)| \le \sum_{m=0}^{\infty} \frac{\lambda^m (k-1)!}{(k+m)!} \le \sum_{m=0}^{\infty} \frac{\lambda^m}{k(k+1)^m} = \frac{1}{k(1-(\lambda/(k+1)))} = \frac{k+1}{k(k+1-\lambda)}. \tag{2.8}$$

Using (2.7) for $k \le \lambda + \frac{1}{5}$, and (2.8) for $k \ge \lambda + \frac{1}{5}$, we get $|f(k)| \le \frac{5}{4}$ for $k \ge 2$. Also, for $k = 1$, (2.4) gives us $f(1) = \lambda^{-1}(h(0) - Eh(Z))$, which is maximized over all choices of A by taking $A = \{0\}$, and minimized by taking $A = \{1, 2, 3, \ldots\}$, so that $|f(1)| \le \lambda^{-1}(1 - \mathrm{e}^{-\lambda}) < 1$. Thus for all λ and all $k \in \mathbb{Z}_+$, we have $|f(k)| \le \frac{5}{4}$, and hence for all $k \in \mathbb{Z}_+$, $|f(k+1) - f(k)| \le 3$.

It remains to prove that $f(k+1) - f(k) \le \lambda^{-1}$ for all k. Consider first the special case $A = \{j\}$ with $j \in \mathbb{Z}_+$, $j \ne 0$. Then $E[h(Z)] = \mathrm{e}^{-\lambda}\lambda^j/j!$, and for $k \le j$, (2.5) implies

$$f_{\{j\}}(k) = -\frac{\mathrm{e}^{-\lambda}\lambda^j}{j!} \sum_{w=0}^{k-1} \frac{\lambda^{w-k}(k-1)!}{w!} = -\frac{\mathrm{e}^{-\lambda}\lambda^j}{j!} \sum_{r=1}^{k} \frac{\lambda^{-r}(k-1)!}{(k-r)!}. \tag{2.9}$$

Since each coefficient of λ^{-r} is non-increasing in k, we have $f_{\{j\}}(k+1) - f_{\{j\}}(k) \le 0$ for $k < j$. Also, for $k > j$, by (2.6),

$$f_{\{j\}}(k) = \frac{\mathrm{e}^{-\lambda}\lambda^j}{j!} \sum_{w=k}^{\infty} \frac{\lambda^{w-k}(k-1)!}{w!} = \frac{\mathrm{e}^{-\lambda}\lambda^j}{j!} \sum_{r=0}^{\infty} \frac{\lambda^{r}(k-1)!}{(k+r)!}. \tag{2.10}$$

Again each coefficient of λ^r is decreasing in k so that $f_{\{j\}}(k+1) - f_{\{j\}}(k) < 0$ for $k > j$. Thus, $f_{\{j\}}(k+1) - f_{\{j\}}(k)$ is positive only when $k = j$, and by the middle expression in each of (2.9) and (2.10), its value in this case is given by

$$\begin{aligned} f_{\{j\}}(j+1) - f_{\{j\}}(j) &= \mathrm{e}^{-\lambda} \left(\sum_{w=0}^{j-1} \frac{\lambda^w}{w!j} + \sum_{w=j+1}^{\infty} \frac{\lambda^{w-1}}{w!} \right) \\ &\le \frac{\mathrm{e}^{-\lambda}}{\lambda} \left(\sum_{w=0}^{j-1} \frac{\lambda^{w+1}}{(w+1)!} + \sum_{w=j+1}^{\infty} \frac{\lambda^w}{w!} \right) \le \frac{1}{\lambda}. \end{aligned}$$

Also, note by (2.10) that $f_{\{0\}}(k+1) - f_{\{0\}}(k) \le 0$ for all k.

Now consider general $A \subseteq \mathbb{Z}_+$. By (2.5), f is linear in the input function h, so that $f_A = \sum_{j \in A} f_{\{j\}}$, and so by the above, $f_A(k+1) - f_A(k) \le \lambda^{-1}$ for all k and all A. Also, $f_A + f_{A^c} = f_{\mathbb{Z}_+} = 0$, so that $-(f_A(k+1) - f_A(k)) = f_{A^c}(k+1) - f_{A^c}(k) \le \lambda^{-1}$, and thus $|f_A(k+1) - f_A(k)| \le \lambda^{-1}$, which completes the proof of (2.3). □

Proof of Theorem 2.1 Let $A \subseteq \mathbb{Z}_+$, let $h : \mathbb{Z}_+ \to \mathbb{R}$ be the indicator function $\mathbf{1}_A$, and let $f : \mathbb{Z}_+ \to \mathbb{R}$ be the solution to (2.2) with $f(0) = 0$. Then

$$\begin{aligned} P[Z \in A] - P[W \in A] &= E[Wf(W) - \lambda f(W+1)] \\ &= \sum_{i \in I} E[\xi_i f(W) - p_i f(W+1)] := \Delta_A. \end{aligned}$$

Let $W_i := W - \xi_i$ and $V_i := \sum_{j \notin \mathcal{N}_i} \xi_j$. Then $\xi_i f(W) = \xi_i f(W_i + 1)$, so that

$$\begin{aligned} \Delta_A &= \sum_{i \in I} \left(E[(\xi_i - p_i) f(W_i + 1)] + p_i E[f(W_i + 1) - f(W+1)] \right) \\ &= \sum_{i \in I} \left(E[(\xi_i - p_i)(f(W_i + 1) - f(V_i + 1))] \right. \\ &\quad \left. + p_i E[f(W_i + 1) - f(W+1)] \right), \end{aligned} \tag{2.11}$$

where the last line follows by independence of ξ_i and V_i. By Lemma 2.2, $|f(W_i + 1) - f(W+1)| \le \min(3, \lambda^{-1}) \xi_i$, so that

$$|p_i E[f(W_i + 1) - f(W+1)]| \le \min(3, \lambda^{-1}) p_i^2.$$

Also, $f(W_i + 1) - f(V_i + 1)$ can be written as a telescoping sum over $j \in \mathcal{N}_i \backslash \{i\}$ of terms of the form $f(U + \xi_j) - f(U)$, each of which has modulus bounded by $\min(3, \lambda^{-1}) \xi_j$. Hence,

$$\begin{aligned} |E[(\xi_i - p_i)(f(W_i + 1) - f(V_i + 1))]| &\le E\left[|\xi_i - p_i| \sum_{j \in \mathcal{N}_i \backslash \{i\}} \min(3, \lambda^{-1}) \xi_j \right] \\ &\le \min(3, \lambda^{-1}) \sum_{j \in \mathcal{N}_i \backslash \{i\}} (p_{ij} + p_i p_j). \end{aligned}$$

Combining all these estimates in (2.11) and using the fact that $A \subseteq \mathbb{Z}_+$ is arbitrary gives us (2.1). □

2.2 Multivariate Poisson approximation

The result in the previous section gives circumstances under which a sum of Bernoulli variables whose weak dependence is formalized by a dependency graph is approximately Poisson. In this section, we give circumstances under which a collection of several such sums, as well as being approximately Poisson, are approximately independent. This result is from Arratia *et al.* (1989).

Theorem 2.3 *Suppose $(\xi_i, i \in I)$ is a finite collection of Bernoulli random variables with dependency graph $(I, \sim)$. Set $p_i := E[\xi_i]$, and set $p_{ij} := E[\xi_i \xi_j]$. Let $(I(1), I(2), \ldots, I(d))$ be a partition of I. For $1 \le j \le d$, let $W_j := \sum_{i \in I(j)} \xi_i$,* *and let* $\lambda_j := E[W_j] = \sum_{i\in I(j)} p_i$. *Let* $Z_1,\dots,Z_d$ *be independent Poisson variables with parameters* $\lambda_1,\dots,\lambda_d$ *respectively. Let* $\boldsymbol{W} := (W_1,\dots,W_d)$ *and let* $\boldsymbol{Z} := (Z_1,\dots,Z_d)$. *Then for any* $A \subseteq (\mathbb{Z}_+)^d$,

$$|P[\boldsymbol{W}\in A] - P[\boldsymbol{Z}\in A]| \le 3\left(\sum_{i\in I}\sum_{j\in\mathcal{N}_i\setminus\{i\}} p_{ij} + \sum_{i\in I}\sum_{j\in\mathcal{N}_i} p_ip_j\right). \tag{2.12}$$

Proof Let $h : (\mathbb{Z}_+)^d \to [0,1]$ be the indicator function of A. Define the unit vectors $\boldsymbol{e}_1 = (1,0,\dots,0)$, $\boldsymbol{e}_2 = (0,1,0,\dots,0)$, and so on. For $1\le k\le d$, take bounded $f_k : (\mathbb{Z}_+)^d \to \mathbb{R}$, satisfying $f_k(\boldsymbol{w}) = 0$ whenever $w_k = 0$, and

$$\lambda_k f_k(\boldsymbol{w}+\boldsymbol{e}_k) - w_k f_k(\boldsymbol{w}) = h(\boldsymbol{w}) - Eh(w_1,\dots,w_{k-1},Z_k,w_{k+1},\dots,w_d).$$

For $i\in I$, let $k(i)\in I$ be such that $I(k(i))$ is the set in the partition of I that contains i. Let $\boldsymbol{W}^i = (W_1^i,\dots,W_d^i)$ be the vector $\boldsymbol{W} - \xi_i\boldsymbol{e}_{k(i)}$, and let $\boldsymbol{V}^i = (V_1^i,\dots,V_d^i)$ be the vector $\boldsymbol{W} - \sum_{j\in\mathcal{N}_i}\xi_j\boldsymbol{e}_{k(j)}$. Making a computation similar to (2.11), we have

$$\begin{aligned}
Eh(\boldsymbol{W}) - Eh(Z_1,W_2,\dots,W_d) &= E[\lambda_1 f_1(\boldsymbol{W}+\boldsymbol{e}_1) - W_1 f_1(\boldsymbol{W})]\\
&= -E\sum_{i\in I(1)}\left(\xi_i f_1(\boldsymbol{W}^i+\boldsymbol{e}_1) - p_i f_1(\boldsymbol{W}+\boldsymbol{e}_1)\right)\\
&= -E\sum_{i\in I(1)}\left((\xi_i-p_i)f_1(\boldsymbol{W}^i+\boldsymbol{e}_1) + p_i(f_1(\boldsymbol{W}^i+\boldsymbol{e}_1) - f_1(\boldsymbol{W}+\boldsymbol{e}_1))\right)\\
&= -\sum_{i\in I(1)} E\left[(\xi_i-p_i)(f_1(\boldsymbol{W}^i+\boldsymbol{e}_1) - f_1(\boldsymbol{V}^i+\boldsymbol{e}_1))\right]\\
&\quad -\sum_{i\in I(1)} p_iE\left[f_1(\boldsymbol{W}^i+\boldsymbol{e}_1) - f_1(\boldsymbol{W}+\boldsymbol{e}_1)\right].
\end{aligned} \tag{2.13}$$

The first difference $f_1(\boldsymbol{W}^i+\boldsymbol{e}_1) - f_1(\boldsymbol{V}^i+\boldsymbol{e}_1)$ can be expressed as a telescoping sum, over $j\in\mathcal{N}_i\setminus\{i\}$, of terms of the form $\xi_j(f_1(\boldsymbol{U}+\boldsymbol{e}_{k(j)}) - f_1(\boldsymbol{U}))$, and since $|f_1(\cdot)|$ is uniformly bounded by 1.25 (by Lemma 2.2), each of these has absolute value bounded by $3\xi_j$. Hence the absolute value of the first sum is bounded by the sum $\sum_{i\in I(1)}\sum_{j\in\mathcal{N}_i\setminus\{i\}} 3(p_ip_j + p_{ij})$.

Since $|f_1(\boldsymbol{W}^i+\boldsymbol{e}_1) - f_1(\boldsymbol{W}+\boldsymbol{e}_1)| \le 3\xi_i$, the second sum on the right-hand side is bounded by $3\sum_{i\in I(1)} p_i^2$, and combining these bounds, we have

$$|Eh(\boldsymbol{W}) - Eh(Z_1,W_2,\dots,W_n)| \le 3\sum_{i\in I(1)}\left(p_i^2 + \sum_{j\in\mathcal{N}(i)\setminus\{i\}}(p_ip_j + p_{ij})\right).$$

Next, note that

$$\begin{aligned}
&Eh(Z_1,W_2,\dots,W_d) - Eh(Z_1,Z_2,W_3,\dots,W_d)\\
&\qquad = E[\lambda_2 f_2(Z_1,W_2+1,W_3,\dots,W_d) - W_2 f_2(Z_1,W_2,\dots,W_d)],
\end{aligned}$$

and by a similar argument to (2.13), this is equal to

$$- \sum_{i\in I(2)} E\left[(\xi_i - p_i)(f_2(Z_1, W_2^i + 1, W_3, \dots, W_d) - f_2(Z_1, V_2^i + 1, V_3^i, \dots, V_d^i))\right]$$
$$- \sum_{i\in I(2)} p_i E\left[f_2(Z_1, W_2^i + 1, W_3, \dots, W_d) - f_2(Z_1, W_2 + 1, \dots, W_d)\right].$$

The ith term in the first of these sums is a telescoping sum over $j \in \mathcal{N}_i \backslash (\{i\} \cup I(1))$ of terms of the form $(\xi_i - p_i)\xi_j(f_2(\boldsymbol{U} + \boldsymbol{e}_{k(j)}) - f_2(\boldsymbol{U}))$, and therefore is bounded by $3\sum_{i\in I(2)}\sum_{j\in\mathcal{N}_i\backslash\{i\}}(p_i p_j + p_{ij})$. The absolute value of the second sum is bounded by $3\sum_{i\in I(2)} p_i^2$, and hence

$$|Eh(Z_1, W_2, \dots, W_d) - Eh(Z_1, Z_2, W_3, \dots, W_d)|$$
$$\leq 3 \sum_{i\in I(2)} \left(p_i^2 + \sum_{j\in\mathcal{N}_i\backslash\{i\}} (p_i p_j + p_{ij}) \right).$$

Repeating the process we may successively change the third, fourth, ..., kth coordinates from Z to W, picking up similar error terms each time whose total is bounded by the right-hand side of (2.12). □

2.3 Normal approximation

The main result of this section is on normal approximation for a sum of weakly dependent variables by Stein's method. Throughout this section, for continuous $g : \mathbb{R} \to \mathbb{R}$ we write $\|g\|_\infty$ for $\sup\{g(x) : x \in \mathbb{R}\}$.

Theorem 2.4 *Suppose $(\xi_i)_{i\in I}$ is a finite collection of random variables with dependency graph $(I, \sim)$ with maximum degree $D - 1$, with $E[\xi_i] = 0$ for each i. Set $W := \sum_{i\in I}\xi_i$, and suppose $E[W^2] = 1$. Let $Z = \mathcal{N}(0, 1)$. Then, for all $t \in \mathbb{R}$,*

$$|P[W \leq t] - P[Z \leq t]| \leq 2(2\pi)^{-1/4}\sqrt{D^2 \sum_{i\in I} E[|\xi_i|^3]}$$
$$+6\sqrt{D^3 \sum_{i\in I} E[|\xi_i|^4]}. \tag{2.14}$$

Let $h : \mathbb{R} \to \mathbb{R}$ be an arbitrary bounded and continuous test function with bounded piecewise continuous derivative. The plan for proving Theorem 2.4 is to show that $Eh(W)$ is close to $Eh(Z)$. The first step is to look for a bounded $g : \mathbb{R} \to \mathbb{R}$ satisfying the differential equation

$$g'(x) - xg(x) = h(x) - Eh(Z). \tag{2.15}$$

This is the analogue, in the normal approximation setting, to the difference equation (2.2) used for Poisson approximation. Once g is found, the idea will be to show that $E[g'(W) - Wg(W)]$ is small.

The left-hand side of (2.15), multiplied by the integrating factor $e^{-x^2/2}$, is the derivative of $e^{-x^2/2}g(x)$. Therefore, (2.15) is solved (with one particular choice of constant of integration) by setting

$$g(x) = e^{x^2/2} \int_{-\infty}^{x} e^{-y^2/2}(h(y) - Eh(Z))dy. \tag{2.16}$$

Since $\int_{-\infty}^{\infty}(h(y) - Eh(Z))e^{-y^2/2}dy = 0$, eqn (2.16) implies the alternative formula

$$g(x) = -e^{x^2/2} \int_{x}^{\infty} e^{-y^2/2}(h(y) - Eh(Z))dy. \tag{2.17}$$

To establish boundedness properties of g and its derivatives, we shall use the following analytical fact.

Lemma 2.5 *Let $w \in \mathbb{R}$. Then*

$$w + (1 + w^2)e^{w^2/2} \int_{-\infty}^{w} e^{-x^2/2}dx > 0. \tag{2.18}$$

Proof Clearly (2.18) holds for $w \geq 0$, so now assume $w < 0$. By an integration by parts, the left-hand side of (2.18) is equal to the expression

$$w + e^{w^2/2} \int_{-\infty}^{w} e^{-x^2/2}dx + w^2 e^{w^2/2}\left(-w^{-1}e^{-w^2/2} - \int_{-\infty}^{w} x^{-2}e^{-x^2/2}dx\right)$$
$$= e^{w^2/2} \int_{-\infty}^{w}\left(1 - \frac{w^2}{x^2}\right)e^{-x^2/2}dx > 0. \qquad \square$$

Lemma 2.6 *Let g be given by (2.16) above. Then $\|g\|_\infty < \infty$ and $\|g'\|_\infty \leq 2\|h - Eh(Z)\|_\infty$.*

Proof Let $K := \|h - Eh(Z)\|_\infty$, that is, $K := \sup_{y\in\mathbb{R}} |h(y) - Eh(Z)|$ (which is finite since h is bounded). First suppose $x > 0$. Then using (2.17) and integrating by parts, we have

$$|g(x)| \leq e^{x^2/2} \int_{x}^{\infty} e^{-y^2/2}K dy = Ke^{x^2/2} \int_{x}^{\infty}\left(-\frac{1}{y}\right)\left(\frac{d}{dy}e^{-y^2/2}\right)dy$$
$$= Ke^{x^2/2}\left(\frac{1}{x}e^{-x^2/2} - \int_{x}^{\infty}\left(\frac{1}{y^2}\right)e^{-y^2/2}dy\right) \leq \frac{K}{x}.$$

For $x < 0$, using (2.16) and setting $z = -y$, we have

$$|g(x)| \leq Ke^{x^2/2} \int_{-\infty}^{x} e^{-y^2/2}dy = Ke^{|x|^2/2} \int_{|x|}^{\infty} e^{-z^2/2}dz \leq \frac{K}{|x|}.$$

Thus $\sup_{x\in\mathbb{R}}\{|xg(x)|\} \leq K$, and g is continuous so also $\sup_{|x|\leq 1}\{|g(x)|\} < \infty$. Hence $\sup_{x\in\mathbb{R}}\{|g(x)|\} < \infty$. Applying (2.15), we have

$$\sup_{x\in\mathbb{R}} |g'(x)| \leq \sup_{x\in\mathbb{R}} |xg(x)| + \sup_{x\in\mathbb{R}} |h(x) - Eh(Z)| \leq 2K. \qquad \square$$

Lemma 2.7 *With g as above, $\|g''\|_\infty \leq 2\|h'\|_\infty$.*

Proof Set $\phi(y) := (2\pi)^{-1/2}\exp(-y^2/2)$ and $\Phi(x) := \int_{-\infty}^x \phi(y)\mathrm{d}y$, the standard normal density and distribution functions respectively. Then

$$\begin{aligned} h(x) - Eh(Z) &= \int_{-\infty}^{\infty} (h(x) - h(y))\phi(y)\mathrm{d}y \\ &= \int_{-\infty}^{x}\int_{y}^{x} h'(z)\phi(y)\mathrm{d}z\mathrm{d}y - \int_{x}^{\infty}\int_{x}^{y} h'(z)\phi(y)\mathrm{d}z\mathrm{d}y, \end{aligned} \tag{2.19}$$

and by Fubini's theorem,

$$h(x) - Eh(Z) = \int_{-\infty}^{x} h'(z)\Phi(z)\mathrm{d}z - \int_{x}^{\infty} h'(z)(1-\Phi(z))\mathrm{d}z. \tag{2.20}$$

Substituting (2.19) in (2.16) gives us

$$\begin{aligned} g(w) &= \mathrm{e}^{w^2/2}\Bigg(\iiint\limits_{y<z<x<w} h'(z)\phi(y)\mathrm{e}^{-x^2/2}\mathrm{d}z\mathrm{d}y\mathrm{d}x \\ &\quad - \iiint\limits_{x<z<y,x<w} h'(z)\phi(y)\mathrm{e}^{-x^2/2}\mathrm{d}z\mathrm{d}y\mathrm{d}x\Bigg) \\ &= \sqrt{2\pi}\mathrm{e}^{w^2/2}\Bigg(\int_{-\infty}^{w} h'(z)\Phi(z)[\Phi(w)-\Phi(z)]\mathrm{d}z \\ &\quad - \int_{-\infty}^{w} h'(z)(1-\Phi(z))\Phi(z)\mathrm{d}z - \int_{w}^{\infty} h'(z)(1-\Phi(z))\Phi(w)\mathrm{d}z\Bigg) \\ &= -\sqrt{2\pi}\mathrm{e}^{w^2/2}\Bigg((1-\Phi(w))\int_{-\infty}^{w} h'(z)\Phi(z)\mathrm{d}z \\ &\quad + \Phi(w)\int_{w}^{\infty} h'(z)(1-\Phi(z))\mathrm{d}z\Bigg). \end{aligned} \tag{2.21}$$

By definition, $g'(w) - wg(w) = h(w) - Eh(Z)$, and hence, differentiating, we have

$$\begin{aligned} g''(w) &= wg'(w) + g(w) + h'(w) \\ &= w(wg(w) + h(w) - Eh(Z)) + g(w) + h'(w). \end{aligned}$$

Hence, substituting from (2.20) and (2.21), we obtain

$$
\begin{aligned}
g''(w) &= h'(w) + (1+w^2)g(w) + w(h(w) - Eh(Z)) \\
&= h'(w) - (1+w^2)\sqrt{2\pi}\mathrm{e}^{w^2/2} \\
&\quad \times \left((1-\Phi(w)) \int_{-\infty}^{w} h'(z)\Phi(z)\mathrm{d}z + \Phi(w) \int_{w}^{\infty} h'(z)(1-\Phi(z))\mathrm{d}z \right) \\
&\quad + w \left(\int_{-\infty}^{w} h'(z)\Phi(z)\mathrm{d}z - \int_{w}^{\infty} h'(z)(1-\Phi(z))\mathrm{d}z \right) \\
&= h'(w) + \left(w - \sqrt{2\pi}(1+w^2)\mathrm{e}^{w^2/2}(1-\Phi(w)) \right) \int_{-\infty}^{w} h'(z)\Phi(z)\mathrm{d}z \\
&\quad + \left(-w - \sqrt{2\pi}(1+w^2)\mathrm{e}^{w^2/2}\Phi(w) \right) \int_{w}^{\infty} h'(z)(1-\Phi(z))\mathrm{d}z. \qquad (2.22)
\end{aligned}
$$

For all $w \in \mathbb{R}$, by (2.18) applied to w and to $-w$ we have

$$
\begin{aligned}
w + \sqrt{2\pi}(1+w^2)\mathrm{e}^{w^2/2}\Phi(w) &> 0, \\
-w + \sqrt{2\pi}(1+w^2)\mathrm{e}^{w^2/2}(1-\Phi(w)) &> 0.
\end{aligned}
$$

Therefore, by (2.22),

$$
\begin{aligned}
|g''(w)| \le \|h'\|_\infty \Bigg(1 &+ (-w + \sqrt{2\pi}(1+w^2)\mathrm{e}^{w^2/2}(1-\Phi(w))) \int_{-\infty}^{w} \Phi(x)\mathrm{d}x \\
&+ (w + \sqrt{2\pi}(1+w^2)\mathrm{e}^{w^2/2}\Phi(w)) \int_{w}^{\infty} (1-\Phi(x))\mathrm{d}x \Bigg).
\end{aligned}
$$

Carrying out the integrals $\int \Phi(x)\mathrm{d}x$ and $\int (1-\Phi(x))\mathrm{d}x$ by parts, and using also the fact that $x\phi(x) = -\phi'(x)$, we find that for all $w \in \mathbb{R}$,

$$
\begin{aligned}
\frac{|g''(w)|}{\|h'\|_\infty} &\le 1 + \left(-w + \sqrt{2\pi}(1+w^2)\mathrm{e}^{w^2/2}(1-\Phi(w)) \right) (w\Phi(w) + \phi(w)) \\
&\quad + \left(w + \sqrt{2\pi}(1+w^2)\mathrm{e}^{w^2/2}\Phi(w) \right) (-w(1-\Phi(w)) + \phi(w)) \\
&= 1 - w^2 + \sqrt{2\pi}(1+w^2)\mathrm{e}^{w^2/2}\phi(w) = 2,
\end{aligned}
$$

as asserted. □

Proof of Theorem 2.4 Let $h : \mathbb{R} \to \mathbb{R}$ be bounded and continuous with bounded, piecewise continuous derivative. Let g be given by (2.16) above. We first prove that

$$
\begin{aligned}
|E[g'(W) - Wg(W)]| &\le D^2 \|h'\|_\infty \sum_{i \in I} E[|\xi_i|^3] \\
&\quad + 6\|h - Eh(Z)\|_\infty \sqrt{D^3 \sum_{i \in I} E[|\xi_i|^4]}. \qquad (2.23)
\end{aligned}
$$

For each i, set $W_i = \sum_{j \notin \mathcal{N}_i} \xi_j$, which is independent of ξ_i. We have the following:

$$\begin{aligned} E[g'(W) - Wg(W)] &= E[g'(W)] - \sum_i E[\xi_i g(W)] \\ &= E[g'(W)] + \sum_i E[\xi_i(g(W_i) - g(W))] \quad \text{(independence)} \\ &= \tau + \rho \quad \text{(Taylor expansion)}, \end{aligned} \tag{2.24}$$

where we set

$$\tau = E[g'(W)] + \sum_i E[\xi_i(W_i - W)g'(W)]$$

and

$$\rho = \sum_i E[\xi_i(g(W_i) - g(W) - (W_i - W)g'(W))].$$

We need to show that τ and ρ are small. First consider ρ. By Taylor's theorem, the quantity $|g(W_i) - g(W) - (W_i - W)g'(W)|$ is bounded by $\frac{1}{2}\|g''\|_\infty (W - W_i)^2$, and so, using Lemma 2.7 and taking expectations, we obtain

$$|\rho| \le \sum_i \|h'\|_\infty E[|\xi_i|(W - W_i)^2] \le \|h'\|_\infty \sum_i \sum_{j,k \in \mathcal{N}_i} E[|\xi_i||\xi_j||\xi_k|],$$

and so by the arithmetic–geometric mean inequality.

$$|\rho| \le \frac{1}{3}\|h'\|_\infty \sum_i \sum_{j,k \in \mathcal{N}_i} (E[|\xi_i|^3] + E[|\xi_j|^3] + E[|\xi_k|^3]).$$

The number of pairs (j, k) with $j \in \mathcal{N}_i, k \in \mathcal{N}_i$ is at most D^2, as is the number of pairs (i, k) with $i \in \mathcal{N}_j, k \in \mathcal{N}_i$. Thus,

$$|\rho| \le D^2 \|h'\|_\infty \sum_i E[|\xi_i|^3]. \tag{2.25}$$

Next look at the other remainder term τ. Let $\sigma_{ij} = E[\xi_i \xi_j]$ for each pair (i, j). By the conditions in the statement of the theorem, $\sum_i \sum_{j \in \mathcal{N}_i} \sigma_{ij} = 1$. Hence,

$$\tau = E\left[g'(W) \sum_i \sum_{j \in \mathcal{N}_i} (\sigma_{ij} - \xi_i \xi_j) \right],$$

so that

$$\tau^2 \le \|g'\|_\infty^2 E\left[\left(\sum_i \sum_{j \in \mathcal{N}_i} (\sigma_{ij} - \xi_i \xi_j) \right)^2 \right] = \|g'\|_\infty^2 \operatorname{Var}\left[\sum_i \sum_{j \in \mathcal{N}_i} \xi_i \xi_j \right]$$

$$= \|g'\|_\infty^2 \left(E\left[\left(\sum_i \sum_{j \in \mathcal{N}_i} \xi_i \xi_j \right)^2 \right] - 1 \right).$$

Expanding the square in the last line above, we get a quadruple sum of terms $E[\xi_i\xi_j\xi_k\xi_l]$ over (i,j,k,l) with $j \in \mathcal{N}_i$ and $l \in \mathcal{N}_k$. We split this into a sum $\sum'$ over quadruples (i,j,k,l) with $j \in \mathcal{N}_i$ and $l \in \mathcal{N}_k$ and $\{k,l\} \cap (\mathcal{N}_i \cup \mathcal{N}_j) \neq \emptyset$, and a sum $\sum''$ over (i,j,k,l) with $j \in \mathcal{N}_i$ and $l \in \mathcal{N}_k$ and $\{k,l\} \cap (\mathcal{N}_i \cup \mathcal{N}_j) = \emptyset$. This gives us

$$\begin{aligned} E\left[\left(\sum_i \sum_{j \in \mathcal{N}_i} \xi_i\xi_j\right)^2\right] &= \sum{}' \; E[\xi_i\xi_j\xi_k\xi_l] + \sum{}'' \; E[\xi_i\xi_j\xi_k\xi_l] \\ &= \sum{}' \; E[\xi_i\xi_j\xi_k\xi_l] + \sum{}'' \; \sigma_{ij}\sigma_{kl}. \end{aligned}$$

Since $(\sum_i \sum_{j \in \mathcal{N}_i} \sigma_{ij})^2 = 1$, we have $\sum' \sigma_{ij}\sigma_{kl} + \sum'' \sigma_{ij}\sigma_{kl} = 1$, so that

$$E\left[\left(\sum_i \sum_{j \in \mathcal{N}_i} \xi_i\xi_j\right)^2\right] - 1 = \sum{}' \; (E[\xi_i\xi_j\xi_k\xi_l] - \sigma_{ij}\sigma_{kl}).$$

For each i the number of (j,k,l) in the sum $\sum'$ is at most $4D^3$. Similarly, for each j the number of (i,k,l) in the sum $\sum'$ is at most $4D^3$, and so on. By the arithmetic–geometric mean inequality the absolute value of the first term $\sum' E[\xi_i\xi_j\xi_k\xi_l]$ is bounded by $4D^3 \sum_i E[\xi_i^4]$. The other term is also bounded by $4D^3 \sum_i E[\xi_i^4]$ for similar reasons, since $\sigma_{ij}\sigma_{kl} = E\xi_i\xi_j\xi'_k\xi'_l$ where (ξ'_k, ξ'_l) is an independent copy of (ξ_k, ξ_l). Hence,

$$\tau^2 \leq 8\|g'\|_\infty^2 D^3 \sum_i E[\xi_i^4] \leq 32\|h - Eh(Z)\|_\infty^2 D^3 \sum_i E[\xi_i^4].$$

Combining this with (2.25) in (2.24) gives us (2.23). It is immediate from this and (2.15) that

$$\begin{aligned} |Eh(W) - Eh(Z)| \leq{} & D^2\|h'\|_\infty \sum_{i \in I} E[|\xi_i|^3] \\ & +6\|h - Eh(Z)\|_\infty \sqrt{D^3 \sum_{i \in I} E[|\xi_i|^4]}. \end{aligned} \tag{2.26}$$

It remains to deduce (2.14) from (2.26) by choosing h in a suitable way. Given t, we make the following rather obvious choice of h: set $h(x) = 1$ for $x \leq t$ and $h(x) = 0$ for $x \geq t + \Delta$, and take h to be continuous everywhere and linear on $[x, x + \Delta]$. The constant Δ will be selected below.

Set $A_3 = D^2 \sum_{i \in I} E[|\xi_i|^3]$ and $A_4 = D^3 \sum_{i \in I} E[|\xi_i|^4]$. Then, by (2.26),

$$\begin{aligned} P[W \leq t] &\leq Eh(W) \leq Eh(Z) + \Delta^{-1}A_3 + 6\sqrt{A_4} \\ &\leq P[Z \leq t + \Delta] + \Delta^{-1}A_3 + 6\sqrt{A_4} \\ &\leq P[Z \leq t] + (2\pi)^{-1/2}\Delta + \Delta^{-1}A_3 + 6\sqrt{A_4}, \end{aligned}$$

and setting $\Delta = (2\pi)^{1/4}\sqrt{A_3}$, we obtain

$$P[W \le t] \le P[Z \le t] + 2(2\pi)^{-1/4}\sqrt{A_3} + 6\sqrt{A_4}.$$

Similarly, applying (2.26) to the function $\tilde{h}(x) = h(x + \Delta)$, we obtain

$$P[W \le t] \ge P[Z \le t] - 2(2\pi)^{-1/4}\sqrt{A_3} - 6\sqrt{A_4}.$$

Combining these bounds gives us (2.14). □

2.4 Martingale theory

If $(M_1, M_2, \dots, M_n)$ is a martingale with respect to a filtration $(\mathcal{F}_1, \mathcal{F}_2, \dots, \mathcal{F}_n)$, then the variables $D_1, \dots, D_n$ defined by $D_i = M_i - M_{i-1}$ (with $D_1 = M_1 - EM_1$) are said to form a *martingale difference sequence.* The following result can be very useful in proving the concentration of the distribution of variables arising in geometrical probability.

Theorem 2.8 (Azuma's inequality) *Let $(M_1, \dots, M_n)$ be a martingale with corresponding martingale difference sequence $D_1, \dots, D_n$. Then for any $a > 0$,*

$$P\left[\left|\sum_{i=1}^n D_i\right| > a\right] \le 2\exp\left(-\frac{a^2}{2\sum_{i=1}^n \|D_i\|_\infty^2}\right),$$

where, as usual, $\|D_i\|_\infty$ denotes the infimum of all b such that $P[|D_i| \le b] = 1$.

For a proof, see, for example, Williams (1991), Steele (1997), or Yukich (1998). The latter two references demonstrate many applications in geometric probability.

Sometimes Azuma's inequality on its own is not useful because the numbers $\|D_i\|_\infty$ are insufficiently small; one can retrieve the situation sometimes in cases where there is some 'sufficiently small' b with $P[|D_i| \ge b]$ also small.

Theorem 2.9 (Chalker *et al.* 1999) *Let $M_1, \dots, M_n$ be a martingale with corresponding martingale difference sequence $D_1, \dots, D_n$. Then for any $a > 0$ and for any $b > 0$,*

$$P\left[\left|\sum_{i=1}^n D_i\right| > a\right] \le 2\exp\left(-\frac{a^2}{32nb^2}\right) + \left(1 + \frac{2\sup_i \|D_i\|_\infty}{a}\right)\sum_{i=1}^n P[|D_i| > b].$$

Proof Let $D_i^* := D_i \mathbf{1}_{\{|D_i| \le b\}}$. Then

$$P\left[\left|\sum_{i=1}^n D_i\right| > a\right] \le P\left[\bigcup_{i=1}^n \{|D_i| > b\}\right] + P\left[\left|\sum_{i=1}^n D_i^*\right| > a\right],$$

and setting $D_i' = D_i^* - E[D_i^*|\mathcal{F}_{i-1}]$, we have

$$P\left[\left|\sum_{i=1}^n D_i^*\right| > a\right] \le P\left[\left|\sum_{i=1}^n D_i'\right| > a/2\right] + P\left[\left|\sum_{i=1}^n E[D_i^*|\mathcal{F}_{i-1}]\right| > a/2\right].$$

Since the D'_i form a martingale difference sequence with $\|D'_i\|_\infty \le 2b$, Azuma's inequality can be applied to the first of these probabilities, and Markov's inequality to the second, to obtain

$$P\left[\left|\sum_{i=1}^n D_i^*\right| > a\right] \le 2\exp(-a^2/(32nb^2)) + (2/a)E\left[\left|\sum_{i=1}^n E[D_i^*|\mathcal{F}_{i-1}]\right|\right].$$

By the martingale property,

$$|E[D_i^*|\mathcal{F}_{i-1}]| = |E[D_i \mathbf{1}_{\{|D_i|>b\}}|\mathcal{F}_{i-1}]| \le \|D_i\|_\infty P[|D_i| > b|\mathcal{F}_i],$$

and therefore

$$E\left[\left|\sum_{i=1}^n E[D_i^*|\mathcal{F}_{i-1}]\right|\right] \le \sup_i \|D_i\|_\infty \sum_{i=1}^n P[|D_i| > b].$$

Combining all this gives us the result. □

Also of use to us is the following central limit theorem of McLeish (1974).

Theorem 2.10 (Central limit theorem for martingale difference arrays). *Suppose that $k_n, n \ge 1$, is an $\mathbb{N}$-valued sequence with $k_n \to \infty$ as $n \to \infty$. Suppose that for each $n \in \mathbb{N}$, the sequence $(M_{n,1}, \ldots, M_{n,k_n})$ is a martingale with respect to some filtration, let $M_{n,0} = E[M_{n,1}]$, and let $D_{n,1}, \ldots, D_{n,n}$ be the corresponding sequence of martingale differences $D_{n,i} = M_{n,i} - M_{n,i-1}$. Suppose that*

$$\sup_{n\ge 1} E\left(\max_{j\le k_n} |D_{n,j}|^2\right) < \infty, \tag{2.27}$$

$$\max_{j\le k_n} |D_{n,j}| \xrightarrow{P} 0 \quad \text{as } n \to \infty, \tag{2.28}$$

and for some $\sigma > 0$,

$$\sum_{j=1}^{k_n} D_{n,j}^2 \xrightarrow{P} \sigma^2. \tag{2.29}$$

Then $M_{n,k_n} - M_{n,0} \xrightarrow{\mathcal{D}} \mathcal{N}(0,\sigma^2)$ as $n \to \infty$.

Proof For each n set $D'_{n,1} := D_{n,1}$ and for $j = 2, 3, \ldots, k_n$ set

$$D'_{n,j} := D_{n,j} \mathbf{1}_{\{\sum_{l=1}^{j-1} D_{n,l}^2 \le 2\sigma^2\}}$$

and set $M'_{n,j} = \sum_{l=1}^j D'_{n,j}$. Then $(M'_{n,1}, \ldots, M'_{n,k_n})$ is also a martingale and

$$P\left[\bigcup_{j=1}^{k_n} \{M_{n,j} \ne M_{n,j} - M_{n,0}\}\right] \le P\left[\sum_{l=1}^{k_n} D_{n,l}^2 \ge 2\sigma^2\right] \to 0$$

as $n \to \infty$. Hence, it suffices to show that $M'_{n,k_n} \xrightarrow{\mathcal{D}} \mathcal{N}(0,\sigma^2)$.

Let $\mathrm{i} = \sqrt{-1}$. Given $t \in \mathbb{R}$, define complex random variables $T_n := \prod_{j=1}^{k_n}(1 + \mathrm{i}tD'_{n,j})$ and $Y_n := \mathrm{e}^{\mathrm{i}tM'_{n,k_n}}$. For real x, define

$$r(x) := -(\log(1 + \mathrm{i}x) - \mathrm{i}x - x^2/2). \tag{2.30}$$

Then $|r(x)| \leq 1$ for $|x| \leq \frac{1}{2}$ since for $|z| \leq 1$ the absolute value of the complex power series $f(z) = \log(1-z)+z+z^2/2$ is bounded by $|f(|z|)|$, and $|f(t)| \leq 1$ for $0 < t \leq \frac{1}{2}$. By (2.30) for real x it is the case that $\mathrm{e}^{\mathrm{i}x} = (1+\mathrm{i}x)\exp(-x^2/2+r(x))$, so that

$$Y_n = \prod_{j=1}^{k_n} \mathrm{e}^{\mathrm{i}tD'_{n,j}} = T_n U_n,$$

where we set

$$U_n = \exp\left(-\frac{1}{2}\sum_{j=1}^{k_n} {D'_{n,j}}^2 + \sum_{j=1}^{k_n} r(D'_{n,j})\right).$$

By Lévy's theorem on the equivalence of convergence in distribution and convergence of characteristic functions (see, e.g., Williams (1991)), it suffices to prove that $E[Y_n] \to \exp(-t^2\sigma^2/2)$ for all real t. Observe first that $E[T_n] = 1$ by definition of T_n and the martingale property. Also, by (2.28), except on an event with probability tending to zero

$$\left|\sum_{j=1}^{k_n} r(tD'_{n,j})\right| \leq |t|^3 \sum_{j=1}^{k_n} |D'_{n,j}|^3 \leq |t|^3 \max_{j \leq k_n} |D_{n,j}| \sum_{j=1}^{k_n} D_{n,j}^2,$$

which tends to 0 in probability by (2.28) and (2.29). Thus $U_n \to \exp(-\sigma^2 t^2/2)$ in probability, so $T_n(U_n - \mathrm{e}^{-\sigma^2t^2/2}) \xrightarrow{P} 0$, and so it suffices to prove that the variables $T_n(U_n - \mathrm{e}^{-\sigma^2t^2/2}) = Y_n - T_n\mathrm{e}^{-\sigma^2t^2/2}$ are uniformly integrable. Since $|Y_n| = 1$ for all n, it suffices to prove that the variables T_n are uniformly integrable.

Define $J_n := \min\{l : \sum_{j=1}^{l} D_{n,l}^2 > 2\sigma^2\}$, if this set is non-empty, and $J_n = k_n$ otherwise. Then $D'_{n,l} = 0$ for $J_n < l \leq k_n$, and

$$\begin{aligned} E[|T_n|^2] &= E\left[\prod_{j=1}^{k_n}(1 + t^2 {D'_{n,j}}^2)\right] \\ &\leq E\exp\left(t^2\sum_{j=1}^{J_n-1} D_{n,j}^2\right)(1 + t^2 D_{n,J_n}^2) \leq \mathrm{e}^{2\sigma^2t^2}(1 + t^2 E[D_{n,J_n}^2]), \end{aligned}$$

and by (2.27), this is uniformly bounded, so that the variables T_n are uniformly integrable. □

To conclude this section, we give a further application of Azuma's inequality. This will not be used until Chapter 12. Suppose W_i are independent identically distributed Poisson variables and $\varepsilon > 0$. We shall require estimates on the rate of exponential decay of $P[\sum_{i=1}^n (W_i^2 - E[W_i^2]) > \varepsilon n]$, which is not amenable to standard methods because the square of a Poisson variable does not have a well-behaved moment generating function. The following result is essentially the best possible of this type.

Lemma 2.11 *Suppose that* $W_1, W_2, W_3, \ldots$ *are independent* $\mathrm{Po}(\lambda)$ *random variables with* $\lambda \in (0, \infty]$. *Let* $\varepsilon > 0$. *Then*

$$\limsup_{n\to\infty} \left(n^{-1/2} \log P \left[\sum_{i=1}^n (W_i^2 - EW_i^2) > \varepsilon n \right] \right) < 0. \tag{2.31}$$

Proof Define a sequence of integers $(\beta_n)_{n\geq 2}$ by

$$P[W_1 \geq \beta_n] > n^{-1} \geq P[W_1 \geq \beta_n + 1], \qquad n \geq 2. \tag{2.32}$$

By (1.12), $P[W_1 \geq \log n] \leq n^{-1}$ for large enough n, and hence $\beta_n + 1 \leq \log n$ for large enough n. Hence,

$$P[W_1 = \beta_n] = P[W_1 = \beta_n + 1](\beta_n + 1)/\lambda \leq (\log n)/(\lambda n). \tag{2.33}$$

By Azuma's inequality (Theorem 2.8) applied to the martingale with successive increments given by the independent variables $W_i^2 \mathbf{1}_{\{W_i < \beta_n\}} - E[W_i^2 \mathbf{1}_{\{W_i < \beta_n\}}]$, which are uniformly bounded by β_n^2, we obtain for large enough n that

$$\begin{aligned} P\left[\sum_{i=1}^n (W_i^2 \mathbf{1}_{\{W_i < \beta_n\}} - EW_i^2) \geq \varepsilon n \right] &\leq 2 \exp\left\{ -\frac{(\varepsilon n)^2}{2n\beta_n^4} \right\} \\ &\leq 2\exp\left\{ -\frac{\varepsilon^2 n}{2(\log n)^4} \right\}. \end{aligned} \tag{2.34}$$

Choose $\delta \in (0, \varepsilon^{1/2})$. By (2.32) and (2.33), the mean of the binomial variable $\sum_{i=1}^n \mathbf{1}_{\{W_i \geq \beta_n\}}$ is bounded by $1 + \lambda^{-1} \log n$. Hence by (1.7), for large n we have

$$\begin{aligned} P\left[\sum_{i=1}^n \mathbf{1}_{\{W_i \geq \beta_n\}} > \frac{\delta n^{1/2}}{\log n} \right] &\leq \exp\left(-\left(\frac{\delta n^{1/2}}{2 \log n} \right) \log \left(\frac{\delta n^{1/2}}{(1 + \lambda^{-1} \log n) \log n} \right) \right) \\ &\leq \exp(-\delta n^{1/2}/5). \end{aligned} \tag{2.35}$$

For each n, let $(Z_{i,n}, i \geq 1)$ be independent identically distributed variables with the conditional distribution of W_1 given that $W_1 \geq \beta_n$, that is, with $P[Z_{i,n} \leq t] = P[W_1 \leq t | W_i \geq \beta_n]$ for all real t. Then by (2.32),

$$E[\exp(Z_{i,n})] \leq \frac{E[\exp(W_i)]}{P[W_i \geq \beta_n]} \leq \exp(\lambda(\mathrm{e} - 1) + \log n),$$

so that

$$P\left[\sum_{i=1}^{n} W_i^2 \mathbf{1}_{\{W_i \geq \beta_n\}} > \varepsilon n \middle| \sum_{i=1}^{n} \mathbf{1}_{\{W_i \geq \beta_n\}} \leq \delta n^{1/2}/\log n\right]$$
$$\leq P\left[\sum_{i=1}^{\lfloor \delta n^{1/2}/\log n \rfloor} Z_{i,n}^2 > \varepsilon n\right] \leq P\left[\sum_{i=1}^{\lfloor \delta n^{1/2}/\log n \rfloor} Z_{i,n} > \varepsilon^{1/2} n^{1/2}\right]$$
$$\leq \exp\{(\delta n^{1/2}/\log n)(\lambda(\mathrm{e}-1)+\log n) - \varepsilon^{1/2} n^{1/2}\}, \tag{2.36}$$

which decays exponentially in $n^{1/2}$. Combining (2.34)–(2.36) yields (2.31). □

2.5 De-Poissonization

The techniques described in the preceding sections for proving central limit theorems are most naturally applied to geometric graphs on the Poissonized (and therefore spatially independent) point process $\mathcal{P}_n$ described in Section 1.7. We now give a result on recovering central limit theorems for $\mathcal{X}_n$ from those obtained for $\mathcal{P}_n$. It is stated in general terms, in terms of a sequence of functionals $(H_n)_{n \geq 1}$ defined on finite point sets in $\mathbb{R}^d$ with the property that the increment

$$R_{m,n} := H_n(\mathcal{X}_{m+1}) - H_n(\mathcal{X}_m) \tag{2.37}$$

is close in mean to a constant α, when m is close to n.

Theorem 2.12 *Suppose that for each $n \in \mathbb{N}$ the real-valued functional $H_n(\mathcal{X})$ is defined for all finite sets $\mathcal{X} \subset \mathbb{R}^d$. Suppose that for some $\sigma^2 \geq 0$ we have $n^{-1}\mathrm{Var}(H_n(\mathcal{P}_n)) \to \sigma^2$ and*

$$n^{-1/2}(H_n(\mathcal{P}_n) - EH_n(\mathcal{P}_n)) \xrightarrow{\mathcal{D}} \mathcal{N}(0, \sigma^2),$$

as $n \to \infty$. Suppose also that there are constants $\alpha \in \mathbb{R}$ and $\gamma > \frac{1}{2}$ such that the increments $R_{m,n}$ defined by (2.37) satisfy

$$\lim_{n\to\infty}\left(\sup_{n-n^\gamma \leq m \leq n+n^\gamma} |E[R_{m,n}] - \alpha|\right) = 0, \tag{2.38}$$

$$\lim_{n\to\infty}\left(\sup_{n-n^\gamma \leq m < m' \leq n+n^\gamma} \left|E[R_{m,n}R_{m',n}] - \alpha^2\right|\right) = 0, \tag{2.39}$$

and

$$\lim_{n\to\infty}\left(n^{-1/2} \sup_{n-n^\gamma \leq m < m' \leq n+n^\gamma} E[R_{m,n}^2]\right) = 0. \tag{2.40}$$

Finally assume that $H_n(\mathcal{X}_m)$ is uniformly bounded by a polynomial in n, m in the sense that there exists a constant $\beta > 0$ such that

$$|H_n(\mathcal{X}_m)| \le \beta(n+m)^\beta, \qquad \text{a.s.} \tag{2.41}$$

Then $\alpha^2 \le \sigma^2$, *and as* $n \to \infty$ *we have* $n^{-1}\mathrm{Var}(H_n(\mathcal{X}_n)) \to \sigma^2 - \alpha^2$, *and*

$$n^{-1/2}(H_n(\mathcal{X}_n) - EH_n(\mathcal{X}_n)) \xrightarrow{\mathcal{D}} \mathcal{N}(0, \sigma^2 - \alpha^2). \tag{2.42}$$

Typical applications will be to random geometric graphs $G(\mathcal{X}_n; r_n)$, with r_n some given sequence of parameters; in these applications we shall normally take $H_n(\mathcal{X}) = H_0(r_n^{-1}\mathcal{X})$, where $H_0(\mathcal{X})$ is some specified functional of $G(\mathcal{X}; 1)$ and $r_n^{-1}\mathcal{X} := \{r_n^{-1}x : x \in \mathcal{X}\}$ (see, e.g., Theorem 2.16).

Proof of Theorem 2.12 Let $\xi_n := H_n(\mathcal{X}_n)$ and $\xi'_n := H_n(\mathcal{P}_n)$. Assume $\mathcal{P}_n$ is coupled to $\mathcal{X}_n$ as described in Section 1.7, with N_n denoting the number of points of $\mathcal{P}_n$. The first step is to prove that as $n \to \infty$,

$$E\left[(n^{-1/2}(\xi'_n - \xi_n - (N_n - n)\alpha))^2\right] \to 0. \tag{2.43}$$

To prove this, note that the expectation on the left-hand side of (2.43) is equal to

$$\sum_{m:|m-n|\le n^\gamma} E\left[n^{-1}(H_n(\mathcal{X}_m) - H_n(\mathcal{X}_n) - (m-n)\alpha)^2\right] P[N_n = m]$$
$$+ n^{-1}E\left[(\xi'_n - \xi_n - (N_n - n)\alpha)^2 \mathbf{1}_{\{|N_n - n| > n^\gamma\}}\right]. \tag{2.44}$$

Let $\varepsilon > 0$. By definition of $R_{m,n}$ and by conditions (2.38)–(2.40), for large enough n and all m with $n \le m \le n + n^\gamma$,

$$E[(H_n(\mathcal{X}_m) - H_n(\mathcal{X}_n) - (m-n)\alpha)^2] = E\left[\left(\sum_{l=n}^{m-1}(R_{l,n} - \alpha)\right)^2\right]$$
$$\le \varepsilon(m-n)^2 + \varepsilon n^{1/2}(m-n),$$

where the bound comes from expanding out the double sum arising from the expectation of the squared sum. A similar argument applies when $n - n^\gamma \le m \le n$, and hence the first term in (2.44) is bounded by

$$n^{-1}E[\varepsilon(N_n - n)^2 + \varepsilon n^{1/2}|N_n - n|],$$

which is bounded by 2ε since $E|N_n - n| \le \sqrt{E|N_n - n|^2} = n^{1/2}$.

By the polynomial bound (2.41), the value of $|\xi'_n - \xi_n - (N_n - n)\alpha|$ is bounded by a constant times $N_n^\beta + n^\beta$, so its fourth moment is bounded by a constant times $n^{4\beta}$. By the Cauchy–Schwarz inequality, there is a constant β_1 such that the second term in (2.44) is bounded by $\beta_1 n^{2\beta-1}(P[|N_n - n| > n^\gamma])^{1/2}$. By Lemma 1.4, $P[|N_n - n| > n^\gamma]$ decays exponentially in $n^{2\gamma-1}$, so the second term in (2.44) tends to zero. This completes the proof of (2.43).

To prove convergence of $n^{-1}\mathrm{Var}(\xi_n)$, we use the identity

$$n^{-1/2}\xi'_n = n^{-1/2}\xi_n + n^{-1/2}(N_n - n)\alpha + n^{-1/2}(\xi'_n - \xi_n - (N_n - n)\alpha).$$

On the right-hand side, the third term has variance tending to zero by (2.43), while the second term has variance α^2 and is independent of the first term. Therefore by assumption,

$$\sigma^2 = \lim_{n\to\infty} n^{-1}\mathrm{Var}(\xi'_n) = \lim_{n\to\infty}(n^{-1}\mathrm{Var}(\xi_n) + \alpha^2),$$

so that $\sigma^2 \geq \alpha^2$ and $n^{-1}\mathrm{Var}(\xi_n) \to \sigma^2 - \alpha^2$.

By assumption, $n^{-1/2}(\xi'_n - E\xi'_n) \xrightarrow{\mathcal{D}} \mathcal{N}(0, \sigma^2)$. Combined with (2.43) and Slutsky's theorem, this yields

$$n^{-1/2}(\xi_n - E\xi'_n + (N_n - n)\alpha) \xrightarrow{\mathcal{D}} \mathcal{N}(0, \sigma^2),$$

and since $n^{-1/2}(N_n - n)\alpha$ is independent of ξ_n and converges in distribution to $\mathcal{N}(0, \alpha^2)$, it follows by an argument using characteristic functions that

$$n^{-1/2}(\xi_n - E\xi'_n) \xrightarrow{\mathcal{D}} \mathcal{N}(0, \sigma^2 - \alpha^2). \tag{2.45}$$

By (2.43), the expectation of $n^{-1/2}(\xi'_n - \xi_n - (N_n - n)\alpha)$ tends to zero, so in (2.45) we can replace $E\xi'_n$ by $E\xi_n$, which gives us (2.42). □

In many cases we check the conditions (2.38)–(2.40) by coupling arguments providing an estimate on the total variation distance between the random 2-vector $(R_{m,n}, R_{m',n})$ and the random 2-vector (Δ, Δ'), where Δ and Δ' are a pair of independent identically distributed random variables.

Lemma 2.13 *Suppose there is a pair of independent identically distributed random variables (Δ, Δ'), such that for any $(\mathbb{N} \times \mathbb{N})$-valued sequence $((\nu(n), \nu'(n)), n \geq 1)$ satisfying $\nu(n) < \nu'(n)$ for all n and $n^{-1}\nu(n) \to 1$ and $n^{-1}\nu'(n) \to 1$ as $n \to \infty$ we have*

$$(R_{\nu(n),n}, R_{\nu'(n),n}) \xrightarrow{\mathcal{D}} (\Delta, \Delta') \qquad \text{as } n \to \infty. \tag{2.46}$$

Suppose also that for some $p > 2$ and some $\eta > 0$ we have

$$\sup_{n\in\mathbb{N}} \sup_{m\in[(1-\eta)n,(1+\eta)n]} \{E[|R_{m,n}|^p]\} < \infty. \tag{2.47}$$

Then $E[\Delta]$ is finite, and conditions (2.38)–(2.40) hold with $\alpha := E[\Delta]$ and $\gamma := \frac{3}{4}$.

Proof It follows from (2.46) that if $\nu(n) < \nu'(n)$ and $n^{-1}\nu(n) \to 1, n^{-1}\nu'(n) \to 1$, then as $n \to \infty$, $R_{\nu(n),n} \xrightarrow{\mathcal{D}} \Delta$ and

$$R_{\nu(n),n} R_{\nu'(n),n} \xrightarrow{\mathcal{D}} \Delta\Delta'. \tag{2.48}$$

By the assumption (2.47) of bounded pth moments and the Cauchy–Schwarz inequality, there exists $n_0 \in \mathbb{N}$ such that

$$\sup_{n \geq n_0} E[|R_{\nu(n),n} R_{\nu'(n),n}|^{p/2}] < \infty$$

and therefore the variables $R_{\nu(n),n} R_{\nu'(n),n}$, defined for each $n \geq n_0$, are uniformly integrable, so that the convergence (2.48) also holds in the sense of convergence of means, that is,

$$\lim_{n \to \infty} E[R_{\nu(n),n} R_{\nu'(n),n}] = E[\Delta\Delta'] = E[\Delta]^2. \tag{2.49}$$

Also the limit in (2.49) is finite so Δ has finite mean. Also, by a similar (simpler) argument $\lim_{n\to\infty} E[R_{\nu(n),n}] = E[\Delta]$. Since the choice of $\nu(n), \nu'(n)$ is arbitrary subject to $\nu(n) \sim \nu'(n) \sim n$, the conditions (2.38) and (2.39) follow. The condition (2.40) also follows from (2.47). □

Often when applying Theorem 2.12 we have no *a priori* guarantee that the limiting variance $\sigma^2 - \alpha^2$ is non-zero. However, a set of conditions similar to those of Lemma 2.13, with the extra condition that Δ have a non-degenerate distribution (i.e. one that is not concentrated on a single value), can be used to ensure that this is the case. As well as the increment $R_{m,n}$ defined earlier at (2.37), we consider the increments $G_{i,n}$ and $\tilde{G}_{i,n}$ defined for $i \leq n$ by

$$G_{i,n} := H_n(\mathcal{X}_n) - H_n(\mathcal{X}_n \setminus \{X_i\}), \tag{2.50}$$

$$\tilde{G}_{i,n} := H_n(\mathcal{X}_{n+1} \setminus \{X_i\}) - H_n(\mathcal{X}_n \setminus \{X_i\}), \tag{2.51}$$

both of which have the same distribution as $R_{n-1,n}$.

Lemma 2.14 *Suppose that there is a random variable Δ with non-degenerate distribution, such that if Δ' denotes an independent copy of Δ, then for any $\mathbb{N}$-valued sequence $(\nu(n), n \geq 1)$ satisfying $\nu(n) \leq n$ for all n and $n^{-1}\nu(n) \to 1$ as $n \to \infty$ we have*

$$(R_{\nu(n)-1,n}, \tilde{G}_{\nu(n),n}) \xrightarrow{\mathcal{D}} (\Delta, \Delta') \quad \text{as } n \to \infty, \tag{2.52}$$

and

$$P[R_{\nu(n)-1,n} \neq G_{\nu(n),n}] \to 0 \quad \text{as } n \to \infty. \tag{2.53}$$

Suppose also for some $p > 2$ and some $\eta > 0$ that (2.47) holds. Then

$$\liminf_{n \to \infty} n^{-1} \mathrm{Var}(H_n(\mathcal{X}_n)) > 0.$$

Proof Set $\alpha = E[\Delta]$. Since $R_{n-1,n} \xrightarrow{\mathcal{D}} \Delta$ by (2.52) and the variables $R_{n-1,n}, n \geq 1$ are uniformly integrable by the moments condition (2.47), α is finite.

Given n, construct a filtration as follows. Let $\mathcal{F}_0$ be the trivial σ-field, let $\mathcal{F}_i := \sigma(X_1, \ldots, X_i)$ and write E_i for conditional expectation given $\mathcal{F}_i$. Define martingale differences $D_{i,n} := E_i H_n(\mathcal{X}_n) - E_{i-1} H_n(\mathcal{X}_n)$. Then $H_n(\mathcal{X}_n) - EH_n(\mathcal{X}_n) = \sum_{i=1}^n D_{i,n}$, and by orthogonality of martingale differences,

$$\mathrm{Var}(H_n(\mathcal{X}_n)) = \sum_{i=1}^{n} E[D_{i,n}^2]. \tag{2.54}$$

We seek lower bounds for $E[D_{i,n}^2]$. Given $i \leq n$, by (2.50) and (2.51) we have $D_{i,n} = E_i[G_{i,n} - \tilde{G}_{i,n}]$.

Let $i(n), n \geq 1$ be an arbitrary $\mathbb{N}$-valued sequence satisfying $i(n) \leq n$ for all n and $n^{-1} i(n) \to 1$ as $n \to \infty$. In what follows we write simply i for $i(n)$.

We approximate to $G_{i,n}$ by $R_{i-1,n}$ which is a good approximation when i is close to n. By (2.53), and uniform integrability of $(G_{i,n} - R_{i-1,n})^2$ which follows from (2.47),

$$E[(E_i[G_{i,n} - R_{i-1,n}])^2] \leq E[(G_{i,n} - R_{i-1,n})^2] \to 0. \tag{2.55}$$

Since $R_{i-1,n} \xrightarrow{\mathcal{D}} \Delta$ by (2.52), and $E_i G_{i,n} = R_{i-1,n} + E_i[G_{i,n} - R_{i-1,n}]$, it follows by Slutsky's theorem that $E_i G_{i,n} \xrightarrow{\mathcal{D}} \Delta$, and hence, by the assumed non-degeneracy of Δ, we can choose $\delta > 0$ such that

$$\liminf_{n \to \infty} P[E_i G_{i,n} \geq \alpha + 2\delta] > 2\delta. \tag{2.56}$$

Define $g : \mathbb{R} \to \mathbb{R}$ by $g(t) = 0$ for $t \leq \alpha + \delta$ and $g(t) = 1$ for $t \geq \alpha + 2\delta$, interpolating linearly between $\alpha + \delta$ and $\alpha + 2\delta$. Set $Y_i := g(E_i G_{i,n})$. Then $Y_i E_i(G_{i,n} - \alpha)$ is a non-negative random variable, and (2.56) implies that for large enough n,

$$E[(G_{i,n} - \alpha) Y_i] = E[Y_i E_i (G_{i,n} - \alpha)] \geq 4\delta^2. \tag{2.57}$$

Next consider $E[(\tilde{G}_{i,n} - \alpha) Y_i]$, writing the second factor as the sum of $g(R_{i-1,n})$ and $g(E_i G_{i,n}) - g(R_{i-1,n})$. By (2.52), we have

$$(\tilde{G}_{i,n} - \alpha) g(R_{i-1,n}) \xrightarrow{\mathcal{D}} (\Delta' - \alpha) g(\Delta)$$

and also the variables $(\tilde{G}_{i,n} - \alpha) g(R_{i-1,n}), n \geq 1$ are uniformly integrable by (2.47). Therefore

$$E[(\tilde{G}_{i,n} - \alpha) g(R_{i-1,n})] \to E[(\Delta' - \alpha) g(\Delta)] = 0. \tag{2.58}$$

By (2.47), there is a constant K such that $E[(\tilde{G}_{i,n} - \alpha)^2] \leq K^2$ for all n. By the Cauchy–Schwarz inequality and the fact that g' is bounded by δ^{-1}, and that $R_{i-1,n}$ is $\mathcal{F}_i$-measurable,

$$\begin{aligned}
&|E[(\tilde{G}_{i,n} - \alpha)(g(E_i G_{i,n}) - g(R_{i-1,n}))]| \\
&\quad \leq E[(\tilde{G}_{i,n} - \alpha)^2]^{1/2} E[(g(E_i G_{i,n}) - g(R_{i-1,n}))^2]^{1/2} \\
&\quad \leq K\delta^{-1} E[(E_i G_{i,n} - R_{i-1,n})^2]^{1/2} = K\delta^{-1} E[(E_i[G_{i,n} - R_{i-1,n}])^2]^{1/2},
\end{aligned}$$

which tends to zero by (2.55). Combining this with (2.58), for n large, we have

$$E[(\tilde{G}_{i,n} - \alpha)Y_i] = E[(\tilde{G}_{i,n} - \alpha) g(E_i G_{i,n})] \to 0.$$

Combined with (2.57) this implies that for large n

$$E[(G_{i,n} - \tilde{G}_{i,n})Y_i] \geq 2\delta^2.$$

Since Y_i is $\mathcal{F}_i$-measurable and lies in the range $[0,1]$, we obtain for large n that

$$2\delta^2 \leq E[Y_i E_i(G_{i,n} - \tilde{G}_{i,n})] \leq E[|E_i(G_{i,n} - \tilde{G}_{i,n})|] = E[|D_{i,n}|],$$

and hence, $\liminf_{n\to\infty} E[D_{i,n}^2] \geq \liminf_{n\to\infty} E[|D_{i,n}|]^2 \geq 4\delta^4$, for $i = i(n)$ an arbitrary sequence satisfying $i(n) \leq n$ and $n^{-1} i(n) \to 1$.

It follows by a diagonal argument that there exists $n_1 \in \mathbb{N}$ and $\varepsilon_1 > 0$ such that $E[D_{i,n}^2] \geq \delta^4$ for all $n \geq n_1$ and $i \in [n(1-\varepsilon_1), n]$; if not, there would be a sequence of integers $n' \to \infty$ and a sequence $i(n')$ with $i(n')/n' \to 1$ and $i(n') \leq n'$, such that $E[D_{i(n'),n'}^2] < \delta^4$ for all n', a contradiction.

Thus, using (2.54), we have for all large enough n that $\mathrm{Var}(H_n(\mathcal{X}_n)) \geq (\varepsilon_1 n - 1)\delta^4$, and the conclusion $\liminf n^{-1}\mathrm{Var}(H_n(\mathcal{X}_n)) > 0$ follows. □

For functions of random geometric graphs in the thermodynamic limit $nr_n^d \to$ const., the conditions (2.46), (2.52) and (2.53) can often be checked using the following notion of *stabilization.* Let H_0 be a real-valued measurable functional defined for all finite subsets $\mathcal{X}$ of $\mathbb{R}^d$. Assume that H_0 is translation-invariant, meaning that $H_0(\mathcal{X} \oplus y) = H_0(\mathcal{X})$ for all finite $\mathcal{X} \subset \mathbb{R}^d$ and all $y \in \mathbb{R}^d$ (here $\mathcal{X} \oplus y := \{x + y : x \in \mathcal{X}\}$). Define the associated 'add one cost' $\Delta(\mathcal{X})$ to be the increment of H_0 if we insert a point at the origin, that is, define

$$\Delta(\mathcal{X}) := H_0(\mathcal{X} \cup \{\mathbf{0}\}) - H_0(\mathcal{X}).$$

As in Section 1.7, let $\mathcal{H}_\lambda$ be a homogeneous Poisson process of intensity λ on $\mathbb{R}^d$.

Definition 2.15 *The functional H_0 is strongly stabilizing on $\mathcal{H}_\lambda$ if there exist a.s. finite random variables S (a radius of stabilization of H_0) and $\Delta(\mathcal{H}_\lambda)$ (the limiting add one cost) such that with probability* 1, *$\Delta(\mathcal{A}) = \Delta(\mathcal{H}_\lambda)$ for all finite $\mathcal{A} \subset \mathbb{R}^d$ satisfying $\mathcal{A} \cap B(\mathbf{0}; S) = \mathcal{H}_\lambda \cap B(\mathbf{0}; S)$.*

Thus, S is a radius of stabilization if the add one cost for $\mathcal{H}_\lambda$ is unaffected by changes in the configuration outside the ball $B(\mathbf{0}; S)$.

Given a strongly stabilizing functional H_0, and given any almost surely strictly positive random variable Λ, define the d-dimensional point process $\mathcal{H}_\Lambda$ and the variable $\Delta(\mathcal{H}_\Lambda)$ as follows. First take a random variable Λ' with the distribution of Λ, and then given $\Lambda' = \lambda$, take $\mathcal{H}_\Lambda$ to be a homogeneous Poisson process on $\mathbb{R}^d$ with intensity λ and take $\Delta(\mathcal{H}_\Lambda)$ to be its limiting add one cost. Note that $\mathcal{H}_\Lambda$ is a Cox process, that is, a Poisson process whose intensity is itself random (see, e.g., Stoyan *et al.* (1995)).

Our interest is mainly in the special case where $\Lambda := \mu f(X)$ with X defined to be a random d-vector with density f. Note that in this case

$$E[\Delta(\mathcal{H}_{\mu f(X)})] = \int_{\mathbb{R}^d} E[\Delta(\mathcal{H}_{\mu f(x)})] f(x) \mathrm{d}x. \tag{2.59}$$

Theorem 2.16 *Suppose that* $\lim_{n\to\infty}(nr_n^d) = \mu \in (0,\infty)$. *Suppose for all* $\lambda > 0$ *that* H_0 *is strongly stabilizing on* $\mathcal{H}_\lambda$ *with limiting add one cost* $\Delta(\mathcal{H}_\lambda)$. *For each finite* $\mathcal{X} \subset \mathbb{R}^d$ *and each* $n \in \mathbb{N}$ *set* $H_n(\mathcal{X}) := H_0(r_n^{-1}\mathcal{X})$.

Suppose there exists $\sigma \geq 0$ *such that as* $n \to \infty$ *we have* $n^{-1}\mathrm{Var}H_n(\mathcal{P}_n) \to \sigma^2$ *and* $n^{-1/2}(H_n(\mathcal{P}_n) - EH_n(\mathcal{P}_n)) \xrightarrow{\mathcal{D}} \mathcal{N}(0,\sigma^2)$. *Suppose also that* $H_n(\cdot)$ *satisfies the polynomial bound* (2.41) *and the moments condition* (2.47) *for some* $\beta > 0, p > 2, \eta > 0$. *Set* $\tau^2 := \sigma^2 - (E[\Delta(\mathcal{H}_{\mu f(X)})])^2$.

Then $\tau^2 \geq 0$ *and as* $n \to \infty$ *we have* $n^{-1}\mathrm{Var}H_n(\mathcal{X}_n) \to \tau^2$ *and* $n^{-1/2}(H_n(\mathcal{X}_n) - EH_n(\mathcal{X}_n)) \xrightarrow{\mathcal{D}} \mathcal{N}(0,\tau^2)$. *Moreover, if the distribution of* $\Delta(\mathcal{H}_{\mu f(X)})$ *is non-degenerate, then* $\tau^2 > 0$ *and* $\sigma^2 > 0$.

Proof By Lemmas 2.13 and 2.14, and Theorem 2.12, it suffices to prove that if $((\nu(n), \nu'(n)), n \geq 1)$ is an arbitrary $(\mathbb{N} \times \mathbb{N})$-valued sequence satisfying $\nu(n) < \nu'(n)$ for all n and $n^{-1}\nu(n) \to 1$ and $n^{-1}\nu'(n) \to 1$ as $n \to \infty$, the condition (2.46) holds, and if also $\nu(n) < n$ then the conditions (2.52) and (2.53) hold, all with $\Delta := \Delta(\mathcal{H}_{\mu f(X)})$.

To prove (2.52), we produce an explicit coupling. That is, we find a family of variables $D_n, D'_n, \rho_n, \rho'_n$, all defined on the same probability space for each n, with the following properties:

- D_n and D'_n are independent and each have the same distribution as $\Delta(\mathcal{H}_{\mu f(X)})$;
- (ρ_n, ρ'_n) have the same joint distribution as $(R_{\nu(n),n}, R_{\nu'(n),n})$, with $R_{m,n}$ defined at (2.37);
- $\lim_{n\to\infty} P[\{D_n \neq \rho_n\} \cup \{D'_n \neq \rho'_n\}] = 0$.

To do this we find a coupling of a realization of the binomial process $\mathcal{X}_n$ to a Cox process with the distribution of $\mathcal{H}_{\mu f(X)}$. Assume on a suitable probability space that we have, independently, a sequence of independent identically distributed random d-vectors $(X, Y, V_1, V_2, V_3, \ldots)$ with common density f, and two homogeneous $(d+1)$-dimensional Poisson processes $\mathcal{P}$, $\mathcal{Q}$, both of unit intensity on $\mathbb{R}^d \times [0,\infty)$.

Given n, define coupled point processes $\mathcal{X}_m^n, m \geq 1$ (a sequence of binomial processes) and $\mathcal{H}_n^X$, $\mathcal{H}_n^Y$ (both Cox processes) and variables ζ_n^X and ζ_n^Y, all in terms of $\mathcal{P}, \mathcal{Q}, X, Y, V_1, V_2, \ldots$, as follows.

Let $\mathcal{P}(n)$ be the image of the restriction of $\mathcal{P}$ to the set $\{(x,t) \in \mathbb{R}^d \times [0,\infty) : t \leq nf(x)\}$, under the projection $(x,t) \mapsto x$, and let $N(n)$ be the number of points of $\mathcal{P}(n)$. Choose an ordering on the points of $\mathcal{P}(n)$, uniformly at random from all $N(n)!$ possible such orderings. Use this ordering to list the points of $\mathcal{P}(n)$ as $W_1, W_2, \ldots, W_N$. Also, set $W_{N+1} = V_1, W_{N+2} = V_2, W_{N+3} = V_3$ and so on. The resulting random d-vectors $W_1, W_2, \ldots$ have common density function f, and are independent of each other and of (X,Y). Define the sequence $(X_m^n)_{m\geq 1}$ by replacing the $(\nu(n)+1)$st and $(\nu'(n)+1)$st terms in the sequence (W_m) by X, Y, respectively, that is, set $X^n_{\nu(n)+1} := X, X^n_{\nu'(n)+1} := Y$, and $X^n_m := W_m$ for $m \notin \{\nu(n)+1, \nu'(n)+1\}$.

Set $\mathcal{X}^n_m := \{X^n_1, \ldots, X^n_m\}$ for each m. Let $\rho_n := H_n(\mathcal{X}^n_{\nu(n)+1}) - H_n(\mathcal{X}^n_{\nu(n)})$. Let $\rho'_n := H_n(\mathcal{X}^n_{\nu'(n)+1}) - H_n(\mathcal{X}^n_{\nu'(n)})$. Since $(X^n_1, X^n_2, \ldots)$ is a sequence of independent identically distributed random d-vectors with common density f, the point process $\mathcal{X}^n_m$ has the same distribution as $\mathcal{X}_m$, and (ρ_n, ρ'_n) have the same joint distribution as $(R_{\nu(n),n}, R_{\nu'(n),n})$ defined at (2.37).

By definition of H_n and translation invariance, we have

$$\rho_n = H_0(r_n^{-1}(\mathcal{X}^n_{\nu(n)} \cup \{X\})) - H_0(r_n^{-1}\mathcal{X}^n_{\nu(n)}) = \Delta(r_n^{-1}(\mathcal{X}^n_{\nu(n)} \oplus (-X)))$$

and

$$\rho'_n = H_0(r_n^{-1}(\mathcal{X}^n_{\nu'(n)+1} \cup \{Y\})) - H_0(r_n^{-1}\mathcal{X}^n_{\nu'(n)}) = \Delta(r_n^{-1}(\mathcal{X}^n_{\nu'(n)} \oplus (-Y))).$$

Let F_X be the half-space of points in $\mathbb{R}^d$ closer to X than to Y, and let $F_Y := \mathbb{R}^d \setminus F_X$. Let $\mathcal{P}^n_X$ be the restriction of $\mathcal{P}$ to the set $F_X \times [0, \mu r_n^{-d} f(X)]$; let $\mathcal{Q}^n_X$ be the restriction of $\mathcal{Q}$ to the set $F_Y \times [0, \mu r_n^{-d} f(X)]$. Let $\mathcal{H}^X_n$ be the image of the point process $\mathcal{P}^n_X \cup \mathcal{Q}^n_X$ under the mapping

$$(x,t) \mapsto r_n^{-1}(x - X).$$

Given $X = x$, the point process $\mathcal{P}^n_X \cup \mathcal{Q}^n_X$ is a homogeneous Poisson process of intensity 1 on $\mathbb{R}^d \times [0, \mu r_n^{-d} f(x)]$. Hence, given $X = x$, $\mathcal{H}^X_n$ is a homogeneous Poisson process on $\mathbb{R}^d$ of intensity $\mu f(x)$; let D_n be the associated limiting add one cost $\Delta(\mathcal{H}^X_n)$.

Construct $\mathcal{H}^Y_n$ in the following analogous manner. Let $\mathcal{P}^n_Y$ be the restriction of $\mathcal{P}$ to the set $F_Y \times [0, \mu r_n^{-d} f(Y)]$; let $\mathcal{Q}^n_Y$ be the restriction of $\mathcal{Q}$ to the set $F_X \times [0, \mu r_n^{-d} f(Y)]$. Let $\mathcal{H}^Y_n$ be the image of the point process $\mathcal{P}^n_Y \cup \mathcal{Q}^n_Y$ under the mapping

$$(x,t) \mapsto r_n^{-1}(x - Y).$$

By an argument similar to that used for $\mathcal{H}^X_n$, the point process $\mathcal{H}^Y_n$, given $Y = y$, is a homogeneous Poisson process on $\mathbb{R}^d$ of intensity $\mu f(y)$; set $D'_n := \Delta(\mathcal{H}^Y_n)$.

Then $\mathcal{H}^X_n$ is a Cox process, where the randomness of the intensity measure comes from the value of $f(X)$; also $\mathcal{H}^Y_n$ is a Cox process. Moreover, the distributions of the Cox processes $\mathcal{H}^X_n$ and H^Y_n are identical to that of $\mathcal{H}_{\mu f(X)}$, for all n.

Finally, we assert that $\mathcal{H}_n^X$ and $\mathcal{H}_n^Y$ are *independent*, which can be seen by conditioning on the values of X, Y; the point processes $\mathcal{H}_n^X$ and $\mathcal{H}_n^Y$ are conditionally independent given (X, Y), with the conditional distribution of $\mathcal{H}_n^X$ determined by X and the conditional distribution of $\mathcal{H}_n^Y$ determined by Y, and integration over possible values of (X, Y) yields the independence asserted. Therefore, for each n, the variables D_n and D'_n are independent, and each have the distribution of $\Delta(\mathcal{H}_{\mu f(X)})$.

Given $\varepsilon > 0$, choose $K > 0$ so that the probability that $\mathcal{H}_{\mu f(X)}$ has a radius of stabilization greater than K, is less than ε. If the radius of stabilization of $\mathcal{H}_n^X$ is at most K, and if also the point processes $r_n^{-1}(\mathcal{X}_{\nu(n)}^n \oplus (-X))$ and $\mathcal{H}_n^X$ are identical on $B(\mathbf{0}; K)$, then ρ_n is equal to $R_{\nu(n),n}$. Arguing similarly for ρ'_n, and using Lemma 2.17 below, we see that for all large enough n we have $P[(\rho_n, \rho'_n) \neq (R_{\nu(n),n}, R_{\nu'(n),n})] \leq 3\varepsilon$, for n large enough. This completes the proof of (2.46).

The condition (2.52) follows by a slight modification of the coupling construction just given, which we omit. Likewise, the condition (2.53) holds by the above coupling construction, and Lemma 2.17 below, and we omit the details for this too. □

The last lemma concludes the preceding proof, and notation from that proof is carried over into this lemma.

Lemma 2.17 *Given* $K > 0$, *we have*

$$\lim_{n\to\infty} P[r_n^{-1}(\mathcal{X}_{\nu(n)}^n \oplus (-X)) \cap B(\mathbf{0}; K) = \mathcal{H}_n^X \cap B(\mathbf{0}; K)] = 1, \qquad (2.60)$$

$$\lim_{n\to\infty} P[r_n^{-1}(\mathcal{X}_{\nu'(n)}^n \oplus (-Y)) \cap B(\mathbf{0}; K) = \mathcal{H}_n^Y \cap B(\mathbf{0}; K)] = 1. \qquad (2.61)$$

Proof Note first that

$$\lim_{n\to\infty} P[B(X; Kr_n) \subseteq F_X] = \lim_{n\to\infty} P[B(Y; Kr_n) \subseteq F_Y] = 1.$$

Suppose $x \in \mathbb{R}^d$ is a Lebesgue point of f (see Section 1.6). Given $X = x$ and given that $B(x; Kr_n) \subseteq F_X$, the expected number of points of $\mathcal{P}$ in $B(x; Kr_n) \times [0, \infty)$ that contribute to $\mathcal{H}_n^X$ but not to $\mathcal{P}(n)$ is

$$\int_{B(x;Kr_n)} \max(\mu r_n^{-d} f(x) - nf(y), 0) \mathrm{d}y,$$

while the expected number of points of $\mathcal{P}$ in $B(x; Kr_n) \times [0, \infty)$ that contribute to $\mathcal{P}(n)$ but not to $\mathcal{H}_n^X$ is

$$\int_{B(x;Kr_n)} \max(nf(y) - \mu r_n^{-d} f(x), 0) \mathrm{d}y.$$

Each of these integrals tends to zero, because their sum is bounded by

$$\int_{B(x;Kr_n)} |\mu r_n^{-d}(f(x) - f(y))| \mathrm{d}y + \int_{B(x;Kr_n)} |(\mu r_n^{-d} - n)| f(y) \mathrm{d}y$$

which tends to zero because x is a Lebesgue point of f and $nr_n^d \to \mu$. Finally the probability that $\mathcal{P}(n) \cap B(x; Kr_n) \neq \mathcal{X}_{\nu(n)}^n \cap B(x; Kr_n)$ tends to zero as $n \to \infty$, since $|N(n) - \nu(n)|$ is $o(n)$ in probability. Integrating over possible values of X and using the dominated convergence theorem, we obtain (2.60). The proof of (2.61) is similar. □

2.6 Notes

Section 2.3. Theorem 2.4 is adapted from a result in Baldi and Rinott (1989), which is based on a more general result of Stein (1986). Its usefulness in geometric probability was recognized by Avram and Bertsimas (1993), who applied it to problems concerning nearest-neighbour and other graphs.

Section 2.4. Lemma 2.11 is a slight improvement on a lemma in Penrose (2000*b*).

Section 2.5. The results in this section are new in the generality given, but use ideas which have been used elsewhere for de-Poissonization in geometric settings such as minimal spanning tree and nearest-neighbour graph; see Kesten and Lee (1996), Lee (1997), and Penrose and Yukich (2001). The notion of stabilization along the lines of Definition 2.15 was introduced by Lee (1997) in the context of minimal spanning trees, and has been applied to many random geometrical problems (not only for de-Poissonization, but also for proving laws of large numbers and central limit theorems) by Penrose and Yukich (2001, 2003).

3

SUBGRAPH AND COMPONENT COUNTS

The number of edges is a fundamental quantity for the random geometric graph $G(\mathcal{X}_n; r_n)$, and its properties have been considered in various guises by numerous authors. In this chapter, it is treated as a special case in the following more general context.

Let Γ be a fixed connected graph on k vertices, $k \geq 2$. Consider the number of subgraphs of $G(\mathcal{X}_n; r_n)$ isomorphic to Γ. Some care is needed in defining this quantity. For example, if Γ is the *3-path*, that is, the connected graph with two edges and three vertices, then each copy in $G(\mathcal{X}_n; r_n)$ of the complete graph K_3 on three vertices could be considered to contribute three copies of Γ, there being three ways to select the two edges.

With this in mind, let $G_n = G_n(\Gamma)$ denote the number of *induced* subgraphs of $G(\mathcal{X}_n; r)$ isomorphic to Γ (or *induced* Γ*-subgraphs* for short), that is, the number of subsets $\mathcal{Y}$ of $\mathcal{X}_n$ such that $G(\mathcal{Y}; r_n)$ is isomorphic to Γ. Clearly the edge count is the simplest special case of this quantity.

One could also consider the quantity $\tilde{G}_n$, defined to be the number of (unlabelled) subgraphs of $G(\mathcal{X}_n; r_n)$ isomorphic to Γ. This is a linear combination of those $G_n(\Gamma')$ for which Γ' is a graph on k vertices having Γ as a subgraph; for example, if Γ is the 3-path, then $\tilde{G}_n(\Gamma) = 3G_n(K_3) + G_n(\Gamma)$. The asymptotic theory for $\tilde{G}_n$ follows readily enough from that for G_n which is to be developed here.

A related concept is the number of Γ*-components* of $G(\mathcal{X}_n; r_n)$ (i.e. components isomorphic to Γ), which we denote by J_n or $J_n(\Gamma)$. To be a component, an induced Γ-subgraph must additionally be disconnected from the rest of $\mathcal{X}_n$; hence, $J_n(\Gamma) \leq G_n(\Gamma)$. Components are usually referred to as 'clusters' in the percolation literature; we steer clear of this nomenclature since the word 'cluster' has somewhat wider connotations in statistical cluster analysis, as described in Section 1.2. Even when Γ is of degree 1, the value of J_n (unlike that of G_n) is of interest since it is the number of isolated vertices.

For some choices of Γ, there is *never* an induced Γ-subgraph of a geometric graph, for example, if Γ is star-shaped (see below) with a sufficiently large degree of its central vertex. In these cases, $G_n(\Gamma) = 0$ almost surely for all n, although $\tilde{G}_n(\Gamma)$ can still be non-zero, for example, because of Γ-graphs arising as subgraphs of induced subgraphs isomorphic to the complete graph on k vertices, where k is the order of Γ. We shall say that Γ is *feasible* if $P[G(\mathcal{X}_k; r) \cong \Gamma] > 0$ for some $r > 0$. For example, if $d = 2$ with the Euclidean norm, the star-shaped graph with one vertex of degree $k-1$ and the other $k-1$ vertices of degree 1 is feasible for $k \leq 6$ but not for $k \geq 7$, since if XY and XZ are edges of the geometric graph $G(\mathcal{X}; r)$ making an angle less than $60°$ at vertex X, then YZ is also an edge of $G(\mathcal{X}; r)$.

The results of this chapter are summarized as follows. For arbitrary feasible connected Γ with k vertices, the Γ-subgraph count G_n satisfies a Poisson limit theorem (in the case where $n^k r_n^{d(k-1)}$ tends to a finite constant) and a normal limit theorem (in the case where $n^k r_n^{d(k-1)} \to \infty$ but $r_n \to 0$, or when r_n is a constant). Moreover, *multivariate* Poisson and normal limit theorems hold for the joint distribution of the subgraph counts associated with two or more feasible graphs. Also, the Γ-subgraph count satisfies strong laws of large numbers. Finally, similar results hold for Γ-component count J_n in the thermodynamic limit.

As well as $G_n(\Gamma)$ and $J_n(\Gamma)$, also of interest are the Γ-subgraph and Γ-component counts in the Poisson process $\mathcal{P}_n$ defined at (1.16); let these be denoted $G'_n(\Gamma)$ and $J'_n(\Gamma)$, respectively. For technical reasons we also consider subgraphs located in some specific region of $\mathbb{R}^d$. Given a finite point set $\mathcal{Y} \subset \mathbb{R}^d$, let the first element of $\mathcal{Y}$ according to the lexicographic ordering on $\mathbb{R}^d$ be called the *left-most point* of $\mathcal{Y}$, and denoted $\mathrm{LMP}(\mathcal{Y})$. For $A \subseteq \mathbb{R}^d$, let $G_{n,A}$ (respectively, $G'_{n,A}, J_{n,A}, J'_{n,A}$) be the number of induced Γ-subgraphs of $G(\mathcal{X}_n; r_n)$ (respectively, induced Γ-subgraphs of $G(\mathcal{P}_n, r_n)$, Γ-components of $G(\mathcal{X}_n, r_n)$, Γ-components of $G(\mathcal{P}_n, r_n)$) for which the left-most point of the vertex set lies in A.

The type of set A that we consider is open, and has $\mathrm{Leb}(\partial A) = 0$, where ∂A denotes the intersection of the closure of A with that of its complement, and $\mathrm{Leb}(\cdot)$ is Lebesgue measure. If the subscript A in $G_{n,A}, G'_{n,A}, J_{n,A}$, or $J'_{n,A}$ is omitted, it is to be understood that $A = \mathbb{R}^d$ (the main case of interest). When wishing to emphasize dependence on the graph Γ, we write them as $G_{n,A}(\Gamma), G'_{n,A}(\Gamma), J_{n,A}(\Gamma)$, and $J'_{n,A}(\Gamma)$.

3.1 Expectations

This section contains asymptotic results for the means of the Γ-subgraph counts G_n and G'_n, and the Γ-component counts J_n and J'_n. Given a connected graph Γ on k vertices, and given $A \subseteq \mathbb{R}^d$, define the indicator functions $h_\Gamma(\mathcal{Y})$ and $h_{\Gamma,n,A}(\mathcal{Y})$ for all finite $\mathcal{Y} \subset \mathbb{R}^d$ by

$$h_\Gamma(\mathcal{Y}) := \mathbf{1}_{\{G(\mathcal{Y};1) \cong \Gamma\}}, \qquad h_{\Gamma,n,A}(\mathcal{Y}) := \mathbf{1}_{\{G(\mathcal{Y};r_n) \cong \Gamma\} \cap \{\mathrm{LMP}(\mathcal{Y}) \in A\}}, \tag{3.1}$$

and set $h_{\Gamma,n}(\mathcal{Y}) := h_{\Gamma,n,\mathbb{R}^d}(\mathcal{Y})$ (i.e. omit the third subscript in the case $A = \mathbb{R}^d$). Observe that $h_\Gamma(\mathcal{Y}) = h_{\Gamma,n,A}(\mathcal{Y}) = 0$ unless $\mathcal{Y}$ has k elements. Set

$$\mu_{\Gamma,A} := k!^{-1} \int_A f(x)^k \mathrm{d}x \int_{(\mathbb{R}^d)^{k-1}} h_\Gamma(\{\mathbf{0}, x_1, \ldots, x_{k-1}\}) \mathrm{d}(x_1, \ldots, x_{k-1}), \tag{3.2}$$

and write μ_Γ for $\mu_{\Gamma,\mathbb{R}^d}$.

Proposition 3.1 *Suppose that Γ is a feasible connected graph of order $k \geq 2$, that $A \subseteq \mathbb{R}^d$ is open with* $\mathrm{Leb}(\partial A) = 0$, *and that* $\lim_{n\to\infty}(r_n) = 0$. *Then*

$$\lim_{n\to\infty} r_n^{-d(k-1)} n^{-k} E[G'_{n,A}(\Gamma)] = \lim_{n\to\infty} r_n^{-d(k-1)} n^{-k} E[G_{n,A}(\Gamma)] = \mu_{\Gamma,A}. \quad (3.3)$$

Proof Clearly $E[G_{n,A}(\Gamma)] = \binom{n}{k} E[h_{\Gamma,n,A}(\mathcal{X}_k)]$. Hence,

$$\begin{aligned} E[G_{n,A}(\Gamma)] &= \binom{n}{k} \int_{\mathbb{R}^d} \cdots \int_{\mathbb{R}^d} h_{\Gamma,n,A}(\{x_1, \ldots, x_k\}) f(x_1)^k \mathrm{d}x_k \cdots \mathrm{d}x_1 \\ &\quad + \binom{n}{k} \int_{\mathbb{R}^d} \cdots \int_{\mathbb{R}^d} h_{\Gamma,n,A}(\{x_1, \ldots, x_k\}) \\ &\quad \times \left(\prod_{i=1}^{k} f(x_i) - f(x_1)^k \right) \prod_{i=1}^{k} \mathrm{d}x_i. \end{aligned} \quad (3.4)$$

By the change of variables $x_i = x_1 + r_n y_i$ for $2 \leq i \leq k$, and $x_1 = x$, the first term on the right-hand side of (3.4) equals

$$\binom{n}{k} r_n^{d(k-1)} \int_{\mathbb{R}^d} \cdots \int_{\mathbb{R}^d} h_{\Gamma,n,A}(\{x, x + r_n y_2, \ldots, x + r_n y_k\}) \mathrm{d}y_k \cdots \mathrm{d}y_2 f(x)^k \mathrm{d}x.$$

Since A is open, for $x \in A$ the function $h_{\Gamma,n,A}(\{x, x+r_n y_2, \ldots, x+r_n y_k\})$ equals $h_\Gamma(\{\mathbf{0}, y_2, \ldots, y_k\})$ for all large enough n, while for $x \notin A \cup \partial A$ it equals zero for all n. Also, $h_{\Gamma,n,A}(\{x, x+r_n y_2, \ldots, x+r_n y_k\})$ is zero except for $(y_2, \ldots, y_k)$ in a bounded region of $(\mathbb{R}^d)^{k-1}$, while $f(x)^k$ is integrable over $x \in \mathbb{R}^d$ since f is assumed bounded. Therefore by the dominated convergence theorem for integrals, the first term on the right-hand side of (3.4) is asymptotic to $n^k r_n^{d(k-1)} \mu_{\Gamma,A}$.

On the other hand, the absolute value of the second term on the right-hand side of (3.4) multiplied by $n^{-k} r_n^{-d(k-1)}$ is bounded by $\int_{\mathbb{R}^d} w_n(x_1) f(x_1) \mathrm{d}x_1$, where we set

$$w_n(x) := \int_{B(x;kr_n)} \cdots \int_{B(x;kr_n)} r_n^{-d(k-1)} \left| \prod_{i=2}^{k} f(x_i) - f(x)^{k-1} \right| \mathrm{d}x_2 \cdots \mathrm{d}x_k.$$

If f is continuous at x, then clearly $w_n(x)$ tends to zero. Even if f is not almost everywhere continuous, we assert that $w_n(x)$ still tends to zero if x is a Lebesgue point of f. This is proved by an induction on k; the inductive step is to bound the integrand by

$$r_n^{-d(k-1)} \left(|f(x_k) - f(x)| \prod_{i=2}^{k-1} f(x_i) \right) + r_n^{-d(k-1)} \left| \prod_{i=2}^{k-1} f(x_i) - f(x)^{k-2} \right| f(x).$$

The integral of the first expression over $B(x; kr_n)^{k-1}$ tends to zero by the definition of a Lebesgue point (and boundedness of f), while that of the second tends to zero by the inductive hypothesis. Hence, by the Lebesgue density theorem and the dominated convergence theorem, $\int_{\mathbb{R}^d} w_n(x_1)f(x_1)\mathrm{d}x_1$ tends to zero, which proves the second equality in (3.3).

By Palm theory (Theorem 1.6), we have

$$E[G'_{n,A}(\Gamma)] = E \sum_{\mathcal{Y}\subseteq\mathcal{P}_n} h_{\Gamma,n,A}(\mathcal{Y}) = \left(\frac{n^k}{k!}\right) E[h_{\Gamma,n,A}(\mathcal{X}_k)],$$

whereas $E[G_{n,A}] = \binom{n}{k}E[h_{\Gamma,n,A}(\mathcal{X}_k)]$. Hence $E[G'_{n,A}]/E[G_{n,A}]$ tends to 1 as $n \to \infty$, and the first equality in (3.3) follows. □

Now consider J_n, the number of Γ-components of $G(\mathcal{X}_n; r_n)$. In the sparse limiting regime $nr_n^d \to 0$, the asymptotic behaviour of J_n is much the same as that of G_n. This is because, given that a collection of k vertices of $\mathcal{X}_n$ form the vertices of a Γ-graph, the probability that they do *not* form a component is $O(nr_n^d)$, and so is close to zero. The next result illustrates this; in fact, many subsequent asymptotic results in this chapter for G_n in the sparse limit are also true for J_n, but are not spelt out in the latter case.

Proposition 3.2 *Suppose that $A \subseteq \mathbb{R}^d$ is open with* $\mathrm{Leb}(\partial A) = 0$, *that Γ is a feasible connected graph of order $k \geq 2$, and that $nr_n^d \to 0$. Then, with $\mu_{\Gamma,A}$ defined at* (3.2), $\lim_{n\to\infty} E[J_{n,A}(\Gamma)]/(n^k r_n^{d(k-1)}) = \mu_{\Gamma,A}$.

Proof Recall that θ denotes the volume of the unit ball $B(\mathbf{0}; 1)$. Let B_n be the event that $G(\mathcal{X}_k; r_n)$ is a component of $G(\mathcal{X}_n; r_n)$ isomorphic to Γ with its left-most vertex in A. Given that $G(\mathcal{X}_k; r_n)$ is isomorphic to Γ, with its left-most vertex in A, the conditional probability of event B_n is the conditional probability that no point of $\mathcal{X}_n \setminus \mathcal{X}_k$ is connected to any point of $\mathcal{X}_k$, and this conditional probability is bounded below by $(1 - f_{\max}\theta(kr_n)^d)^{n-k}$, a lower bound which tends to 1 since we assume $nr_n^d \to 0$. Hence,

$$E[J_{n,A}(\Gamma)] = \binom{n}{k} P[B_n] = (1 + o(1))E[G_{n,A}(\Gamma)],$$

and the result follows from Proposition 3.1. □

Next, consider J_n in the thermodynamic limit where nr_n^d tends to a constant. Given $\lambda > 0$, and given a feasible connected graph Γ of order $k \geq 2$, define $p_\Gamma(\lambda)$ by

$$p_\Gamma(\lambda) := \frac{\lambda^{k-1}}{(k-1)!}\int_{(\mathbb{R}^d)^{k-1}} h_\Gamma(\{\mathbf{0}, x_1, \ldots, x_{k-1}\}) \times \exp(-\lambda V(\mathbf{0}, x_1, \ldots, x_{k-1}))\mathrm{d}(x_1, \ldots, x_{k-1}), \tag{3.5}$$

where $V(y_1, \ldots, y_m)$ denotes the Lebesgue measure (volume) of the union of balls of unit radius (in the chosen norm) centred at $y_1, \ldots, y_m$. If Γ consists of a single point (i.e. if $k = 1$), set $p_\Gamma(\lambda) := \exp(-\lambda\theta)$.

The quantity $p_\Gamma(\lambda)$ has the following interpretation. Let $\mathcal{H}_\lambda$ denote a homogeneous Poisson process of intensity λ in $\mathbb{R}^d$. Then $p_\Gamma(\lambda)$ is the probability that the component of $G(\mathcal{H}_\lambda \cup \{\mathbf{0}\}; 1)$ containing the origin is isomorphic to Γ. This can be proved using Theorem 1.6; we omit its proof. See Theorem 9.23 for a proof of a closely related fact.

Proposition 3.3 *Suppose that* $A \subseteq \mathbb{R}^d$ *is open with* $\mathrm{Leb}(\partial A) = 0$, *that* Γ *is a feasible connected graph of order* $k \in \mathbb{N}$, *and that* $nr_n^d \to \rho \in (0, \infty)$. *Then*

$$\lim_{n\to\infty} \left(\frac{E[J'_{n,A}(\Gamma)]}{n} \right) = \lim_{n\to\infty} \left(\frac{E[J_{n,A}(\Gamma)]}{n} \right) = k^{-1} \int_A p_\Gamma(\rho f(x)) f(x) \mathrm{d}x. \quad (3.6)$$

Proof For $x_1, \ldots, x_k$ in $\mathbb{R}^d$, let $I_n(x_1, \ldots, x_k)$ be the integral

$$I_n(x_1, \ldots, x_k) := F\left(\bigcup_{j=1}^m B(x_j; r_n) \right) = \int_{\cup_{j=1}^m B(x_j; r_n)} f(x) \mathrm{d}x. \quad (3.7)$$

Then with $h_{\Gamma,n,A}(\cdot)$ defined at (3.1),

$$\begin{aligned} n^{-1} E J_{n,A}(\Gamma) = n^{-1} \binom{n}{k} \int_{\mathbb{R}^d} \cdots \int_{\mathbb{R}^d} & h_{\Gamma,n,A}(\{x_1, \ldots, x_k\}) \\ & \times (1 - I_n(x_1, \ldots, x_k))^{n-k} f(x_1)^k \prod_{i=1}^k \mathrm{d}x_i \\ & + n^{-1} \binom{n}{k} \int_{\mathbb{R}^d} \cdots \int_{\mathbb{R}^d} h_{\Gamma,n,A}(\{x_1, \ldots, x_k\}) \\ & \times (1 - I_n(x_1, \ldots, x_k))^{n-k} \left(\prod_{i=1}^n f(x_i) - f(x_1)^k \right) \prod_{i=1}^k \mathrm{d}x_i. \end{aligned} \quad (3.8)$$

By the change of variables $y_i = r_n^{-1}(x_i - x_1)$ $(2 \le i \le k)$, the first term on the right-hand side of (3.8) is asymptotic to

$$\begin{aligned} \frac{\rho^{k-1}}{k!} \int_{\mathbb{R}^d} \mathrm{d}x_1 \int_{(\mathbb{R}^d)^{k-1}} & \mathrm{d}(y_2, \ldots, y_k) h_{\Gamma,n,A}(\{x_1, x_1 + r_n y_2, \ldots, x_1 + r_n y_k\}) f(x_1)^k \\ & \times \exp\{(n-k) \log(1 - I_n(x_1, x_1 + r_n y_2, \ldots, x_1 + r_n y_k))\}. \end{aligned} \quad (3.9)$$

As a consequence of the definition of a Lebesgue point (see Rudin (1987, Theorem 7.10)), for each Lebesgue point x_1 of f, and each $y_2, \ldots, y_n$, it is the case that

$$r_n^{-d} I_n(x_1, x_1 + r_n y_2, \ldots, x_1 + r_n y_k) \to f(x_1) V(\mathbf{0}, y_2, \ldots, y_k),$$

so that in the preceding expression (3.9), the exponent converges to $-\rho f(x_1) \times V(\mathbf{0}, y_2, \ldots, y_k)$. Also, $h_{\Gamma,n,A}(\{x_1, x_1 + r_n y_2, \ldots, x_1 + r_n y_k\})$ converges to $h_\Gamma(\{\mathbf{0}, y_2, \ldots, y_k\})$ for $x_1 \in A$ and to 0 for $x_1 \notin A \cup \partial A$. Hence by the Lebesgue density theorem and the dominated convergence theorem, the expression (3.9) converges to the right-hand side of (3.6).

Now consider the second term on the right-hand side of (3.8). By the crude bound

$$h_{\Gamma,n,A}(\{x_1,\ldots,x_k\})(1-I_n(x_1,\ldots,x_k))^{n-k} \leq \prod_{i=2}^{n} \mathbf{1}_{B(x_1;kr_n)}(x_i),$$

and the fact that $n^{-1}\binom{n}{k} \leq c r_n^{d(1-k)}$ for some constant c, the absolute value of this last term in (3.8) is bounded by $c\int_{\mathbb{R}^d} f(x_1)w_n(x_1)\mathrm{d}x_1$ where we set

$$w_n(x) := \int_{B(x;kr_n)} \cdots \int_{B(x;kr_n)} \mathrm{d}x_2 \cdots \mathrm{d}x_n r_n^{d(1-k)} \left| \prod_{i=1}^{n} f(x_i) - f(x)^{k-1} \right|,$$

which tends to zero for each Lebesgue point x, as in the proof of Proposition 3.1. Hence, by the Lebesgue density theorem and dominated convergence theorem, $\int_{\mathbb{R}^d} f(x_1)w_n(x_1)\mathrm{d}x_1 \to 0$ as $n \to \infty$, completing the proof of the second equality in (3.6).

For finite point sets $\mathcal{Y} \subseteq \mathcal{X}$, let $g_{\Gamma,n,A}(\mathcal{Y},\mathcal{X})$ be the indicator of the event that $G(\mathcal{Y};r_n)$ is a Γ-component of $G(\mathcal{X};r_n)$ with its left most vertex in A. Then $J_n'(\Gamma,A) = \sum_{\mathcal{Y}\subseteq\mathcal{P}_n} g_{\Gamma,n,A}(\mathcal{Y};\mathcal{P}_n)$ and so by Theorem 1.6,

$$\begin{aligned} n^{-1}E[J'_{n,A}(\Gamma)] &= \frac{n^{k-1}}{k!}E[g_{\Gamma,n,A}(\mathcal{X}_k';\mathcal{X}_k'\cup\mathcal{P}_n)] \\ &= \frac{n^{k-1}}{k!}\int_{\mathbb{R}^d}\cdots\int_{\mathbb{R}^d} h_{\Gamma,n,A}(\{x_1,\ldots,x_k\}) \\ &\quad \times \exp(-nI_n(x_1,\ldots,x_k))\prod_{i=1}^{k}(f(x_i)\mathrm{d}x_i). \end{aligned}$$

This expression is quite similar to the one at (3.8) and, by an argument similar to the one used before, can be shown to converge to the same limit. This gives us (3.6). □

3.2 Poisson approximation

The basic Poisson approximation theorem for the induced Γ-subgraph count G_n goes as follows. As well as convergence in distribution of G_n to the Poisson when EG_n tends to a finite limit, it also yields convergence to the normal when $EG_n \to \infty$ and $nr_n^d \to 0$, and provides error bounds for these convergence results. Recall that $\mu_\Gamma = \mu_{\Gamma,\mathbb{R}^d}$ is defined at (3.2).

Theorem 3.4 *Let Γ be a feasible connected graph of order $k \geq 2$, and let $G_n := G_n(\Gamma)$. Suppose $(nr_n^d)_{n\geq 1}$ is a bounded sequence. Let Z_n be Poisson with parameter $E[G_n]$. Then there is a constant c such that for all n,*

$$d_{\mathrm{TV}}(G_n, Z_n) \le c n r_n^d. \tag{3.10}$$

If $n^k r_n^{d(k-1)} \to \alpha \in (0,\infty)$, *then* $G_n \xrightarrow{\mathcal{D}} \mathrm{Po}(\lambda)$ *with* $\lambda = \alpha \mu_\Gamma$. *If* $n^k r_n^{d(k-1)} \to \infty$ *and* $n r_n^d \to 0$, *then* $(n^k r_n^{d(k-1)} \mu_\Gamma)^{-1/2}(G_n - EG_n) \xrightarrow{\mathcal{D}} \mathcal{N}(0,1)$.

Proof We have $G_n = \sum_{\boldsymbol{i} \in \mathcal{I}_n} \xi_{\boldsymbol{i},n}$, where $\boldsymbol{i}$ runs through the index set $\mathcal{I}_n$ of all k-subsets $\boldsymbol{i} = \{i_1, \ldots, i_k\}$ of $\{1, 2, \ldots, n\}$, and $\xi_{\boldsymbol{i},n} := h_{\Gamma,n}(\{X_i : i \in \boldsymbol{i}\})$ as defined at (3.1).

For each index $\boldsymbol{i} \in \mathcal{I}_n$, let $\mathcal{N}_{\boldsymbol{i}}$ be the set of $\boldsymbol{j} \in \mathcal{I}_n$ such that $\boldsymbol{i}$ and $\boldsymbol{j}$ have at least one element in common. Let $\sim$ be the associated adjacency relation on $\mathcal{I}_n$, that is, let $\boldsymbol{i} \sim \boldsymbol{j}$ if $\boldsymbol{j} \in \mathcal{N}_{\boldsymbol{i}}$ but $\boldsymbol{j} \neq \boldsymbol{i}$. Then $\xi_{\boldsymbol{i},n}$ is independent of $\xi_{\boldsymbol{j},n}$ except when $\boldsymbol{j} \in \mathcal{N}_{\boldsymbol{i}}$, and the graph $(\mathcal{I}_n, \sim)$ is a dependency graph for $(\xi_{\boldsymbol{i},n}, \boldsymbol{i} \in \mathcal{I}_n)$. The plan is to use Theorem 2.1.

By connectedness all vertices of any Γ-subgraph of $G(\mathcal{X}_n; r_n)$ lie within a distance $(k-1)r_n$ of one another, and hence, with θ denoting the volume of the unit ball, $E\xi_{\boldsymbol{i},n} \le (f_{\max}\theta(kr_n)^d)^{k-1}$. Also,

$$\mathrm{card}(\mathcal{N}_{\boldsymbol{i}}) = \binom{n}{k} - \binom{n-k}{k} = k!^{-1} k^2 n^{k-1} + O(n^{k-2}), \tag{3.11}$$

so that

$$\sum_{\boldsymbol{i} \in \mathcal{I}_n} \sum_{\boldsymbol{j} \in \mathcal{N}_{\boldsymbol{i}}} E\xi_{\boldsymbol{i},n} E\xi_{\boldsymbol{i},n} \le c n^{2k-1} r_n^{2d(k-1)} = c n^{k+1} r_n^{dk} (n r_n^d)^{k-2}. \tag{3.12}$$

Next we bound $E[\xi_{\boldsymbol{i},n}\xi_{\boldsymbol{j},n}]$ when $\boldsymbol{i} \sim \boldsymbol{j}$ but $\boldsymbol{i} \neq \boldsymbol{j}$. In this case the number of elements of $\boldsymbol{i} \cap \boldsymbol{j}$, which we denote h, lies in the range $\{1, \ldots, k-1\}$. We have

$$E[\xi_{\boldsymbol{i},n}\xi_{\boldsymbol{j},n}] \le (f_{\max}\theta k^d r_n^d)^{2k-h-1}.$$

Given $h \in \{1, 2, \ldots, k-1\}$, the number of pairs $(\boldsymbol{i}, \boldsymbol{j}) \in \mathcal{I}_n \times \mathcal{I}_n$ with h elements in common is $\binom{n}{k}\binom{k}{h}\binom{n-k}{k-h}$ which is bounded by a constant times n^{2k-h}. Thus,

$$\begin{aligned}\sum_{\boldsymbol{i} \in \mathcal{I}_n} \sum_{\boldsymbol{j} \in \mathcal{N}_{\boldsymbol{i}} \setminus \{\boldsymbol{i}\}} E[\xi_{\boldsymbol{i},n}\xi_{\boldsymbol{j},n}] &\le c \sum_{h=1}^{k-1} n^{2k-h} r_n^{d(2k-h-1)} \\ &= c \sum_{h=1}^{k-1} n^{k+1} r_n^{dk} (n r_n^d)^{k-h-1}. \end{aligned} \tag{3.13}$$

By the bounds (3.12) and (3.13) and Theorem 2.1,

$$d_{\mathrm{TV}}(G_n, Z_n) \le \min(3, 1/EG_n) c n^{k+1} r_n^{dk},$$

and by Proposition 3.1, this is bounded by a constant times $n r_n^d$. This gives us (3.10), and the remaining assertions of the theorem follow at once from Proposition 3.1, and the convergence of the standardized $\mathrm{Po}(\lambda)$ distribution to the normal as $\lambda \to \infty$. □

Given two or more non-isomorphic connected graphs $\Gamma_1, \ldots, \Gamma_m$, each of order k, it is of interest, in the case $\lim_{n\to\infty}(n^k r_n^{d(k-1)}) \in (0,\infty)$, to know not only that each of the variables $G_n(\Gamma_i)$ is asymptotically Poisson (as shown by the preceding theorem), but also that they are asymptotically independent. The next result demonstrates that this is true.

Theorem 3.5 *Let $k \in \mathbb{N}$ with $k \geq 2$. Let $\Gamma_1, \ldots, \Gamma_m$ be non-isomorphic feasible connected graphs, each with k vertices. Suppose* $\limsup_{n\to\infty}(nr_n^d) < \infty$. *Given $n \in \mathbb{N}$, let $Z_{1,n}, \ldots, Z_{m,n}$ be independent Poisson variables with $EZ_{j,n} = EG_n(\Gamma_j)$. Then there is a constant c such that for all $A \subseteq \mathbb{Z}^m$, and $n \in \mathbb{N}$,*

$$|P[(G_n(\Gamma_1), \ldots, G_n(\Gamma_m)) \in A] - P[(Z_{1,n}, \ldots, Z_{m,n}) \in A]| \leq cn^{k+1}r_n^{dk}. \quad (3.14)$$

Remark The main case of interest occurs when $n^k r_n^{d(k-1)}$ converges to a finite positive limit. Then the above result, along with Proposition 3.1, shows that $(G_n(\Gamma_1), \ldots, G_n(\Gamma_m))$ converge in distribution to independent Poisson variables, and gives a bound on the rate of convergence. In cases where $n^k r_n^{d(k-1)}$ tends to infinity, the result does not give such a good error bound as in the univariate case.

Proof of Theorem 3.5 We have $G_n(\Gamma_j) = \sum_{\boldsymbol{i}} \xi_{\boldsymbol{i},j}$, where $\boldsymbol{i}$ runs through the index set $\mathcal{I}_n$ of all k-subsets $\boldsymbol{i} = \{i_1, \ldots, i_k\}$ of $\{1,2,\ldots,n\}$; and $\xi_{\boldsymbol{i},j} := h_{\Gamma,n}(\{X_i : i \in \boldsymbol{i}\})$.

Set $J := \{1,2,\ldots,m\}$. For each $(\boldsymbol{i},j) \in \mathcal{I}_n \times J$ let $\mathcal{N}_{(\boldsymbol{i},j)}$ be the set of $(\boldsymbol{i}',j') \in \mathcal{I}_n \times J$ such that $\boldsymbol{i}$ and $\boldsymbol{i}'$ have at least one element in common. Let $\sim$ be the associated adjacency relation on $\mathcal{I}_n \times J$, that is, set $(\boldsymbol{i},j) \sim (\boldsymbol{i}',j')$ if $(\boldsymbol{i}',j') \in \mathcal{N}_{(\boldsymbol{i},j)}$ and $(\boldsymbol{i}',j') \neq (\boldsymbol{i},j)$.

Then $\xi_{\boldsymbol{i},j}$ is independent of $\xi_{\boldsymbol{i}',j'}$ except when $(\boldsymbol{i}',j') \in \mathcal{N}_{(\boldsymbol{i},j)}$, and the graph $(\mathcal{I}_n \times J, \sim)$ is a dependency graph for $(\xi_{\boldsymbol{i},j}, (\boldsymbol{i},j) \in \mathcal{I}_n \times J)$. The plan is to use Theorem 2.3.

By (3.11), for each $(\boldsymbol{i},j) \in \mathcal{I}_n \times J$ the cardinality of $\mathcal{N}_{(\boldsymbol{i},j)}$ is equal to $m(k!^{-1}k^2n^{k-1} + O(n^{k-2}))$, so that, since m is fixed,

$$\sum_{(\boldsymbol{i},j)\in\mathcal{I}_n\times J}\ \sum_{(\boldsymbol{i}',j')\in\mathcal{N}_{(\boldsymbol{i},j)}} E\xi_{\boldsymbol{i},j}E\xi_{\boldsymbol{i}',j'} \leq cn^{2k-1}r_n^{2d(k-1)} = cn^{k+1}r_n^{dk}(nr_n^d)^{k-2}. \quad (3.15)$$

Next we bound $E[\xi_{\boldsymbol{i},j}\xi_{\boldsymbol{i}',j'}]$ when $(\boldsymbol{i}',j') \in \mathcal{N}_{(\boldsymbol{i},j)} \setminus \{(\boldsymbol{i}',j')\}$. In this case the number of common elements of $\boldsymbol{i}$ and $\boldsymbol{i}'$, which we denote h, lies in the range $\{1,\ldots,k\}$. If $h = k$ we must have $j \neq j'$ so that $E\xi_{\boldsymbol{i},j}\xi_{\boldsymbol{i}',j'} = 0$. If $1 \leq h \leq k-1$, then

$$E[\xi_{\boldsymbol{i},j}\xi_{\boldsymbol{i}',j'}] \leq (f_{\max}\theta k^d r_n^d)^{2k-h-1}.$$

Given $h \in \{1,2,\ldots,k-1\}$, the number of pairs $((\boldsymbol{i},j),(\boldsymbol{i}',j')) \in (\mathcal{I}_n \times J)^2$, such that $\boldsymbol{i}$ and $\boldsymbol{i}'$ have h elements in common, is $m^2\binom{n}{k}\binom{k}{h}\binom{n-k}{k-h}$ which is bounded by a constant times n^{2k-h}. Thus, as at (3.13), there is a constant c' such that

$$\sum_{(i,j)\in\mathcal{I}_n\times J}\ \sum_{(i',j')\in\mathcal{N}_{(i,j)}\setminus\{(i,j)\}} E[\xi_{i,j}\xi_{i',j'}] \le c'\sum_{h=1}^{k-1} n^{k+1}r_n^{dk}(nr_n^d)^{k-h-1}. \quad (3.16)$$

By the bounds (3.15) and (3.16) and Theorem 2.3, along with the assumption that nr_n^d is bounded, we obtain (3.14). □

Corollary 3.6 *Let $k \in \mathbb{N}$ with $k \ge 2$. Let $\Gamma_1, \ldots, \Gamma_m$ be a collection of non-isomorphic feasible connected graphs, each with k vertices. Suppose for some $\alpha \in (0, \infty)$ that $\lim_{n\to\infty}(n^k r_n^{d(k-1)}) = \alpha$. Let $Z_1, \ldots, Z_m$ be independent Poisson variables with $EZ_j = \alpha\mu_{\Gamma_j}$. Then as $n \to \infty$,*

$$(G_n(\Gamma_1), \ldots, G_n(\Gamma_m)) \xrightarrow{\mathcal{D}} (Z_1, \ldots, Z_m) \quad (3.17)$$

and

$$(J_n(\Gamma_1), \ldots, J_n(\Gamma_m)) \xrightarrow{\mathcal{D}} (Z_1, \ldots, Z_m). \quad (3.18)$$

Proof The first result (3.17) is immediate from Theorem 3.5 and Proposition 3.1. To deduce (3.18), observe that if $\mathcal{Y} \subseteq \mathcal{X}_n$ has k elements and $G(\mathcal{Y}, r_n)$ is an induced Γ_j-subgraph of $G(\mathcal{X}_n, r_n)$, but is not a component, then there exists a point set $\mathcal{U}$ with $k+1$ elements such that $\mathcal{Y} \subset \mathcal{U} \subseteq \mathcal{X}_n$ and $G(\mathcal{U}; r_n)$ is connected. Hence if R_n denotes the number of sets $\mathcal{U} \subseteq \mathcal{X}_n$ of cardinality $k+1$ such that $G(\mathcal{U}; r_n)$ is connected, we have

$$0 \le G_n(\Gamma_j) - J_n(\Gamma_j) \le (k+1)R_n.$$

Since $E[R_n] = O(n^{k+1}r_n^{dk})$, it follows that $E[R_n] \to 0$, and hence $P[G_n(\Gamma_j) \ne J_n(\Gamma_j)]$ tends to zero. Combined with (3.17), this gives us (3.18). □

Example Let $k = 3$. Let Γ_1 be the 3-path with three vertices and two edges, and let Γ_2 be the triangle, that is, the complete graph K_3. Suppose also that $n^3 r_n^{2d}$ tends to a finite constant. If $G(\mathcal{X}_n; r_n)$ has no component of order greater than 3 (an event of probability tending to 1), then the number of vertices of degree 2 is equal to $G_n(\Gamma_1) + 3G_n(\Gamma_2)$, and so converges in distribution to $Z_1 + 3Z_2$, described in Corollary 3.6. Hence, the distribution number of vertices of degree 2 is asymptotically *compound Poisson*, not asymptotically Poisson, as would be the case in the analogous setting for Erdös–Rényi random graphs.

More generally, for $k \ge 3$, suppose $n^k r_n^{d(k-1)}$ converges to a constant. Enumerate the non-isomorphic feasible graphs on k vertices as $\Gamma_1, \ldots, \Gamma_\nu$. The number of vertices of degree k is asymptotically compound Poisson, since it is a linear combination of the variables $G_n(\Gamma_1), \ldots, G_n(\Gamma_\nu)$, $1 \le j \le \nu$, with the coefficient of $G_n(\Gamma_j)$ given by the number of vertices of degree k in Γ_j, and the variables $G_n(\Gamma_1), \ldots, G_n(\Gamma_\nu)$ are asymptotically independent Poisson variables by Corollary 3.6.

3.3 Second moments in a Poisson process

Let Γ, Γ' be fixed, feasible, connected graphs of order k, k', respectively. Let $A \subset \mathbb{R}^d$ be a fixed open set (possibly $\mathbb{R}^d$ itself) with $\mathrm{Leb}(\partial A) = 0$ and $F(A) > 0$.

Recall that $G'_{n,A}(\Gamma)$ denotes the number of induced Γ-subgraphs of $G(\mathcal{P}_n; r_n)$ with left-most vertex in A. This section contains asymptotic expressions for the covariance of $G'_{n,A}(\Gamma)$ and $G'_{n,A}(\Gamma')$.

Recall the definition of $h_\Gamma(\cdot)$ at (3.1). For $(x_1, \ldots, x_{k+k'-j}) \in (\mathbb{R}^d)^{k+k'-j}$, with $1 \leq j \leq \min(k, k')$, define the indicator function $h^j_{\Gamma,\Gamma'}((x_1, \ldots, x_{k+k'-j}))$ by

$$h^j_{\Gamma,\Gamma'}((x_1, \ldots, x_{k+k'-j})) := h_\Gamma(\{x_1, \ldots, x_k\}) \times h_{\Gamma'}(\{x_1, \ldots, x_j, x_{k+1}, \ldots, x_{k+k'-j}\}),$$

and set

$$h^j_{\Gamma,\Gamma',n,A}((x_1, \ldots, x_{k+k'-j})) := h_{\Gamma,n,A}(\{x_1, \ldots, x_k\}) \times h_{\Gamma',n,A}(\{x_1, \ldots, x_j, x_{k+1}, \ldots, x_{k+k'-j}\}).$$

For $j = 1, 2, \ldots, \min(k, k')$, let $\Phi_{j,A} = \Phi_{j,A}(\Gamma, \Gamma')$ be defined by

$$\Phi_{j,A}(\Gamma, \Gamma') := \left(\frac{\int_A f(x)^{k+k'-j} \mathrm{d}x}{j!(k-j)!(k'-j)!} \right) \times \int_{\mathbb{R}^d} \mathrm{d}x_2 \cdots \int_{\mathbb{R}^d} \mathrm{d}x_{k+k'-j} h^j_{\Gamma,\Gamma'}(\mathbf{0}, x_2, \ldots, x_{k+k'-j}). \quad (3.19)$$

Proposition 3.7 *Suppose* $\min(k, k') \geq 2$. *Suppose* $r_n \to 0$, *and set* $\rho_n = nr_n^d$. *Then as* $n \to \infty$,

$$\mathrm{Cov}(G'_{n,A}(\Gamma), G'_{n,A}(\Gamma')) \sim n \sum_{j=1}^{\min(k,k')} \rho_n^{k+k'-j-1} \Phi_{j,A}(\Gamma, \Gamma'), \quad (3.20)$$

where $\sim$ *here means that the ratio of the two sides tends to* 1.

Remarks Note that $\Phi_{k,A} = 0$ when $k = k'$ but Γ, Γ' are not isomorphic. When $\Gamma = \Gamma'$, $\Phi_{k,A} > 0$, and in this case the expression (3.20) describes the asymptotic behaviour of $\mathrm{Var}(G'_{n,A}(\Gamma))$; moreover in this case $\Phi_{k,A} = \mu_{\Gamma,A}$, defined at (3.2).

The dominant term in the asymptotic expression for the covariance depends on the limiting regime. For example, since $\Phi_{k,A}(\Gamma, \Gamma) = \mu_{\Gamma,A}$ we have

$$\mathrm{Var}(G'_{n,A}(\Gamma)) \sim n\rho_n^{k-1} \mu_{\Gamma,A} \quad \text{if } \rho_n \to 0. \quad (3.21)$$

If $k = k'$ but Γ, Γ' are not isomorphic, in the sparse limit we have

$$\mathrm{Cov}(G'_{n,A}(\Gamma), G'_{n,A}(\Gamma')) \sim n\rho_n^k \Phi_{k-1,A}(\Gamma, \Gamma') \quad \text{if } \rho_n \to 0. \quad (3.22)$$

Also, whenever $k = k'$ we have

$$\mathrm{Cov}(G'_{n,A}(\Gamma), G'_{n,A}(\Gamma')) \sim n\rho_n^{2k-2} \Phi_{1,A}(\Gamma, \Gamma') \quad \text{if } \rho_n \to \infty, \ r_n \to 0. \quad (3.23)$$

In the thermodynamic limit $\rho_n \to$ const., all terms in the sum on the right-hand side of (3.20) tend to positive finite limits. Also, the rate of growth of $\mathrm{Var}(G'_{n,A}(\Gamma))$ is independent of k for the thermodynamic limit but not for the sparse or dense limit.

Proof of Proposition 3.7 Without loss of generality, assume that $k \leq k'$. Then

$$E[G'_{n,A}(\Gamma)G'_{n,A}(\Gamma')] = \sum_{j=0}^{k} E \sum_{\mathcal{Y}\subseteq\mathcal{P}_n} \sum_{\mathcal{Y}'\subseteq\mathcal{P}_n} h_{\Gamma,n,A}(\mathcal{Y})h_{\Gamma',n,A}(\mathcal{Y}')\mathbf{1}_{\{|\mathcal{Y}\cap\mathcal{Y}'|=j\}}, \quad (3.24)$$

and by Theorem 1.7 the $j = 0$ term in this sum equals $E[G'_{n,A}(\Gamma)]E[G'_{n,A}(\Gamma')]$. For $1 \leq j \leq k$ the jth term equals $E\sum_{\mathcal{X}\subseteq\mathcal{P}_n} h'_{n,j}(\mathcal{X})$, with the function $h'_{n,j}(\mathcal{X})$ given by

$$h'_{n,j}(\mathcal{X}) := \mathbf{1}_{\{|\mathcal{X}|=k+k'-j\}} \sum_{\mathcal{Y}\subseteq\mathcal{X}} \sum_{\mathcal{Y}'\subseteq\mathcal{X}} h_{\Gamma,n,A}(\mathcal{Y})h_{\Gamma',n,A}(\mathcal{Y}')\mathbf{1}_{\{\mathcal{Y}\cup\mathcal{Y}'=\mathcal{X}\}}.$$

By Theorem 1.6, for $j > 0$ the jth term in (3.24) equals

$$\left(\frac{n^{k+k'-j}}{(k+k'-j)!}\right) Eh'_{n,j}(\mathcal{X}_{k+k'-j}),$$

and since the number of ways of partitioning $\mathcal{X}_{k+k'-j}$ into an ordered triple of sets of cardinality $j, k-j, k'-j$ respectively is $(k+k'-j)!/(j!(k-j)!(k'-j)!)$, this is equal to

$$\begin{aligned}
&\left(\frac{n^{k+k'-j}}{j!(k-j)!(k'-j)!}\right) E[h_{\Gamma,n,A}(\mathcal{X}_k)h_{\Gamma',n,A}(\mathcal{X}_j \cup (\mathcal{X}_{k+k'-j} \setminus \mathcal{X}_k))] \\
&= \frac{n^{k+k'-j}}{j!(k-j)!(k'-j)!} \int_{\mathbb{R}^d} \cdots \int_{\mathbb{R}^d} h^j_{\Gamma,\Gamma',n,A}(x_1,\ldots,x_{k+k'-j}) \\
&\qquad \times f(x_1)\mathrm{d}x_1 \cdots f(x_{k+k'-1})\mathrm{d}x_{k+k'-j} \\
&= \frac{n(nr_n^d)^{k+k'-j-1}}{j!(k-j)!(k'-j)!} \int_{\mathbb{R}^d} f(x_1)\mathrm{d}x_1 \int_{\mathbb{R}^d} \cdots \int_{\mathbb{R}^d} \\
&\times h^j_{\Gamma,\Gamma',n,A}(x_1, x_1+r_ny_2,\ldots,x_1+r_ny_{k+k'-j}) \prod_{i=2}^{k+k'-j} (f(x_1+r_ny_i)\mathrm{d}y_i).
\end{aligned}$$

We assert that the integral in this expression tends to

$$\int_A f(x)^{k+k'-j}\mathrm{d}x \int_{\mathbb{R}^d} \cdots \int_{\mathbb{R}^d} h^j_{\Gamma,\Gamma'}(\mathbf{0}, y_2,\ldots,y_{k+k'-j})\mathrm{d}y_2 \cdots \mathrm{d}y_{k+k'-j}.$$

If f is almost everywhere continuous, this follows from the dominated convergence theorem; if not, an extra argument using the Lebesgue density theorem, similar to that in the proof of Proposition 3.1, is needed and is left as an exercise.

It follows that the jth term in the sum (3.24) is asymptotic to $n(nr_n^d)^{k+k'-j-1}\Phi_{j,A}$, and the result follows. □

Now consider $J'_{n,A}(\Gamma)$, the number of *components* of $G(\mathcal{P}_n; r_n)$ isomorphic to Γ with left-most vertex in A, and $J'_{n,A}(\Gamma')$ defined likewise. In this case, we consider only the thermodynamic limit, but now allow for the possibility that $k = 1$ or $k' = 1$. Given $\lambda > 0$, recall that $p_\Gamma(\lambda)$ is defined at (3.5) and denotes the probability that $\mathbf{0}$ lies in a Γ-component of the graph $G(\mathcal{H}_\lambda \cup \{\mathbf{0}\}; 1)$, and that $V(x_1, \ldots, x_m)$ denotes the Lebesgue measure of $\cup_{i=1}^m B(x_i; 1)$. For $y \in \mathbb{R}^d$ and $\lambda > 0$, define $q_{\Gamma,\Gamma'}(y, \lambda)$ (in the case with $\min(k, k') > 1$) by

$$
\begin{aligned}
&q_{\Gamma,\Gamma'}(y,\lambda) \\
&:= \left(\frac{\lambda^{k+k'-2}}{(k-1)!(k'-1)!}\right) \int_{(\mathbb{R}^d)^{k+k'-2}} h_\Gamma(\{\mathbf{0}, x_1, \ldots, x_{k-1}\}) h_{\Gamma'}(\{y, y_1, \ldots, y_{k'-1}\}) \\
&\quad \times \mathbf{1}\{G(\{\mathbf{0}, x_1, \ldots, x_{k-1}, y, y_1, \ldots, y_{k-1}\}; 1) \text{ is disconnected}\} \\
&\quad \times \exp(-\lambda V(\mathbf{0}, y, x_1, \ldots, x_{k-1}, y_1, \ldots, y_{k-1})) \mathrm{d}(x_1, \ldots, x_{k-1}, y_1, \ldots, y_{k-1}).
\end{aligned}
$$

If $1 = k < k'$, then set $B(\mathbf{0}; 1)^{\mathrm{c}} := \mathbb{R}^d \setminus B(\mathbf{0}; 1)$, and set

$$
\begin{aligned}
q_{\Gamma,\Gamma'}(y,\lambda) := \mathbf{1}_{B(\mathbf{0};1)^{\mathrm{c}}}(y) \left(\frac{\lambda^{k'-1}}{(k'-1)!}\right) \int_{(B(\mathbf{0};1)^{\mathrm{c}})^{k'-1}} h_{\Gamma'}(\{y, y_1, \ldots, y_{k'-1}\}) \\
\times \exp(-\lambda V(\mathbf{0}, y, y_1, \ldots, y_{k-1})) \mathrm{d}(y_1, \ldots, y_{k-1}).
\end{aligned}
$$

Define $q_{\Gamma,\Gamma'}(y, \lambda)$ analogously when $1 = k' < k$, and if $1 = k = k'$, set $q_{\Gamma,\Gamma'}(y, \lambda) := \mathbf{1}_{B(\mathbf{0};1)^{\mathrm{c}}}(y) \exp(-\lambda V(\mathbf{0}, y))$.

It can be shown by Palm theory (Theorem 1.6; we leave this as an exercise) that $q_{\Gamma,\Gamma'}(y, \lambda)$ is the probability that in $G(\mathcal{H}_\lambda \cup \{\mathbf{0}, y\}; 1)$ there are distinct components C, C', such that $\mathbf{0} \in C$ and $y \in C'$, and such that $C \cong \Gamma$ and $C' \cong \Gamma'$.

Proposition 3.8 *Suppose that* $\lim_{n\to\infty}(nr_n^d) = \rho \in (0, \infty)$. *Set*

$$
\Psi_A(\Gamma, \Gamma') := \frac{\rho}{k'k} \int_A f(x)^2 \int_{\mathbb{R}^d} (q_{\Gamma,\Gamma'}(y, \rho f(x)) - p_\Gamma(\rho f(x)) p_{\Gamma'}(\rho f(x))) \mathrm{d}y \mathrm{d}x.
$$

If Γ *and* Γ' *are non-isomorphic, then*

$$
\lim_{n\to\infty} \left(n^{-1} \mathrm{Cov}(J'_{n,A}(\Gamma), J'_{n,A}(\Gamma'))\right) = \Psi_A(\Gamma, \Gamma'), \tag{3.25}
$$

while

$$
\lim_{n\to\infty} \left(n^{-1} (\mathrm{Var}(J'_{n,A}(\Gamma)) - E[J'_{n,A}(\Gamma)])\right) = \Psi_A(\Gamma, \Gamma). \tag{3.26}
$$

Proof For any finite set $\mathcal{X} \subset \mathbb{R}^d$ and any $x \in \mathcal{X}$, let $v_n(x; \mathcal{X})$ be the indicator function of the event that x lies in a component of $G(\mathcal{X}; r_n)$ isomorphic to Γ with left-most vertex in A. Then $J'_{n,A}(\Gamma) = k^{-1} \sum_{X \in \mathcal{P}_n} v_n(X; \mathcal{P}_n)$, and hence by Theorem 1.6,

$$
E[J'_{n,A}(\Gamma)] = \frac{n}{k} \int_{\mathbb{R}^d} P[B_x^\Gamma] f(x) \mathrm{d}x, \tag{3.27}
$$

where B_x^Γ denotes the event that x is a vertex of a Γ-component of $G(\mathcal{P}_n \cup \{x\}; r_n)$ with left-most vertex in A.

Suppose Γ, Γ' are non-isomorphic. For any finite set $\mathcal{X} \subset \mathbb{R}^d$ and any $\{x, y\} \subseteq \mathcal{X}$, let $w_{\Gamma,\Gamma',n}(\{x, y\}, \mathcal{X})$ be the indicator function of the event that $G(\mathcal{X}; r_n)$ contains two distinct components C, C', with one of x, y a vertex of C and the other a vertex of C', with $C \cong \Gamma$ and $C' \cong \Gamma'$, and with the left-most vertex of C in A and the left-most vertex of C' in A. Then

$$J'_{n,A}(\Gamma)J'_{n,A}(\Gamma') = \frac{1}{k'k} \sum_{\{X,Y\} \subseteq \mathcal{P}_n} w_{\Gamma,\Gamma',n}(\{X, Y\}, \mathcal{P}_n), \tag{3.28}$$

so that by Theorem 1.6,

$$\begin{aligned} E[J'_{n,A}(\Gamma)J'_{n,A}(\Gamma')] &= \left(\frac{n^2}{2k'k}\right) E[w_{\Gamma,\Gamma',n}(\mathcal{X}'_2, \mathcal{X}'_2 \cup \mathcal{P}_n)] \\ &= \frac{n^2}{k'k} \int_{\mathbb{R}^d} \int_{\mathbb{R}^d} P[F_{x,y}] f(y) \mathrm{d}y f(x) \mathrm{d}x, \end{aligned} \tag{3.29}$$

where $F_{x,y}$ denotes the event that there are distinct components $C_x \cong \Gamma$ and $C_y \cong \Gamma'$ in $G(\mathcal{P}_n \cup \{x, y\}; r_n)$, with x and y being vertices of C_x, C_y, respectively, and with the left-most vertex of C_x in A and the left-most vertex of C_y in A.

It follows from (3.29) and (3.27), followed by a change of variable, that

$$\begin{aligned} \mathrm{Cov}(J'_n(\Gamma), J'_n(\Gamma')) &= \frac{n^2}{k'k} \int_{\mathbb{R}^d} \int_{\mathbb{R}^d} (P[F_{x,y}] - P[B_x^\Gamma]P[B_y^{\Gamma'}]) f(y) \mathrm{d}y f(x) \mathrm{d}x \\ &= \frac{n^2 r_n^d}{k'k} \int_{\mathbb{R}^d} \int_{\mathbb{R}^d} (P[F_{x,x+r_n z}] - P[B_x^\Gamma]P[B_{x+r_n z}^{\Gamma'}]) f(x + r_n z) \mathrm{d}z f(x) \mathrm{d}x. \end{aligned} \tag{3.30}$$

By the independence properties of the Poisson process, $P[F_{xy}] = P[B_x^\Gamma]P[B_y^{\Gamma'}]$ for $\|x - y\| > (k + k')r_n$, and hence in the last expression the integrand is zero for $\|z\| > k + k'$.

Suppose $\min(k, k') > 1$. With $h_{\Gamma,n,A}(\cdot)$ defined at (3.1) and $I_n(\cdot)$ at (3.7), by Theorem 1.6 we have

$$\begin{aligned} P[B_x^\Gamma] &= \frac{n^{k-1}}{(k-1)!} \int_{\mathbb{R}^d} \cdots \int_{\mathbb{R}^d} h_{\Gamma,n,A}(\{x, x_1, \ldots, x_{k-1}\}) \\ &\quad \times \exp(-nI_n(x, x_1, \ldots, x_{k-1})) \prod_{i=1}^{k-1} (f(x_i) \mathrm{d}x_i) \\ &= \frac{(nr_n^d)^{k-1}}{(k-1)!} \int_{\mathbb{R}^d} \cdots \int_{\mathbb{R}^d} h_{\Gamma,n,A}(\{x, x + r_n y_1, \ldots, x + r_n y_{k-1}\}) \\ &\quad \times \exp(-nI_n(x, x + r_n y_1, \ldots, x + r_n y_{k-1})) \prod_{i=1}^{k-1} (f(x + r_n y_i) \mathrm{d}y_i). \end{aligned}$$

Suppose $x \in A$ and x is a continuity point of f. Then by the dominated convergence theorem, this expression for $P[B_x^\Gamma]$ tends to

$$\frac{(\rho f(x))^{k-1}}{(k-1)!} \int_{\mathbb{R}^d} \cdots \int_{\mathbb{R}^d} h_\Gamma(\{\mathbf{0}, y_1, \ldots, y_{k-1}\}) \times \exp(-\rho f(x) V(\mathbf{0}, y_1, \ldots, y_{k-1})) \mathrm{d}y_1 \cdots \mathrm{d}y_{k-1},$$

that is, $P[B_x^\Gamma] \to p_\Gamma(\rho f(x))$. Similarly, again by a change of variable and the dominated convergence theorem, we obtain $P[B_{x+r_n z}^{\Gamma'}] \to p_{\Gamma'}(\rho f(x))$, and also

$$P[F_{x,x+r_n z}] \to q_\Gamma(z, \rho f(x)).$$

On the other hand, if $x \notin A \cup \partial A$, then each of $P[B_x^\Gamma]$, $P[B_{x+r_n z}^{\Gamma'}]$, and $P[F_{x,x+r_n z}]$ tends to zero. Moreover, all of these limiting statements are also valid when $k = 1$ or $k' = 1$.

Using these limits and the dominated convergence theorem in the expression (3.30) for $\mathrm{Cov}(J'_{n,A}(\Gamma), J'_{n,A}(\Gamma'))$ gives us the limit (3.25), for the special case where f is almost everywhere continuous. The general case can be dealt with by using the Lebesgue density theorem, in a similar manner to that used in the proofs of Propositions 3.1 and 3.3.

The proof of (3.26) is similar, except that in the case where $\Gamma = \Gamma'$, eqn (3.28) must be modified to

$$(J'_{n,A}(\Gamma))^2 = J'_{n,A}(\Gamma) + \frac{2}{k^2} \sum_{\{X,Y\} \subseteq \Gamma} w_{\Gamma,\Gamma,n}(\{X,Y\}, \mathcal{P}_n).$$

The extra term $J'_{n,A}(\Gamma)$ is accounted for on the left-hand side of (3.26), and the extra factor of 2 is lost at (3.29). □

3.4 Normal approximation for Poisson processes

Suppose Γ, Γ' are non-isomorphic connected graphs of order k. The goal now is to prove that appropriately scaled and centred versions of $G_n(\Gamma)$ and $G_n(\Gamma')$, are asymptotically *bivariate normal.* If $n^k r_n^{d(k-1)} \to \infty$ but $nr_n^d \to 0$ as $n \to \infty$, then $G_n(\Gamma)$, suitably scaled and centred, is asymptotically normal as already seen from the Poisson approximation in Theorem 3.4; however this is insufficient to show a bivariate normal limit, and Theorem 3.5 does not help unless $n^{k+1} r_n^{dk} \to 0$. Also, one might expect a central limit theorem to hold for $G_n(\Gamma)$ even in the dense limit. Therefore we take a different approach, proceeding via the Poissonized setting. Attention is restricted here to cases with $r_n \to 0$; when $r_n =$ const., Hoeffding's classical theory of U-statistics (Lee 1990, p. 76) yields a central limit theorem for $G_n(\Gamma)$, but we shall not discuss this case further.

Throughout this section, assume $A \subseteq \mathbb{R}^d$ is open with $\mathrm{Leb}(\partial A) = 0$; we give central limit theorems for $G'_{n,A}(\Gamma)$ and for $J'_{n,A}(\Gamma)$. The main case of interest is when $A = \mathbb{R}^d$. The first result for $G'_{n,A}$ includes both sparse and dense limiting regimes for r_n (the thermodynamic limit is considered later on).

Theorem 3.9 *Let $k \in \mathbb{N}$ with $k \geq 2$. Let $\Gamma_1, \ldots, \Gamma_m$ be non-isomorphic feasible connected graphs, each with k vertices. Suppose that $r_n \to 0$ and $n^k r_n^{d(k-1)} \to \infty$*

as $n \to \infty$. Suppose also that $\rho_n := nr_n^d$ tends either to 0 or to ∞ as $n \to \infty$. If $\rho_n \to 0$ then set $\tau_n := (n\rho_n^{k-1})^{1/2}$, but if $\rho_n \to \infty$ then set $\tau_n := (n\rho_n^{2(k-1)})^{1/2}$.

Then as $n \to \infty$, the joint distribution of the variables $\tau_n^{-1}(G'_{n,A}(\Gamma_j) - EG'_{n,A}(\Gamma_j))$ $1 \le j \le m$, converges to a centred multivariate normal distribution with the following covariance matrix $\Sigma'(A) = (\sigma'_{ij}(A))$. In the case $\rho_n \to 0$, Σ' is a diagonal matrix with $\sigma'_{ii}(A) = \mu_{\Gamma_i,A}$ defined at (3.2), while in the case $\rho_n \to \infty$, we have $\sigma'_{ij}(A) = \Phi_{1,A}(\Gamma_i, \Gamma_j)$, with $\Phi_{1,A}$ defined at (3.19).

Proof Let $a_1, \dots, a_m$ be arbitrary constants. Let $S'_{n,A} = \sum_{j=1}^m a_j G'_{n,A}(\Gamma_j)$. By (3.21)–(3.23), we obtain

$$\tau_n^{-2}\mathrm{Var}(S'_{n,A}) \to \sum_{i=1}^m \sum_{j=1}^m a_i a_j \sigma'_{ij}(A) := \gamma^2(A). \tag{3.31}$$

By the Cramér–Wold device, it suffices to prove that $\tau_n^{-1}(S'_{n,A} - ES'_{n,A})$ converges in distribution to $\mathcal{N}(0, \gamma^2(A))$. If $\gamma^2(A) = 0$, then this is clearly true (we do not require that $\Sigma'(A)$ be strictly positive definite), so let us assume $\gamma^2(A) > 0$.

First suppose that A *is bounded.* Given n, divide $\mathbb{R}^d$ into little cubes of side r_n, denoted $Q_{i,n}$, $i \in \mathbb{N}$. Let $V_n := \{i \in \mathbb{N} : Q_{i,n} \cap A \neq \emptyset\}$. Recalling the definition of $h_{\Gamma,n,A}(\cdot)$ at (3.1), set

$$\xi_{i,n} := \sum_{\mathcal{Y} \subseteq \mathcal{P}_n} \sum_{j=1}^m a_j h_{\Gamma_j,n,A\cap Q_{i,n}}(\mathcal{Y}). \tag{3.32}$$

Then $S'_{n,A} = \sum_{i \in V_n} \xi_{i,n}$. Also, if we make V_n into the vertex set of a graph by setting $i \sim i'$ if and only if the minimum distance between points in $Q_{i,n}$ and $Q_{i',n}$ is at most $2kr_n$, it is evident that $(V_n, \sim)$ is a dependency graph for $\{\xi_{i,n} : i \in V_n\}$, with $O(r_n^{-d})$ vertices (since A is assumed to be bounded) and with degree bounded uniformly by a constant that does not depend on n. Therefore by Theorem 2.4, it suffices to show that as $n \to \infty$,

$$b_{p,n} := r_n^{-d} \max_i \left(\tau_n^{-p} E[|\xi_{i,n} - E\xi_{i,n}|^p]\right) \to 0, \quad p = 3, 4. \tag{3.33}$$

Consider first *the case where $\rho_n \to 0$*. For any positive integer m let us write $(m)_k$ for the descending factorial $m(m-1)\cdots(m-k+1)$. To estimate the moments of $\xi_{i,n}$, observe that $|\xi_{i,n}|$ is bounded by a constant times $(Z_{i,n})_k$, where $Z_{i,n}$ denotes the number of points of $\mathcal{P}_n$ lying within distance kr_n of the cube $Q_{i,n}$. Then $Z_{i,n}$ is stochastically dominated by a Poisson variable with parameter $c\rho_n$, where c is a constant depending only on f and the choice of norm. Since $(m)_k$ is zero for $m < k$, and $(m)_k^4$ is a polynomial in m, we have

$$E[(Z_{i,m})_k^4] \le \sum_{m=k}^\infty (m)_k^4 \left(\frac{\mathrm{e}^{-c\rho_n}(c\rho_n)^m}{m!} \right) \le c'\rho_n^k,$$

for some constant c'. Similarly $E[(Z_{i,m})_k^3]$ and $E[Z_{i.n}]$ are also bounded by a constant times ρ_n^k. Hence there is a constant c'' such that in the case $\rho_n \to 0$, for $p = 3, 4$ we have

$$b_{p,n} \le c''(n\rho_n^{k-1})^{-p/2} r_n^{-d} \rho_n^k = c''(n^k r_n^{d(k-1)})^{1-p/2} \to 0,$$

which gives us (3.33) as required in the case where $\rho_n \to 0$ (and A is bounded).

Now consider *the case where* $\rho_n \to \infty$ (still with A bounded), for which more care is needed. Consider first $b_{4,n}$, expressing $E[(\xi_{i,n} - E\xi_{i,n})^4]$ as a linear combination of moments of $\xi_{i,n}$:

$$E[(\xi_{i,n} - E\xi_{i,n})^4] = \sum_{j=0}^{4} \binom{4}{j} (-E[\xi_{i,n}])^{4-j} E[\xi_{i,n}^j]. \tag{3.34}$$

For finite $\mathcal{Y} \subset \mathbb{R}^d$, let $g_{n,i}(\mathcal{Y}) := \sum_{j=1}^m a_j h_{\Gamma_j,n,A\cap Q_{i,n}}(\mathcal{Y}) I_{i,n}(\mathcal{Y})$. Then $\xi_{i,n}$ equals $\sum_{\mathcal{Y}\subseteq\mathcal{P}_n} g_{n,i}(\mathcal{Y})$, and

$$E[\xi_{i,n}^4] = E \sum_{\mathcal{Y}\subseteq\mathcal{P}_n} \sum_{\mathcal{Y}'\subseteq\mathcal{P}_n} \sum_{\mathcal{Y}''\subseteq\mathcal{P}_n} \sum_{\mathcal{Y}'''\subseteq\mathcal{P}_n} g_{n,i}(\mathcal{Y}) g_{n,i}(\mathcal{Y}') g_{n,i}(\mathcal{Y}'') g_{n,i}(\mathcal{Y}'''), \tag{3.35}$$

with similar expressions for lower moments of $\xi_{i,n}$.

The leading-order term in the expression (3.35) for $E[\xi_{i,n}^4]$ comes from the contribution from ordered 4-tuples $\mathcal{Y}, \mathcal{Y}', \mathcal{Y}'', \mathcal{Y}'''$ with *no* elements in common. By Theorem 1.7, this term is equal to $(E[\xi_{i,n}])^4$. Similarly the leading-order term in $E[\xi_{i,n}^3]$ is equal to $(E[\xi_{i,n}])^3$, and the leading-order term in $E[\xi_{i,n}^2]$ is equal to $(E[\xi_{i,n}])^2$. Combining all these we find that the sum of the leading-order terms on the right-hand side of (3.34) is zero.

The second-order term in (3.35) comes from 4-tuples of subsets $\mathcal{Y}, \mathcal{Y}', \mathcal{Y}'', \mathcal{Y}'''$ of $\mathcal{P}_n$ with *one* element in common between them, that is, with a total of $4k-1$ elements. For example, the contribution from $\mathcal{Y}$ and $\mathcal{Y}'$ having precisely one element in common but $\mathcal{Y}''$ and $\mathcal{Y}'''$ having no element in common with each other or with $\mathcal{Y}$ or $\mathcal{Y}'$ is equal, by Theorem 1.7, to

$$\left(E \sum_{\mathcal{Y}\subseteq\mathcal{P}_n} \sum_{\mathcal{Y}'\subseteq\mathcal{P}_n} g_{n,i}(\mathcal{Y}) g_{n,i}(\mathcal{Y}') \mathbf{1}_{\{|\mathcal{Y}\cap\mathcal{Y}'|=1\}} \right) (E[\xi_{i,n}])^2 . \tag{3.36}$$

There are six such terms, according to which two out of $\mathcal{Y}, \mathcal{Y}', \mathcal{Y}'', \mathcal{Y}'''$ have one element in common, so the overall second order term in $E[\xi_{i,n}^4]$ is equal to six times the expression at (3.36). Similarly, the second order term in $E[\xi_{i,n}^3]$ is equal to

$$3 \left(E \sum_{\mathcal{Y}\subseteq\mathcal{P}_n} \sum_{\mathcal{Y}'\subseteq\mathcal{P}_n} g_{n,i}(\mathcal{Y}) g_{n,i}(\mathcal{Y}') \mathbf{1}_{\{|\mathcal{Y}\cap\mathcal{Y}'|=1\}} \right) E[\xi_{i,n}],$$

so that the overall second-order contribution from $j = 3$ to the right-hand side of (3.34) is equal to -12 times the expression (3.36). Moreover, by Theorem

1.7, the overall second-order contribution from $j = 2$ to the right-hand side of (3.34) is equal to six times the expression (3.36), and there is no second-order contribution from $j = 1$ or from $j = 0$. Since $6 - 12 + 6 = 0$, the total of all second-order contributions to the right-hand side of (3.34) is zero.

Thus the lowest-order non-zero term on the right-hand side of (3.34) is (at worst) the third-order term from 4-tuples $(\mathcal{Y}, \mathcal{Y}', \mathcal{Y}'', \mathcal{Y}''')$ having a total of $4k-2$ elements. We assert that this term is bounded by a constant times $(nr_n^d)^{4k-2}$. For example, the contribution from $\mathcal{Y}$ and $\mathcal{Y}'$ having precisely two elements in common but $\mathcal{Y}''$ and $\mathcal{Y}'''$ having no element in common with each other or with $\mathcal{Y}$ or $\mathcal{Y}'$ is equal, by Theorem 1.7, to

$$\left(E \sum_{\mathcal{Y} \subseteq \mathcal{P}_n} \sum_{\mathcal{Y}' \subseteq \mathcal{P}_n} g_{n,i}(\mathcal{Y}) g_{n,i}(\mathcal{Y}') \mathbf{1}_{\{|\mathcal{Y} \cap \mathcal{Y}'| = 2\}} \right) (E[\xi_{i,n}])^2 .$$

By Theorem 1.6, $E[\xi_{i,n}] = (n^k/k!) E[g_{n,i}(\mathcal{X}_k)]$, so that $E[\xi_{i,n}]$ is bounded by a constant times $(nr_n^d)^k$. Also, by Theorem 1.6, there is a constant c such that

$$E \sum_{\mathcal{Y} \subseteq \mathcal{P}_n} \sum_{\mathcal{Y}' \subseteq \mathcal{P}_n} g_{n,i}(\mathcal{Y}) g_{n,i}(\mathcal{Y}') \mathbf{1}_{\{|\mathcal{Y} \cap \mathcal{Y}'| = 2\}} \leq c n^{2k-2} E[h_{n,i,2}(\mathcal{X}_{2k-2})], \quad (3.37)$$

where $h_{n,i,2}(\mathcal{X})$ is the indicator of the event that $\mathcal{X}$ has $2k-2$ elements, all lying within distance $2kr_n$ of $Q_{i,n}$. Since $E[h_{n,i,2}(\mathcal{X}_{2k-2})]$ is bounded by a constant times $r_n^{d(2k-2)}$, the expression (3.37) is $O(nr_n^d)^{2k-2}$; by this estimate and similar ones for other contributions to the third-order term in (3.34), the assertion follows. Similarly, fourth- and higher-order terms are all $o(nr_n^d)^{4k-2}$.

Therefore, there is a constant c such that $E[(\xi_{i,n} - E\xi_{i,n})^4] \leq c(nr_n^d)^{4k-2}$, and hence

$$b_{4,n} \leq \frac{cr_n^{-d}(nr_n^d)^{4k-2}}{(n^{1/2}(nr_n^d)^{k-1})^4} = cr_n^d,$$

which tends to zero by assumption.

Turning to $b_{3,n}$, observe that by Jensen's inequality and the preceding bound for $E[|\xi_{i,n} - E\xi_{i,n}|^4]$, there is a constant c such that

$$E[|\xi_{i,n} - E\xi_{i,n}|^3] \leq c(nr_n^d)^{3(4k-2)/4} = c(nr_n^d)^{3k-3/2},$$

so that for some constant c',

$$b_{3,n} \leq \frac{c' r_n^{-d}(nr_n^d)^{3k-3/2}}{(n^{1/2}(nr_n^d)^{k-1})^3} = c' r_n^{d/2},$$

which tends to zero. Thus (3.33) holds in the case $\rho_n \to \infty$, too, and this completes the proof for the case where A is bounded.

Now suppose that A *is unbounded* (e.g. $A = \mathbb{R}^d$). Set $\zeta_n(A) := \tau_n^{-1}(S'_n(A) - ES'_n(A))$. Set $A_K := A \cap (-K, K)^d$, and $A^K := A \setminus [-K, K]^d$. Then A_K is open and bounded with $\text{Leb}(\partial A_K) = 0$, so that by the case considered already,

$$\zeta_n(A_K) \xrightarrow{\mathcal{D}} \mathcal{N}(0, \gamma^2(A_K)) \quad \text{as } n \to \infty. \tag{3.38}$$

Given $w \in \mathbb{R}$ and $\varepsilon > 0$,

$$\begin{aligned} P[\zeta_n(A) \le w] = &\ P[\zeta_n(A_K) \le w - \varepsilon] - P[\{\zeta_n(A_K) \le w - \varepsilon\} \cap \{\zeta_n(A) > w\}] \\ &+ P[\{|\zeta_n(A_K) - w| < \varepsilon\} \cap \{\zeta_n(A) \le w\}] \\ &+ P[\{\zeta_n(A_K) \ge w + \varepsilon\} \cap \{\zeta_n(A) \le w\}]. \end{aligned}$$

Hence, since $\zeta_n(A) = \zeta_n(A_K) + \zeta_n(A^K)$ a.s.,

$$|P[\zeta_n(A) \le w] - P[\zeta_n(A_K) \le w - \varepsilon]| \le P[|\zeta_n(A^K)| \ge \varepsilon] + P[|\zeta_n(A_K) - w| < \varepsilon].$$

By Chebyshev's inequality, (3.31), and (3.38),

$$\begin{aligned} &\limsup_{n\to\infty} |P[\zeta_n(A) \le w] - P[\zeta_n(A_K) \le w - \varepsilon]| \\ &\qquad \le \frac{\gamma^2(A^K)}{\varepsilon^2} + \frac{2\varepsilon}{\sqrt{2\pi\gamma^2(A_K)}}. \end{aligned} \tag{3.39}$$

Set $\Phi(t) := P[\mathcal{N}(0,1) \le t]$, $t \in \mathbb{R}^d$. As $K \to \infty$, $\gamma^2(A_K)$ tends to $\gamma^2(A)$, and $\gamma^2(A^K)$ tends to zero. Hence, by taking ε sufficiently small and K sufficiently large, we can make right-hand side of (3.39) arbitrarily small, and also make $\Phi\left((w - \varepsilon)/\gamma(A_K)\right)$ arbitrarily close to $\Phi\left(w/\gamma(A)\right)$. Then by (3.38), it follows that $P[\zeta_n(A) \le w]$ tends to $\Phi\left(w/\gamma(A)\right)$, that is, $\zeta_n(A) \xrightarrow{\mathcal{D}} \mathcal{N}(0, \gamma^2(A))$, completing the proof. □

Now consider the thermodynamic limit. In this case we consider J'_n as well as G'_n. The argument is just the same as in the sparse limit (the easier case in the result just given), except that now the asymptotic covariance of $n^{-1/2}G'_n(\Gamma_i)$ and $n^{-1/2}G'_n(\Gamma_j)$ is non-zero, even if Γ_i, Γ_j have a different number of vertices.

Theorem 3.10 *Suppose $m \in \mathbb{N}$ and for $j \in \{1, 2, \ldots, m\}$, Γ_j is a feasible connected graph of order $k_j \in [2, \infty)$, with $\Gamma_1, \ldots, \Gamma_m$ non-isomorphic. Suppose $\lim_{n\to\infty}(nr_n^d) = \rho \in (0, \infty)$. Then the joint distribution of the variables $n^{-1/2}(G'_{n,A}(\Gamma_j) - EG'_{n,A}(\Gamma_j))$, $1 \le j \le m$, converges, as $n \to \infty$, to a centred multivariate normal with covariance matrix whose (i, l)th entry is*

$$\sum_{j=1}^{\min(k_i, k_l)} \rho^{k_i + k_l - j - 1} \Phi_{j,A}(\Gamma_i, \Gamma_l),$$

with $\Phi_{j,A}(\Gamma_i, \Gamma_l)$ defined at (3.19).

Proof The proof is just the same as for the case $\rho_n \to 0$ of the preceding result, except that now the limiting covariances come directly from eqn (3.20). □

The same argument (with details therefore omitted) yields the following multivariate central limit theorem for component counts in the thermodynamic limit. This time the limiting covariance structure comes from Propositions 3.8 and 3.3, and $\Psi_A(\Gamma, \Gamma')$ is as defined in the statement of Proposition 3.8.

Theorem 3.11 *Let $\Gamma_1, \ldots, \Gamma_m$ be a collection of non-isomorphic feasible connected finite graphs. Suppose $\lim_{n\to\infty}(nr_n^d) = \rho \in (0, \infty)$. Then the joint distribution of the variables $n^{-1/2}(J'_{n,A}(\Gamma_j) - EJ'_{n,A}(\Gamma_j))$, $1 \leq j \leq m$, converges to a centred multivariate normal as $n \to \infty$, with covariance matrix whose (i,j)th entry equals $\Psi_A(\Gamma_i, \Gamma_j)$ for $i \neq j$, and equals $\Psi_A(\Gamma_i, \Gamma_i) + k^{-1} \int_A p_{\Gamma_i}(\rho f(x))dx$ for $i = j$.*

3.5 Normal approximation: de-Poissonization

This section contains central limit theorems for G_n and J_n, which are deduced from those obtained in the preceding section for G'_n and J'_n, using the de-Poissonization techniques from Section 2.5. As in the Poissonized case above, the results for sparse and dense limits are stated together, with the results for the thermodynamic limit given later.

Theorem 3.12 *Let $\Gamma_1, \ldots, \Gamma_m$ be non-isomorphic feasible connected graphs, each of order k, with $2 \leq k < \infty$. Suppose that $r_n \to 0$ and $n^k r_n^{d(k-1)} \to \infty$ as $n \to \infty$. Suppose also that $\rho_n := nr_n^d$ tends either to 0 or to ∞ as $n \to \infty$. If $\rho_n \to 0$ then set $\tau_n := (n\rho_n^{k-1})^{1/2}$, but if $\rho_n \to \infty$ then set $\tau_n := (n\rho_n^{2(k-1)})^{1/2}$. Then as $n \to \infty$, the joint distribution of the variables $\tau_n^{-1}(G_n(\Gamma_j) - EG_n(\Gamma_j))$ $1 \leq j \leq m$, converges to a centred multivariate normal distribution with the following covariance matrix $\Sigma = (\sigma_{ij})$. In the case $\rho_n \to \infty$, $\sigma_{ij} = \Phi_1(\Gamma_i, \Gamma_j) - k^2 \mu_{\Gamma_i}\mu_{\Gamma_j}$, with $\Phi_1 := \Phi_{1,\mathbb{R}^d}$ defined at (3.19) and $\mu_\Gamma := \mu_{\Gamma,\mathbb{R}^d}$ defined at (3.2). In the case $\rho_n \to 0$, Σ is a diagonal matrix with $\sigma_{ii} = \mu_{\Gamma_i}$.*

Moreover, $\mathrm{Cov}(\tau_n^{-1}G_n(\Gamma_i), \tau_n^{-1}G_n(\Gamma_j))$ converges to σ_{ij} for each i, j.

Proof Let $(a_1, \ldots, a_m) \in \mathbb{R}^m$. By the Cramér–Wold device, it suffices to prove that

$$\tau_n^{-1} \sum_{j=1}^m a_j \left(G_n(\Gamma_j) - EG_n(\Gamma_j)\right) \xrightarrow{\mathcal{D}} \mathcal{N}\left(0, \sum_{i=1}^m \sum_{j=1}^m a_i a_j \sigma_{ij}\right), \tag{3.40}$$

and that the variance of the left-hand side of (3.40) converges to that of the right-hand side. The aim is to use Theorem 2.12.

Suppose $1 \leq j \leq m$. Recall the definition of $h_{\Gamma,n}$ at (3.1). For $s \in \mathbb{N}$, let $D_{s,n}^j$ be the increment

$$D_{s,n}^j := \left(\sum_{\mathcal{Y} \subseteq \mathcal{X}_{s+1}} h_{\Gamma_j,n}(\mathcal{Y})\right) - \sum_{\mathcal{Y} \subseteq \mathcal{X}_s} h_{\Gamma_j,n}(\mathcal{Y}).$$

Then $D^j_{s,n}$ is the number of induced Γ_j-subgraphs with one vertex at X_{s+1} in the graph $G(\mathcal{X}_{s+1}, r_n)$, and therefore $ED^j_{s,n} = \binom{s}{k-1} Eh_{\Gamma_j,n}(\mathcal{X}_k)$. By the proof of Proposition 3.1, this is asymptotic to $(k/n)E[G_n(\Gamma_j)]$, and hence to $k(nr_n^d)^{k-1}\mu_{\Gamma_j}$, uniformly over $n - n^{2/3} \leq s \leq n + n^{2/3}$, as $n \to \infty$. In other words,

$$\lim_{n\to\infty} \sup_{s\in[n-n^{2/3},n+n^{2/3}]} |(nr_n^d)^{1-k} ED^j_{s,n} - k\mu_{\Gamma_j}| = 0. \tag{3.41}$$

Also, for $s, t \in \mathbb{N}$, and for $i, j \in \{1, 2, \ldots, m\}$,

$$E[D^i_{s,n} D^j_{t,n}] = E \sum_{\mathcal{Y}\subseteq\mathcal{X}_s} \sum_{\mathcal{Y}'\subseteq\mathcal{X}_t} h_{\Gamma_i,n}(\mathcal{Y} \cup \{X_{s+1}\}) h_{\Gamma_j,n}(\mathcal{Y}' \cup \{X_{t+1}\}). \tag{3.42}$$

Suppose $s < t$. The leading-order term in (3.42) comes from pairs $(\mathcal{Y}, \mathcal{Y}')$ with $\mathcal{Y} \cup \{X_{s+1}\}$ and $\mathcal{Y}'$ disjoint, and so is equal to the expression

$$\binom{s}{k-1}\binom{t-k}{k-1} E[h_{\Gamma_i,n}(\mathcal{X}_k)] E[h_{\Gamma_j,n}(\mathcal{X}_k)].$$

Again by the proof of Proposition 3.1 as before, this expression is asymptotic to

$$(k(nr_n^d)^{k-1}\mu_{\Gamma_i})(k(nr_n^d)^{k-1}\mu_{\Gamma_j}), \tag{3.43}$$

uniformly over $s, t \in [n - n^{2/3}, n + n^{2/3}]$, as $n \to \infty$.

The second- and higher-order terms in $E[D^i_{s,n} D^j_{t,n}]$, that is, those coming from $\mathcal{Y}, \mathcal{Y}'$ such that $\mathcal{Y}' \cap (\mathcal{Y} \cup \{X_t\})$ is non-empty, are bounded by $\binom{s}{k-1} k \binom{t-k}{k-2}$ times the probability that $G(\mathcal{X}_{2k-1}; 2kr_n)$ is a complete graph. Therefore, these terms are bounded by a constant times $n^{2k-3} r_n^{d(2k-2)}$, which is negligible compared to the expression (3.43). The upshot is that for all $i, j \in \{1, \ldots, m\}$ we have

$$\lim_{n\to\infty} \sup_{n-n^{2/3}\leq s<t\leq n+n^{2/3}} |(nr_n^d)^{2(1-k)} ED^i_{s,n} D^j_{t,n} - k^2 \mu_{\Gamma_i}\mu_{\Gamma_j}| = 0. \tag{3.44}$$

For $s = t$, the leading-order term in (3.42) is equal to

$$\binom{s}{k-1}\binom{s-(k-1)}{k-1} E[h_{\Gamma_i,n}(\mathcal{X}_k) h_{\Gamma_j,n}(\mathcal{X}_{2k-1} \setminus \mathcal{X}_{k-1})] \leq cn^{2(k-1)} r_n^{2d(k-1)},$$

so that

$$E[D^i_{s,n} D^j_{s,n}] \leq c(nr_n^d)^{2(k-1)}. \tag{3.45}$$

For each $n \in \mathbb{N}$, and for finite $\mathcal{X} \subset \mathbb{R}^d$, define the functional

$$H_n(\mathcal{X}) := n^{1/2}\tau_n^{-1} \sum_{j=1}^m a_j \sum_{\mathcal{Y}\subseteq\mathcal{X}} h_{\Gamma_j,n}(\mathcal{Y}).$$

Consider first the case with $\rho_n \to \infty$. By Theorem 3.9 and Proposition 3.7, together with the estimates (3.41), (3.44), and (3.45), the functional $H_n(\cdot)$ satisfies all the conditions for Theorem 2.12, with $\alpha = k^2 \sum_{i=1}^m \sum_{j=1}^m \mu_{\Gamma_i}\mu_{\Gamma_j}$, and that result yields the desired conclusion at eqn (3.40), in the case $\rho_n \to \infty$.

Now consider the case with $\rho_n \to 0$. In this case we may deduce from (3.41), (3.44), and (3.45) that

$$\lim_{n\to\infty} \sup_{s\in[n-n^{2/3},n+n^{2/3}]} |(nr_n^d)^{(1-k)/2} ED_{s,n}^j| = 0,$$

$$\lim_{n\to\infty} \sup_{n-n^{2/3}\leq s\leq t\leq n+n^{2/3}} |(nr_n^d)^{1-k} ED_{s,n}^i D_{t,n}^j| = 0.$$

By these estimates, together with Theorem 3.9, the functional $H_n(\cdot)$ satisfies all the conditions for Theorem 2.12 (with $\alpha = 0$), and that result yields the desired conclusion, for the case $\rho_n \to 0$. □

In the case of the thermodynamic limit, we can check the stabilization criterion for de-Poissonization given at Definition 2.15, and then use Theorem 2.16.

Theorem 3.13 *Let $\Gamma_1, \ldots, \Gamma_m$ be a collection of non-isomorphic feasible connected graphs, with Γ_i of order $k_i \in [2,\infty)$ for each i. Suppose $nr_n^d \to \rho \in (0,\infty)$. Then the joint distribution of the variables $n^{-1/2}(G_n(\Gamma_j) - EG_n(\Gamma_j))$, $1 \leq j \leq m$, is asymptotically centred multivariate normal with covariance matrix whose (i,l)th entry is*

$$\left(\sum_{j=1}^{\min(k_i,k_l)} \rho^{k_i+k_l-j-1}\Phi_j(\Gamma_i,\Gamma_l)\right) - k_i k_l \rho^{k_i+k_l-2}\mu_{\Gamma_i}\mu_{\Gamma_l},$$

with $\Phi_j(\Gamma,\Gamma') := \Phi_{j,\mathbb{R}^d}(\Gamma,\Gamma')$ as defined at (3.19).

Proof Let $(a_1,\ldots,a_m) \in \mathbb{R}^m$. By the Cramér–Wold device, it suffices to prove that the linear combination $n^{-1/2}\sum_{j=1}^m a_j(G_n(\Gamma_j) - EG_n(\Gamma_j))$ converges in distribution to a normal variable with mean zero, and its variance converges to the variance of that normal variable. By Theorem 3.10, this condition holds with G_n replaced by G'_n. In order to use Theorem 2.16, we need to check that the functional

$$H_0(\mathcal{X}) := \sum_{j=1}^m a_j \sum_{\mathcal{Y}\subseteq\mathcal{X}} h_{\Gamma_j}(\mathcal{Y})$$

is strongly stabilizing (see Definition 2.15). This is rather obvious since the effect of an inserted point at the origin has only finite range. The associated limiting add one cost $\Delta(\mathcal{H}_\lambda)$ is the number of induced Γ_j-graphs in $G(\mathcal{H}_\lambda \cup \{\mathbf{0}\};1)$ with one vertex at the origin, multiplied by a_j and summed over j. By an application of Palm theory (Theorem 1.6), the expectation of this is given by

$$E\Delta(\mathcal{H}_\lambda) = \sum_{j=1}^m \left(\frac{a_j\lambda^{k_j-1}}{(k_j-1)!}\right)\int_{(\mathbb{R}_d)^{k_j-1}} h_\Gamma(\{\mathbf{0},y_1,\ldots,y_{k_j-1}\})\mathrm{d}(y_1,\ldots,y_{k-1})$$

and hence, by (2.59) and the definition of μ_Γ at (3.2),

$$E\Delta(\mathcal{H}_{\rho f(X)}) = \sum_{j=1}^{m} a_j k_j \rho^{k_j-1} \mu_{\Gamma_j}.$$

Set $H_n(\mathcal{X}) := H_0(r_n^{-1}\mathcal{X})$, and set $k_{\max} := \max(k_1, \ldots, k_m)$. Then $|H_n(\mathcal{X}_{n+1}) - H_n(\mathcal{X}_n)|$ is bounded by a constant times $(\mathcal{X}_n(B(X_{n+1}; k_{\max} r_n)))^{k_{\max}-1}$, which is stochastically dominated by $(\mathrm{Bi}(n, f_{\max}\theta(k_{\max}r_n)^d))^{k_{\max}-1}$, which has uniformly bounded fourth moment, confirming the moments condition (2.47) in this setting. Therefore, all conditions for Theorem 2.16 apply, and that result gives the required convergence of $H_n(\mathcal{X}_n)$ to a normal. □

Next we give an analogous central limit theorems for component counts in the thermodynamic limit, now allowing for components of order 1 (i.e. isolated points). Recall the definition of $p_\Gamma(\cdot)$ at (3.5), and set $\Psi(\Gamma, \Gamma') := \Psi_{\mathbb{R}^d}(\Gamma, \Gamma')$ as defined in the statement of Proposition 3.8.

Theorem 3.14 *Let $\Gamma_1, \ldots, \Gamma_m$ be a collection of non-isomorphic feasible connected graphs, set k_j to be the order of Γ_j and assume $1 \leq k_j < \infty$ for each j. Suppose $nr_n^d \to \rho \in (0, \infty)$. For $1 \leq j \leq m$, set*

$$\begin{aligned} u_j := & \int_{\mathbb{R}^d} p_{\Gamma_j}(\rho f(x)) f(x) \mathrm{d}x \\ & - \int_{\mathbb{R}^d} \mathrm{d}x \left(\frac{\rho^{k_j} f(x)^{k_j+1}}{k_j!} \right) \int_{(\mathbb{R}^d)^k \setminus (\mathbb{R}^d \setminus B(\mathbf{0};1))^k} h_\Gamma(\{y_1, \ldots, y_k\}) \\ & \times \exp(-\rho f(x) V(y_1, \ldots, y_k)) \mathrm{d}(y_1, \ldots, y_k). \end{aligned}$$

Then the joint distribution of the variables $n^{-1/2}(J_n(\Gamma_j) - EJ_n(\Gamma_j))$, $1 \leq j \leq m$, is asymptotically centred multivariate normal with covariance matrix whose (i,j)th entry equals $\Psi(\Gamma_i, \Gamma_i) + k^{-1} \int_{\mathbb{R}^d} p_{\Gamma_i}(\rho f(x)) \mathrm{d}x - u_i^2$ for $i = l$, and equals $\Psi(\Gamma_i, \Gamma_j) - u_i u_l$ for $i \neq l$.

Proof The proof is similar to that given for the preceding result, except that this time we use Theorem 3.11 instead of Theorem 3.10. In the present case, define

$$H_0(\mathcal{X}) := \sum_{j=1}^{m} a_j \sum_{\mathcal{Y} \subseteq \mathcal{X}} h_{\Gamma_j}(\mathcal{Y}) \mathbf{1}_{\{\mathcal{Y} \text{ is a component of} \mathcal{X}\}}.$$

Consider first the case where $m = 1$ and $a_1 = 1$, $\Gamma_1 = \Gamma$. Then the limiting add one cost $\Delta(\mathcal{H}_\lambda)$ is the indicator of the event that an inserted point at $\mathbf{0}$ lies in a Γ-component of $G(\mathcal{H}_\lambda \cup \{\mathbf{0}\}; 1)$ minus the number of Γ-components of $G(\mathcal{H}_\lambda; 1)$ having at least one vertex within unit distance of the origin. By Theorem 1.6 we obtain

$$\begin{aligned} \Delta(\mathcal{H}_\lambda) = p_\Gamma(\lambda) - \frac{\lambda^k}{k!} & \int_{(\mathbb{R}^d)^k \setminus (\mathbb{R}^d \setminus B(\mathbf{0};1))^k} h_\Gamma(\{y_1, \ldots, y_k\}) \\ & \times \exp(-\lambda V(y_1, \ldots, y_k)) \mathrm{d}(y_1, \ldots, y_k), \end{aligned}$$

and it follows for general $m, a_1, \ldots, a_m$ that $E\Delta(\mathcal{H}_{\rho f(X)}) = \sum_{j=1}^m a_j u_j$. We may then deduce the result using Theorem 2.16. □

3.6 Strong laws of large numbers

The central limit theorems for the Γ-subgraph count $G_n(\Gamma)$ and the Γ-component count $J_n(\Gamma)$, described in the preceding section, imply that these quantities satisfy a weak law of large numbers. In the present section we improve this to a strong law of large numbers. The first of these is for the number of Γ-components $J_n(\Gamma)$ in the thermodynamic limit where nr_n^d tends to a constant.

Theorem 3.15 *Suppose that Γ is a connected feasible graph of order k, $k \in \mathbb{N}$, and that $nr_n^d \to \rho \in (0,\infty)$. Then with $p_\Gamma(\cdot)$ defined at (3.5), $J_n = J_n(\Gamma)$ satisfies*

$$\lim_{n\to\infty} (J_n/n) = k^{-1} \int_{\mathbb{R}^d} p_\Gamma(\rho f(x)) f(x) \mathrm{d}x \quad \text{c.c.}$$

Proof To deduce complete convergence from the convergence of means established in Proposition 3.3, we use Azuma's inequality. With $\mathcal{F}_0$ denoting the trivial σ-field, define σ-fields $\mathcal{F}_i = \sigma(X_1, \ldots, X_i)$, and write $J_n - EJ_n$ as the sum of a series of martingale differences $\sum_{i=1}^n D_{i,n}$, where $D_{i,n} := E[J_n|\mathcal{F}_i] - E[J_n|\mathcal{F}_{i-1}]$. Let $\tilde{J}_{n,i} = \tilde{J}_{n,i}(\Gamma)$ denote the number of Γ-components in $G(\mathcal{X}_{n+1}\setminus\{X_i\}; r_n)$; then $D_{i,n} = E[(J_n - \tilde{J}_{n,i})|\mathcal{F}_i]$. Given a set $\mathcal{X}$ of points in $\mathbb{R}^d$, the addition of a point x to $\mathcal{X}$ can cause the number of Γ-components to increase by at most 1, and can cause it to decrease by a geometric constant K depending only on d, namely the maximum number of distinct points that it is possible to have in the unit ball without any two of them lying within unit distance of one another. Therefore $|J_n - \tilde{J}_{n,i}| \leq 2K$ a.s., and $|D_{i,n}| \leq 2K$ a.s. By Azuma's inequality,

$$P[|J_n - EJ_n| > \varepsilon n] \leq 2 \exp\left(-\frac{\varepsilon^2 n^2}{4nK}\right),$$

which is summable in n for any $\varepsilon > 0$. The complete convergence then follows from Proposition 3.3. □

Theorem 3.16 *Suppose that Γ is a connected feasible graph of order $k \in \mathbb{N}$. Suppose that $\lim_{n\to\infty}(nr_n^d) = 0$, and that $\lim_{n\to\infty}(n^{2k-1}r_n^{d(2k-2)}/\log n) = \infty$. Then $\lim_{n\to\infty}(J_n/(n^k r_n^{d(k-1)})) = \mu_\Gamma$, c.c., with μ_Γ defined at (3.2).*

Proof By the same argument using Azuma's inequality as in the proof of the preceding result,

$$P[|J_n - EJ_n| \geq \varepsilon n^k r_n^{d(k-1)}] \leq 2 \exp\left(-\frac{\varepsilon^2 n^{2k} r_n^{2d(k-1)}}{4nK}\right),$$

which is summable in n by the second condition on the limiting behaviour of $(r_n)_{n\geq 1}$. Thus the result follows using Proposition 3.2. □

The next result is a strong law of large numbers for the number of induced Γ-subgraphs $G_n = G_n(\Gamma)$, analogous to the strong law just given for J_n. The range of application includes both the thermodynamic limit $nr_n^d \to$ const., the dense limit $nr_n^d \to \infty$, and some cases of sparse limit $nr_n^d \to 0$.

Theorem 3.17 *Suppose that Γ is a connected graph on k vertices, $k \geq 2$. Suppose f has bounded support. Suppose that $r_n \to 0$, and there exists $\eta > 0$ such that* $\liminf_{n\to\infty}(n^{2k-1-\eta}r_n^{d(2k-2)}) > 0$. *Then* $\lim_{n\to\infty}(G_n/(n^k r_n^{d(k-1)})) = \mu_\Gamma$, *c.c., with μ_Γ defined at* (3.2).

Proof The basic idea is the same as for Theorem 3.15. However, direct application of Azuma's inequality no longer works because there is no uniform bound on the change in the number of induced Γ-subgraphs when a single point is added or removed. To get around this difficulty, we shall use the refinement of Azuma's inequality in Theorem 2.9.

Recall from (1.7) in Lemma 1.1 that for $\lambda > \mathrm{e}^2 mp$, we have

$$P[\mathrm{Bi}(m,p) > \lambda] \leq \exp(-\lambda). \tag{3.46}$$

Let $\gamma := \min(1,\eta)/(4(k-1))$. Divide $\mathbb{R}^d$ into cubes of side r_n, denoted $Q_{n,i}, i \in \mathbb{N}$, and let A be the set of n-point configurations $\mathcal{X}$ such that $\mathcal{X}(Q_{n,i}) \leq n^\gamma \max(nr_n^d, 1)$ for every cube $Q_{n,i}$ intersecting the support of f. Since for each cube the variable $\mathcal{X}_n(Q_{n,i})$ is binomial with mean at most $nr_n^d f_{\max}$, by (3.46), the probability that $\mathcal{X}_n(Q_{n,i}) \geq n^\gamma \max(nr_n^d, 1)$ is bounded by $\exp(-n^\gamma)$, for n large enough. Since f is assumed to have bounded support, there is a constant c_1 such that

$$P[\mathcal{X}_n \in A] \leq c_1 n \exp(-n^\gamma). \tag{3.47}$$

Define σ-fields $\mathcal{F}_i = \sigma(X_1, \ldots, X_i)$, with $\mathcal{F}_0$ denoting the trivial σ-field, and write $G_n - EG_n$ as the sum of a series of martingale differences $\sum_{i=1}^n D_{i,n}$, where $D_{i,n} := E[G_n|\mathcal{F}_i] - E[G_n|\mathcal{F}_{i-1}]$.

Let $G_n^{(i)}$ denote the number of induced Γ-subgraphs in $G(\mathcal{X}_{n+1}\backslash\{X_i\}; r_n)$; then $D_{i,n} = E[G_n - G_n^{(i)}|\mathcal{F}_i]$. Suppose two configurations of n points both lie in A, and they differ in the position of a single point. Then there exists a constant c_2 such that the difference in the number of induced Γ-graphs for these two configurations is at most $c_2(n^\gamma \max(nr_n^d, 1))^{k-1}$. Therefore, if $A_{i,n}$ denotes the event that both $\mathcal{X}_n$ and $\mathcal{X}_{n+1}\backslash\{X_i\}$ lie in A, we have $|G_n - G_n^{(i)}| \leq c_2(n^\gamma \max(nr_n^d, 1))^{k-1}$ on event $A_{i,n}$. In any event we always have $|G_n - G_n^{(i)}| \leq n^k$. Hence,

$$\begin{aligned}|D_{i,n}| &\leq E[(|G_n - G_n^{(i)}|\mathbf{1}_{A_{i,n}})|\mathcal{F}_i] + E[(|G_n - G_n^{(i)}|\mathbf{1}_{A_{i,n}^{\mathrm{c}}})|\mathcal{F}_i] \\ &\leq c_2(n^\gamma \max(nr_n^d, 1))^{k-1} + n^k P[A_{i,n}^{\mathrm{c}}|\mathcal{F}_i].\end{aligned} \tag{3.48}$$

Define the event $B_{i,n} := \{P[A_{i,n}^{\mathrm{c}}|\mathcal{F}_i] \leq n^{-k}\}$. By Markov's inequality, and (3.47),

$$P[B_{i,n}^{\mathrm{c}}] \leq n^k E[P[A_{i,n}^{\mathrm{c}}|\mathcal{F}_i]] = n^k P[A_{i,n}^{\mathrm{c}}] \leq 2c_1 n^{k+1} \exp(-n^\gamma).$$

Set $c_3 := c_2 + 1$. On the event $B_{i,n}$ (which is in $\mathcal{F}_i$), by (3.48) we have $|D_{i,n}| \leq c_3(n^\gamma \max(nr_n^d, 1))^{k-1}$. Hence, by Theorem 2.9,

$$
\begin{aligned}
P[|G_n - EG_n| > \varepsilon n^k r_n^{d(k-1)}] &\leq 2\exp\left(-\frac{\varepsilon^2 n^{2k} r_n^{2d(k-1)}}{32c_3^2 n(n^\gamma \max(nr_n^d, 1))^{2(k-1)}}\right) \\
&+ \left(1 + \frac{2n^k}{\varepsilon n^k r_n^{d(k-1)}}\right) 2c_1 n^{k+2} \exp(-n^\gamma) \\
&\leq 2\exp\left(-\min(n^{1-2\gamma(k-1)}, \varepsilon^2 n^{2k-1-2\gamma(k-1)} r_n^{2d(k-1)})/(32c_3^2)\right) \\
&+ c' n^{2k+2} \exp(-n^\gamma),
\end{aligned}
$$

which is summable in n for any $\varepsilon > 0$, by the conditions on r_n and the definition of γ in terms of η. The required complete convergence follows by Proposition 3.1. □

The preceding results establish strong laws for the number of induced Γ-subgraphs or Γ-components in $G(\mathcal{X}_n; r_n)$, when r_n^d decays more slowly than $n^{-1-1/(2k-2)}$. Even for more rapidly decaying r_n, as long as r_n^d decays more slowly than $n^{-1-1/(k-1)}$, then $n^k r_n^{d(k-1)}$ tends to infinity, so that $E[G_n] \to \infty$ and one might hope for a strong law. This is true, but without complete convergence, if one imposes an extra condition of *regular variation* on the sequence $(r_n)_{n\geq 1}$, which encapsulates the idea that we usually think of r_n as behaving roughly like a power of n. A sequence $(r_n)_{n\geq 1}$ is *regularly varying* if, for all $t > 0$, $\lim_{n\to\infty} r_{[nt]}/r_n$ exists and is finite and strictly positive. In this case the limit is always of the form t^ρ for some $\rho \in \mathbb{R}$ (the *index of regular variation*) (see Bingham *et al.* (1987, Theorem 1.9.5)).

Theorem 3.18 *Suppose that Γ is a connected graph of order k, $2 \leq k < \infty$. Suppose that $\lim_{n\to\infty}(nr_n^d) = 0$, and that there exists $\eta > 0$ such that $n^k r_n^{d(k-1)} > (\log n)^{1+\eta}$ for all large enough n. Suppose also that $(r_n)_{n\geq 1}$ is a regularly varying sequence. Then $(G_n/(n^k r_n^{d(k-1)})) \to \mu_\Gamma$ a.s., as $n \to \infty$, with μ_Γ defined at (3.2).*

Proof First assume Γ is the complete graph on k vertices. By Proposition 3.1 and Theorem 3.12, we have $EG_n \sim \mu_\Gamma n^k r_n^{d(k-1)}$ and $\mathrm{Var}(G_n) \sim \mu_\Gamma n^k r_n^{d(k-1)}$. Let $\varepsilon > 0$, and for $m \in \mathbb{N}$, set $\nu(m) := \lfloor (1+\varepsilon)^m \rfloor$. Let $s_m := \max\{r_l : \nu(m) \leq l < \nu(m+1)\}$, and let $t_m := \min\{r_l : \nu(m) \leq l < \nu(m+1)\}$. Let $T_m^\uparrow$ be the number of induced Γ-subgraphs in $G(\mathcal{X}_{\nu(m+1)}; s_m)$ and let $T_m^\downarrow$ be the number of induced Γ-subgraphs in $G(\mathcal{X}_{\nu(m)}; t_m)$.

Suppose $n, m \in \mathbb{N}$, with $\nu(m) \leq n < \nu m + 1$. By the assumption that Γ is the complete graph, the number of induced Γ-subgraphs is a monotone increasing graph function, and hence $T_m^\downarrow \leq G_n \leq T_m^\uparrow$ with probability 1.

Let $-\rho$ be the index of regular variation of the sequence $(r_n)_{n\geq 1}$. Then $\rho \geq 0$, and by Bingham *et al.* (1987, Theorems 1.9.5 and 1.5.2), $r_{\lfloor \lambda n \rfloor}/r_n$ converges to

$\lambda^{-\rho}$ uniformly in λ in the range $[1, 1+2\varepsilon]$. Hence $\limsup_{m\to\infty}(s_m/t_m) \leq (1+\varepsilon)^\rho$. For large m, we have

$$\begin{aligned}\max(ET_m^\uparrow, \mathrm{Var}(T_m^\uparrow)) &\leq (1+\varepsilon)\mu_\Gamma \nu(m+1)^k s_m^{d(k-1)} \\ &\leq (1+\varepsilon)^{2+k+\rho d(k-1)} \mu_\Gamma \nu(m)^k t_m^{d(k-1)}.\end{aligned}$$

Hence by Chebyshev's inequality, there exist constants c, c' such that

$$P[T_m^\uparrow > (1+\varepsilon)^{3+k+\rho d(k-1)} \mu_\Gamma \nu(m)^k t_m^{d(k-1)}] \leq \frac{c}{\nu(m)^k t_m^{d(k-1)}} \leq \frac{c}{(\log \nu(m))^{1+\eta}},$$

which is summable in m. Hence by the Borel–Cantelli lemma, with probability 1,

$$\limsup_{n\to\infty}\left(\frac{G_n}{n^k r_n^{d(k-1)}}\right) \leq \limsup_{m\to\infty}\left(\frac{T_m^\uparrow}{\nu(m)^k t_m^{d(k-1)}}\right) \leq (1+\varepsilon)^{3+k+\rho d(k-1)}\mu_\Gamma.$$

By a similar argument,

$$\liminf_{n\to\infty}\left(\frac{G_n}{n^k r_n^{d(k-1)}}\right) \geq \liminf_{n\to\infty}\left(\frac{T_m^\downarrow}{(\nu(m)+1)^k s_m^{d(k-1)}}\right) \geq (1+\varepsilon)^{-3-k-\rho d(k-1)}\mu_\Gamma.$$

Since ε is arbitrarily small, it follows that $G_n/(n^k r_n^{d(k-1)}) \to \mu_\Gamma$ a.s., completing the proof for the case where Γ is the complete graph on k vertices.

Now suppose Γ has k vertices but is not the complete graph. The above argument fails only because we lose the monotonicity from which we were able to deduce that $T_m^\downarrow \leq G_n \leq T_m^\uparrow$. To recover monotonicity, recall from the start of this chapter that $\tilde{G}_n(\Gamma)$ denotes the number of Γ-subgraphs (induced or not) in $G(\mathcal{X}_n; r_n)$. Then $\tilde{G}_n$ is monotone under the addition of edges or vertices, so by the same argument as given above for the case where Γ is the complete graph, we obtain a.s. convergence of $\tilde{G}_n/(n^k r_n^{d(k-1)})$ to a limit, given by an appropriate linear combination of the expressions $\mu_{\Gamma'}$, $\Gamma' \in \mathcal{G}(\Gamma)$, where $\mathcal{G}(\Gamma)$ denotes the set of all non-isomorphic graphs Γ' on k vertices having Γ as a subgraph.

It is not hard to see that $G_n(\Gamma)$ is a linear combination of the variables $\tilde{G}_n(\Gamma')$, $\Gamma' \in \mathcal{G}(\Gamma)$. Therefore the almost sure convergence of G_n, divided by $n^k r_n^{d(k-1)}$, follows from that of each of the variables $\tilde{G}_n(\Gamma')$. □

Theorem 3.19 *Suppose that Γ is a connected graph of order $k \geq 2$. Suppose that $\lim_{n\to\infty}(nr_n^d) = 0$, and that there exists $\eta > 0$ such that $n^k r_n^{d(k-1)} > (\log n)^{1+\eta}$ for all large enough n, and $nr_n^d < n^{-\eta}$ for all large enough n. Suppose also that $(r_n)_{n\geq 1}$ is a regularly varying sequence. Then $J_n/(n^k r_n^{d(k-1)}) \to \mu_\Gamma$ almost surely, with μ_Γ defined at (3.2).*

Proof Choose $\gamma \in \mathbb{Z}$ so that $\gamma\eta > 1$. For $m \in \mathbb{Z}$, set $\nu(m) := m^\gamma$. Set $s_m := \max\{r_l : \nu(m) \leq l \leq \nu(m+1)\}$ and $t_m := \min\{r_l : \nu(m) \leq l \leq \nu(m+1)\}$.

Let R_n denote the number of subsets S of size $k+1$ of $\mathcal{X}_n$ such that $G(S;r_n)$ is connected, and let $R_m^\uparrow$ be the number of subsets S of size $k+1$ in $\mathcal{X}_{\nu(m+1)}$ such that $G(S;m_n)$ is connected. By the regular variation property, s_m/t_m is bounded (and in fact tends to 1). Also, $\nu(m+1)/\nu(m) \to 1$. We have

$$\frac{E[R_m^\uparrow]}{\nu(m)^k t_m^{d(k-1)}} \leq \frac{c\nu(m)^{k+1} r_{\nu(m)}^{dk}}{\nu(m)^k r_{\nu(m)}^{d(k-1)}} \leq c'\nu(m)^{-\eta},$$

which is summable in m by the choice of γ. Hence, with probability 1, the sum $\sum_m R_m^\uparrow/(\nu(m)^k t_m^{d(k-1)})$ converges (since it has finite mean) and consequently $R_m^\uparrow/(\nu(m)^k t_{\nu(m)}^{d(k-1)})$ tends to zero. Since for $\nu(m) \leq n \leq \nu(m+1)$ we have

$$0 \leq \frac{G_n - J_n}{n^k r_n^{d(k-1)}} \leq \frac{(k+1)R_n}{n^k r_n^{d(k-1)}} \leq \frac{cR_m^\uparrow}{\nu(m)^k t_m^{d(k-1)}},$$

it follows that $R_n/(n^k r_n^{d(k-1)})$ tends to zero almost surely, and then the result follows from Theorem 3.18. □

3.7 Notes

In the case of the sparse limiting regime Hafner (1972) proved Poisson and normal limit theorems for $J_n(\Gamma)$, the number of components isomorphic to a given graph Γ. A number of results in the literature on U-statistics are applicable to $G_n(\Gamma)$, the number of induced Γ-subgraphs; early papers of this type include Silverman and Brown (1978) for Poisson limit theorems, and Weber (1983) for normal limit theorems. Subsequent papers include Jammalamadaka and Janson (1986), Bhattacharya and Ghosh (1992), and have demonstrated a variety of different ways of obtaining results of this sort, under various conditions on f and r_n. For example, Bhattacharya and Ghosh (1992) proved a result similar to Theorem 3.12 by different methods using the martingale central limit theorem, but required stronger conditions on r_n than those assumed here. Another set of results having some overlap with those of this chapter (for the uniform case only) appears in Yang (1995, Chapter III).

In the sparse limit ($nr_n^d \to 0$), Hall (1988, p. 252) has a result along the lines of Theorem 3.9, and also a Poissonized version of Theorem 3.4, but restricts attention to the uniform distribution. Hall (1986) also has the Poissonized version of the case $k=1$ of Theorem 3.15 above, for uniformly distributed points.

The methods of proof of limit theorems used here, based on the Stein–Chen method, are not particularly closely related to those in the works cited above. They are more closely related to Barbour and Eagleson (1984) in the case of Section 3.2, and to Avram and Bertsimas (1993) in the case of Section 3.4 (at least for the thermodynamic limit), although neither of these works is specifically concerned with random geometric graphs.

For an account of results for Erdös–Rényi random graphs analogous to those given here for $G_n(\Gamma)$, see Bollobás (1985, Chapter 4).

4

TYPICAL VERTEX DEGREES

This chapter is concerned with the following question. Given $k \in \mathbb{N}$, and $r > 0$, how many vertices of $G(\mathcal{X}_n; r)$ have degree at least k? Equivalently, how many of the points of $\mathcal{X}_n$ have their kth nearest neighbour at a distance of at most r? Here we take the second question as our starting point, and investigate the asymptotic empirical distribution of the *k-nearest-neighbour distances* in the point set $\mathcal{X}_n$ or $\mathcal{P}_n$ (these point processes are defined in Sections 1.5 and 1.7 respectively). These are a multivariate analogue to *k-spacings* in one dimension; k-spacings and k-nearest-neighbour distances have been studied in a variety of contexts, but especially in the context of goodness of fit tests for a null hypothesis of a uniform underlying distribution of points, or some other specified underlying distribution or family of distributions (see the notes in Section 4.7).

Given a finite point set $\mathcal{X} \subset \mathbb{R}^d$ (typically the random point set $\mathcal{X}_n$ or $\mathcal{P}_n$), and given $x \in \mathcal{X}$, let $R_k(x; \mathcal{X})$ denote the distance from x to its kth nearest neighbour in $\mathcal{X}$, that is, the minimal r for which x has degree at least k in $G(\mathcal{X}; r)$. The empirical process of k-nearest-neighbour distances in $\mathcal{X}$ is the integer-valued stochastic process $(\zeta(t), t \geq 0)$ defined by $\zeta(t) := \sum_{x \in \mathcal{X}} \mathbf{1}_{\{R_k(x;\mathcal{X}) \leq t\}}$, and is our object of study here, after appropriate renormalization of space and time parameters. For fixed r, $\zeta(r)$ is simply the answer to the question posed at the start of this chapter, but we shall also consider weak convergence of the entire process $\zeta(\cdot)$, suitably renormalized. This is a standard approach in empirical process theory (see Shorack and Wellner (1986)), and its application to multivariate nearest-neighbour statistics, with a goodness-of-fit test in mind, dates back at least to Bickel and Breiman (1983), who took $k = 1$. We consider here asymptotic regimes, either with k is fixed or with k growing with n, as is often appropriate in non-parametric density estimation based on distances from points of $\mathcal{X}_n$ to k-nearest neighbours (see, e.g., Silverman (1986)).

Gaussian processes feature in this chapter, and are defined as follows. Suppose T is an abstract set and $\sigma : T \times T \to \mathbb{R}$ is non-negative definite, that is, suppose σ satisfies $\sum_{i=1}^{n} \sum_{j=1}^{n} a_i a_j \sigma(t_i, t_j) \geq 0$ for any finite subset $\{t_1, \ldots, t_k\}$ of T and any $(a_1, \ldots, a_k) \in \mathbb{R}^k$. A *centred Gaussian process* with covariance function σ is a family of random variables $(X(t), t \in T)$ with the property that for any finite subset $\{t_1, \ldots, t_k\}$ of T and any $(a_1, \ldots, a_k) \in \mathbb{R}^k$, the linear combination $\sum_{i=1}^{n} a_i X(t_i)$ has the normal $\mathcal{N}(0, \sum_{i=1}^{n} \sum_{j=1}^{n} a_i a_j \sigma(t_i, t_j))$ distribution. Such a process exists for any T and any non-negative definite σ. See, for example, the discussion of 'Gaussian systems' in Karlin and Taylor (1975).

4.1 The setup

We consider k_n-nearest-neighbour distances, in two different types of limiting regime. In the first regime, we specify a value $k \in \mathbb{N}$, and take $k_n = k$ for all n; in this case we shall say that k_n is *fixed.* The other regime is to let $(k_n)_{n\geq 1}$ be a sequence with $k_n \to \infty$ but

$$\lim_{n\to\infty} n^{-2/(d+2)} k_n = 0. \tag{4.1}$$

Since there are some similarities between them, some of the notation used will be common to both types of limiting regimes for k_n. First of all, we choose a sequence of distance parameters r_n in such a way that k_n is a 'typical' vertex degree in the sense that the expected proportion of vertices of degree at least k_n tends to a non-trivial limit. In the case where k_n is fixed, for $t \in (0,\infty)$ define $r_n = r_n(t)$ by

$$nr_n(t)^d = t.$$

In the case where $(k_n)_{n\geq 1}$ is a sequence tending to infinity (and satisfying (4.1)), for $s > 0$ and $t \in \mathbb{R}$, define $r_n = r_n(t)$ by

$$nr_n(t)^d = s(k_n + tk_n^{1/2}). \tag{4.2}$$

In either of the limiting regimes under consideration, let $Z_n(t)$ be the number of vertices of $G(\mathcal{X}_n; r_n(t))$ of degree at least k_n, and let $Z'_n(t)$ be the number of vertices of $G(\mathcal{P}_n; r_n(t))$ of degree at least k_n. Note that with this definition, in the first regime (k_n fixed) the dependence on the parameter k is suppressed, while in the second regime ($k_n \to \infty$) the dependence on the parameter s and on the the sequence $(k_n)_{n\geq 1}$ is suppressed.

We shall consider the asymptotic distribution of $Z_n(t)$ and $Z'_n(t)$, suitably scaled and centred, as n tends to infinity, and show they are each asymptotically normal, for any fixed t. More generally, we consider the asymptotic behaviour of $Z_n(\cdot)$ and $Z'_n(\cdot)$ (scaled and centred). We shall see that the finite-dimensional distributions converge to those of a Gaussian process, and at least in the case of $Z'_n(\cdot)$, this can be extended to convergence in the space of Skorohod functions with the Skorohod topology.

We use the following notation. For $\lambda > 0$, let $\pi_\lambda(\cdot)$ denote the Poisson probability function with parameter λ. That is, let $\pi_\lambda(k) := P[\text{Po}(\lambda) = k]$ and for $A \subseteq \mathbb{R}$ let $\pi_\lambda(A) := P[\text{Po}(\lambda) \in A]$. For $x \in \mathbb{R}^d$, and any point process $\mathcal{X}$, let $\mathcal{X}^x$ denote the point process $\mathcal{X} \cup \{x\}$ (e.g. $\mathcal{X}_n^x = \mathcal{X}_n \cup \{x\}$). Let $\phi(t) := (2\pi)^{-1/2}\exp(-x^2/2)$, and let $\Phi(t) := \int_{-\infty}^t \phi(s)ds$, the standard normal density and distribution function respectively. Recall also that θ denotes the volume of the unit ball in the chosen norm. Given $x \in \mathbb{R}^d$, define the ball

$$B_n(x;t) := B(x; r_n(t)). \tag{4.3}$$

The definition of $B_n(x;t)$ depends on which limiting regime is taken for (k_n). In either case, for Borel $A \subseteq \mathbb{R}^d$, define

$$Z_n(t;A) := \sum_{i=1}^{n} \mathbf{1}_{\{\mathcal{X}_n(B_n(X_i;t)) \geq k_n+1\} \cap \{X_i \in A\}},$$

$$Z'_n(t;A) := \sum_{i=1}^{N_n} \mathbf{1}_{\{\mathcal{P}_n(B_n(X_i;t)) \geq k_n+1\} \cap \{X_i \in A\}}.$$

The main concern here is with the case where A is $\mathbb{R}^d$; observe that $Z_n(t)$ defined earlier is equal to $Z_n(t;\mathbb{R}^d)$ and $Z'_n(t) = Z_n(t;\mathbb{R}^d)$. It will be convenient in the sequel to approximate $Z_n(t)$ by $Z_n(t;A)$ for bounded A. In the second limiting regime ($k_n \to \infty$) we write $Z_n(s,t;A)$ for $Z_n(t;A)$ and $Z'_n(s,t;A)$ for $Z'_n(t;A)$ when we wish to emphasize the dependence on s as well as t.

In the second limiting regime with $r_n \to \infty$, given $s > 0$, define the level set

$$L_s := \{x \in \mathbb{R}^d : s\theta f(x) = 1\}, \tag{4.4}$$

and also set $L_s^+ := \{x \in \mathbb{R}^d : s\theta f(x) > 1\}$. The limiting normal distribution for $Z_n(t)$, scaled and centred, will be non-degenerate only when the parameter s at (4.2) is chosen so that $F(L_s) > 0$. For example, if F is a uniform distribution there is just one such choice of s.

We require a mild technical condition on the underlying probability density function f of the points X_i. This concerns the 'region of regularity' R defined by

$$R := \{x \in \mathbb{R}^d : f(x) > 0 \text{ and } \limsup_{y \to x}(|f(y) - f(x)|/\|y - x\|) < \infty\}. \tag{4.5}$$

Throughout this chapter, we assume that $F(R) = 1$. This assumption holds, for example, if f is differentiable almost everywhere.

4.2 Laws of large numbers

This section is concerned with the asymptotic first-order behaviour of $Z_n(t)$ as $n \to \infty$. The first result concerns the mean of $Z_n(t)$ in the two regimes under consideration.

Theorem 4.1 *Suppose $A \subseteq \mathbb{R}^d$ is a Borel set. If k_n is fixed (i.e. $k_n = k$ for all n), then*

$$\lim_{n \to \infty} \left(n^{-1} E[Z_n(t;A)]\right) = \int_A \pi_{\theta t f(x)}([k,\infty)) f(x) \mathrm{d}x, \tag{4.6}$$

$$\lim_{n \to \infty} \left(n^{-1} E[Z'_n(t;A)]\right) = \int_A \pi_{\theta t f(x)}([k,\infty)) f(x) \mathrm{d}x. \tag{4.7}$$

If instead $k_n \to \infty$ but (4.1) holds, then

$$\lim_{n \to \infty} \left(n^{-1} E[Z_n(s,t;A)]\right) = F(L_s^+ \cap A) + \Phi(t) F(L_s \cap A), \tag{4.8}$$

and likewise for $Z'_n(s,t;A)$.

Proof With $B_n(x;t)$ defined at (4.3), let $p_n(x;t) := F(B_n(x;t))$. Then

$$E[Z_n(t;A)] = n \int_A P[\text{Bi}(n-1, p_n(x;t)) \geq k_n] f(x) \text{d}x, \tag{4.9}$$

and by Palm theory (Theorem 1.6),

$$E[Z'_n(t;A)] = n \int_A P[\text{Po}(np_n(x;t)) \geq k_n] f(x) \text{d}x. \tag{4.10}$$

Suppose k_n is fixed. If x is in the region R defined at (4.5), then f is continuous at x and $np_n(x;t) \to \theta t f(x)$, in which case by binomial convergence to the Poisson distribution, the probability $P[\text{Bi}(n-1, p_n(x;t)) \geq k]$ tends to $\pi_{\theta t f(x)}([k, \infty))$. Then by (4.9) and the dominated convergence theorem for integrals, we obtain (4.6). The proof of (4.7) is similar, using (4.10).

Next suppose $k_n \to \infty$ and (4.1) holds. If f is continuous at x, then

$$np_n(x;t) \sim n\theta r_n^d f(x) \sim s\theta f(x) k_n, \tag{4.11}$$

so that by Lemma 1.1,

$$\lim_{n\to\infty} P[\text{Bi}(n-1, p_n(x;t)) \geq k_n] = \begin{cases} 1 & \text{if } s\theta f(x) > 1 \\ 0 & \text{if } s\theta f(x) < 1. \end{cases}$$

Now suppose that $x \in R \cap L_s$. Then

$$\begin{aligned} np_n(x;t) &= n\theta r_n^d f(x) + n \int_{B_n(x;t)} (f(y) - f(x)) \text{d}F(y) \\ &= k_n + t k_n^{1/2} + n \int_{B_n(x;t)} (f(y) - f(x)) \text{d}F(y). \end{aligned}$$

Since the radius of $B_n(x;t)$ is bounded by a constant times $(k_n/n)^{1/d}$, and $x \in R$, the remainder term satisfies the bound

$$n \int_{B_n(x;t)} |f(y) - f(x)| \text{d}F(y) \leq cn(k_n/n)^{1+1/d} \ll k_n^{1/2},$$

the last comparison coming from the condition (4.1) on k_n. Hence,

$$nF(B_n(x;t)) = k_n + t k_n^{1/2} + o(k_n^{1/2}), \quad x \in R \cap L_s. \tag{4.12}$$

Suppose $x \in R \cap L_s$. Let $Y = \text{Bi}(n-1, p_n)$ with $p_n = p_n(x;t)$. Then $(Y - EY)/\sqrt{np_n}$ is approximately standard normal, and

$$P[Y \geq k_n] = P\left[\frac{Y - EY}{\sqrt{np_n}} \geq \frac{k_n - (n-1)p_n}{\sqrt{np_n}}\right],$$

which converges to $\Phi(t)$ by (4.12). The convergence of expectations (4.8) follows from (4.9) by the dominated convergence theorem. The proof of the analogous result for $Z'_n(s,t;A)$ is similar, using (4.10). □

Theorem 4.2 *Suppose $A \subseteq \mathbb{R}$ is a Borel set. If k_n takes the fixed value k, then as $n \to \infty$,*

$$n^{-1} Z_n(t; A) \to \int_A \pi_{\theta t f(x)}([k, \infty)) f(x) \mathrm{d}x, \qquad \text{c.c.}$$

If $k_n \to \infty$ and $n^{-1} k_n^2 / \log n \to 0$, then as $n \to \infty$,

$$n^{-1} Z_n(s, t; A) \to F(L_s^+ \cap A) + \Phi(u) F(L_s \cap A), \qquad \text{c.c.}$$

The proof of this uses the following lemma.

Lemma 4.3 *There is a constant c depending only on the dimension d such that for all $k \in \mathbb{N}$ and any finite $\mathcal{X} \subset \mathbb{R}^d$ and any $x \in \mathcal{X}$, the number of $y \in \mathcal{X}$ having x as kth nearest neighbour is at most ck.*

Proof Consider an infinite cone with a point at x, subtending an angle less than $60°$. There cannot be more than k points in the cone having x as one of their k nearest neighbours. One can take finitely many such cones to cover $\mathbb{R}^d$, completing the proof. □

Proof of Theorem 4.2 With the aim of using Azuma's inequality (Theorem 2.8), define σ-fields $\mathcal{F}_0 = \{\emptyset, \Omega\}$ and $\mathcal{F}_i = \sigma(X_1, \ldots, X_i)$, $1 \leq i \leq n$. Write $Z_n(t; A) - EZ_n(t; A)$ as the sum of a series of martingale differences $\sum_{i=1}^n D_{i,n}$, with $D_{i,n} := E[Z_n(t; A)|\mathcal{F}_i] - E[Z_n(t; A)|\mathcal{F}_{i-1}]$.

Let $Z_{n,i}(t; A)$ denote the number of vertices in $G(\mathcal{X}_{n+1} \backslash \{X_i\}; r_n)$ having degree at least k_n and located in A. Then

$$D_{i,n} = E[Z_n(t; A) - Z_{n,i}(t; A)|\mathcal{F}_i].$$

By Lemma 4.3, there is a constant c such that $|Z_n(t; A) - Z_{n,i}(t; A)| \leq ck_n$, a.s., so that $|D_{i,n}| \leq ck_n$, a.s. Let $\varepsilon > 0$. By Azuma's inequality,

$$P[|Z_n(t; A) - EZ_n(t; A)| \geq \varepsilon n] \leq 2 \exp\left(-\frac{\varepsilon^2 n^2}{2nck_n^2}\right).$$

By the condition $n^{-1} k_n^2 \log n \to 0$, this is summable in n for any $\varepsilon > 0$. Combined with Theorem 4.1 this yields the complete convergence asserted. □

4.3 Asymptotic covariances

This section is concerned with the second-order behaviour of $Z_n'(t)$, and contains results on asymptotic covariance structure of the process $Z_n'(\cdot)$, for both the case where k_n is fixed and the case where $k_n \to \infty$. This is a step towards the eventual goal of obtaining a Gaussian limit process for $Z_n'(\cdot)$.

As in Section 1.7, $\mathcal{H}_\lambda$ denotes a homogeneous Poisson process of intensity λ on $\mathbb{R}^d$, and for $z \in \mathbb{R}^d$ let $\mathcal{H}_\lambda^z$ be the point process $\mathcal{H}_\lambda \cup \{z\}$. Also, let $\mathcal{W}$ denote homogeneous white noise of intensity θ^{-1} on $\mathbb{R}^d$, that is, a centred Gaussian process indexed by the bounded Borel sets in $\mathbb{R}^d$, with covariance function given by $\mathrm{Cov}(\mathcal{W}(A), \mathcal{W}(B)) = \theta^{-1}|A \cap B|$; here $|\cdot|$ denotes Lebesgue measure.

Proposition 4.4 *Suppose k_n is fixed and takes the value k. Suppose A is an open set in $\mathbb{R}^d$. Then for $0 \le t \le u$, as $n \to \infty$,*

$$n^{-1}\mathrm{Cov}(Z'_n(t;A), Z'_n(u;A)) \to \int_A \pi_{\theta t f(x)}([k,\infty))F(\mathrm{d}x) + \int_A \int_{\mathbb{R}^d} \psi_\infty(z; f(x))\mathrm{d}z f(x)^2 \mathrm{d}x, \quad (4.13)$$

with $\psi_\infty(z;\lambda)$ defined for $z \in \mathbb{R}^d$ and $\lambda > 0$ by

$$\psi_\infty(z;\lambda) := P[\{\mathcal{H}_\lambda^z(B(\mathbf{0};u^{1/d})) \ge k\} \cap \{\mathcal{H}_\lambda^{\mathbf{0}}(B(z;t^{1/d})) \ge k\}] - P[\mathcal{H}_\lambda(B(\mathbf{0};u^{1/d})) \ge k]P[\mathcal{H}_\lambda(B(\mathbf{0};t^{1/d})) \ge k]. \quad (4.14)$$

Proposition 4.5 *Suppose $k_n \to \infty$ but (4.1) holds. Let A be an open set in $\mathbb{R}^d$. Let $s,t,u \in \mathbb{R}$ with $s > 0$ and $t \le u$. Then*

$$\lim_{n\to\infty} (nk_n)^{-1}\mathrm{Cov}(Z'_n(t;A), Z'_n(u;A)) = \left(\frac{|L_s \cap A|}{\theta^2 s}\right) \int_{\mathbb{R}^d} \mathrm{Cov}(\mathbf{1}_{\{\mathcal{W}(B(\mathbf{0};1)) \le t\}}, \mathbf{1}_{\{\mathcal{W}(B(z;1)) \le u\}})\mathrm{d}z. \quad (4.15)$$

The proofs of these two results start in the same way. For $x, y \in \mathbb{R}^d$, define the ball $B_n(x;t)$ by (4.3), set $W_n(x;t) := \mathcal{P}_n(B_n(x;t))$, and set $W_n^y(x;t) := \mathcal{P}_n^y(B_n(x;t))$. By Palm theory for the Poisson process (Theorem 1.6),

$$E[Z'_n(t;A)] = n \int_A P[W_n(x;t) \ge k_n]F(\mathrm{d}x) \quad (4.16)$$

and

$$E[Z'_n(t;A)(Z'_n(t;A) - 1)] = n^2 \int_A \int_A P[\{W_n^y(x;t) \ge k_n\} \cap \{W_n^x(y;t) \ge k_n\}]F(\mathrm{d}y)F(\mathrm{d}x). \quad (4.17)$$

Similarly, for $t \le u$,

$$E[Z'_n(t;A)(Z'_n(u;A) - Z'_n(t;A))] = n^2 \int_A \int_A P[\{W_n^y(x;t) \ge k_n\} \cap \{W_n^x(y;t) < k_n \le W_n^x(y;u)\}]F(\mathrm{d}y)F(\mathrm{d}x). \quad (4.18)$$

Define $\psi_n(x,y)$ (also dependent on t and u) by

$$\psi_n(x,y) := P[\{W_n^y(x;t) \ge k_n\} \cap \{W_n^x(y;u) \ge k_n\}] - P[W_n(x;t) \ge k_n]P[W_n(y;u) \ge k_n]. \quad (4.19)$$

By (4.16)–(4.18),

$$\mathrm{Cov}(Z_n'(t;A), Z_n'(u;A)) - EZ_n'(t;A) = n^2 \int_A \int_A \psi_n(x,y) f(x) f(y) \mathrm{d}y \mathrm{d}x. \quad (4.20)$$

Given x and y in $\mathbb{R}^d$, define random variables

$$U_n(x,t,y,u) := \mathcal{P}_n(B_n(x;t) \backslash B_n(y;u)), \quad (4.21)$$

$$V_n(x,t,y,u) := \mathcal{P}_n(B_n(x;t) \cap B_n(y;u)). \quad (4.22)$$

Then $U_n(x,t,y,u)$, $U_n(y,u,x,t)$ and $V_n(x,t,y,u)$ are independent Poisson variables. Also, $W_n(x;t) = U_n(x,t,y,u)+V_n(x,t,y,u)$ and $W_n(y;u) = U_n(y,u,x,t)+ V_n(x,t,y,u)$.

Lemma 4.6 *Suppose k is fixed. Suppose that $x \in R$, $z \in \mathbb{R}^d$, and $-\infty < t \leq u < \infty$. Then with $\psi_\infty(z;\lambda)$ given by (4.14),*

$$\lim_{n\to\infty} \psi_n\left(x, x + n^{-1/d}z\right) = \psi_\infty(z; f(x)). \quad (4.23)$$

Proof Set $y_n := x + n^{-1/d}z$. Then $n|B_n(x;t) \setminus B_n(y_n;u)| = |B(\mathbf{0};t^{1/d}) \setminus B(z;u^{1/d})|$ so that by the continuity of f at x,

$$\begin{aligned} E[U_n(x,t,y_n,u)] &= nf(x)|B_n(x;t)\backslash B_n(y_n;u)|(1+o(1)) \\ &\to f(x)|B(\mathbf{0};t^{1/d})\backslash B(z;u^{1/d})| \qquad \text{as } n \to \infty. \end{aligned}$$

Likewise $EU_n(y_n,u,x,t)$ tends to $f(x)|B(z;u^{1/d})\backslash B(\mathbf{0};t^{1/d})|$ and $EV_n(x,t,y_n,u)$ tends to $f(x)|B(\mathbf{0};t^{1/d}) \cap B(z;u^{1/d})|$. Then (4.23) follows from the definition of $\psi_n(x,y)$ and the remarks following eqn (4.22). □

Lemma 4.7 *Suppose $k_n \to \infty$ and* (4.1) *holds. Let $x \in R$ and $z \in \mathbb{R}^d$. Let $s > 0$ and set $y_n := x + (sk_n/n)^{1/d}z$. Then*

$$\lim_{n\to\infty} \psi_n(x,y_n) = \begin{cases} 0, & s\theta f(x) \neq 1, \\ \mathrm{Cov}(\mathbf{1}_{\{\mathcal{W}(B(\mathbf{0};1))\leq t\}}, \mathbf{1}_{\{\mathcal{W}(B(z;1))\leq u\}}), & s\theta f(x) = 1. \end{cases} \quad (4.24)$$

Proof By (4.11), $EW_n(x;t) \sim s\theta f(x)k_n$ and $\mathrm{Var}W_n(x;t) \sim s\theta f(x)k_n$, so that if $s\theta f(x) > 1$, then $P[W_n(x;t) \geq k_n] \to 1$ by Chebyshev's inequality. Similarly $P[W_n(y_n;u) \geq k_n] \to 1$. If instead $s\theta f(x) < 1$, then $P[W_n(x;t) \geq k_n] \to 0$ and $P[W_n(y_n;u) \geq k_n] \to 0$, and the first case of (4.24) follows.

Now assume $s\theta f(x) = 1$. If $t > 0$, then by (4.12), $W_n(x;t) - W_n(x;0)$ is Poisson with mean $tk_n^{1/2} + o(k_n^{1/2})$, so that $k_n^{-1/2}(W_n(x;t) - W_n(x;0))$ converges in probability to t. Similarly, when $t < 0$, $k_n^{-1/2}(W_n(x;0) - W_n(x;t))$ converges in probability to $-t$. Hence, for all t,

$$\lim_{n\to\infty} P[\{W_n(x;t) \geq k_n\} \triangle \{W_n(x;0) \geq k_n - tk_n^{1/2}\}] = 0,$$

and likewise for W_n^y with $y = y_n(x,z)$. Similarly,

$$\lim_{n\to\infty} P[\{W_n(y_n;u) \geq k_n\} \triangle \{W_n(y_n;0) \geq k_n - uk_n^{1/2}\}] = 0,$$

and likewise for W_n^x. Hence,

$$\begin{aligned}\psi_n(x,y_n) = P[\{W_n^{y_n}(x;0) \geq k_n - tk_n^{1/2}\} \cap \{W_n^x(y_n;0) \geq k_n - uk_n^{1/2}\}] \\ -P[W_n^{y_n}(x;0) \geq k_n - tk_n^{1/2}]P[W_n^x(y_n;0) \geq k_n - uk_n^{1/2}] + o(1).\end{aligned}$$

Since we assume $x \in R$ so f is well behaved at x, we have $|f(\cdot) - f(x)| = O(k_n/n)^{1/d}$ on $B_n(x;t)$, so with $U_n(\cdot)$ defined at (4.21),

$$\begin{aligned}E[U_n(x,0,y_n,0)] &= n\left(f(x)|B_n(x;0)\backslash B_n(y_n;0)| + O(k_n/n)^{1+1/d}\right) \\ &= nf(x)(sk_n/n)|B(\mathbf{0};1)\backslash B(z;1)| + o(k_n^{1/2}) \qquad \text{by (4.1)} \\ &= k_n\theta^{-1}|B(\mathbf{0};1)\backslash B(z;1)| + o(k_n^{1/2}).\end{aligned}$$

Likewise,

$$EU_n(y_n,0,x,0) = k_n\theta^{-1}|B(\mathbf{0};1)\backslash B(z;1)| + o(k_n^{1/2}),$$

and $EV_n(x,0,y_n,0) = k_n\theta^{-1}|B(\mathbf{0};1) \cap B(z;1)| + o(k_n^{1/2})$. Set

$$\begin{aligned}U_n^{(1)}(x,z) &:= k_n^{-1/2}(k_n\theta^{-1}|B(\mathbf{0};1)\backslash B(z;1)| - U_n(x,0,y_n,0)), \\ U_n^{(2)}(x,z) &:= k_n^{-1/2}(k_n\theta^{-1}|B(\mathbf{0};1)\backslash B(z;1)| - U_n(y_n,0,x,0)),\end{aligned}$$

and

$$V_n'(x,z) := k_n^{-1/2}(k_n\theta^{-1}|B(\mathbf{0};1) \cap |B(z;1)| - V_n(x,0,y_n,0)).$$

Then $U_n^{(1)}(x,z) + V_n'(x,z) = k_n^{-1/2}(k_n - W_n(x;0))$, so that

$$\{W_n(x;0) \geq k_n - tk_n^{1/2}\} = \{U_n^{(1)}(x,z) + V_n'(x,z) \leq t\}.$$

Similarly,

$$\{W_n(y_n;0) \geq k_n - uk_n^{1/2}\} = \{U_n^{(2)}(x,z) + V_n'(x,z) \leq u\}.$$

Also, $U_n^{(1)}(x,z), U_n^{(2)}(x,z)$, and $V_n'(x,z)$ are independent and asymptotically normal with mean zero and variances $\theta^{-1}|B(\mathbf{0};1)\backslash B(z;1)|, \theta^{-1}|B(\mathbf{0};1)\backslash B(z;1)|$, and $\theta^{-1}|B(\mathbf{0};1) \cap B(z;1)|$, respectively; the result follows. □

Define the function $f\mathbf{1}_A(\cdot)$ by $f\mathbf{1}_A(x) := f(x)\mathbf{1}_A(x)$, $x \in \mathbb{R}^d$. This is used in the next two proofs.

Proof of Proposition 4.4 Suppose k_n is fixed. By (4.20), the change of variable $y = y_n(x,z) = x + n^{-1/d}z$, and the definition (4.19) of $\psi_n(\cdot)$,

$$\begin{aligned}&\text{Cov}(Z_n'(t;A), Z_n'(u;A)) - EZ_n'(t;A) \\ &\quad = n\int_A\int_{\mathbb{R}^d} \psi_n(x, y_n(x,z))\, f\mathbf{1}_A(y_n(x,z))\text{d}z f(x)\text{d}x. \qquad (4.25)\end{aligned}$$

If $\|y - x\| > 2(u/n)^{1/d}$, then $B_n(x;t) \cap B_n(y;u) = \emptyset$ and $\psi_n(x,y) = 0$. Hence,

$$|\psi_n(x, y_n(x,z))| f(y_n(x,z)) \leq \mathbf{1}_{B(\mathbf{0};2u^{1/d})}(z) f_{\max}, \qquad \forall\, z, x \in \mathbb{R}^d.$$

Hence, by Lemma 4.6, the assumption that $F(R) = 1$, and the dominated convergence theorem,

$$\begin{aligned}\lim_{n\to\infty} \int_A \int_{\mathbb{R}^d} \psi_n(x, y_n(x,z)) f \mathbf{1}_A(y_n(x,z)) \mathrm{d}z f(x) \mathrm{d}x \\ = \int_A \int_{\mathbb{R}^d} \psi_\infty(z; f(x)) \mathrm{d}z f(x)^2 \mathrm{d}x.\end{aligned}$$

Combined with (4.25) and (4.7), this gives us the result (4.13). □

Proof of Proposition 4.5 Suppose that $k_n \to \infty$ and (4.1) holds. By (4.20), and the change of variable $y = y_n(x,z) = x + (sk_n/n)^{1/d} z$,

$$\begin{aligned}\mathrm{Cov}(Z_n'(t;A), Z_n'(u;A)) - EZ_n'(t;A) = snk_n \int\int \psi_n\,(x, y_n(x,z)) \\ \times f\mathbf{1}_A(y_n(x,z)) \mathrm{d}z f(x) \mathrm{d}x. \quad (4.26)\end{aligned}$$

For n large, if $\|y - x\| > 3(sk_n/n)^{1/d}$ then $B_n(x;u) \cap B_n(y;t) = \emptyset$ and $\psi_n(x,y) = 0$. Hence $|\psi_n(x, y_n(x,z))| \leq \mathbf{1}_{B(\mathbf{0};3s^{1/d})}(z)$. Hence by Lemma 4.7, the assumption that $F(R) = 1$, and the dominated convergence theorem for integrals,

$$\begin{aligned}\lim_{n\to\infty} \int_A \int_{\mathbb{R}^d} \psi_n\,(x, y_n(x,z))\, f\mathbf{1}_A(y_n(x,z)) \mathrm{d}z f(x) \mathrm{d}x \\ = \frac{|L_s \cap A|}{(s\theta)^2} \int_{\mathbb{R}^d} \mathrm{d}z \mathrm{Cov}(\mathbf{1}_{\{\mathcal{W}(B(\mathbf{0};1)) \leq t\}}, \mathbf{1}_{\{\mathcal{W}(B(z;1)) \leq u\}}),\end{aligned}$$

and then (4.15) follows by (4.26). □

4.4 Moments for de-Poissonization

This section is concerned only with the limiting regime with $k_n \to \infty$. Take s in (4.2) to be fixed. As at (4.3), set $B_n(x;t) := B(x; r_n(t))$. For $n, m \in \mathbb{N}$, set

$$T_{m,n}(t) := \sum_{i=1}^{m} \mathbf{1}_{\{\mathcal{X}_m(B_n(X_i;t) \setminus \{X_i\}) \geq k_n\}},$$

so that $T_{n,n} = Z_n(t)$ and $T_{N_n,n}(t) = Z_n'(t)$. Set $\tilde{D}_{m,n}(t) := T_{m+1,n}(t) - T_{m,n}(t)$.

This section contains a series of results, which show that for distinct m and m' close to n, the mean of $\tilde{D}_{m,n}(t)$ is close to $\phi(t)F(L_s)$, its second moment is uniformly bounded, and the covariance of $\tilde{D}_{m,n}(t)$ and $\tilde{D}_{m',n}(u)$ is close to zero. In the next section, these will be used to apply the de-Poissonization technique from Section 2.5 to deduce the central limit theorem for Z_n from the central limit theorem for Z'_n. Observe that $\tilde{D}_{m,n}(t)$ is equal to $D_{m,n}(t) + D'_{m,n}(t)$, where we set

$$D_{m,n}(t) := \sum_{i=1}^{m} \mathbf{1}_{\{\mathcal{X}_m(B_n(X_i;t)\setminus\{X_i\})=k_n-1\}\cap\{X_{m+1}\in B_n(X_i)\}}, \tag{4.27}$$

$$D'_{m,n}(t) := \mathbf{1}_{\{\mathcal{X}_m(B_n(X_{m+1};t))\geq k_n\}}. \tag{4.28}$$

For $n \in \mathbb{N}$, $p \in (0,1)$, and $k \in \{0,1,\ldots,n\}$, define the binomial probability

$$\beta_{n,p}(k) := P[\mathrm{Bi}(n,p) = k] = \binom{n}{k} p^k (1-p)^{n-k}.$$

The proofs in this section use the following facts about the binomial distribution. The first is a matter of simple calculus, while the second is a local central limit theorem, and can be proved by the argument in Shiryayev (1984, p. 56).

Lemma 4.8 (a) *Suppose $n, k \in \mathbb{N}$ with $k < n$. Then $\beta_{n,p}(k)$ is maximized over $p \in (0,1)$ by setting $p = k/n$, and $p\beta_{n,p}(k)$ is maximized over $p \in (0,1)$ by setting $p = (k+1)/(n+1)$.*

(b) *Suppose $(j_n)_{n\geq 1}$ is a sequence of integers satisfying $j_n \to \infty$ and $(j_n/n) \to 0$ as $n \to \infty$. Suppose $t \in \mathbb{R}$ and $(p_n)_{n\geq 1}$ is a sequence in $(0,1)$ satisfying $(j_n - np_n)/(np_n)^{1/2} \to t$ as $n \to \infty$. Then*

$$j_n^{1/2} \beta_{n,p_n}(j_n) \to \phi(t) \qquad \text{as } n \to \infty.$$

Lemma 4.9 *Suppose $k_n \to \infty$ and (4.1) holds. Then*

$$\lim_{n\to\infty} \sup_{\{m:|m-n|\leq n^{2/3}\}} |k_n^{-1/2} E\tilde{D}_{m,n}(t) - \phi(t)F(L_s)| = 0. \tag{4.29}$$

Proof Take an arbitrary $\mathbb{N}$-valued sequence $(m_n)_{n\geq 1}$ with $|m_n - n| \leq n^{2/3}$ for each n. By (4.27),

$$k_n^{-1/2} ED_{m_n,n}(t) = \int_{\mathbb{R}^d} m_n k_n^{-1/2} P[\mathcal{X}_{m_n-1}(B_n(x;t)) = k_n - 1] \times F(B_n(x;t))F(\mathrm{d}x). \tag{4.30}$$

Let $x \in R \cap L_s$. Then $\mathcal{X}_{m_n-1}(B_n(x;t))$ is binomial with parameters $m_n - 1$ and $F(B_n(x;t))$, and by (4.12) its mean is given by

$$m_n F(B_n(x;t)) = (1 + O(n^{-1/3}))(k_n + tk_n^{1/2} + o(k_n^{1/2})),$$

and since $k_n = o(n^{2/3})$ by (4.1), this implies that

$$m_n F_n(B_n(x;t)) = k_n + tk_n^{1/2} + o(k_n^{1/2}), \qquad x \in R \cap L_s. \tag{4.31}$$

By Lemma 4.8,

$$\lim_{n\to\infty} k_n^{1/2} P[\mathcal{X}_{m_n-1}(B_n(x;t)) = k_n - 1] = \phi(t), \qquad x \in R \cap L_s.$$

Also, by (4.11) and Lemma 1.1,

$$\lim_{n\to\infty} k_n^{1/2} P[\mathcal{X}_{m_n-1}(B_n(x;t)) = k_n - 1] = 0, \qquad x \in R \setminus L_s.$$

Thus, for $x \in R$, the integrand on the right-hand side of (4.30) tends to $\phi(t)\mathbf{1}_{L_s}(x)$. Also, $(m_n/k_n)F(B_n(x))$ is bounded uniformly in x and n, and by Lemma 4.8, $k_n^{1/2} \sup_{0\le p\le 1} \beta_{m_n-1,p}(k_n - 1)$ is also uniformly bounded. Hence by the dominated convergence theorem, $k_n^{-1/2} ED_{m_n,n}(t)$ tends to $\phi(t)F(L_s)$. Also, $0 \le D'_{m_n,n}(t) \le 1$ so $k_n^{-1/2} D'_{m_n,n}(t)$ tends to zero. Since the choice of sequence (m_n) was arbitrary, subject to $|m_n - n| \le n^{2/3}$, (4.29) follows. □

Lemma 4.10 *Suppose $k_n \to \infty$ and* (4.1) *holds. Let $t \in \mathbb{R}, u \in \mathbb{R}$. Then*

$$\lim_{n\to\infty} \sup_{n-n^{2/3}\le l<m\le n+n^{2/3}} |k_n^{-1} E\tilde{D}_{l,n}(t)\tilde{D}_{m,n}(u) - \phi(t)\phi(u)F(L_s)^2| = 0. \quad (4.32)$$

Proof For $l \le m$, it is the case that

$$\begin{aligned} ED_{l,n}(t)D_{m,n}(u) &= \sum_{i\le l}\sum_{j\le m} P[\{\mathcal{X}_l(B_n(X_i;t)) = k_n\} \cap \{\mathcal{X}_m(B_n(X_j;u)) = k_n\} \\ &\qquad \cap \{X_{l+1} \in B_n(X_i;t)\} \cap \{X_{m+1} \in B_n(X_j;u)\}] \\ &= l(l-1) \int\int g_{n,l,m}(x,y)F(\mathrm{d}y)F(\mathrm{d}x) \\ &\quad + l(m-l) \int\int g'_{n,l,m}(x,y)F(\mathrm{d}y)F(\mathrm{d}x) \\ &\quad + l \int g''_{n,l,m}(x)F(\mathrm{d}x), \end{aligned} \quad (4.33)$$

where all integrals are over $\mathbb{R}^d$, and where we set

$$\begin{aligned} g_{n,l,m}(x,y) := P[&\{\mathcal{X}^y_{l-2}(B_n(x;t)) = k_n - 1\} \cap \{X_{l-1} \in B_n(x;t)\} \\ &\cap \{\mathcal{X}^x_{m-2}(B_n(y;u)) = k_n - 1\} \cap \{X_{m-1} \in B_n(y;u)\}], \end{aligned} \quad (4.34)$$

$$\begin{aligned} g'_{n,l,m}(x,y) := P[&\{\mathcal{X}_{l-1}(B_n(x;t)) = k_n - 1\} \cap \{X_l \in B_n(x;t)\} \\ &\cap \{\mathcal{X}^x_{m-2}(B_n(y;u)) = k_n - 1\} \cap \{X_{m-1} \in B_n(y;u)\}], \end{aligned} \quad (4.35)$$

and

$$\begin{aligned} g''_{n,l,m}(x) := P[&\{\mathcal{X}_{l-1}(B_n(x;t)) = k_n - 1\} \cap \{X_l \in B_n(x;t)\} \\ &\cap \{\mathcal{X}^x_{m-1}(B_n(x;u)) = k_n - 1\} \cap \{X_m \in B_n(x;u)\}]. \end{aligned} \quad (4.36)$$

Take x and y in R with $x \ne y$. Choose arbitrary $\mathbb{N}$-valued sequences $(l_n)_{n\ge1}$ and $(m_n)_{n\ge1}$ with $n - n^{2/3} \le l_n < m_n \le n + n^{2/3}$. Then by (4.11), as $n \to \infty$,

$$(n/k_n)P[X_{l_n-1} \in B_n(x;t)] \to s\theta f(x), \tag{4.37}$$
$$(n/k_n)P[X_{m_n-1} \in B_n(y;u)] \to s\theta f(x). \tag{4.38}$$

By Lemma 4.8 and (4.31),

$$k_n^{1/2} P[\mathcal{X}^x_{m_n-2}(B_n(y;u)) = k_n - 1 | X_{l_n-1} \in B_n(x;t)] \to \phi(u) \quad y \in L_s. \tag{4.39}$$

Let $x, y \in \mathbb{R}^d$ with $y \notin B(x; r_n(t) + r_n(u))$, so that $B_n(x;t) \cap B_n(y;u) = \emptyset$. If $\mathcal{X}_{m_n-2}(B_n(y;u)) = k_n - 1$, then $\mathcal{X}_{l_n-2}(B_n(y;u)) = j$ for some j with $0 \leq j \leq k_n - 1$. Given $\mathcal{X}_{l_n-2}(B_n(y;u)) = j$ for such a j, the conditional distribution of $\mathcal{X}_{l_n-2}(B_n(x;t))$ is binomial with parameters $l_n - 2 - j$ and $F(B_n(x;t))/(1 - F(B_n(y;u)))$. For all such j, if also x lies in L_s then by (4.31) the mean of this distribution satisfies

$$\begin{aligned} \frac{(l_n - 2 - j)F(B_n(x;t))}{1 - F(B_n(y;u))} &= (k_n + tk_n^{1/2} + o(k_n^{1/2}))\left(1 + O\left(\frac{k_n}{n}\right)\right) \\ &= k_n + tk_n^{1/2} + o(k_n^{1/2}), \end{aligned} \tag{4.40}$$

where the last line follows from the fact that $k_n = o(n^{2/3})$ by (4.1). Hence, by Lemma 4.8, for $x \in L_s$, and for any $y \neq x$,

$$\begin{aligned} k_n^{1/2} P[\mathcal{X}_{l_n-2}(B_n(x;t)) = k_n - 1 | \{\mathcal{X}_{m_n-2}(B_n(y;u)) = k_n - 1\} \\ \cap \{X_{l_n-1} \in B_n(x;t)\}] \to \phi(t). \end{aligned}$$

Combining this with (4.37)–(4.39), we obtain

$$(n^2/k_n) g_{n,l_n,m_n}(x,y) \to \phi(t)\phi(u), \qquad x, y \in R \cap L_s,\ x \neq y. \tag{4.41}$$

On the other hand, by (4.11) and Lemma 1.1,

$$k_n P[\mathcal{X}^x_{m_n-2}(B_n(y;u)) = k_n - 1 | X_{l_n-1} \in B_n(x;t)] \to 0 \quad y \in R \setminus L_s,$$

and similarly

$$k_n P[\mathcal{X}^y_{l_n-2}(B_n(x;t)) = k_n - 1] \to 0, \quad x \in R \setminus L_s.$$

Combined with (4.37) and (4.38), these imply that

$$(n^2/k_n) g_{n,l_n,m_n}(x,y) \to 0, \qquad (x,y) \in (R \times R) \setminus (L_s \times L_s). \tag{4.42}$$

Set $B_n^2(x) := B(x; r_n(t) + r_n(u))$. If $y \notin B_n^2(x)$, then, setting $p_1 := F(B_n(y;u))$ and $p_2 := F(B_n(x;t))/(1 - p_1)$, we have

$$g_{n,l_n,m_n}(x,y) \leq \max_{0 \leq j \leq k_n - 1} p_1 p_2 \beta_{m_n-3,p_1}(k_n - 1)\beta_{l_n-2-j,p_2}(k_n - 1),$$

so by Lemma 4.8, there is a constant c such that

$$g_{n,l_n,m_n}(x,y) \le c\left(k_n^{-1/2}(k_n/n)\right)^2 = ck_n/n^2. \tag{4.43}$$

It follows by (4.41), (4.42), and the dominated convergence theorem that

$$\lim_{n\to\infty}\left(\frac{n^2}{k_n}\right)\int_{\mathbb{R}^d}\int_{\mathbb{R}^d\setminus B_n^2(x)} g_{n,l_n,m_n}(x,y)F(\mathrm{d}y)F(\mathrm{d}x) = \phi(t)\phi(u)F(L_s)^2. \tag{4.44}$$

To deal with x and y close together, observe that

$$\begin{aligned} g_{n,l_n,m_n}(x,y) &\le P[\mathcal{X}^y_{l_n-2}(B_n(x;t)) = k_n - 1]F(B_n(x;t))F(B_n(y;u)) \\ &= O(k_n^{-1/2}(k_n/n)^2). \end{aligned} \tag{4.45}$$

Since $F(B_n^2(x))$ is bounded by a constant times k_n/n, and since $k_n = o(n^{2/3})$ by (4.1), it follows that there is a constant c such that

$$\left(\frac{n^2}{k_n}\right)\int_{\mathbb{R}^d}\int_{B_n^2(x)} g_{n,l_n,m_n}(x,y)F(\mathrm{d}y)F(\mathrm{d}x) \le c\left(\frac{n^2}{k_n}\right)\left(\frac{k_n}{n}\right)\left(\frac{k_n^{3/2}}{n^2}\right) \to 0.$$

Thus (4.44) holds with the region of integration modified to $\mathbb{R}^d \times \mathbb{R}^d$. The asymptotics for g'_{n,l_n,m_n} are just the same. Also, by similar arguments there is a constant c' such that

$$l_n \sup_{x\in\mathbb{R}} g''_{n,l}(x) \le c' n k_n^{-1/2}(k_n/n)^2 \to 0.$$

Therefore (4.33) yields

$$k_n^{-1} E D_{l_n,n}(t) D_{m_n,n}(u) \to \phi(t)\phi(u)F(L_s)^2. \tag{4.46}$$

We need to show that (4.46) still holds with $D_{l_n,n}$ replaced by $\tilde{D}_{l_n,n} := D_{l_n,n} + D'_{l_n,m}$, and with $D_{m_n,n}$ replaced by $\tilde{D}_{m_n,n}$. By definition, $0 \le D'_{l_n,n}(t) \le 1$, and by the proof of Lemma 4.9, $E[D_{l_n,n}(t)] = O(k_n^{1/2})$; likewise for $D'_{m_n,n}(u)$ and $D_{m_n,n}(u)$. It follows that $E[D_{l_n,n}(t)D'_{m_n,n}(u)]$, $E[D'_{l_n,n}(t)D'_{m_n,n}(u)]$, and $E[D'_{l_n,n}(t)D_{m_n,n}(u)]$ are all $O(k_n^{1/2})$, so (4.46) indeed still holds with $D_{l_n,n}(t)$ and $D_{m_n,n}(u)$ replaced by $\tilde{D}_{l_n,n}(t)$ and $\tilde{D}_{m_n,n}(u)$, respectively. Since the sequences (l_n) and (m_n) are arbitrary, (4.32) follows. □

Lemma 4.11 *Suppose that $k_n \to \infty$ and (4.1) holds. Let $t \in \mathbb{R}$ and $u \in \mathbb{R}$. Then*

$$\limsup_{n\to\infty}\left(k_n^{-3/2} \sup_{\{m:|m-n|\le n^{2/3}\}} E\tilde{D}_{m,n}(t)^2\right) < \infty. \tag{4.47}$$

Proof Since $0 \leq D'_{m,n}(t) \leq 1$, it suffices to show that there is a constant c such that $k_n^{-3/2} ED_{m_n,n}(t)^2 \leq c$ for any sequence (m_n) satisfying $|m_n - n| \leq n^{2/3}$ for all n. Choose such a sequence. By (4.33) with $l = m = m_n$,

$$E[D_{m_n,n}(t)^2] = m_n(m_n - 1) \int_{\mathbb{R}^d} \int_{\mathbb{R}^d} g_{n,m_n,m_n}(x,y) F(\mathrm{d}x) F(\mathrm{d}y) + E[D_{m_n,n}(t)], \tag{4.48}$$

with $g_{n,m,m}(x,y)$ defined by (4.34) (with $u = t$). By Lemma 4.8, there is a constant c such that

$$g_{n,m_n,m_n}(x,y) \leq \sup_{0 \leq p \leq 1} p \max(\beta_{m_n,p}(k_n - 1), \beta_{m_n,p}(k_n - 2)) \leq c \left(\frac{k_n}{n} \right) k_n^{-1/2}.$$

Also, $g_{n,m,m}(x,y) = 0$ unless $y \in B(x; 2r_n(t))$. Hence, there is a constant c' such that

$$\int_{\mathbb{R}^d} \int_{\mathbb{R}^d} g_{n,m_n,m_n}(x,y) F(\mathrm{d}x) F(\mathrm{d}y) \leq \frac{c' k_n^{3/2}}{n^2}. \tag{4.49}$$

The factor $m_n(m_n - 1)$ in (4.48) is $O(n^2)$, while $ED_{m_n,n}$ is $O(k_n^{1/2})$ by Lemma 4.9. So $E[D_{m_n,n}(t)^2] = O(k_n^{3/2})$, as required. □

4.5 Finite-dimensional central limit theorems

This section contains Gaussian limit theorems for the finite-dimensional distributions of the empirical processes $Z'_n(\cdot)$ and $Z_n(\cdot)$ of re-scaled k_n-nearest-neighbour distances. In the case of $Z'_n(\cdot)$, that is, in the case where the number of points is Poisson, the results go as follows. They are stated as limit theorems for $Z'_n(\cdot; A)$, the main interest being in the special case with $A = \mathbb{R}^d$.

Theorem 4.12 *Suppose that k_n is fixed, and that A is an open set in $\mathbb{R}^d$. The finite-dimensional distributions of the process*

$$n^{-1/2}[Z'_n(t; A) - EZ'_n(t; A)], \qquad t \geq 0$$

converge to those of a centred Gaussian process $(Z'_\infty(t; A), t > 0)$ with covariance $E[Z'_\infty(t; A) Z'_\infty(u; A)]$ given by the right-hand side of (4.13).

Theorem 4.13 *Suppose that $k_n \to \infty$, that (4.1) holds, and that A is an open set in $\mathbb{R}^d$. Let $s > 0$ and suppose $F(A \cap L_s) > 0$. The finite-dimensional distributions of the process*

$$(nk_n)^{-1/2}[Z'_n(t; A) - EZ'_n(t; A)], \qquad t \in \mathbb{R}$$

converge to those of a centred Gaussian process $(Z'_\infty(t; A), t \in \mathbb{R})$ with covariance $E[Z'_\infty(t; A) Z'_\infty(u; A)]$ given by the right-hand side of (4.15).

Proof We prove these two theorems together. Set $\tilde{k}_n := 1$ in the case when k_n is fixed, and set $\tilde{k}_n := k_n$ in the case when $k_n \to \infty$. Let $M \in \mathbb{N}$, $\boldsymbol{b} = (b_1, \ldots, b_M) \in \mathbb{R}^M$, and $\boldsymbol{t} = (t_1, \ldots, t_M) \in \mathbb{R}^M$. In the case of k_n fixed, assume each t_j is positive. Set $t_{\max} := \max(t_1, \ldots, t_m)$ and

$$Z'_n(\boldsymbol{t}, \boldsymbol{b}; A) := \sum_{m=1}^{M} b_m Z'_n(t_m; A). \tag{4.50}$$

By Proposition 4.4 (when k_n is fixed) or Proposition 4.5 (when $k_n \to \infty$),

$$\lim_{n\to\infty} (n\tilde{k}_n)^{-1} \mathrm{Var}\, Z'_n(\boldsymbol{t}, \boldsymbol{b}; A) = \mathrm{Var}\left(\sum_{m=1}^{M} b_m Z'_\infty(t_m; A)\right) := \sigma'(\boldsymbol{t}, \boldsymbol{b}; A). \tag{4.51}$$

Assume first that A is bounded. Given $n \in \mathbb{N}$, divide $\mathbb{R}^d$ into cubes $Q_{j,n}, j \geq 1$, of volume $\tilde{k}_n/n$, and let $Y_{j,n}$ be the contribution to $Z'_n(\boldsymbol{t}, \boldsymbol{b}; A)$ from points of $\mathcal{P}_n$ in $Q_{j,n}$; that is, let $Y_{j,n}$ be the sum over $m \in \{1, \ldots, M\}$ of b_m times the number of vertices of $G(\mathcal{P}_n; r_n(t_m))$ having degree at least k_n and lying in $Q_{j,n} \cap A$.

Let G_n be a graph with vertex set $V_n := \{j : Q_{j,n} \cap A \neq \emptyset\}$, and with vertices j and j' linked by an edge if and only if $\mathrm{dist}(Q_{j,n}, Q_{j',n}) \leq 3r_n(t_{\max})$. Then G_n is a dependency graph for the variables $Y_{j,n}, j \in V_n$, since $Y_{j,n}$ is determined by the positions of the points of $\mathcal{P}_n$ distant at most $r_n(t_{\max})$ from the set $Q_{j,n}$. Moreover, since $r_n(t_{\max}) = O(\tilde{k}_n/n)^{1/d}$ in both limiting regimes for k_n, the degrees of vertices of G_n are uniformly bounded.

For each j, n, let $N_{j,n} := \mathcal{P}_n(Q_{j,n})$, a Poisson variable with mean bounded by $\tilde{k}_n f_{\max}$. Then $0 \leq Y_{j,n} \leq \left(\sum_{m=1}^{M} |b_m|\right) N_{j,n}$, and hence for some constant c, and $p = 3$ or $p = 4$, $E|Y_{j,n}|^p \leq c\tilde{k}_n^p$.

Now, $Z'_n(\boldsymbol{t}, \boldsymbol{b}; A) = \sum_{j \in V_n} Y_{j,n}$, and since V_n has $O(n/\tilde{k}_n)$ elements, by (4.51), if $\sigma'(\boldsymbol{t}, \boldsymbol{b}; A) > 0$, then there is a constant c such that for $p = 3, 4$,

$$(\mathrm{Var} Z'_n(\boldsymbol{t}, \boldsymbol{b}; A))^{-p/2} \sum_{j \in V_n} E|Y_{j,n}|^p \leq c(n\tilde{k}_n)^{-p/2} \tilde{k}_n^p \left(\frac{n}{\tilde{k}_n}\right) = c\left(\frac{n}{\tilde{k}_n}\right)^{1-p/2}.$$

This tends to zero, so by Theorem 2.4 on normal approximation, setting

$$\xi_n(A) := (n\tilde{k}_n)^{-1/2}(Z'_n(\boldsymbol{t}, \boldsymbol{b}; A) - EZ'_n(\boldsymbol{t}, \boldsymbol{b}; A)),$$

we have $\xi_n(A) \xrightarrow{\mathcal{D}} \mathcal{N}(0, \sigma'(\boldsymbol{t}, \boldsymbol{b}; A))$. This also holds for $\sigma'(\boldsymbol{t}, \boldsymbol{b}; A) = 0$.

Now suppose that A is unbounded (e.g. $A = \mathbb{R}^d$). Set $A_K := A \cap (-K, K)^d$, and $A^K := A \setminus [-K, K]^d$. Then A_K is bounded, so that

$$\xi_n(A_K) \xrightarrow{\mathcal{D}} \mathcal{N}(0, \sigma'(\boldsymbol{t}, \boldsymbol{b}; A_K)). \tag{4.52}$$

Given $w \in \mathbb{R}$ and $\varepsilon > 0$,

$$\begin{aligned}P[\xi_n(A) \le w] = P[\xi_n(A_K) \le w - \varepsilon] - P[\{\xi_n(A_K) \le w - \varepsilon\} \cap \{\xi_n(A) > w\}] \\ +P[\{|\xi_n(A_K) - w| < \varepsilon\} \cap \{\xi_n(A) \le w\}] \\ +P[\{\xi_n(A_K) \ge w + \varepsilon\} \cap \{\xi_n(A) \le w\}].\end{aligned}$$

Hence, since $\xi_n(A) = \xi_n(A_K) + \xi_n(A^K)$ a.s.,

$$|P[\xi_n(A) \le w] - P[\xi_n(A_K) \le w - \varepsilon]| \le P[|\xi_n(A^K)| \ge \varepsilon] + P[|\xi_n(A_K) - w| < \varepsilon].$$

By Chebyshev's inequality, (4.51) and (4.52),

$$\begin{aligned}\limsup_{n\to\infty} |P[\xi_n(A) \le w] - P[\xi_n(A_K) \le w - \varepsilon]| \\ \le \frac{\sigma'(\boldsymbol{t}, \boldsymbol{b}; A^K)}{\varepsilon^2} + \frac{2\varepsilon}{\sqrt{2\pi\sigma'(\boldsymbol{t}, \boldsymbol{b}; A_K)}}.\end{aligned} \tag{4.53}$$

As $K \to \infty$, $\sigma'(\boldsymbol{t}, \boldsymbol{b}; A_K)$ tends to $\sigma'(\boldsymbol{t}, \boldsymbol{b}; A)$, and $\sigma'(\boldsymbol{t}, \boldsymbol{b}; A^K)$ tends to zero. Hence, by taking ε sufficiently small and K large, we can make the right-hand side of (4.53) arbitrarily small, and also make $\Phi\left((w-\varepsilon)/\sigma'(\boldsymbol{t}, \boldsymbol{b}; A_K)^{1/2}\right)$ arbitrarily close to $\Phi\left(w/\sigma'(\boldsymbol{t}, \boldsymbol{b}; A)^{1/2}\right)$. Then by (4.52), it follows that $P[\xi_n(A) \le w]$ tends to $\Phi\left(w/\sigma'(\boldsymbol{t}, \boldsymbol{b}; A)^{1/2}\right)$, that is, $\xi_n(A) \xrightarrow{\mathcal{D}} \mathcal{N}(0, \sigma'(\boldsymbol{t}, \boldsymbol{b}; A))$. The results then follow by the Cramér–Wold device. □

The next two results are central limit theorems for the finite-dimensional distributions of the process $Z_n(\cdot)$, and are obtained by de-Poissonizing Theorems 4.12 and 4.13, using results from Section 2.5 and in particular the notion of stabilization given at Definition 2.15.

Theorem 4.14 *Suppose k_n takes the fixed value k. The finite-dimensional distributions of the process*

$$n^{-1/2}[Z_n(t) - EZ_n(t)], \qquad t > 0$$

converge to those of a centred Gaussian process $(Z_\infty(t), t > 0)$ with

$$E[Z_\infty(t)Z_\infty(u)] = E[Z'_\infty(t)Z'_\infty(u)] - h(t)h(u), \tag{4.54}$$

with $(Z'_\infty(t)) = Z'_\infty(t; \mathbb{R}^d)$ as given in Theorem 4.12, *and with $h(t) = h(t; k)$ defined by*

$$h(t) := \int \{\pi_{f(x)\theta t}(k-1) f(x)\theta t^d + \pi_{f(x)\theta t}([k, \infty))\} f(x) \mathrm{d}x. \tag{4.55}$$

Proof Let $\boldsymbol{t} \in (0, \infty)^M$ and $\boldsymbol{b} \in \mathbb{R}^M$. For each finite $\mathcal{X} \subset \mathbb{R}^d$, let $H_0(\mathcal{X}) := \sum_{\mu=1}^M b_\mu \sum_{x\in\mathcal{X}} \mathbf{1}_{\{\mathcal{X}(B(x;t_\mu^{1/d})) \ge k_n+1\}}$ and let $H_n(\mathcal{X}) := H_0(n^{1/d}\mathcal{X})$. Then using the notation at (4.50) and (4.51), we have $H_n(\mathcal{P}_n) = Z'_n(\boldsymbol{t}, \boldsymbol{b}; \mathbb{R}^d)$, and by Theorem 4.12, $n^{-1/2}(H_n(\mathcal{P}_n) - EH_n(\mathcal{P}_n))$ is asymptotically normal $\mathcal{N}(0, \sigma'(\boldsymbol{t}, \boldsymbol{b}; \mathbb{R}^d))$.

The functional H_0 stabilizes because it has finite range. Also, the expected value of the associated limiting add one cost on a homogeneous Poisson process is

$$E[\Delta(\mathcal{H}_\lambda)] = \sum_{\mu=1}^{M} b_\mu(\pi_{\lambda\theta t_\mu}([k,\infty)) + \lambda\theta\pi_{\lambda\theta t_\mu}(k-1)),$$

where the first term $\pi_{\lambda\theta t_\mu}([k,\infty))$ is the probability that an inserted point at the origin has degree at least k, and the second term $\lambda\theta\pi_{\lambda\theta t_\mu}(k-1)$ is the expected number of points whose degrees go up from $k-1$ to k as a result of an insertion at the origin into the Poisson process $\mathcal{H}_\lambda$, for the geometric graph $G(\cdot; t_\mu^{1/d})$. Hence, with the Cox process $\mathcal{H}_{f(X)}$ defined just after Definition 2.15, and with $h(t)$ defined at (4.55), we have $E[\Delta(\mathcal{H}_{f(X)})] = \sum_{\mu=1}^{m} b_\mu h(t_\mu)$.

Also, $|H_n(\mathcal{X}_m)| \leq \sum_{\mu=1}^{M} |b_\mu| m$, and setting $t_{\max} := \max(t_1, \ldots, t_M)$, we have

$$|H_n(\mathcal{X}_{m+1}) - H_n(\mathcal{X}_m)| \leq \left(\sum_{\mu=1}^{M} |b_\mu|\right)(\mathcal{X}_m(B_n(X_{m+1}; t_{\max})) + 1),$$

which is stochastically dominated by a constant times $(\mathrm{Bi}(m, f_{\max}\theta r_n^d) + 1)$, and so has a fourth moment that is bounded uniformly over $m, n \in \mathbb{N}$ with $m \leq 2n$. Therefore the functional H satisfies all the conditions for Theorem 2.16, so that $n^{-1/2}(H_n(\mathcal{X}_n) - EH_n(\mathcal{X}_n))$ is asymptotically $\mathcal{N}(0, \tau^2)$ with $\tau^2 = \sigma'(\boldsymbol{t}, \boldsymbol{b}; \mathbb{R}^d) - (E\Delta(\mathcal{H}_{f(X)}))^2$. The result then follows by the Cramér–Wold device. □

Theorem 4.15 *Suppose $k_n \to \infty$ and (4.1) holds. Let $s > 0$ and suppose $F(L_s) > 0$. The finite-dimensional distributions of the process*

$$(nk_n)^{-1/2}[Z_n(t) - EZ_n(t)], \quad t \in \mathbb{R}$$

converge to those of a centred Gaussian process $(Z_\infty(t), t \in \mathbb{R})$, satisfying (4.54) for all t, u but with $Z'_\infty(t)$ now given by Theorem 4.13 and with $h(\cdot)$ now defined by $h(t) := \phi(t)F(L_s)$.

Proof Let $\boldsymbol{t} \in \mathbb{R}^M$ and $\boldsymbol{b} \in \mathbb{R}^M$, and for finite $\mathcal{X} \subset \mathbb{R}^d$ set

$$H_n(\mathcal{X}) := k_n^{-1/2} \sum_{\mu=1}^{M} b_\mu \sum_{x \in \mathcal{X}} \mathbf{1}_{\{\mathcal{X}(B_n(x;t)) \geq k_n + 1\}}.$$

Using notation from (4.50) and (4.51), we have $H_n(\mathcal{P}_n) = k_n^{-1/2} Z'_n(\boldsymbol{t}, \boldsymbol{b}; \mathbb{R}^d)$, and by Theorem 4.13,

$$n^{-1/2}(H_n(\mathcal{P}_n) - EH_n(\mathcal{P}_n)) \xrightarrow{\mathcal{D}} \mathcal{N}(0, \sigma'(\boldsymbol{t}, \boldsymbol{b}; \mathbb{R}^d)). \tag{4.56}$$

Set $\alpha := \sum_{\mu=1}^{M} b_\mu \phi(t_\mu) F(L_s)$, and define the increments $R_{m,n} := H_n(\mathcal{X}_{m+1}) - H_n(\mathcal{X}_m)$. Then $R_{m,n} = \sum_{\mu=1}^{M} b_\mu D_{m,n}(t_\mu)$ and, by Lemma 4.9,

$$\lim_{n\to\infty}\left(\sup_{n-n^{2/3}\leq m\leq n+n^{2/3}} |E[R_{m,n}]-\alpha|\right)=0,$$

while by Lemma 4.10

$$\lim_{n\to\infty}\left(\sup_{n-n^{2/3}\leq m<m'\leq n+n^{2/3}} \left|E[R_{m,n}R_{m',n}]-\alpha^2\right|\right)=0,$$

and by Lemma 4.11, and the Cauchy–Schwarz inequality,

$$\lim_{n\to\infty}\left(n^{-1/2}\sup_{n-n^{2/3}\leq m<m'\leq n+n^{2/3}} E[R_{m,n}^2]\right)=0.$$

Also, $|H_n(\mathcal{X}_m)| \leq \sum_{\mu=1}^M |b_\mu| m$, a.s. Hence the functional H_n satisfies all the conditions of Theorem 2.12, and that result gives us

$$n^{-1/2}\left(H_n(\mathcal{X}_n)-EH_n(\mathcal{X}_n)\right) \xrightarrow{\mathcal{D}} \mathcal{N}(0,\sigma(\boldsymbol{t},\boldsymbol{b})),$$

with $\sigma(\boldsymbol{t},\boldsymbol{b}) := \sigma'(\boldsymbol{t},\boldsymbol{b};\mathbb{R}^d)-\alpha^2$, that is, $\sigma(\boldsymbol{t},\boldsymbol{b}) = \operatorname{Var}\sum_{\mu=1}^M b_\mu Z_\infty(t_\mu)$. The result then follows by the Cramér–Wold device. □

4.6 Convergence in Skorohod space

The preceding section contains weak convergence of the finite-dimensional distributions of the process $Z'_n(\cdot)$, suitably scaled and centred, to a Gaussian limit process $Z'_\infty(\cdot)$. The present section contains an extension of this to weak convergence of the stochastic process $Z'_n(\cdot)$ in the standard function space for processes of this type, namely the Skorohod space, as described in Billingsley (1968), and extended to non-compact time intervals in Whitt (1980). Convergence in Skorohod space can be important in the construction of statistical tests; see Bickel and Breiman (1983) or more generally, Shorack and Wellner (1986).

In brief, the notion of weak convergence in this setting goes as follows. For $a<b$, let $D[a,b]$ denote the space of all right-continuous real-valued functions on $[a,b]$ with left limits. Let $\Lambda[a,b]$ be the class of strictly increasing continuous mappings of $D[a,b]$ onto itself and, for $x,y\in D[a,b]$, let $d(x,y)$ be the infimum of the set of $\varepsilon>0$ for which there exists $\lambda\in\Lambda$ such that $\sup_{a\leq x\leq b}|\lambda(t)-t|\leq\varepsilon$ and $\sup_{a\leq x\leq b}|x(t)-y(\lambda(t))|\leq\varepsilon$. Then d is a metric on $D[a,b]$ and generates the so-called Skorohod J_1 topology. This topology induces a notion of weak convergence (convergence in distribution) for any sequence of stochastic processes $(\xi_n(t), a\leq t\leq b)_{n\geq 1}$.

Let T be either the interval $[0,\infty)$ or the interval $(-\infty,\infty)$. We say a sequence of stochastic processes $(\xi_n(t), t\in T)_{n\geq 1}$ converges weakly in $D(T)$ to a limit process $(\xi(t), t\in T)$ if (and only if) for any $a<b$ with $a,b\in T$ the restrictions to time interval $[a,b]$ of the processes ξ_n converge weakly to the restriction to $[a,b]$ of the process ξ_∞. This is equivalent to convergence in distribution using an appropriate topology on $D(T)$; see Whitt (1980, Theorem 2.8).

Theorem 4.16 *Suppose that $d \geq 2$, and that $\|\cdot\|$ is the Euclidean norm. Suppose k_n is fixed. Then the sequence of processes*

$$n^{-1/2}[Z'_n(t) - EZ'_n(t)], \quad t \geq 0$$

converges weakly in $D[0,\infty)$ to a zero-mean Gaussian process $(Z'_\infty(t), t > 0)$ with $E[Z'_\infty(t)Z'_\infty(u)]$ given by the right-hand side of (4.13).

Theorem 4.17 *Suppose $d \geq 2$, and $\|\cdot\|$ is the Euclidean norm. Suppose $k_n \to \infty$ and (4.1) holds. Let $s > 0$ and suppose $F(L_s) > 0$. Then the sequence of processes*

$$(nk_n)^{-1/2}[Z'_n(t) - EZ'_n(t)], \quad t \in \mathbb{R}$$

converges weakly in $D(-\infty,\infty)$ to a zero-mean Gaussian process $(Z'_\infty(t), t \in \mathbb{R})$ with $E[Z'_\infty(t)Z'_\infty(u)]$ given by the right-hand side of (4.15).

Proof of Theorems 4.16 and 4.17 Set $\tilde{k}_n = 1$ in the case when k_n is fixed, $\tilde{k}_n = k_n$ in the case where $k_n \to \infty$. By Theorems 4.12 and 4.13, we have convergence of the finite-dimensional distributions of $(n\tilde{k}_n)^{-1/2}(Z'_n(\cdot) - EZ_n(\cdot))$ to those of $Z'_\infty(\cdot)$. Therefore by Billingsley (1968, Theorem 15.6), it suffices to prove that given $K > 0$, there are constants $c > 0$ and $\alpha > 1$ such that for $-K \leq t < u < v \leq K$,

$$M_{t,u,v} := (n\tilde{k}_n)^{-2}E[(Z'_n(v) - Z'_n(u))^2(Z'_n(u) - Z'_n(t))^2] \leq c(v-t)^\alpha. \quad (4.57)$$

Since $\tilde{k}_n$ is within a constant of k_n it suffices to prove (4.57) with $\tilde{k}_n$ replaced by k_n.

The proof of (4.57) is essentially identical to that of Penrose (2000*a*, eqn (7.8)), although there are some minor differences in the setup (see Section 4.7 below). The argument in Penrose (2000*a*) is rather long and technical, and we do not repeat it here. However, we do take this opportunity to correct an error in Penrose (2000*a*, Lemma 7.1).

Lemma 4.18 *Suppose $d \geq 2$. Let $A(x;r,\varepsilon)$ denote the annulus $B(x;r+\varepsilon)\setminus B(x;r)$. There exists $\varepsilon_0 > 0$ and $c > 0$ such that for any $r, r' \in (0,1]$, any $\varepsilon, \varepsilon' \in (0,\varepsilon_0)$, and any $x \in \mathbb{R}^d$ with $|x| \geq 5\max(\varepsilon,\varepsilon')^{1/2}$, it is the case that*

$$|A(\mathbf{0};r,\varepsilon) \cap A(x;r',\varepsilon')| \leq c\max(\varepsilon,\varepsilon')^{5/4}.$$

In Penrose (2000*a*) the exponent in this bound was incorrectly given as $\frac{3}{2}$, not $\frac{5}{4}$. This does not affect the argument to prove Penrose (2000*a*, eqn (7.8)) (any exponent strictly greater than 1 suffices).

We sketch a proof of Lemma 4.18, concentrating on the case $d = 2$. We assume without loss of generality that x, the centre of the second annulus, lies on the horizontal axis, to the right of the origin, and also that $r \geq r'$ and $\varepsilon = \varepsilon'$.

Set $\delta := \|x\|$, and assume ε is small with $\delta \geq 5\varepsilon^{1/2}$. Assume first that $r' = 1-\delta$ (the 'worst case'). Then the region $A(\mathbf{0};r,\varepsilon) \cap A(x;r',\varepsilon')$ is the more darkly shaded region in Fig. 4.1.

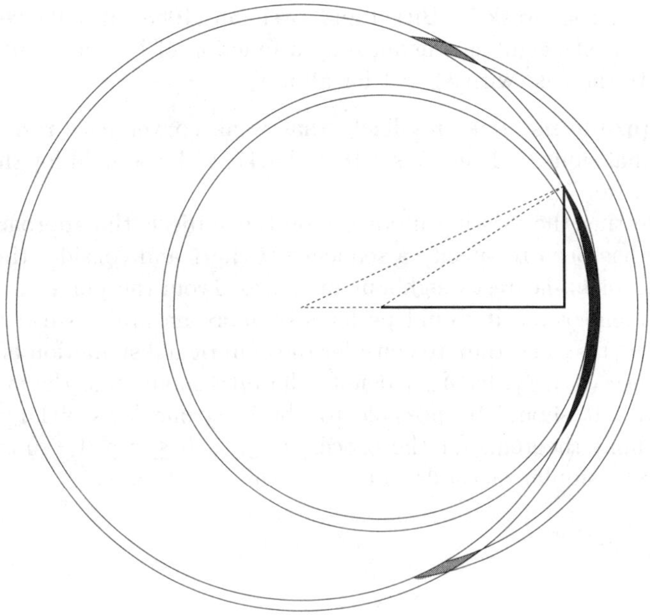

FIG. 4.1. Three annuli (one of them partially obscured) are shown. The largest annulus has radius r and is centred at $\mathbf{0}$, while the others are centred at x.

Some elementary trigonometry shows that the length of the bold horizontal line is at least $r - (r\varepsilon/\delta) - \delta$, and hence the height of the bold vertical line is at most $\sqrt{2\varepsilon/\delta}$, and so is at most a constant times $\varepsilon^{1/4}$.

From this we can deduce that the more darkly shaded region has area bounded by a constant times $\varepsilon^{5/4}$. Other cases where $r' > 1 - \varepsilon$ are illustrated by the more lightly shaded region, which has two components. For either component, the bounding arcs (centred at x) have length less than a constant times $\varepsilon^{1/4}$, by comparison with the 'worst case' already considered. From this we can deduce the result.

4.7 Notes and open problems

Notes The theory of one-dimensional spacings is discussed in Barbour *et al.* (1992); for statistical applications see, for example, Hall (1986), Wells *et al.* (1993), and references therein. For general discussion of applications of multivariate k-nearest-neighbour distances in statistical testing, see Henze (1987), Cressie (1991), Byers and Raftery (1998) and L'Écuyer *et al.* (2000).

The results of this chapter are adapted from those in Penrose (2000*a*), the differences being that (i) only the case with $k_n \to \infty$ is considered there, and (ii) k-nearest-neighbour distances from X are weighted by $f(X)^{1/d}$ in Penrose

(2000a). Earlier work by Bickel and Breiman (1983) also allowed for weighting of the nearest-neighbour distance by a function of location, but was concerned only with the case with $k_n = 1$ for all n.

Open problems It seems likely that weak convergence results in Skorohod space, analogous to Theorems 4.16 and 4.17, will also hold for the process $Z_n(\cdot)$ as well as for $Z'_n(\cdot)$.

Reflecting the existing literature on the subject, the approach taken in this chapter has been to specify a sequence $(k_n)_{n\geq 1}$ and consider the empirical distribution of k_n-nearest-neighbour distances. From the point of view of the rest of this monograph, it would perhaps be more natural instead to specify a sequence $(r_n)_{n\geq 1}$ and then to consider the empirical distribution of degrees. Given a sequence $(r_n)_{n\geq 1}$, let $\delta_{m,n}$ denote the mth smallest of the vertex degrees of $G(\mathcal{X}_n; r_n)$. It should be possible to obtain strong laws of large numbers and central limit theorems for the process $(\delta_{\lfloor an \rfloor, n}, 0 \leq a \leq 1)$, suitably scaled and centred, by similar methods to those of this chapter.

5

GEOMETRICAL INGREDIENTS

The next few chapters are concerned with results on extremal vertex degrees, cliques, and so forth. Further geometrical and measure-theoretic preliminaries are required, and are collected in the present chapter.

Throughout this chapter, we write $|\cdot|$ for Lebesgue measure. Recall that $\theta := |B(\mathbf{0};1)|$, the volume of the unit ball in the chosen norm.

5.1 Consequences of the Lebesgue density theorem

Recall that F is the measure on $\mathbb{R}^d$ associated with the underlying probability density function f, that is, $F(A) = \int_A f(x)\mathrm{d}x$. Recall also that $f_{\max}$ denotes the essential supremum of f, that is, the smallest h such that $P[f(X_1) \leq h] = 1$, and that we assume $f_{\max} < \infty$.

As mentioned in Section 1.6, the Lebesgue density theorem is often of use to us in dispensing with any assumption of continuity on f. The following lemmas are a case in point.

Lemma 5.1 *Suppose $\phi < f_{\max}$. Let $\delta \in (0,1]$. For $r > 0$ let $\sigma(r)$ be the maximum number of points $x_i \in \mathbb{R}^d$ which can be found such that the balls $B(x_i;r)$ are disjoint and satisfy $F(B(x_i;r)) \geq \phi\theta r^d$, while $F(B(x_i;\delta r)) \geq \phi\theta(\delta r)^d$. Then*

$$\liminf_{r\downarrow 0} r^d \sigma(r) > 0. \tag{5.1}$$

Proof Define the number ϕ_1 by

$$\phi_1 := (d!4)^{-1}\delta^d\phi + (1-(d!4)^{-1}\delta^d)f_{\max}.$$

By the Lebesgue density theorem, we can choose $x_0 \in \mathbb{R}^d$ and $r_0 > 0$ such that

$$F(B(x_0;r_0)) > \phi_1\theta r_0^d. \tag{5.2}$$

Set $B := B(x_0;r_0)$.

By convexity, the volume of the unit ball $B(\mathbf{0};1)$ divided by that of the smallest product of intervals containing $B(\mathbf{0};1)$ is a constant, depending on the choice of norm, that is at least $d!^{-1}$ (this minimum value being achieved by the l_1 norm, but in any event the value of the constant is unimportant). Therefore, for small enough r it is possible to pack into the ball B a collection of $n = n(r)$ disjoint balls $B(x_1;r), B(x_2;r), \ldots, B(x_n;r)$, each contained in B and of total volume at least $(d!2)^{-1}|B|$. For each i let B_i (respectively, B'_i) denote the ball $B(x_i;r)$ (respectively, $B(x_i;\delta r)$).

Suppose more than half of the n points x_i satisfied either $F(B_i) \le \phi|B_i|$ or $F(B_i') \le \phi|B_i'|$. Then by taking an appropriate union of balls we could find a set $A \subseteq B$, with

$$|A| \ge (d!4)^{-1}\delta^d|B|$$

and $F(A) \le \phi|A|$. But then we would have

$$F(B) \le \phi|A| + f_{\max}|B \backslash A| \le \phi_1|B|,$$

which contradicts (5.2). Therefore at least half of the n points x_i satisfy $F(B_i) > \phi|B_i|$ and $F(B_i') > \phi|B_i'|$, so that $\sigma(r) \ge n/2$. Since nr^d is bounded away from zero, (5.1) follows. □

We give a similar result for the infimum. Let Ω denote the support of f, that is, the intersection of all closed sets in $\mathbb{R}^d$ with F-measure 1. Let f_0 be the essential infimum of f over Ω, that is, the largest h such that $P[f(X_1) \ge h] = 1$.

Lemma 5.2 *Suppose $\psi > f_0$. Let $\delta \in (0,1]$. For $r > 0$ let $\sigma'(r)$ be the maximum number of points $x_i \in \mathbb{R}^d$ which can be found such that the balls $B(x_i;r)$ are disjoint and satisfy $F(B(x_i;r)) \le \psi\theta r^d$, while $F(B(x_i;\delta r)) \ge (f_0/2)\theta(\delta r)^d$. Then*

$$\liminf_{r \downarrow 0} r^d \sigma'(r) > 0. \tag{5.3}$$

Proof Choose numbers $\varepsilon > 0$ and $\psi_0 \in (f_0, \psi)$ such that $\varepsilon < (d!16)^{-1}\delta^d$ and

$$\psi_0 < (d!8)^{-1}\psi + (1 - (d!8)^{-1})f_0 - \varepsilon f_0.$$

By the Lebesgue density theorem, there exists $x_0 \in \mathbb{R}^d$ and $r_0 > 0$ such that

$$|F(B(x_0;r_0))| < \psi_0\theta r_0^d \tag{5.4}$$

and additionally

$$|B(x_0;r_0) \setminus \Omega| \le \varepsilon\theta r_0^d. \tag{5.5}$$

Set $B := B(x_0;r_0)$. For small enough r it is possible to pack into B a collection of $n = n(r)$ disjoint balls $B(x_1;r), B(x_2;r), \dots, B(x_n;r)$, each contained in B and of total volume at least $(d!2)^{-1}|B|$. For each i let B_i (respectively, B_i') denote the ball $B(x_i;r)$ (respectively, $B(x_i;\delta r)$).

Suppose more than one-quarter of the n balls B_i satisfied $F(B_i) \ge \psi|B_i|$. Then the union A of such balls B_i would satisfy $|A| \ge (d!8)^{-1}|B|$, and $F(A) \ge \psi|A|$, and

$$F(B \setminus A) \ge f_0(|B \setminus A| - \varepsilon|B|),$$

and hence

$$F(B) \ge \psi|A| + f_0|B \setminus A| - \varepsilon f_0|B| > \psi_0|B|,$$

contradicting (5.4).

Suppose more than one-quarter of the n balls B'_i satisfied $F(B'_i) < (f_0/2)|B'_i|$. Then the union A of such balls B'_i would satisfy $|A| \geq (d!8)^{-1}\delta^d|B|$, and $F(A) < (f_0/2)|A|$; the latter condition implies that $|A \setminus \Omega| \geq |A|/2$, and hence $|A \setminus \Omega| \geq (d!16)^{-1}\delta^d|B|$, contradicting (5.5).

Thus, at least one half of the points x_i satisfy both $F(B_i) \leq \psi|B_i|$ and $F(B'_i) \geq (f_0/2)|B'_i|$, so that $\sigma'(r) \geq n/2$. Since nr^d is bounded away from zero, (5.3) follows. □

5.2 Covering, packing, and slicing

For any $U \subseteq \mathbb{R}^d$ and $r > 0$ define the *r-covering number* of U, denoted $\kappa(U;r)$, to be the minimum n such that there exists a collection of n balls of the form $B(x;r)$ with $x \in U$, whose union contains U. Define the *r-packing number*, denoted $\sigma(U;r)$, to be the maximum n such that there exist n disjoint balls of the form $B(x;r)$ with $x \in U$. The next result has a simple proof, which is omitted.

Lemma 5.3 *Suppose $(U_n)_{n\geq 1}$ is a uniformly bounded sequence of subsets of $\mathbb{R}^d$, and $(r_n)_{n\geq 1}$ is a strictly positive sequence with $r_n \to 0$ as $n \to \infty$. Then*

$$\limsup_{n\to\infty} r_n^d \kappa(U_n; r_n) < \infty. \tag{5.6}$$

In the statement of results on random geometric graphs, let Ω denote the support of the underlying density function f of the random points X_i under consideration. Then Ω is a closed subset of $\mathbb{R}^d$; let $\partial\Omega$ denote its topological boundary, that is, its intersection with the closure of its complement.

We shall sometimes assume that $\partial\Omega$ is a *$(d-1)$-dimensional C^2 submanifold of $\mathbb{R}^d$*. By this we mean that there exists a collection of pairs $\{(U_i, \phi_i)\}$, where $\{U_i\}$ is a collection of open sets in $\mathbb{R}^d$, whose union contains $\partial\Omega$, and ϕ_i is a C^2 diffeomorphism of U_i onto an open set in $\mathbb{R}^d$, with the property that $\phi_i(U_i \cap \partial\Omega) = \phi_i(U_i) \cap (\mathbb{R}^{d-1} \times \{0\})$.

Examples where $\partial\Omega$ is a $(d-1)$-dimensional C^2 submanifold of $\mathbb{R}^d$ include cases with $d = 2$ and Ω bounded by a smooth closed curve, and cases with $d = 3$ and Ω bounded by a smooth surface such as a sphere or ellipsoid. On the other hand, if $d \geq 2$ and Ω is polyhedral its boundary is *not* a $(d-1)$-dimensional C^2 submanifold of $\mathbb{R}^d$.

The proof of the following result is fairly simple and is omitted.

Lemma 5.4 *Suppose $\partial\Omega$ is a compact $(d-1)$-dimensional C^2 submanifold of $\mathbb{R}^d$. Then*

$$\limsup_{r\downarrow 0} r^{d-1}\kappa(\partial\Omega; r) < \infty. \tag{5.7}$$

Moreover, for any open $U \subseteq \mathbb{R}^d$ with $U \cap \partial\Omega \neq \emptyset$,

$$\liminf_{r\downarrow 0} r^{d-1}\sigma(U \cap \partial\Omega; r) > 0. \tag{5.8}$$

For $x, y \in \mathbb{R}^d$, write $x \cdot y$ for the usual l_2 inner product $\sum_{i=1}^d x_i y_i$, and recall that $\|x\|_2 = (x \cdot x)^{1/2}$. Let $S^{d-1} := \{x \in \mathbb{R}^d : \|x\|_2 = 1\}$ (the unit sphere). Using the equivalence of norms on $\mathbb{R}^d$, take $\eta_0 \in (0,1)$, depending on the norm $\|\cdot\|$, such that $\|x\| \leq \frac{1}{2}$ whenever $\|x\|_2 \leq \eta_0$.

Suppose $x \in \mathbb{R}^d$ and $r > 0$, $e \in S^{d-1}$, and $\eta \in (0, \eta_0)$. Define $B^*(x; r, \eta, e)$ and $B^*(x; r, \eta, -e)$ to be the two components obtained by starting with the ball $B(x; r)$ and removing a slice of relative l_2 thickness 2η orthogonal to e at the centre of the ball. More precisely, set

$$B^*(x; r, \eta, e) := \{y \in B(x; r) : (y - x) \cdot e > \eta r\}. \tag{5.9}$$

The key geometrical result for dealing with boundary effects is the following lemma, which reflects the fact that $\partial\Omega$ is locally (almost) flat, and is illustrated by Fig. 5.1.

Lemma 5.5 *Suppose $\Omega \subset \mathbb{R}^d$ is compact, and $\partial\Omega$ is a $(d-1)$-dimensional C^2 submanifold of $\mathbb{R}^d$. Suppose $x \in \partial\Omega$, and $\eta > 0$. Then there exist $e \in S^{d-1}$ and $\delta > 0$, such that*

$$B^*(y; r, \eta, e) \subseteq \Omega, \quad \forall y \in B(x; \delta) \cap \Omega, \quad \forall r \in (0, \delta] \tag{5.10}$$

and

$$B^*(y; r, \eta, -e) \subseteq \Omega^c, \quad \forall y \in B(x; \delta) \cap \partial\Omega, \quad \forall r \in (0, \delta]. \tag{5.11}$$

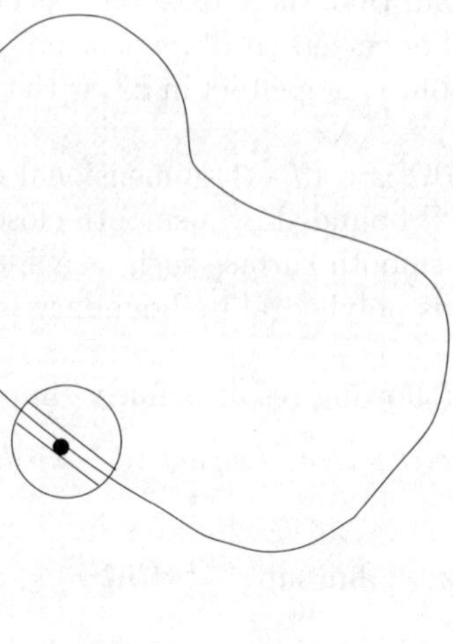

FIG. 5.1. Illustration of Lemma 5.5.

Proof Let $x \in \partial\Omega$. By the definition of a submanifold, there is a C^2 diffeomorphism ϕ from an open neighbourhood U of x to a ball $\phi U \subseteq \mathbb{R}^d$, centred at $\mathbf{0}$, with $\phi(x) = \mathbf{0}$ and

$$\phi(U \cap \partial\Omega) = \{w \in \phi U : \pi_d(w) = 0\}, \tag{5.12}$$

with π_d denoting projection onto the dth coordinate.

If $y, z \in U \backslash \partial\Omega$ and $\pi_d(\phi(y))$ and $\pi_d(\phi(z))$ have the same sign, then there is a path in $\phi(U)$ from $\phi(y)$ to $\phi(z)$ which avoids $\phi(U \cap \partial\Omega)$, so the points y and z must be both in Ω or both in Ω^c. Hence, either

$$U \cap \Omega = (\pi_d \circ \phi)^{-1}([0,\infty)) \text{ and } U \cap \Omega^c = (\pi_d \circ \phi)^{-1}((-\infty, 0)) \tag{5.13}$$

or

$$U \cap \Omega = (\pi_d \circ \phi)^{-1}((-\infty, 0]) \text{ and } U \cap \Omega^c = (\pi_d \circ \phi)^{-1}((0,\infty)). \tag{5.14}$$

In what follows we assume (5.13), but would argue similarly in the case of (5.14).

The derivative $\phi'(x)$ is a linear isomorphism on $\mathbb{R}^d$, and the composition $\pi_d \circ \phi'(x)$ is the l_2 inner product with some vector, which we denote v; set $b := \|v\|_2$ and $e := b^{-1}v$, an l_2 unit vector in the direction of v.

Take $r_1 > 0$ such that $B(x; 3r_1) \subseteq U$, and such that for $y \in B(x; 2r_1)$, and all unit vectors f,

$$\|\phi'(y)(f) - \phi'(x)(f)\|_2 \leq \|y - x\|. \tag{5.15}$$

By Taylor's theorem and the equivalence of norms on $\mathbb{R}^d$, there exists a constant $M \geq 1$ such that if $y, z \in B(x; 2r_1)$, then

$$\|\phi(z) - \phi(y) - \phi'(y)(z - y)\|_2 \leq M\|z - y\|^2,$$

and hence by (5.15),

$$\|\phi(z) - \phi(y) - \phi'(x)(z - y)\|_2 \leq M\left(\|z - y\|^2 + \|y - x\| \times \|z - y\|\right). \tag{5.16}$$

Let $\delta := \min((b\eta/(2M)), r_1)$. Suppose $y \in B(x; \delta)$ and $\|z - y\| \leq r \leq \delta$. Then, by (5.16),

$$\|\phi(z) - \phi(y) - \phi'(x)(z - y)\|_2 \leq b\eta\|z - y\| \leq b\eta r,$$

so that

$$-b\eta r \leq \pi_d \circ \phi(z) - \pi_d \circ \phi(y) - \pi_d \circ \phi'(x)(z - y) \leq b\eta r. \tag{5.17}$$

If also $y \in \Omega$, then $\pi_d \circ \phi(y) \geq 0$ by (5.13), and hence

$$\pi_d \circ \phi(z) - \pi_d \circ \phi'(x)(z - y) \geq -b\eta r. \tag{5.18}$$

Suppose also $b^{-1}\pi_d \circ \phi'(x)(z - y) > \eta r$; then $\pi_d \circ \phi'(x)(z - y) > b\eta r$ and hence by (5.18), $\pi_d \circ \phi(z) > 0$ so $z \in \Omega$ by (5.13). Then (5.10) follows, because $b^{-1}\pi_d \circ \phi'(x)$ is the inner product with the l_2 unit vector e defined earlier.

Similarly, if $y \in B(x;\delta)$ and $\|z-y\| \leq r \leq \delta$, but now we also assume $y \in \partial\Omega$ then $\pi_d \circ \phi(y) = 0$ by (5.13), so that if $b^{-1}\pi_d \circ \phi'(x)(z-y) < -\eta r$, then the second inequality of (5.17) yields $\pi_d \circ \phi(z) < 0$ so that $z \in \Omega^c$. Then (5.11) follows. □

Lemma 5.6 *Suppose $\Omega \subseteq \mathbb{R}^d$ is compact, and $\partial\Omega$ is a $(d-1)$-dimensional C^2 submanifold of $\mathbb{R}^d$. Given $\eta \in (0,\eta_0)$, there exists $\delta = \delta(\eta) > 0$ such that for all $r < \delta$ and all $y \in \partial\Omega$, there exists $e \in S^{d-1}$ such that*

$$B^*(y;r,\eta,e) \subseteq \Omega \quad \text{and} \quad B^*(y;r,\eta,-e) \subseteq \Omega^c. \tag{5.19}$$

Proof Given $x \in \partial\Omega$, by Lemma 5.5 we can find $e = e(x) \in S^{d-1}$ and $\delta = \delta(x) > 0$ such that (5.10) and (5.11) hold. By compactness, it is possible to cover $\partial\Omega$ by finitely many balls of the form $B(x;\delta(x))$, and the minimum of the corresponding numbers $\delta(x)$ is the required number δ. □

Lemma 5.7 *Suppose $\Omega \subseteq \mathbb{R}^d$ is compact, and $\partial\Omega$ is a $(d-1)$-dimensional C^2 submanifold of $\mathbb{R}^d$. Given $\varepsilon > 0$, there exists $\delta > 0$ such that for $x \in \Omega$ and $s \in (0,\delta)$, the Lebesgue measure of $B(x;s) \cap \Omega$ exceeds $(1-\varepsilon)\theta s^d/2$.*

Proof Take $\eta > 0$ such that $|B^*(\mathbf{0};1,4\eta,e)| > (1-\varepsilon)\theta/2$, for all $e \in S^{d-1}$.

Given $x \in \partial\Omega$, by Lemma 5.5 we can find $e = e(x) \in S^{d-1}$ and $\delta = \delta(x) > 0$ such that (5.10) holds. By compactness, it is possible to cover $\partial\Omega$ by finitely many balls of the form $B(x_i;\delta(x_i)/2), 1 \leq i \leq k$, with each $x_i \in \partial\Omega$; let $\delta_0 := \min(\delta(x_1),\ldots,\delta(x_k))/2$.

If $y \in \Omega$ is at a distance at most δ_0 from $\partial\Omega$, then by the triangle inequality y lies in one of the balls $B(x_i;\delta(x_i)), 1 \leq i \leq k$, so that by (5.10) and the choice of η, there exists an l_2 unit vector e such that for any $s < \delta_0$ the set $B^*(y;s,\eta,e)$ is contained in Ω and has Lebesgue measure greater than $(1-\varepsilon)\theta s^d/2$; hence $|B(y;s)| > (1-\varepsilon)\theta s^d/2$.

If $y \in \Omega$ is at a distance greater than δ_0 from $\partial\Omega$ and $0 < s < \delta_0$, then $B(y;s) \subseteq \Omega$ so that $|B(y;s) \cap \Omega| = \theta s^d$. □

Let $f_1 := \inf\{f(x) : x \in \partial\Omega\}$. The following result is analogous to Lemma 5.2, except that it refers to points near the boundary of Ω.

Lemma 5.8 *Suppose that $\partial\Omega$ is a compact $(d-1)$-dimensional C^2 submanifold of $\mathbb{R}^d$, that $f_1 > 0$, and that the restriction of f to Ω is continuous at x for all $x \in \partial\Omega$. Suppose $\psi > f_1$. Let $\delta \in (0,1]$. For $r > 0$ let $\sigma''(r)$ be the maximum number of points $x_i \in \partial\Omega$ which can be found such that the balls $B(x_i;r)$ are disjoint and satisfy $F(B(x_i;r)) \leq \psi\theta r^d/2$, while $F(B(x_i;\delta r)) \geq f_1\theta\delta^d r^d/8$. Then*

$$\liminf_{r\downarrow 0} r^{d-1}\sigma''(r) > 0. \tag{5.20}$$

Proof Choose $f_2 \in (f_1,\psi)$. Then take $x_0 \in \partial\Omega$ and $\varepsilon \in (0,\frac{1}{4})$ such that $\frac{1}{2}f_1 < f(x) < f_2$ for $x \in B(x_0;2\varepsilon) \cap \Omega$, and also $f_2(1+\varepsilon) < \psi$. Set $B_1 := B(x_0;\varepsilon)$. By (5.8) in Lemma 5.4,

$$\liminf_{r \downarrow 0} r^{d-1} \sigma(B_1 \cap \partial\Omega; r) > 0. \tag{5.21}$$

Recall the definition of $B^*(x; r, \eta, e)$ given at (5.9). Pick $\eta_1 > 0$ such that $|B^*(\mathbf{0}; 1, \eta_1, e)| > \theta(1-\varepsilon)/2$ for any $e \in S^{d-1}$. Pick $\delta = \delta(\eta_1)$ by Lemma 5.6. Suppose $y \in B_1 \cap \partial\Omega$ and $r < \min(\delta, \varepsilon)$. Then for some $e \in S^{d-1}$, (5.19) holds so that

$$\begin{aligned} F(B(y;r)) \leq f_2 |B(y;r) \cap \Omega| &\leq f_2 |B(y;r) \setminus B^*(y; r, \eta_1, -e)| \\ &\leq f_2 \theta r^d (1 - (1-\varepsilon)/2) \leq \psi \theta r^d / 2, \end{aligned}$$

and

$$F(B(y; \delta r)) \geq (f_1/2)|B(y; \delta r) \cap \Omega| \geq (f_1/2)|B^*(y; \delta r, \eta_1, e)| \geq f_1 \theta (\delta r)^d / 8.$$

The result follows by Lemma 5.4. □

For $x \in \mathbb{R}^d$ and $e \in S^{d-1}$, let $D(x; r, e)$ denote the cylinder of l_2 height $2r$ and radius r, centred at x, pointing in the direction of e; that is, set

$$D(x; r, e) := \{x + y + \lambda e : y \cdot e = 0, -r < \lambda < r, \|y\|_2 < r\}.$$

For $\eta > 0$, define a cylinder $D^*(x; r, \eta, e)$ analogously to $B^*(x; r, \eta, e)$ by

$$D^*(x; r, \eta, e) := \{y \in D(x; r, e) : (y - x) \cdot e > \eta r\}. \tag{5.22}$$

Also, define the line $L(x; e)$ by $L(x; e) = \{x + \lambda e : \lambda \in \mathbb{R}^d\}$.

It is straightforward to recast Lemma 5.5 in terms of cylinders as follows.

Corollary 5.9 *Suppose $x \in \partial\Omega$, and $\eta \in (0,1)$. Then there exists $e \in S^{d-1}$ and $\delta > 0$, such that*

$$D^*(y; r, \eta, e) \subseteq \Omega, \quad \forall y \in D(x; \delta) \cap \Omega, \quad \forall r \in (0, \delta) \tag{5.23}$$

and

$$D^*(y; r, \eta, -e) \subseteq \Omega^c, \quad \forall y \in D(x; \delta) \cap \partial\Omega, \quad \forall r \in (0, \delta). \tag{5.24}$$

Proposition 5.10 *There exists a constant $\delta_1 > 0$, and a finite collection of pairs $\{(\xi_i, e_i), i = 1, 2, \ldots, \mu\}$ with $\xi_i \in \partial\Omega$ and $e_i \in S^{d-1}$, such that*

$$\partial\Omega \subseteq \bigcup_{i=1}^{\mu} D(\xi_i; \delta_1, e_i),$$

and for $1 \leq i \leq \mu$,

$$D^*(y; r, 0.1, e_i) \subseteq \Omega, \quad \forall y \in D(\xi_i; 10\delta_1, e_i) \cap \Omega, \quad \forall r \in (0, 10\delta_1), \tag{5.25}$$

and

$$D^*(y; r, 0.1, -e_i) \subseteq \Omega^c, \quad \forall y \in D(\xi_i; 10\delta_1, e_i) \cap \partial\Omega, \quad \forall r \in (0, 10\delta_1). \tag{5.26}$$

Moreover, if $x \in D(\xi_i; 10\delta_1, e_i)$, there is a unique point denoted $\psi_i(x)$ of the line $L(x; e_i)$ which is in $D(\xi_i; 10\delta_1, e_i) \cap \partial\Omega$.

Finally, there exists a constant $c_2 > 0$ such that for all $i \leq \mu$, for all $u, v \in D(\xi_i; 10\delta_1, e_i)$ with $\|v - u\|_2 < 5\delta_1$, we have $\|\psi_i(v) - \psi_i(u)\| \leq c_2 \|v - u\|$.

Proof By Corollary 5.9, for all $x \in \partial\Omega$ there exists $e = e(x) \in S^{d-1}$ and $\delta = \delta(x) > 0$, such that (5.23) and (5.24) hold with $\eta = 0.1$. By compactness, we can take a finite collection of points $x_j \in \partial\Omega$ such that the cylinders $D(x_j; \delta(x_j)/10, e(x_j))$ cover $\partial\Omega$. For each j write f_j for the vector $e(x_j)$. Let δ_1 denote the minimum of the numbers $\delta(x_j)/12$.

Again by compactness, for each j there is a finite collection of points $\zeta_{jk} \in \partial\Omega \cap D(x_j; \delta(x_j)/10, f_j)$ such that the cylinders $D(\zeta_{jk}; \delta_1, f_j)$ cover $D(x_j; \delta(x_j)/10, f_j) \cap \partial\Omega$. The (ζ_{jk}, f_j), re-labelled as (ξ_i, e_i), $1 \leq i \leq \mu$, are the pairs required, since if $y \in D(\zeta_{jk}; 10\delta_1, f_j)$, then $y \in D(x_j; (\frac{10}{12} + \frac{1}{10})\delta(x_j), f_j)$.

Let $i \leq \mu$. If $y \in D(\xi_i; 10\delta_1, e_i) \cap \Omega$, then it follows from (5.25) that $y + \lambda e_i \in \Omega$ for all $\lambda \in (0, 10\delta_1)$. Hence, for $x \in D(\xi_i; 10\delta_1, e_i)$ there cannot be more than one point of $L(x; e_i)$ in $D(\xi_i; 10\delta_1, e_i) \cap \partial\Omega$. The existence of such a point follows from the fact that $D^*(\xi_i; 10\delta_1, 0.1, e_i) \subseteq \Omega$, but $D^*(\xi_i; 10\delta_1, 0.1, -e_i) \subseteq \Omega^c$.

Finally, suppose $u, v \in D(\xi_i; 10\delta_1, e_i)$, with $\|v - u\|_2 < 5\delta_1$. Then $v \in D(u; 2\|v - u\|_2, e_i)$, and since $D(\psi_i(u); 2\|v - u\|_2, e_i)$ contains points of the line $L(v; e_i)$ both in Ω and in Ω^c, it must also contain the point $\psi_i(v)$. Hence $\|\psi_i(v) - \psi_i(u)\|_2 \leq 4\|v - u\|_2$, and by the equivalence of norms there exists c_2 such that $\|\psi_i(v) - \psi_i(u)\| \leq c_2\|v - u\|$. □

5.3 The Brunn–Minkowski inequality

Minkowski addition $\oplus$ of sets $A \subseteq \mathbb{R}^d$, $B \subseteq \mathbb{R}^d$ is defined by

$$A \oplus B := \{x + y : x \in A, y \in B\}.$$

We give the following theorem from geometric measure theory without proof (see Burago and Zalgaller (1988) or Ledoux (1996)). We continue to denote Lebesgue measure by $|\cdot|$.

Theorem 5.11 (Brunn–Minkowski inequality) *Suppose A and B are non-empty compact sets in $\mathbb{R}^d$. Then*

$$|A \oplus B| \geq \left(|A|^{1/d} + |B|^{1/d}\right)^d. \tag{5.27}$$

The next result is an isodiametric inequality which says that out of all sets of a given diameter (in the chosen norm), the volume is maximized by taking a ball (in the same norm), and which is derived from the Brunn–Minkowski inequality. Recall from (1.2) that $\mathrm{diam}(A) := \sup\{\|x - y\| : x \in A, y \in A\}$, $A \subset \mathbb{R}^d$.

Corollary 5.12 (Bieberbach inequality) *Suppose A is a Borel set in $\mathbb{R}^d$ with* $\mathrm{diam}(A) = r > 0$. *Then* $|A| \leq 2^{-d}\theta r^d$.

Proof It suffices to consider the case where A is convex and compact. Let $-A := \{-x : x \in A\}$ and let $\frac{1}{2}A := \{\frac{1}{2}x : x \in A\}$. Set $B := \frac{1}{2}A \oplus \frac{1}{2}(-A)$. By the Brunn–Minkowski inequality, $|B| \geq |A|$.

If $x \in B$ then $x = \frac{1}{2}(y - z)$ for some $y, z \in A$, so $\|x\| \leq \mathrm{diam}(A)/2 = r/2$. Therefore $B \subseteq B(\mathbf{0}; r/2)$ so $|B| \leq 2^{-d}\theta r^d$, and the result follows. □

For $A \subseteq [0,1]^2$, let A_r denote the set $(A \oplus B(\mathbf{0};r)) \cap [0,1]^2$. Also let A^c denote the set $[0,1]\backslash A$. The following isoperimetric inequality provides a lower bound for the area of the perimeter region $A_r \setminus A$ in terms of the areas of A and A^c. It is a further consequence of the Brunn–Minkowski inequality, and will not be used until Chapter 12.

Proposition 5.13 (Isoperimetric inequality in $[0,1]^2$) *Suppose* $\|\cdot\|$ *is the* l_∞ *norm on* $\mathbb{R}^2$. *Suppose* A *is a compact subset of* $[0,1]^2$, *and* $r \in (0,1)$. *Then*

$$|A_r \backslash A| \geq \min(2r|A|^{1/2}, 2r|A^c|^{1/2} - 4r^2, r). \tag{5.28}$$

Proof For $x \in [0,1]$, set $S_x(A) := \{y \in [0,1] : (x,y) \in S\}$ (a vertical section through A), and let $|S_x|_1$ denote the one-dimensional Lebesgue measure of $S_x(A)$. Let A' be the set in $[0,1]^2$ obtained by 'pushing each vertical section of A down as far as possible towards the x-axis'; more precisely, let

$$A' := \bigcup_{x \in [0,1]: S_x(A) \neq \emptyset} (\{x\} \times [0, |S_x(A)|_1]).$$

The construction of A' is a form of Steiner symmetrization; see Hadwiger (1957) or Burago and Zalgaller (1988). Indeed, one recipe for constructing A' is to take the Steiner symmetrization about the x-axis of the union of A and its reflection in the x-axis, then take the intersection of the resulting set with $[0,1]^2$.

By Fubini's theorem, $|A'| = |A|$, and moreover (see Burago and Zalgaller (1988, Remark 9.3.2))

$$|A'_r| \leq |A_r|. \tag{5.29}$$

In fact Burago and Zalgaller are working in $\mathbb{R}^2$, but the inequality also holds in the square.

Let A'' be the set in $[0,1]^2$, obtained by 'pushing each horizontal section of A' sideways as far as possible towards the y-axis', in an analogous manner to the construction of A' from A. Then $|A''| = |A'| = |A|$, and $|A''_r| \leq |A'_r| \leq |A_r|$. Moreover, A'' is a *down-set* in $[0,1]^2$, that is, A'' has the property that if $(x,y) \in A''$, then $[0,x] \times [0,y] \subseteq A''$. Hence, without loss of generality, we can (and do) assume from now on that A itself is a down-set. We consider four different cases.

First, suppose $(1-r, 0) \in A$ and $(0, 1-r) \notin A$. Then $S_x(A_r \backslash A)$ contains an interval of length at least r, for each $x \in [0,1]$, so that by Fubini's theorem, $|A_r \backslash A| \geq r$.

Second, suppose $(1-r,0) \notin A$ and $(0,1-r) \in A$. Clearly in this case, $|A_r \backslash A| \geq r$ by an analogous argument using horizontal sections.

Third, suppose $(1-r,0) \notin A$ and $(0,1-r) \notin A$. In this case, $A \subseteq [0,1-r]^2$ so that $A \oplus [0,r]^2 \subset [0,1]^2$, and therefore by the Brunn–Minkowski inequality,

$$|A_r \setminus A| \geq |(A \oplus [0,r]^2) \setminus A| \geq 2|A|^{1/2} r.$$

Fourth, suppose $(1-r,0) \in A$ and $(0,1-r) \in A$. In this case, set $B := [0,1]^2 \backslash A_r$. Then $(B_r \setminus B) \subseteq (A_r \setminus A)$ and $B \subseteq [r,1]^2$ so that $B \oplus [-r,0]^2 \subseteq [0,1]^2$. Hence, by the Brunn–Minkowski inequality,

$$\begin{aligned}|A_r \setminus A| &\geq |B_r \setminus B| \geq |(B \oplus [-r,0]^2) \setminus B| \\ &\geq 2r|B|^{1/2} = 2r\sqrt{|A^c| - |A_r \setminus A|}.\end{aligned} \tag{5.30}$$

If $|A_r \setminus A| \geq 2r|A^c|^{1/2}$, then (5.28) is immediate; if not, then by (5.30),

$$\begin{aligned}|A_r \setminus A| &\geq 2r|A^c|^{1/2}\sqrt{1 - 2r|A^c|^{-1/2}} \\ &\geq 2r|A^c|^{1/2}(1 - 2r|A^c|^{-1/2}) = 2r|A^c|^{1/2} - 4r^2\end{aligned}$$

so (5.28) holds in this case, too. □

5.4 Expanding sets in the orthant

This section contains further lower bounds on the volume of the r-neighbourhood of a set A in $\mathbb{R}^d$. We are sometimes interested in r-neighbourhoods in the unit cube (e.g. when considering points uniformly distributed on that cube), rather than in $\mathbb{R}^d$; in the case where A and r are small, A can be viewed as effectively a subset of the orthant $[0,\infty)^d$. The results of this section that will be used subsequently are a lower bound for the volume of the 1-neighbourhood of A in the orthant, when A is of moderate size (Proposition 5.15), and a lower bound for the volume of the 1-neighbourhood of a two-point set in the orthant (Proposition 5.16). Before proving these we give a lemma that will be used in the proof of Proposition 5.16.

For $A \subseteq \mathbb{R}^d$ and $v \in \mathbb{R}^d$ write $A \oplus v$ for $A \oplus \{v\}$. Also, set $D_2(A) := \sup_{x,y \in A} \|x-y\|_2$, the l_2 diameter of A, and define $D_\infty(A)$ likewise.

Lemma 5.14 *Suppose $d \geq 2$. For any convex $A \subseteq \mathbb{R}^d$ and any vector $v \in \mathbb{R}^d$,*

$$|(A \oplus v) \setminus A| \geq \frac{1}{2}|A| \min(1, \|v\|_2 / D_2(A)).$$

Proof Without loss of generality assume v is of the form he_d with $h > 0$ and e_d denoting the dth coordinate vector $(0,0,\ldots,0,1)$. For $x \in \mathbb{R}^{d-1}$ set $A^x = \{t \in \mathbb{R} : (x,t) \in A\}$. Let $|\cdot|_1$ denote one-dimensional Lebesgue measure. Then

$$|(A \oplus he_d) \setminus A| = \int_{\mathbb{R}^{d-1}} |(A^x \oplus h) \setminus A^x|_1 \, dx.$$

By convexity, for each $x \in \mathbb{R}^{d-1}$ the set A^x is an interval so that

$$|(A^x \oplus h) \setminus A^x|_1 = \min(h, |A^x|_1).$$

Let $\pi : \mathbb{R}^d \to \mathbb{R}^{d-1}$ denote projection onto the first $d-1$ coordinates. Let $A_{>h}$ (respectively, $A_{<h}$) denote the set of $x \in A$ such that the length of $A^{\pi(x)}$ is greater than h (respectively, at most h). Then since A^x has length at most $D_2(A)$ for all x,

$$\begin{aligned}|(A \oplus he_d)\backslash A| &= \int_{\pi(A_{<h})} |A^x|_1 \mathrm{d}x + \int_{\pi(A_{>h})} h\mathrm{d}x \\ &\geq \int_{\pi(A_{<h})} |A^x|_1 \mathrm{d}x + (h/D_2(A)) \int_{\pi(A_{>h})} |A^x|_1 \mathrm{d}x \\ &= |A_{<h}| + (h/D_2(A))|A_{>h}|.\end{aligned}$$

If $|A_{<h}| \geq |A|/2$ then $|(A + he_d) \setminus A| \geq |A|/2$, while if $|A_{<h}| \leq |A|/2$ then $|A_{>h}| \geq |A|/2$ so that $|(A + he_d)\backslash A| \geq (h/D_2(A))|A|/2$. □

The remaining results in this section are concerned with subsets of the orthant $\mathcal{O}^d := [0, \infty)^d$. For $A \subseteq \mathcal{O}^d$, let A_1 denote the 1-neighbourhood of A in $\mathcal{O}^d$, that is, set

$$A_1 := (A \oplus B(\mathbf{0}; 1)) \cap \mathcal{O}^d.$$

In this section, for $x \in \mathbb{R}^d$, we write x^i for the ith coordinate of x.

Proposition 5.15 *Suppose $\|\cdot\|$ is an l_p norm on $\mathbb{R}^d$ with $1 \leq p \leq \infty$ and $d \geq 2$. Let $\varepsilon > 0$. Then there exists $\eta_1 = \eta_1(\varepsilon) > 0$, such that if $A \subseteq \mathcal{O}^d$ is compact with l_∞ diameter $D_\infty(A) \geq \varepsilon$ and $x \in A$ with $\sum_{i=1}^d x^i \leq \sum_{i=1}^d y^i$ for all $y \in A$, then*

$$|A_1| \geq |A| + |\{x\}_1| + \eta_1.$$

Proof Let A and x be as described above. Then $\|y - x\|_\infty \geq \varepsilon/2$ for some $y \in A$, so that

$$\max_{i\in\{1,2,\ldots,d\}} \left(\sup_{y\in A}(y^i - x^i) \right) \geq \varepsilon/(2d). \tag{5.31}$$

Assume without loss of generality that this maximum is achieved at $i = 1$.

First suppose $1 \leq p < \infty$. Let $e := (d^{-1/p}, \ldots, d^{-1/p})$, the unit vector in the direction of $(1, 1, \ldots, 1)$. Then $A \oplus e \subseteq A_1$. Also, we assert that

$$|(A \oplus e) \cap B(x; 1)| = 0. \tag{5.32}$$

Indeed, since $\max\{\|u\|_1 : u \in B(\mathbf{0}; 1)\}$ is achieved at $u = e$, if $y \in B(x; 1)$ then $\sum_{i=1}^d (y^i - x^i) \leq \|e\|_1$, while if $y \in A \oplus e$ then $\sum_{i=1}^d (y^i - x^i) \geq \|e\|_1$, since $\sum_{i=1}^d (y^i - e^i) \geq \sum_{i=1}^d x^i$ by assumption. Hence $(A \oplus e) \cap B(x; 1)$ is contained in a $(d - 1)$-dimensional hyperplane, justifying assertion (5.32).

Let T_0 be the slice from near the right-hand side of $B(\mathbf{0}; 1) \cap \mathcal{O}^d$ given by

$$T_0 := B(\mathbf{0}; 1) \cap \left((\max(1 - \varepsilon/(2d), d^{-1/p}), 1] \times \mathcal{O}^{d-1} \right).$$

Set $\eta_1 = |T_0| > 0$. Let z be a *right-most point* of A, that is, take $z \in A$ such that $y^1 \leq z^1$ for all $y \in A$. By the assumption following (5.31), $z^1 \geq x^1 + \varepsilon/(2d)$.

Let $T := T_0 \oplus z$. Then $T \subseteq \{z\}_1 \subseteq A_1$. If $u \in T$ then $u^1 > z^1 + d^{-1/p}$, so $u \notin A \oplus e$, and $u^1 > x^1 + 1$, so that $u \notin B(x;1)$. Combined with (5.32), this implies that

$$\begin{aligned}|A_1| &\geq |A \oplus e| + |\{x\}_1| + |T| \\ &= |A| + |\{x\}_1| + \eta_1.\end{aligned}$$

Now suppose that $p = \infty$ (in which case the argument above breaks down). Set $\delta := \min(\varepsilon/(2d), 1)$. It suffices to find a partition $\{R^1, \ldots, R^d\}$ of A and a collection of disjoint sets $\{\{x\}_1, T, T^1, T^2, \ldots, T^d\}$ in A_1 such that T has volume at least δ whilst each T^i is a translate of R^i so that $|T^i| = |R^i|$.

Let $\mathcal{O}_x$ denote the set of $y \in \mathcal{O}^d$ such that $\sum_{i=1}^d y^i \geq \sum_{i=1}^d x^i$. For $y \in \mathcal{O}_x$ we have $y^j \geq x^j$ for some $j \leq d$. For $1 \leq i \leq d$, let S^i be the set of $y \in \mathcal{O}_x$ such that i is the first such j, that is,

$$S^i := \{y \in \mathcal{O}_x : y^i \geq x^i \ \text{ and } \ y^j < x^j \ \text{ for } \ j < i\}.$$

The sets S^i are disjoint. Let $R^i = S^i \cap A$. The sets R^i form a partition of A.

Let $T^i := R^i \oplus e_i$, with e_i denoting the unit vector in the direction of the ith coordinate. Then $T^1, \ldots, T^d$ are disjoint subsets of $\mathcal{O}^d$ since $T^i \subseteq S^i$ for each i. Also, each T^i is disjoint from the interior of $\{x\}_1$ since $y^i \geq x^i + 1$ for $y \in T^i$. For each i, $T^i \subseteq A_1$ and T^i is a translate of R^i so that $|T^i| = |R^i|$.

It remains to find a set T (see Fig. 5.2). Let z be a right-most point of A (as defined above). By the assumption following (5.31), $z^1 \geq x^1 + \delta$. Let $W := \{y \in T^1 : y^1 > z^1 + 1 - \delta\}$. Let H be the set:

$$\begin{aligned}H := (z^1 + 1 - \delta, z^1 + 1] \times [z^2, z^2 + 1], \times [z^3, z^3 + 1] \times \cdots \\ \times [z^{d-1}, z^{d-1} + 1] \times \{z^d\}.\end{aligned}$$

Let

$$T := ((H \cup W) \oplus \{\xi e_d : 0 \leq \xi \leq 1\}) \setminus (H \cup W).$$

Then $T \subset A_1$, since $\lambda e_1 + \xi e_d \in B_1$ for all $\lambda, \xi \in [0,1]$. Also, $|T| \geq \delta$ by Fubini's theorem, and T is disjoint from each of $B(x;1), T^1, T^2, \ldots, T^d$. The proof is complete. □

Proposition 5.16 *Suppose* $\|\cdot\|$ *is an* l_p *norm with* $1 < p \leq \infty$. *There exists* $\eta_2 > 0$ *such that if* $u, v \in \mathcal{O}^d$ *with* $\|u - v\| \leq 3$ *and* $\sum_{i=1}^d v^i \geq \sum_{i=1}^d u^i$, *then*

$$|\{u, v\}_1| \geq |\{u\}_1| + \eta_2 \|v - u\|.$$

Proof The result is clearly true for $d = 1$, so assume from now on that $d \geq 2$. Set $B := B(\mathbf{0}; 1)$. Note that $(d^{-1/p}, d^{-1/p}, \ldots, d^{-1/p})$ lies on the boundary of B and its coordinates sum to $d^{1-1/p}$, and $\{x : \sum_{i=1}^d x^i = d^{1-1/p}\}$ is a supporting

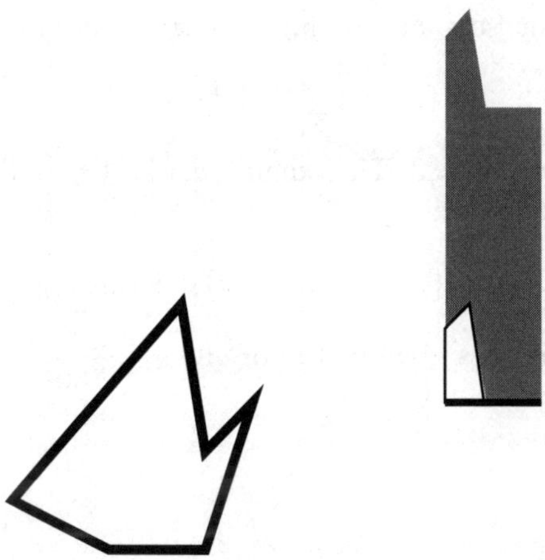

FIG. 5.2. The bold polygon is the boundary of A, while the bold horizontal line (of length δ) is H and the shaded region is T. The smaller polygon is the boundary of W.

hyperplane for B which touches B only at this point (this is why we assume $p > 1$ here). Hence, there exists $\varepsilon > 0$ such that

$$\text{if } x \in B \cap \mathcal{O}^d \text{ with } x^1 < \varepsilon \text{ then } \sum_{i=1}^{d} x^i < d^{1-1/p} - \varepsilon. \tag{5.33}$$

Set

$$A := \left\{ x \in B : \sum_{i=1}^{d} x^i \geq d^{1-1/p} - \varepsilon \right\}.$$

By Lemma 5.14 and the equivalence of all norms on $\mathbb{R}^d$, there exists $\eta \in (0, \varepsilon)$ such that $|(A \oplus y) \backslash A| \geq \eta ||y||$, for any vector y with $||y|| \leq \eta$. If also $\sum_{i=1}^{d} y^i \geq 0$, then $(A \oplus y) \cap (B \backslash A) = \emptyset$, so that

$$|(A \oplus y) \backslash B| \geq \eta ||y||, \quad \text{if } ||y|| \leq \eta, \quad \sum_{i=1}^{d} y^i \geq 0.$$

By (5.33), $A \oplus y$ is contained in $\mathcal{O}^d$ whenever $||y|| \leq \eta$.

Let $u, v \in \mathcal{O}^d$ with $\sum_{i=1}^{d} u^i \leq \sum_{i=1}^{d} v^i$. First suppose $||v - u|| \leq \eta$. Set $y = v - u$. By the above,

$$\begin{aligned}\{u,v\}_1 &= (B(u;1) \cup B(y+u;1)) \cap \mathcal{O}^d \\ &\supseteq (B(u;1) \cap \mathcal{O}^d) \cup (((A \oplus y) \backslash B) \oplus u);\end{aligned}$$

this union is of disjoint sets, and taking Lebesgue measures, we have

$$|\{u,v\}_1| \geq |\{u\}_1| + \eta \|y\|.$$

Next suppose $\eta \leq \|v-u\| \leq 3$. Then $\mathrm{diam}(\{u,v\}) \geq \eta$ and by Proposition 5.15, there exists $\eta_1 > 0$ such that

$$|\{u,v\}_1| \geq |\{u\}_1| + \eta_1 \geq |\{u\}_1| + (\eta_1/3)\|v-u\|.$$

Combining these estimates gives us the result. □

6

MAXIMUM DEGREE, CLIQUES, AND COLOURINGS

Given a graph G with n vertices and with the degrees of the vertices denoted $d_1, \ldots, d_n$, its maximum vertex degree is $\max(d_1, \ldots, d_n)$; in this chapter we study this for random geometric graphs. Given a sequence $(r_n)_{n \geq 1}$, let Δ_n denote the maximum vertex degree of $G(\mathcal{X}_n; r_n)$, and let Δ'_n denote the maximum vertex degree of $G(\mathcal{P}_n; r_n)$, with $\mathcal{X}_n$ and $\mathcal{P}_n$ defined in Sections 1.5 and 1.7, respectively.

Sometimes it is convenient to investigate the maximum degree via the associated threshold distance. Given a finite set $\mathcal{X} \subset \mathbb{R}^d$, and given $k \in \mathbb{N}$, the *smallest k-nearest-neighbour link* of the set $\mathcal{X}$ is the smallest value of r such that the maximum degree of the graph $G(\mathcal{X}; r)$ is at least k. Let $S_k(\mathcal{X})$ denote the smallest k-nearest-neighbour link of $\mathcal{X}$.

A *complete* graph on k vertices is one with $\binom{k}{2}$ edges (i.e. one with each pair of vertices connected by an edge). A *clique* of a graph is a maximal complete subgraph. The *clique number* of a finite graph G, which we shall denote by $\mathcal{C}(G)$ or simply by $\mathcal{C}$, is the order of its largest clique. Given a finite set $\mathcal{X} \subset \mathbb{R}^d$, and given $k > 0$, let $\rho(\mathcal{X}; \mathcal{C} \geq k)$ denote the threshold value of r above which the geometric graph $G(\mathcal{X}; r)$ has clique number at least k. Given a sequence $(r_n)_{n \geq 1}$, set $\mathcal{C}_n := \mathcal{C}(G(\mathcal{X}_n; r_n))$ and $\mathcal{C}'_n := \mathcal{C}(G(\mathcal{P}_n; r_n))$.

In the case where the norm of choice is the l_∞ norm, the clique number of $G(\mathcal{X}; r)$ is the maximal number of points of $\mathcal{X}$ in any 'window' of side r, that is, the maximal number of points in any rectilinear hypercube of side r. This is a form of the multidimensional *scan statistic*, which is of considerable statistical interest. For a comprehensive reference on theory and applications of scan statistics, see Glaz *et al.* (2001); also Glaz and Balakrishnan (1999), Cressie (1991), and references therein.

The *chromatic number* of a graph is the smallest number of colours with which one can colour the vertices in such a way that no two adjacent vertices have the same colour. Colourings of a geometric graph are a natural object of study in connection with *frequency assignment* problem: how does one best assign frequencies to a collection of radio or cellular telephone transmitters located in space, so as to avoid interference between sites less than some specified distance apart, and how many wavebands are required to do this? For an extensive discussion, see Hale (1980) and Leese and Hurley (2002).

As we shall see, the maximum degree, the clique number, and the chromatic number are closely interrelated; we treat them all in this chapter. First we investigate 'focusing' phenomena whereby, under certain limiting regimes for r_n, the distribution of Δ_n or $\mathcal{C}_n$ is asymptotically concentrated on at most two values.

Thereafter, we consider strong laws of large numbers for Δ_n and $\mathcal{C}_n$, and for the chromatic number.

As in previous chapters, the volume of the unit ball is denoted θ in this chapter.

6.1 Focusing

The results in this section show that for $(r_n)_{n\geq 1}$ in subconnective limiting regimes, there is a sequence $(k_n)_{n\geq 1}$ with $P[\Delta_n \in \{k_n - 1, k_n\}] \to 1$; the distribution of Δ_n is focused almost entirely on just two values. At least in the case of sparse regimes where the maximum vertex degree remains bounded in probability, similar results also hold for the clique numbers $\mathcal{C}_n$ and $\mathcal{C}'_n$, and in fact we start with these.

Theorem 6.1 *Let $k \in \mathbb{N}$ with $k \geq 2$, and let Γ_k denote a complete graph on k vertices. Let $\lambda > 0$. If $n^{k/(k-1)} r_n^d \to \lambda^{1/(k-1)}$ as $n \to \infty$, then*

$$\lim_{n\to\infty} P[\mathcal{C}_n = k-1] = 1 - \lim_{n\to\infty} P[\mathcal{C}_n = k]$$
$$= \exp\left(-\frac{\lambda}{k!} \int_{\mathbb{R}^d} f(x)^k \mathrm{d}x \right.$$
$$\left. \times \int_{(\mathbb{R}^d)^k} h_{\Gamma_k}(\{\mathbf{0}, x_1, \ldots, x_{k-1}\}) \mathrm{d}(x_1, \ldots, x_{k-1}) \right), \quad (6.1)$$

where, as in (3.1), *$h_{\Gamma_k}(\mathcal{Y})$ takes the value 1 if $G(\mathcal{Y}; 1) \cong \Gamma_k$, and zero otherwise.*

If $n^{k/(k-1)} r_n^d \to \infty$, then $P[\mathcal{C}_n \geq k] \to 1$ as $n \to \infty$. If $n^{k/(k-1)} r_n^d \to 0$, then $P[\mathcal{C}_n < k] \to 1$ as $n \to \infty$.

Corollary 6.2 *Suppose $k \in \mathbb{N}$ with $k \geq 2$. If $n^{k/(k-1)} r_n^d \to \infty$ and $n^{(k+1)/k} r_n^d \to 0$ as $n \to \infty$, then $P[\mathcal{C}_n = k] \to 1$ as $n \to \infty$. If $n^2 r_n^d \to 0$ then $P[\mathcal{C}_n = 1] \to 1$ as $n \to \infty$.*

Proof of Theorem 6.1 Following the notation from Chapter 3, let $G_n = G_n(\Gamma_k)$ be the number of induced subgraphs of $G(\mathcal{X}_n; r_n)$ isomorphic to Γ_k. Then the events $\{\mathcal{C}_n < k\}$ and $\{G_n = 0\}$ are identical. By Proposition 3.1,

$$\lim_{n\to\infty} n^{-k} r_n^{-d(k-1)} E[G_n] = \frac{1}{k!} \int_{\mathbb{R}^d} f(x)^k \mathrm{d}x$$
$$\times \int_{(\mathbb{R}^d)^k} h_{\Gamma_k}(\{\mathbf{0}, x_1, \ldots, x_{k-1}\}) \mathrm{d}(x_1, \ldots, x_{k-1}). \quad (6.2)$$

Suppose $n r_n^d \to 0$. If Z_n is Poisson with parameter $E[G_n]$ then, by Theorem 3.4,

$$\lim_{n\to\infty} P[\mathcal{C}_n < k] = \lim_{n\to\infty} P[G_n = 0] = \lim_{n\to\infty} P[Z_n = 0], \quad (6.3)$$

whenever the third limit exists.

First suppose that, for some $\lambda \geq 0$, $\lim_{n\to\infty}(n^{k/(k-1)}r_n^d) = \lambda^{1/(k-1)}$, so that $\lim_{n\to\infty}(n^k r_n^{d(k-1)}) = \lambda$. Then by (6.2), $P[Z_n = 0]$ tends to the right-hand side of (6.1), and hence by (6.3), so does $P[\mathcal{C}_n < k]$.

Next suppose $n^{k/(k-1)}r_n^d \to \infty$ and $nr_n^d \to 0$. Then $P[Z_n = 0] \to 0$, so $P[\mathcal{C}_n < k] \to 0$ by (6.3).

Finally, suppose $n^{k/(k-1)}r_n^d \to \infty$ without assuming $nr_n^d \to 0$. Then we can interpolate $s_n \leq r_n$ with $n^{k/(k-1)}s_n^d \to \infty$ and $ns_n^d \to 0$. The clique number of $G(\mathcal{X}_n; r_n)$ is at least as big as that of $G(\mathcal{X}_n; s_n)$, so $P[\mathcal{C}_n < k] \to 0$. The result follows. □

The analogous result for the maximum degree goes as follows.

Theorem 6.3 *Let $k \in \mathbb{N}$. Let $\lambda \geq 0$. If $n^{(k+1)/k}r_n^d \to \lambda^{1/k}$ as $n \to \infty$, then*

$$\lim_{n\to\infty} P[\Delta_n = k-1] = 1 - \lim_{n\to\infty} P[\Delta_n = k]$$
$$= \exp\left(-\frac{\lambda}{(k+1)!}\int_{\mathbb{R}^d} f(x)^{k+1}\mathrm{d}x \times \int_{(\mathbb{R}^d)^k} h^*(\{\mathbf{0}, x_1, \ldots, x_k\})\mathrm{d}(x_1, \ldots, x_k)\right), \quad (6.4)$$

where $h^*(\mathcal{Y})$ takes the value 1 if $G(\mathcal{Y}; 1)$ has at least one vertex of degree k, and zero otherwise.

If $n^{(k+1)/k}r_n^d \to \infty$, then $P[\Delta_n \geq k] \to 1$ as $n \to \infty$. If $n^{(k+1)/k}r_n^d \to 0$, then $P[\Delta_n < k] \to 1$ as $n \to \infty$.

Corollary 6.4 *Suppose $k \in \mathbb{N}$. If $n^{(k+1)/k}r_n^d \to \infty$ and $n^{(k+2)/(k+1)}r_n^d \to 0$ as $n \to \infty$, then $P[\Delta_n = k] \to 1$ as $n \to \infty$. If $n^2 r_n^d \to 0$ as $n \to \infty$ then $P[\Delta_n = 0] \to 1$ as $n \to \infty$.*

Proof of Theorem 6.3 Let $\Gamma'_1, \ldots, \Gamma'_m$ be a maximal collection of non-isomorphic feasible graphs of order $k+1$ all having maximum degree k. Let $G_n(\Gamma'_i)$ be the number of induced Γ'_i-subgraphs of $G(\mathcal{X}_n; r_n)$, as described at the start of Chapter 3, and let the integral $\mu_{\Gamma'_i}$ be as defined at (3.2). Then $\Delta_n < k$ if and only if $G_n(\Gamma'_i) = 0$ for $1 \leq i \leq m$.

Suppose $n^{k+1}r_n^{dk} \to \lambda \in (0, \infty)$ as $n \to \infty$. Then by Corollary 3.6, $P[\Delta_n < k]$ tends to $\exp(-\lambda \sum_{j=1}^m \mu_{\Gamma'_j})$, which is equal to the right-hand side of (6.4).

Also, if $n^{k+1}r_n^{dk} \to 0$, then by Proposition 3.1, for any connected graph Γ on $k+1$ vertices, $E[G_n(\Gamma)]$ tends to zero. Hence, $P[\Delta_n \geq k+1] \to 0$.

Next suppose $n^{k+1}r_n^{dk} \to \infty$. Then $E[G_n(\Gamma'_1)] \to \infty$. If also $nr_n^d \to 0$, then by Theorem 3.4, $P[\Delta_n < k]$ tends to zero.

Finally, suppose $n^{(k+1)/k}r_n^d \to \infty$ without assuming $nr_n^d \to 0$. Then we can interpolate $s_n \leq r_n$ with $n^{(k+1)/k}s_n^d \to \infty$ and $ns_n^d \to 0$. The maximum degree of $G(\mathcal{X}_n; r_n)$ is at least as big as that of $G(\mathcal{X}_n; s_n)$, so by the previous paragraph $P[\Delta_n < k] \to 0$. The result follows. □

It is straightforward to deduce from Theorems 6.1 and 6.3 the corresponding result for the clique number C'_n on a Poisson sample, and the maximum degree Δ'_n on a Poisson sample.

Corollary 6.5 *The statements of Theorems* 6.1 *and* 6.3, *and of Corollaries* 6.2 *and* 6.4, *remain true with C_n replaced by C'_n and Δ_n replaced by Δ'_n.*

Proof Let $\tilde{C}_{m,n}$ denote the clique number of $G(\mathcal{X}_m; r_n)$. For any sequence of integers $(m_n)_{n\geq 1}$ with $\lim_{n\to\infty}(m_n/n) = 1$, the conclusion of Theorem 6.1 remains true with C_n replaced by $\tilde{C}_{m_n,n}$, simply because $m_n^{k/(k-1)} r_n^d$ converges to the same limit as $n^{k/(k-1)} r_n^d$.

If N_n is Poisson with mean n, then $P[|N_n - n| \geq n^{3/4}]$ tends to zero, so that

$$P[C'_n = k] = \sum_{n-n^{3/4}\leq m\leq n+n^{3/4}} P[N_n = m] P[\tilde{C}_{m,n} = k] + o(1)$$

and by the preceding argument, this converges to the same limit as $P[C_n = k]$.

The argument for Δ'_n is similar. □

Now consider a limiting regime with nr_n^d bounded away from zero and with $nr_n^d/\log n$ tending to zero, which includes the thermodynamic limiting regime and goes (almost) all the way up to the connectivity regime. We restrict attention to the case where $f = f_{\rm U}$, defined at (1.1), that is, where the points X_i are uniformly distributed on the unit cube. The main focusing result goes as follows.

Theorem 6.6 *Suppose that $d \geq 2$, and that $f = f_{\rm U}$. Set $\mu_n := n\theta r_n^d$ and suppose that $\inf\{\mu_n : n \geq 1\} > 0$, and that $\lim_{n\to\infty}(\mu_n^{1+\varepsilon}/\log n) = 0$ for some $\varepsilon > 0$. Then there exists a sequence $(j(n))_{n\geq 1}$ such that if we set $\zeta_n := P[\mathrm{Po}(\mu_n) \geq j(n)]$, then as $n \to \infty$,*

$$P[\Delta'_n = j(n) - 1] - \exp(-n\zeta_n) \to 0$$

and

$$P[\Delta'_n = j(n)] - (1 - \exp(-n\zeta_n)) \to 0.$$

Moreover,

$$P[\Delta_n = j(n) - 1] - \exp(-n\zeta_n) \to 0$$

and

$$P[\Delta_n = j(n)] - (1 - \exp(-n\zeta_n)) \to 0.$$

The value of $j(n)$ will be given in course of the proof of this result.

Before proving Theorem 6.6, we give a general estimate which will be used again later on. Let $W_{k,n}(r)$ (respectively, $W'_{k,\lambda}(r)$) be the number of vertices of degree k in $G(\mathcal{X}_n; r)$ (respectively, $G(\mathcal{P}_\lambda; r)$). For $A \subseteq \mathbb{N} \cup \{0\}$, set $W'_{A,\lambda}(r) := \sum_{k\in A} W'_{k,\lambda}(r)$, the number of vertices of $G(\mathcal{P}_\lambda; r)$ whose degree lies in the set A.

Theorem 6.7 *Suppose f is almost everywhere continuous. Let $A \subseteq \mathbb{N} \cup \{0\}$, $r > 0$, and $\lambda > 0$. For $x \in \mathbb{R}^d$, set $B_x := B(x;r)$ and $\mathcal{P}_\lambda^x := \mathcal{P}_\lambda \cup \{x\}$. Define the integrals I_i $(i = 1, 2)$ by*

$$I_1 := \lambda^2 \int_{\mathbb{R}^d} \mathrm{d}F(x) \int_{B(x;3r)} \mathrm{d}F(y) P[\mathcal{P}_\lambda(B_x) \in A] P[\mathcal{P}_\lambda(B_y) \in A], \quad (6.5)$$

$$I_2 := \lambda^2 \int_{\mathbb{R}^d} \mathrm{d}F(x) \int_{B(x;3r)} \mathrm{d}F(y) P[\{\mathcal{P}_\lambda^y(B_x) \in A\} \cap \{\mathcal{P}_\lambda^x(B_y) \in A\}]. \quad (6.6)$$

Then

$$d_{\mathrm{TV}}(W'_{A,\lambda}, \mathrm{Po}(EW'_{A,\lambda})) \leq \min\left(3, \frac{1}{EW'_{A,\lambda}}\right)(I_1 + I_2).$$

Proof Given $m \in \mathbb{N}$, partition $\mathbb{R}^d$ into disjoint hypercubes of side m^{-1}, with corners at the lattice points $\{m^{-1}z : z \in \mathbb{Z}^d\}$. Label these hypercubes $H_{m,1}, H_{m,2}, H_{m,3} \ldots$, and their centres as $a_{m,1}, a_{m,2}, a_{m,3} \ldots$, respectively. For each m, i, define the indicator variable $\xi_{m,i}$ by

$$\xi_{m,i} := \mathbf{1}_{\{\mathcal{P}_\lambda(H_{m,i})=1\} \cap \{\mathcal{P}_\lambda(B_{a_{m,i}} \setminus H_{m,i}) \in A\}}.$$

Set $p_{m,i} := E[\xi_{m,i}]$ and $p_{m,i,j} := E[\xi_{m,i}\xi_{m,j}]$.

For $n \in \mathbb{N}$, let $Q_n := [-n, n]^d$ and $\mathcal{I}_{m,n} := \{i \in \mathbb{N} : H_{m,i} \subseteq Q_n\}$. Define an adjacency relation $\sim_m$ on $\mathbb{N}$ by putting $i \sim_m j$ if and only if $0 < \|a_{m,i} - a_{m,j}\| \leq 3r$, and define the corresponding adjacency neighbourhoods $\mathcal{N}_{m,i}, i \in \mathbb{N}$ by $\mathcal{N}_{m,i} := \{j \in \mathbb{N} : \|a_{m,j} - a_{m,i}\| \leq 3r\}$. Also, for $i \in \mathcal{I}_{m,n}$, set $\mathcal{N}_{m,n,i} := \mathcal{N}_{m,i} \cap \mathcal{I}_{m,n}$. By the spatial independence properties of the Poisson process, this adjacency relation makes $(\mathcal{I}_{m,n}, \sim_m)$ into a dependency graph for the variables $\xi_{m,i}, i \in \mathcal{I}_{m,n}$.

Define $\tilde{W}_{m,n} := \sum_{i \in \mathcal{I}_{m,n}} \xi_{m,i}$; then $W'_{A,\lambda} = \lim_{n\to\infty} \lim_{m\to\infty} \tilde{W}_{m,n}$. By Theorem 2.1,

$$d_{\mathrm{TV}}(\tilde{W}_{m,n}, \mathrm{Po}(E\tilde{W}_{m,n})) \leq \min\left(3, \frac{1}{E\tilde{W}_{m,n}}\right)(b_1(m,n) + b_2(m,n)), \quad (6.7)$$

where we set

$$b_1(m,n) := \sum_{i \in \mathcal{I}_{m,n}} \sum_{j \in \mathcal{N}_{m,n,i}} p_{m,i}p_{m,j}, \qquad b_2(m,n) := \sum_{i \in \mathcal{I}_{m,n}} \sum_{j \in \mathcal{N}_{m,n,i} \setminus \{i\}} p_{m,i,j}.$$

Define the function $(w_m(x), x \in \mathbb{R}^d)$ by $w_m(x) := m^d p_{m,i}$ for $x \in H_{m,i}$. Then $\sum_{i \in \mathcal{I}_{m,n}} p_{m,i} = \int_{Q_n} w_m(x)\mathrm{d}x$. If f is continuous at x, then $\lim_{m\to\infty}(w_m(x)) = \lambda f(x) P[\mathcal{P}_\lambda(B_x) \in A]$. Also, for $x \in H_{m,i}$, we have

$$w_m(x) \leq m^d P[\mathcal{P}_\lambda(H_{m,i}) = 1] \leq \lambda f_{\max}.$$

Therefore, by the dominated convergence theorem for integrals,

$$\lim_{m\to\infty} E\tilde{W}_{m,n} = \lambda \int_{Q_n} P[\mathcal{P}_\lambda(B_x) \in A] f(x) \mathrm{d}x$$

and

$$\lim_{n\to\infty} \lim_{m\to\infty} E\tilde{W}_{m,n} = \lambda \int_{\mathbb{R}^d} P[\mathcal{P}_\lambda(B_x) \in A] f(x) \mathrm{d}x = E[W'_{A,\lambda,n}], \tag{6.8}$$

where the second equality comes from Palm theory (Theorem 1.6).

For $x, y \in \mathbb{R}^d$, define $u_m(x,y)$ and $v_m(x,y)$, if $x \in H_{m,i}$ and $y \in H_{m,j}$, by

$$u_m(x,y) := m^{2d} p_{m,i} p_{m,j} \mathbf{1}_{j \in \mathcal{N}_{m,i}}, \qquad v_m(x,y) := m^{2d} p_{m,i,j} \mathbf{1}_{j \in \mathcal{N}_{m,i} \setminus \{i\}}.$$

Then $b_1(m,n) = \int_{Q_n} \int_{Q_n} u_m(x,y) \mathrm{d}y \mathrm{d}x$ and $b_2(m,n) = \int_{Q_n} \int_{Q_n} v_m(x,y) \mathrm{d}y \mathrm{d}x$.

If x, y are distinct continuity points of f with $\|x-y\| \neq r$ and $\|x-y\| \neq 3r$, then

$$\lim_{m\to\infty} u_m(x,y) = \lambda^2 f(x) f(y) P[\mathcal{P}_\lambda(B_x) \in A] P[\mathcal{P}_\lambda(B_y) \in A] \mathbf{1}_{B(x;3r)}(y),$$
$$\lim_{m\to\infty} v_m(x,y) = \lambda^2 f(x) f(y) P[\{\mathcal{P}^y_\lambda(B_x) \in A\} \cap \{\mathcal{P}^x_\lambda(B_y) \in A\}] \mathbf{1}_{B(x;3r)}(y).$$

Hence, by limiting arguments similar to the one which gave us (6.8),

$$\lim_{n\to\infty} \lim_{m\to\infty} b_1(m,n) = I_1, \qquad \lim_{n\to\infty} \lim_{m\to\infty} b_2(m,n) = I_2.$$

Taking $m \to \infty$ and then $n \to \infty$ in (6.7) gives us the result. □

Proposition 6.8 *Suppose $d \geq 2$ and $f = f_{\mathrm{U}}$. Set $\mu_n := n\theta r_n^d$ and suppose that $\lim_{n\to\infty}(r_n) = 0$ and $\inf\{\mu_n : n \geq 1\} > 0$. Suppose $(k_n)_{n\geq 1}$ is an $\mathbb{N}$-valued sequence such that, for some $\varepsilon > 0$,*

$$\lim_{n\to\infty} (k_n / \mu_n^{1+\varepsilon}) = \infty, \tag{6.9}$$

and set $\zeta_n := \sum_{j=k_n}^{\infty} \mathrm{e}^{-\mu_n} \mu_n^j / j!$. Then

$$\lim_{n\to\infty} (P[\Delta'_n < k_n] - \exp(-n\zeta_n)) = 0. \tag{6.10}$$

Proof Let W'_n be the number of points of $\mathcal{P}_n$ with degree at least k_n in $G(\mathcal{P}_n; r_n)$. By Palm theory (Theorem 1.6), $\lim_{n\to\infty}(E[W'_n]/(n\zeta_n)) = 1$. Hence, by Theorem 6.7, for n large,

$$|P[W'_n = 0] - \exp(-EW'_n)| \leq \min\left(3, \frac{2}{n\zeta_n}\right)(I_1(n) + I_2(n)), \tag{6.11}$$

where for $j = 1, 2$, $I_j(n)$ is the value taken by the integral I_j defined at (6.5) and (6.6), when we take $A = \mathbb{Z} \cap [k_n, \infty)$, $r = r_n$, and $\lambda = n$. We have

$$I_1(n) \leq n^2 \theta (3r_n)^d \zeta_n^2 = 3^d \theta (n\zeta_n)^2 r_n^d,$$

so that by Lemma 1.2 and eqn (6.9),

$$(n\zeta_n)^{-1} I_1(n) \le 3^d \mu_n \zeta_n \to 0. \tag{6.12}$$

Moreover, setting $C := [-\frac{1}{2}, \frac{1}{2}]^d$ (the unit cube), we have

$$I_2(n) = n^2 \int_C \mathrm{d}x \int_{C \cap B(x;3r_n)} \mathrm{d}y P[\{\mathcal{P}_n^y(B(x;r_n)) \ge k_n\} \cap \{\mathcal{P}_n^x(B(y;r_n)) \ge k_n\}].$$

Define $\mathcal{H}_{nr_n^d}$ to be a homogeneous Poisson process of intensity nr_n^d on $\mathbb{R}^d$, and set

$$h_n(z) := P[\{\mathcal{H}_{nr_n^d}(B(\mathbf{0};1)) \ge k_n - 1\} \cap \{\mathcal{H}_{nr_n^d}(B(z;1)) \ge k_n - 1\}].$$

Making the change of variable $z = r_n^{-1}(y - x)$, and using a scaling property of Poisson processes (see Theorem 9.17 below), we have

$$I_2(n) \le n^2 r_n^d \int_{B(\mathbf{0};3)} h_n(z) \mathrm{d}z. \tag{6.13}$$

Using (6.9), choose a positive integer K such that μ_n^{K+1}/k_n^K tends to zero as $n \to \infty$. Then

$$P[\mathcal{H}_{nr_n^d}(B(\mathbf{0};1)) \ge k_n + K] \le \zeta_n (\mu_n/k_n)^K. \tag{6.14}$$

Also, $P[\mathcal{H}_{nr_n^d}(B(\mathbf{0};1)) = k_n - 1] \le (k_n/\mu_n)\zeta_n$, and if n is large enough then $P[\mathcal{H}_{nr_n^d}(B(\mathbf{0};1)) \ge k_n] = \zeta_n \le \zeta_n k_n/\mu_n$, so that

$$P[\mathcal{H}_{nr_n^d}(B(\mathbf{0};1)) \in \{k_n - 1, k_n, \ldots, k_n + K - 1\}] \le 2\zeta_n k_n/\mu_n. \tag{6.15}$$

Since $P[\mathcal{H}_{nr_n^d}(B(z;1)) \ge k_n - 1 | \mathcal{H}_{nr_n^d}(B(\mathbf{0};1)) = j]$ is a nondecreasing function of j, it is maximized over $j \in \{k_n - 1, k_n, \ldots, k_n + K\}$ at $j = k_n + K$. With $|\cdot|$ denoting Lebesgue measure, for $z \ne \mathbf{0}$ let δ_z denote the proportionate volume $|B(\mathbf{0};1) \setminus B(z;1)|/\theta$. The conditional distribution of $\mathcal{H}_{nr_n^d}(B(z;1))$, given that $\mathcal{H}_{nr_n^d}(B(\mathbf{0};1))$ takes the value $k_n + K$, is that of the sum of two independent variables $k_n + K - U$ and V, where U is $\mathrm{Bi}(k_n + K, \delta_z)$ and V is $\mathrm{Po}(nr_n^d \theta \delta_z)$, representing the number of points in $B(\mathbf{0};1) \setminus B(z;1)$ and in $B(z;1) \setminus B(\mathbf{0};1)$, respectively. Provided n is large enough so that $K + 1 < k_n\delta_z/6$, if $U > 2k_n\delta_z/3$ and $V < k_n\delta_z/3$, then $k_n + K - U + V < k_n - 1$. Thus, by Lemmas 1.1 and 1.2, there is a constant $\alpha > 0$ such that, for all large enough n and all $z \in B(\mathbf{0};3)$, if $\delta_z > 6(K+1)/k_n$ then

$$\begin{aligned}
&\max_{j \in \{k_n - 1, k_n, \ldots, k_n + K - 1\}} P[\mathcal{H}_{nr_n^d}(B(z;1)) \ge k_n - 1 | \mathcal{H}_{nr_n^d}(B(\mathbf{0};1)) = j] \\
&\le P[\mathrm{Bi}(k_n + K, \delta_z) \le 2k_n\delta_z/3] + P[\mathrm{Po}(nr_n^d \theta \delta_z) \ge k_n\delta_z/3] \\
&\le 2\exp(-\alpha k_n \delta_z),
\end{aligned}$$

while if $\delta_z \leq 6(K+1)/k_n$ then $e^{-\alpha k_n \delta_z} \geq e^{-6\alpha(K+1)}$, so that $e^{6\alpha(K+1)} e^{-\alpha k_n \delta_z} \geq 1$. Therefore, setting $c_0 := \max(2; e^{6\alpha(K+1)})$, we have for all $z \in \mathbb{R}^d$ that

$$\max_{j \in \{k_n - 1, k_n, \ldots, k_n + K - 1\}} P[\mathcal{H}_{nr_n^d}(B(z;1)) \geq k_n - 1 | \mathcal{H}_{nr_n^d}(B(\mathbf{0};1)) = j]$$
$$\leq c_0 e^{-\alpha k_n \delta_z}.$$

Combining this with (6.14) and (6.15), we obtain for large enough n that

$$h_n(z) \leq \zeta_n \left(\frac{\mu_n}{k_n}\right)^K + 2c_0 \zeta_n \left(\frac{k_n}{\mu_n}\right) \exp(-\alpha k_n \delta_z),$$

and since $\inf\{\delta_z / \|z\|_2 : z \in B(\mathbf{0};3)\} > 0$, there is a constant $\beta > 0$ such that

$$\begin{aligned}\int_{B(\mathbf{0};3)} (h_n(z) - \zeta_n(\mu_n/k_n)^K) \mathrm{d}z &\leq 2c_0 \zeta_n (k_n/\mu_n) \int_{B(\mathbf{0};3)} \exp(-\beta k_n \|z\|_2) \mathrm{d}z \\ &\leq c' \zeta_n (k_n/\mu_n) \int_0^\infty \exp(-\beta k_n s) s^{d-1} \mathrm{d}s \\ &= c' \zeta_n (k_n/\mu_n) \int_0^\infty e^{-t} t^{d-1} \mathrm{d}t / (\beta k_n)^d.\end{aligned}$$

Hence, by (6.13) there is a constant c such that

$$I_2(n) \leq c \mu_n n \zeta_n ((\mu_n/k_n)^K + \mu_n^{-1} k_n^{1-d}).$$

By the choice of K, and the assumption that $d \geq 2$, it follows that

$$(n\zeta_n)^{-1} I_2(n) \to 0,$$

and combining this with (6.12) in (6.11) gives us the result (6.10). □

We now extend the last result from $\mathcal{P}_n$ to $\mathcal{X}_n$, by considering a Poisson process of slightly larger intensity that dominates $\mathcal{X}_n$ with high probability.

Proposition 6.9 *Suppose $f = f_{\mathrm{U}}$. Set $\mu_n := n\theta r_n^d$ and suppose that* $\inf\{\mu_n : n \geq 1\} > 0$ *and* $\lim_{n\to\infty}(\mu_n/n^{1/9}) = 0$. *Suppose $(k_n)_{n \geq 1}$ is an $\mathbb{N}$-valued sequence such that (6.9) holds for some $\varepsilon > 0$, and set $\zeta_n := \sum_{j=k_n}^\infty e^{-\mu_n} \mu_n^j / j!$. Then*

$$\lim_{n\to\infty} (P[\Delta_n < k_n] - \exp(-n\zeta_n)) = 0. \tag{6.16}$$

Proof First suppose that $k_n \geq n^{1/8}$. By Boole's inequality and Lemma 1.1,

$$P[\Delta_n \geq n^{1/8}] \leq nP[\mathrm{Bi}(n-1, \mu_n/n) \geq n^{1/8}] \to 0,$$

so that, if $k_n \geq n^{1/8}$, then both $P[\Delta_n \geq k_n]$ and $\exp(-n\zeta_n)$ tend to zero. Therefore, from now on we may assume without loss of generality that $k_n \leq n^{1/8}$ for all n.

For $n \in \mathbb{N}$, set $\lambda(n) := n + n^{3/4}$. Let $\mathcal{P}_{\lambda(n)}$ be a Poisson process with intensity function $\lambda(n) f_U$, coupled to $\mathcal{X}_n$ as described in Section 1.7. Let Δ_n^+ be the maximum vertex degree in $G(\mathcal{P}_{\lambda(n)}; r_n)$. Also let $\mu_n^+ := \lambda(n)\theta r_n^d$ and $\zeta_n^+ := \sum_{j=k_n}^{\infty} \mathrm{e}^{-\mu_n^+}(\mu_n^+)^j/j!$. By Proposition 6.8,

$$\lim_{n\to\infty} \left(P[\Delta_n^+ < k_n] - \exp(-\lambda(n)\zeta_n^+) \right) = 0. \tag{6.17}$$

Since $(\mu_n^+/\mu_n)^{k_n} = (1 + n^{-1/4})^{k_n} \le \exp(-k_n n^{-1/4})$, which tends to 1 by the assumption that $k_n \le n^{1/8}$, and since $\mu_n^+ - \mu_n = n^{-1/4}\mu_n$, which tends to zero, we have

$$1 \le \frac{\zeta_n^+}{\zeta_n} \le \frac{\mathrm{e}^{-\mu_n^+}(\mu_n^+)^{k_n}(1 + (\mu_n^+/k_n) + (\mu_n^+/k_n)^2 + \cdots)}{\mathrm{e}^{-\mu_n}\mu_n^{k_n}} \to 1. \tag{6.18}$$

Set $t(n) := n\zeta_n$ and $s_n := (\lambda(n)\zeta_n^+/(n\zeta_n)) - 1$. Then $s_n > 0$, and by (6.18), we have $s_n \to 0$ as $n \to \infty$. Also,

$$\exp(-n\zeta_n) - \exp(-\lambda(n)\zeta_n^+) = \mathrm{e}^{-t(n)}(1 - \exp(-s_n t(n))). \tag{6.19}$$

If $t(n) \le s_n^{-1/2}$ then $1 - \mathrm{e}^{-s_n t(n)} \le 1 - \exp(-s_n^{1/2})$, which tends to zero as $n \to \infty$. If $t(n) \ge s_n^{-1/2}$ then $\mathrm{e}^{-t(n)} \le \exp(-s_n^{-1/2})$, which also tends to zero. These estimates show that as $n \to \infty$ the expression (6.19) tends to zero, and hence by (6.17),

$$\lim_{n\to\infty} \left(P[\Delta_n^+ < k_n] - \exp(-n\zeta_n) \right) = 0. \tag{6.20}$$

Let $N_{\lambda(n)}$ be the number of points of $\mathcal{P}_{\lambda(n)}$, a Poisson variable with parameter $n + n^{3/4}$. By Chebyshev's inequality, $P[n < N_{\lambda(n)} < n + 2n^{3/4}] \to 1$. Hence, in view of (6.20), to prove (6.16) it suffices to prove that

$$\lim_{n\to\infty} P[\Delta_n^+ \ge k_n > \Delta_n | n \le N_{\lambda(n)} \le n + 2n^{3/4}] = 0. \tag{6.21}$$

Suppose $\Delta_n^+ \ge k_n$ and $n \le N_{\lambda(n)} \le n + 2n^{3/4}$; then there is at least one point of $\mathcal{P}_{\lambda(n)}$ of degree at least k in $\mathrm{G}(\mathcal{P}_{\lambda(n)}; r_n)$. Pick one such point X, and some collection of k_n points $Y_1, \ldots, Y_{k_n}$ adjacent to X in $G(\mathcal{P}_{\lambda(n)}; r_n)$, uniformly at random from all possibilities. Then the probability that some point of $\{X, Y_1, \ldots, Y_{k_n}\}$ lies in $\mathcal{P}_{\lambda(n)} \setminus \mathcal{X}_n$ is bounded by $(k_n + 1)(2n^{3/4}/n)$, which tends to zero by the assumption that $k_n \le n^{1/8}$. Then (6.21) follows. □

Proof of Theorem 6.6 For each k let $\zeta_n(k) := P[\mathrm{Po}(\mu_n) \ge k]$. For each n take k_n such that

$$n\zeta_n(k_n - 1) > 1 \ge n\zeta_n(k_n).$$

Set $j(n) := k_n - 1$ if $n\zeta_n(k_n) \le \sqrt{\zeta_n(k_n)/\zeta_n(k_n - 1)}$, and set $j(n) := k_n$ otherwise.

Set $i_n := \lfloor \mu_n (\log n)^\delta \rfloor$, with $\delta > 0$ chosen so that $i_n/(\log n)^{1-\delta} \to 0$ as $n \to \infty$ (this is possible by the assumption about μ_n). Then, by (1.15) in Lemma 1.3, there is a constant $c > 0$ such that

$$n\zeta_n(i_n) \geq c n i_n^{-1/2} \exp(-i_n \log(i_n/\mu_n)) \to \infty,$$

so that $k_n \geq i_n$ for large n, and hence $k_n/\mu_n \to \infty$. Thus,

$$\frac{\zeta_n(k_n)}{\zeta_n(k_n-1)} \leq \frac{\mu_n}{k_n} \to 0, \qquad \frac{\zeta_n(k_n+1)}{\zeta_n(k_n)} \to 0, \qquad \frac{\zeta_n(k_n-1)}{\zeta_n(k_n-2)} \to 0.$$

Hence $n\zeta_n(j(n)+1) \to 0$ and $n\zeta_n(j(n)-1) \to \infty$, as $n \to \infty$. Hence, by Proposition 6.8, we have

$$P[\Delta'_n < j(n)+1] \to 1, \qquad P[\Delta'_n < j(n)-1] \to 0,$$
$$P[\Delta'_n < j(n)] - \exp(-n\zeta_n(j(n))) \to 0.$$

By (1.12) in Lemma 1.2, $\zeta_n(\lfloor \log n \rfloor) \leq n^{-1}$ for large n, and hence $j(n) \leq \log n \leq n^{1/9}$. Thus, by Proposition 6.9,

$$P[\Delta_n < j(n)+1] \to 1, \qquad P[\Delta_n < j(n)-1] \to 0,$$
$$P[\Delta_n < j(n)] - \exp(-n\zeta_n(j(n))) \to 0,$$

completing the proof. □

6.2 Subconnective laws of large numbers

This section contains a law of large numbers for Δ_n, for general underlying density functions, for cases with $nr_n^d/\log n \to 0$. It is of interest both to consider cases with $nr_n^d \to \infty$, and cases where nr_n^d remains bounded, or even where $nr_n^d \to 0$ provided $nr_n^d = n^{o(1)}$, that is, $\log(1/(nr_n^d))/\log n \to 0$. If $nr_n^d \to 0$ faster than this, that is, if $nr_n^d \leq n^{-\varepsilon}$ for some $\varepsilon > 0$, then Theorem 6.3 shows that there exists $k > 0$ such that $P[\Delta_n \leq k] \to 1$, and there will not be any interesting law of large numbers for Δ_n.

For the limiting regime considered here, we content ourselves with a weak law of large numbers with convergence in probability rather than convergence almost surely. Hence, there is some overlap of the next result for Δ_n with Theorem 6.6. However, in the next result we consider arbitrary underlying densities, not just the case $f = f_U$, and we also consider here the clique number $\mathcal{C}_n := \mathcal{C}(G(\mathcal{X}_n; r_n))$.

Theorem 6.10 *Suppose that* $(r_n)_{n\geq 1}$ *satisfies both* $nr_n^d/\log n \to 0$ *and* $\log(1/(nr_n^d))/\log n \to 0$ *as* $n \to \infty$. *Then*

$$\lim_{n\to\infty} \left(\frac{\Delta_n \log\left((\log n)/(nr_n^d)\right)}{\log n} \right) = 1 \quad \textit{in probability} \tag{6.22}$$

and

$$\lim_{n\to\infty} \left(\frac{\mathcal{C}_n \log\left((\log n)/(nr_n^d)\right)}{\log n} \right) = 1 \quad \textit{in probability.} \tag{6.23}$$

Proof Set

$$k_n := \frac{\log n}{\log((\log n)/(nr_n^d))} = \frac{\log n}{\log\log n + \log(1/(nr_n^d))},$$

which tends to infinity by the assumed asymptotic behaviour of r_n. We need to show that $\Delta_n \sim k_n$ in probability. Set $y_n := (\log n)/k_n = \log((\log n)/(nr_n^d))$, which tends to infinity by assumption. We have

$$\begin{aligned} k_n \log\left(\frac{k_n}{nr_n^d}\right) &= k_n \log\left(\frac{\log n}{nr_n^d \log((\log n)/(nr_n^d))}\right) \\ &= \frac{\log n}{y_n} \log\left(\frac{e^{y_n}}{y_n}\right) = (\log n)\left(1 - \frac{\log y_n}{y_n}\right). \end{aligned} \tag{6.24}$$

Let $\varepsilon > 0$. Then by Boole's inequality and Lemma 1.1, and the fact that $H(a) := 1 - a + a\log a$ satisfies $H(a) \geq a(\log a - 1)$,

$$\begin{aligned} P[\Delta_n > (1+\varepsilon)k_n] &\leq nP[\mathrm{Bi}(n, f_{\max}\theta r_n^d) \geq (1+\varepsilon)k_n] \\ &\leq n\exp\left(-nf_{\max}\theta r_n^d H\left(\frac{(1+\varepsilon)k_n}{nf_{\max}\theta r_n^d}\right)\right) \\ &\leq n\exp\left(-(1+\varepsilon)k_n\left(\log\left(\frac{(1+\varepsilon)k_n}{nf_{\max}\theta r_n^d}\right) - 1\right)\right), \end{aligned}$$

and by (6.24) this bound is equal to

$$n\exp\left(-(1+\varepsilon)\left(\log n\left(1 - \frac{\log y_n}{y_n}\right) + k_n \times \text{const.}\right)\right).$$

Since $k_n/\log n$ tends to zero and y_n tends to infinity, the above expression is $n\exp(-(1+\varepsilon+o(1))\log n)$, which tends to zero, and so

$$\lim_{n\to\infty} P[\Delta_n > (1+\varepsilon)k_n] = 0. \tag{6.25}$$

Since $\mathcal{C}_n \leq \Delta_n + 1$, we also have

$$\lim_{n\to\infty} P[\mathcal{C}_n > (1+\varepsilon)k_n + 1] = 0. \tag{6.26}$$

For an inequality the other way, for each n, choose a non-random set $\{x_1^n, x_2^n, \ldots, x_{\sigma_n}^n\}$, of maximal cardinality, such that the balls $B(x_i^n; r_n/2)$, $1 \leq i \leq \sigma_n$, are disjoint and satisfy $F(B(x_i^n; r_n/2)) \geq (f_{\max}/2)\theta(r_n/2)^d$. By assumption, $r_n^d < (\log n)/n$ for n large; so, by Lemma 5.1,

$$\liminf_{n\to\infty} \sigma_n(\log n)/n \geq \liminf_{n\to\infty}(\sigma_n r_n^d) > 0. \tag{6.27}$$

For each $n \in \mathbb{N}$, set $\lambda(n) := n - n^{3/4}$. With $\mathcal{P}_\lambda$ and N_λ described in Section 1.7, for $x \in \mathbb{R}^d$ define event $E_n(x)$ by

$$E_n(x) := \{\mathcal{P}_{\lambda(n)}(B(x; r_n/2)) \geq (1-\varepsilon)k_n\}.$$

By the triangle inequality, if $E_n(x)$ occurs, there is a point X of $\mathcal{P}_{\lambda(n)}$ in $B(x; r_n/2)$ with at least $k_n(1-\varepsilon)-1$ other points of $\mathcal{P}_{\lambda(n)}$ in $B(X; r_n)$. Moreover, if $E_n(x)$ occurs then $\mathcal{C}(G(\mathcal{P}_{\lambda(n)}; r_n)) \geq (1-\varepsilon)k_n$. Therefore, since $\mathcal{P}_{\lambda(n)} \subseteq \mathcal{X}_n$ except when $N_{\lambda(n)} > n$, we have the event inclusions

$$\{\Delta_n < (1-\varepsilon)k_n - 1\} \subseteq \{N_{\lambda(n)} > n\} \bigcup \left(\bigcup_{i=1}^{\sigma_n} E_n(x_i^n) \right)^{\mathrm{c}} \tag{6.28}$$

and

$$\{\mathcal{C}_n < (1-\varepsilon)k_n\} \subseteq \{N_{\lambda(n)} > n\} \bigcup \left(\bigcup_{i=1}^{\sigma_n} E_n(x_i^n) \right)^{\mathrm{c}}. \tag{6.29}$$

Set $\gamma := 2^{-(d+2)}\theta f_{\max}$. Then, by (1.15) in Lemma 1.3, and the fact that $k_n/(nr_n^d)$ tends to infinity, we have

$$\begin{aligned} \min_{i \leq \sigma_n} P[E_n(x_i^n)] &\geq P[\mathrm{Po}(\gamma n r_n^d) \geq (1-\varepsilon)k_n] \\ &\geq \left(\frac{\mathrm{e}^{-1/(12k_n)}}{\sqrt{2\pi k_n}} \right) \exp\left(-(1-\varepsilon)k_n \log\left(\frac{(1-\varepsilon)k_n}{\gamma n r_n^d} \right) \right) := \eta_n, \end{aligned}$$

and, by (6.24),

$$\eta_n = \left(\frac{\mathrm{e}^{1/(12k_n)}}{\sqrt{2\pi k_n}} \right) \exp\left(-(1-\varepsilon)\log n \left(1 - \frac{\log y_n}{y_n} \right) + \text{const.} \times k_n \right).$$

Since $k_n/\log n \to 0$, this lower bound equals $\exp((\varepsilon - 1 + o(1))\log n)$.

The events $E_n(x_i^n), 1 \leq i \leq \sigma_n$ are independent, so for large enough n

$$P\left[\left(\bigcup_{i=1}^{\sigma_n} E_n(x_i^n) \right)^{\mathrm{c}} \right] \leq (1 - n^{(\varepsilon/2)-1})^{\sigma_n} \leq \exp(-\sigma_n n^{(\varepsilon/2)-1}), \tag{6.30}$$

which tends to zero by (6.27). Therefore, by (6.28), $P[\Delta_n < k_n(1-\varepsilon) - 1] \to 0$, and combined with (6.25) this gives us (6.22). Also, by (6.29), $P[\mathcal{C}_n < k_n(1-\varepsilon)] \to 0$, and combined with (6.26) this gives us (6.23). □

Remark Since the right-hand side of (6.30) is summable in n, and also $P[N_{n-n^{3/4}} > n]$ is summable in n (see Lemma 1.4), application of the Borel–Cantelli lemma in the above proof of the lower bounds in (6.22) and (6.23) shows that they hold in the stronger sense that

$$\liminf_{n\to\infty}(\Delta_n/k_n) \geq 1 \quad \text{and} \quad \liminf_{n\to\infty}(\mathcal{C}_n/k_n) \geq 1, \quad \text{a.s.} \tag{6.31}$$

6.3 More laws of large numbers for maximum degree

This section contains a strong law of large numbers for the smallest k_n-nearest-neighbour link $S_{k_n}(\mathcal{X}_n)$, when k_n grows at least logarithmically in n. Re-formulating this result in terms of the maximum degree of the geometric graph with specified distance parameter, we shall obtain a law of large numbers for the maximum degree Δ_n of $G(\mathcal{X}_n; r_n)$ when $nr_n^d/\log n$ is bounded away from zero (Theorem 6.14). This adds to the law of large numbers (Theorem 6.10) already given for Δ_n in the case where $nr_n^d/\log n \to 0$.

Central to the statement of our laws of large numbers is the large deviations rate function $H : [0, \infty) \to \mathbb{R}$, defined, as at (1.4), by $H(0) = 1$ and

$$H(a) = 1 - a + a \log a, \qquad a > 0. \tag{6.32}$$

As noted earlier, $H(1) = 0$ and the unique turning point of H is the minimum at 1. Also $H(a)/a$ is increasing on $(1, \infty)$. Let $H_-^{-1} : [0, 1] \to [0, 1]$ be the unique inverse of the restriction of H to $[0, 1]$, and let $H_+^{-1} : [0, \infty) \to [1, \infty)$ be the inverse of the restriction of H to $[1, \infty)$.

In what follows, we use the convention $1/\infty = 0$ to cover cases where $k_n/\log n \to \infty$.

Theorem 6.11 *Suppose f has compact support. Suppose $b \in (0, \infty]$ and suppose the sequence $(k_n)_{n \geq 1}$ satisfies $k_n/\log n \to b$ and $k_n/n \to 0$ as $n \to \infty$. Define $a \geq 1$ by $a/H(a) = b$ (so $a = 1$ if $b = \infty$). Then, with probability* 1,

$$\lim_{n \to \infty} \left(\frac{n\theta(S_{k_n}(\mathcal{X}_n))^d}{k_n} \right) = \frac{1}{f_{\max} a}.$$

Before going into details, we sketch the approach underlying the proofs of strong laws such as this one, and those to be seen in Chapter 7 for the *minimum* degree. Consider first the simplest possible distribution of points, namely, uniform on the unit torus in $\mathbb{R}^d$, and suppose nr_n^d grows logarithmically in n (the connectivity regime). Then the mean number of points in a given r_n-ball also grows logarithmically in n. If also k_n grows logarithmically in n with a coefficient greater (less) than this mean, then the probability that the r_n-ball contains more (fewer) points than k_n decays exponentially in $\log n$, that is, polynomially in n, with an exponent determined precisely by the function H. Suppose with the ball we associate a 'core' of radius εr_n. Provided there is at least one point in the core, the presence of more than (fewer than) k_n other points in a slightly shrunken (expanded) version of this ball ensures that the maximum (minimum) degree is at least (at most) k_n, by the triangle inequality.

The number of disjoint balls of the above type that can be fitted into the unit torus is $O(n/\log n)$, as is the number of cores required to cover the unit torus. Finding k_n so that the maximum (minimum) degree is at least (at most) k_n with high probability is a matter of balancing the number of such balls against the polynomially decaying probability of the event of interest mentioned above happening for any single ball.

The behaviour of the maximum (minimum) degree for non-uniform density functions is determined by the maximum (minimum) value of the density.

As a first step in the proof of Theorem 6.11, we obtain an upper bound on the smallest k-nearest-neighbour link.

Proposition 6.12 *Suppose* $\lim_{n\to\infty}(k_n/\log n) = b \in (0,\infty]$, *and suppose* $a \geq 1$ *satisfies* $a/H(a) = b$. *Let* $\beta > 1$. *For* $u > 0$, *and* $n \in \mathbb{N}$, *define* $\rho_n(u)$ *by*

$$\rho_n(u) := (k_n u/(n\theta f_{\max} a))^{1/d}. \tag{6.33}$$

Then, with probability 1, $S_{k_n}(\mathcal{X}_n)^d \leq \rho_n(\beta)$, *for all large enough* n.

Proof Pick a number $\varepsilon_1 > 0$ such that

$$\varepsilon_1^{1/d} + (1+3\varepsilon_1)^{1/d} < \beta^{1/d}. \tag{6.34}$$

Also, in the case $a > 1$, assume that $1 + \varepsilon_1 < a$. Since $x^{-1}H(x)$ is increasing in x on $x \geq 1$, when $a > 1$ we can (and do) pick $\varepsilon_2 > 0$ such that

$$\left(\frac{1+\varepsilon_1}{a}\right) H\left(\frac{a}{1+\varepsilon_1}\right) < \frac{1-2\varepsilon_2}{b}. \tag{6.35}$$

Set $\phi_1 := f_{\max}(1+2\varepsilon_1)/(1+3\varepsilon_1)$. For each $n \in \mathbb{N}$ and $x \in \mathbb{R}^d$, set $U_n(x) := B(x;\rho_n(1+3\varepsilon_1))$ and $V_n(x) := B(x;\rho_n(\varepsilon_1))$. For each n, choose a non-random set $\{x_1^n, x_2^n, \ldots, x_{\sigma_n}^n\}$, of maximal cardinality, such that the balls $U_n(x_i^n)$, $1 \leq i \leq \sigma_n$, are disjoint, and such that $F(B(x_i^n;\rho_n(1+3\varepsilon_1))) \geq \phi_1\theta\rho_n(1+3\varepsilon_1)^d$ and $F(B(x_i^n;\rho_n(\varepsilon_1))) \geq \phi_1\theta\rho_n(\varepsilon_1)^d$. By Lemma 5.1,

$$\liminf_{n\to\infty}(\sigma_n k_n/n) > 0. \tag{6.36}$$

For $\lambda > 0$, let $\mathcal{P}_\lambda$ and N_λ be as described in Section 1.7. For $n \in \mathbb{N}$ set $\lambda(n) := n - n^{3/4}$, and for $x \in \mathbb{R}^d$ define events $E_n(x)$ and $E'_n(x)$ by

$$E_n(x) := \{\mathcal{P}_{\lambda(n)}(U_n(x)) \geq k_n + 1\};$$
$$E'_n(x) := \{\mathcal{P}_{\lambda(n)}(V_n(x)) \geq 1\} \cap E_n(x).$$

If $\|X - x\| \leq \rho_n(\varepsilon_1)$ and $\|Y - x\| \leq \rho_n(1+3\varepsilon_1)$, then by (6.34) and the triangle inequality, $\|X - Y\| \leq \rho_n(\beta)$. So, if $E'_n(x)$ occurs there is a point X of $\mathcal{P}_{\lambda(n)}$ in $B(x;\rho_n(\varepsilon_1))$ with at least k_n other points of $\mathcal{P}_{\lambda(n)}$ in $B(X;\rho_n(\beta))$, and hence $S_{k_n}(\mathcal{P}_{\lambda(n)}) \leq \rho_n(\beta)$. Therefore, since $\mathcal{P}_{\lambda(n)} \subseteq \mathcal{X}_n$ except when $N_{\lambda(n)} > n$,

$$\{S_{k_n}(\mathcal{X}_n) > \rho_n(\beta)\} \subseteq \{N_{\lambda(n)} > n\} \bigcup \left(\bigcup_{i=1}^{\sigma_n} E'_n(x_i^n)\right)^{\mathrm{c}}. \tag{6.37}$$

For large enough n, and each $x \in \{x_1^n, \ldots, x_{\sigma_n}^n\}$, it is the case that

$$E[\mathcal{P}_{\lambda(n)}(U_n(x))] \geq \lambda(n)\theta\phi_1(\rho_n(1+3\varepsilon_1))^d \geq \frac{k_n(1+\varepsilon_1)}{a}. \tag{6.38}$$

First consider the case with $b < \infty$, so $a > 1$. Then, by Lemma 1.3, for n large enough,

$$\begin{aligned}\min_{i \le \sigma_n} P[E_n(x_i^n)] &\ge P\left[\mathrm{Po}\left(\frac{k_n(1+\varepsilon_1)}{a}\right) \ge k_n + 1\right] \\ &\ge \left(\frac{\mathrm{e}^{-1/(12k_n)}}{\sqrt{2\pi(k_n+1)}}\right) \exp\left(-\left(\frac{k_n(1+\varepsilon_1)}{a}\right) H\left(\frac{a(1+k_n^{-1})}{1+\varepsilon_1}\right)\right) \\ &\ge \exp\left(-(1-\varepsilon_2)\log n\right) = n^{\varepsilon_2 - 1},\end{aligned}$$

where the last inequality is from (6.35) and the assumption that $k_n/\log n \to b$.

Given the number of points of $\mathcal{P}_{\lambda(n)}$ in $U_n(x_i^n)$, the conditional distribution of the number of points of $\mathcal{P}_{\lambda(n)}$ in $V_n(x_i^n)$ is binomial with parameter $F(V_n(x_i^n))/F(U_n(x_i^n))$, which remains bounded away from zero. Therefore,

$$\lim_{n\to\infty}\left(\min_{1\le i\le\sigma_n} P[E'_n(x_i^n)|E_n(x_i^n)]\right) = 1.$$

Hence for n large and $1 \le i \le \sigma_n$,

$$P[E'_n(x_i^n)] \ge \frac{1}{2} n^{\varepsilon_2 - 1}. \qquad (6.39)$$

The events $E'_n(x_i^n), 1 \le i \le \sigma_n$ are independent, so by (6.39), for large enough n,

$$P\left[\left(\bigcup_{i=1}^{\sigma_n} E'_n(x_i^n)\right)^{\mathrm{c}}\right] \le \left(1 - \frac{1}{2}n^{\varepsilon_2-1}\right)^{\sigma_n} \le \exp\left(-\frac{1}{2}\sigma_n n^{\varepsilon_2-1}\right),$$

which is summable in n by (6.36). Also $P[N_{\lambda(n)} > n]$ is summable in n by Lemma 1.4. The result follows, for the case $b < \infty$, by (6.37) and the Borel–Cantelli lemma.

The case with $b = \infty$, so that $a = 1$, is simpler. By (6.38) and by Lemma 1.2, there exists $\delta > 0$ such that in this case we have, for large n, that

$$P[E'_n(x_1^n)^{\mathrm{c}}] \le \exp(-\delta k_n),$$

and since $k_n/\log n \to \infty$ in this case, this implies that $P[(E'_n(x_1^n))^{\mathrm{c}}]$ is summable in n, so again the result follows, by (6.37) and the Borel–Cantelli lemma. □

The proof of an inequality the other way uses a *subsequence trick* that will come up repeatedly. We show that a certain probability under consideration for $(\mathcal{X}_n)$ tends to zero sufficiently fast along a certain subsequence of $\mathcal{X}_n$ to ensure that, by the Borel–Cantelli lemma, the event in question occurs for only finitely many n in the subsequence, almost surely; we shall then fill in the gaps between numbers in the subsequence using the geometric structure of $G(\mathcal{X}_n; r)$.

Proposition 6.13 *Suppose f has compact support. Suppose $(k_n)_{n\ge1}$ satisfies $\lim_{n\to\infty}(k_n/\log n) = b \in (0,\infty]$, and suppose $a \ge 1$ satisfies $a/H(a) = b$. Let $\beta < 1$, and let $\rho_n(\cdot)$ be defined by (6.33). Then, with probability 1, $S_{k_n}(\mathcal{X}_n) \ge \rho_n(\beta)$ for all large enough n.*

Proof Pick $\varepsilon_3 > 0$ such that

$$\varepsilon_3^{1/d} + \beta^{1/d} < (1-\varepsilon_3)^{1/d}. \tag{6.40}$$

Since $x^{-1}H(x)$ is increasing in x on $x \geq 1$, and $a^{-1}H(a) = 1/b$, we have $((1-\varepsilon_3)/a)H(a/(1-\varepsilon_3)) > 1/b$. Pick $\varepsilon_4 \in (0, \varepsilon_3)$ such that

$$\left(\frac{1-\varepsilon_3}{a}\right) H\left(\frac{a(1-\varepsilon_4)}{1-\varepsilon_3}\right) > \frac{1+2\varepsilon_4}{b}. \tag{6.41}$$

Let Ω denote the support of f. With $\rho_n(u)$ defined at (6.33), let κ_n be the smallest number of balls of radius $\rho_n(\varepsilon_3)$ needed to cover Ω. Then

$$\limsup_{n\to\infty}((\kappa_n \log n)/n) < \infty. \tag{6.42}$$

For each n take a deterministic set $\{x_1^n, \ldots, x_{\kappa_n}^n\} \subset \mathbb{R}^d$, with the property that $\Omega \subseteq \cup_i B(x_i^n, \rho_n(\varepsilon_3))$. Given $x \in \mathbb{R}^d$, let $F_n(x)$ be the event

$$F_n(x) := \{\mathcal{X}_n\left[B(x;\rho_n(1-\varepsilon_3))\right] \geq k_n(1-\varepsilon_4)\}. \tag{6.43}$$

Then, for all n and all $x \in \mathbb{R}^d$,

$$\begin{aligned} nF(B(x;\rho_n(1-\varepsilon_3))) &\leq n f_{\max}\theta(\rho_n(1-\varepsilon_3))^d \\ &= k_n(1-\varepsilon_3)/a := \mu_n. \end{aligned}$$

Consider first the case with $b < \infty$. By Lemma 1.1 and eqn (6.41), for all large enough n and all $x \in \mathbb{R}^d$ we have

$$\begin{aligned} P[F_n(x)] &\leq \exp\left(-k_n\left(\frac{1-\varepsilon_3}{a}\right)H\left(\frac{a(1-\varepsilon_4)}{1-\varepsilon_3}\right)\right) \\ &\leq \exp(-(1+\varepsilon_4)\log n) = n^{-(1+\varepsilon_4)}. \end{aligned}$$

Set $G_n := \cup_{i=1}^{\kappa_n} F_n(x_i^n)$. By Boole's inequality and (6.42), for large enough n,

$$P[G_n] \leq n^{-\varepsilon_4} \quad (b < \infty). \tag{6.44}$$

For the case with $b = \infty$ we have $a = 1$ and $\mu_n = k_n(1-\varepsilon_3)$, so by Lemma 1.1 and (6.42), there is a constant $\gamma > 0$ such that

$$P[G_n] \leq n\exp(-\gamma k_n) \qquad (b = \infty), \tag{6.45}$$

which is summable in n since $k_n/\log n \to \infty$.

Pick a positive integer K such that (for the case $b < \infty$) $K\varepsilon_4 > 1$ (or in the case $b = \infty$, take $K = 1$). For $m \in \mathbb{N}$, let $\nu(m) := m^K$ (this is the subsequence trick). By (6.44) and (6.45) we have in either case that $\sum_{m=1}^{\infty} P[G_{\nu(m)}] < \infty$, so by the Borel–Cantelli lemma, $G_{\nu(m)}$ occurs for only finitely many m, with probability 1.

Given $n \in \mathbb{N}$, let $m = m(n) \in \mathbb{N}$ be chosen so that $(m-1)^K < n \leq m^K$. Then since $k_n/\log n \to b$ and $\log n/\log(\nu(m(n))) \to 1$ as $n \to \infty$, we have

$$\lim_{n\to\infty}\left(\frac{k_{\nu(m(n))}}{k_n}\right) = 1, \tag{6.46}$$

so that for all large enough n we have $k_{\nu(m(n))}(1-\varepsilon_4) \leq k_n$.

Pick $n \in \mathbb{N}$, and take $m = m(n)$. Suppose $S_{k_n}(\mathcal{X}_n) < \rho_n(\beta)$. Then there exists a point X of $\mathcal{X}_n$ such that $\mathcal{X}_n(B(X;\rho_n(\beta))) > k_n$. Choose $i \leq \kappa_{\nu(m)}$ such that $X \in B(x_i^{\nu(m)};\rho_{\nu(m)}(\varepsilon_3))$. By (6.40) and (6.46), provided m is large enough,

$$\rho_{\nu(m)}(\varepsilon_3) + \rho_n(\beta) \leq \rho_{\nu(m)}(1-\varepsilon_3),$$

so that for i as just described there are more than k_n points of $\mathcal{X}_{\nu(m)}$ in $B(x_i^{\nu(m)};\rho_{\nu(m)}(1-\varepsilon_3))$. Therefore, by (6.43), $F_{\nu(m)}(x_i^{\nu(m)})$ and hence $G_{\nu(m)}$ occurs. Thus, since $G_{\nu(m)}$ occurs for only finitely many m, $S_{k_n}(\mathcal{X}_n) < \rho_n(\beta)$ for only finitely many values of n, almost surely. □

Proof of Theorem 6.11 Immediate from Propositions 6.12 and 6.13. □

The re-formulation of Theorem 6.11, in terms of the maximum degree Δ_n of $G(\mathcal{X}_n;r_n)$, goes as follows. The inverse function H_+^{-1} is as defined at the start of this section.

Theorem 6.14 *Suppose f has compact support. Suppose that $\alpha \in (0,\infty]$, and that $(r_n)_{n\geq 1}$ satisfies $r_n \to 0$ and $n\theta r_n^d/\log n \to \alpha$ as $n \to \infty$. Then*

$$\lim_{n\to\infty}\left(\frac{\Delta_n}{n\theta r_n^d}\right) = f_{\max} H_+^{-1}\left(\frac{1}{f_{\max}\alpha}\right) \quad \text{a.s.,} \tag{6.47}$$

with the convention $1/\infty = 0$, so the limit is $f_{\max}$ if $\alpha = \infty$.

Proof First suppose $\alpha < \infty$. Given $b > 0$, define $a > 1$ by $a/H(a) = b$, and set $\psi(b) = (f_{\max}H(a))^{-1}$. If $(k_n)_{n\geq 1}$ is a sequence with $k_n/\log n \to b$, then by Theorem 6.11, with probability 1,

$$\lim_{n\to\infty}\left(\frac{n\theta(S_{k_n}(\mathcal{X}_n))^d}{\log n}\right) = \frac{b}{f_{\max}a} = \psi(b). \tag{6.48}$$

Observe that $\psi(b)$ is a continuous, strictly increasing function of b. Choose b, b' with $b < \psi^{-1}(\alpha) < b'$, and choose sequences $(k_n)_{n\geq 1}$ and $(k'_n)_{n\geq 1}$ with $k_n/\log n \to b$ and $k'_n/\log n \to b'$. Then, by (6.48), and a similar limiting expression for $S_{k'_n}$, we have

$$\lim_{n\to\infty}\left(n\theta(S_{k_n}(\mathcal{X}_n))^d/\log n\right) < \alpha < \lim_{n\to\infty}\left(n\theta(S_{k'_n}(\mathcal{X}_n))^d/\log n\right),$$

so that for n large enough, $S_{k_n} < r_n < S_{k'_n}$ and hence $k_n \leq \Delta_n \leq k'_n$. It follows that, with probability 1,

$$b \le \liminf_{n\to\infty}(\Delta_n/\log n) \le \limsup_{n\to\infty}(\Delta_n/\log n) \le b'.$$

By taking $b \uparrow \psi^{-1}(\alpha)$ and $b' \downarrow \psi^{-1}(\alpha)$, we may deduce that $(\Delta_n/\log n) \to \psi^{-1}(\alpha)$ almost surely.

Now set $b = \psi^{-1}(\alpha)$. Suppose $a > 1$ satisfies $a/H(a) = b$. Then, by definition of the function ψ, we have $H(a) = (f_{\max}\alpha)^{-1}$, and therefore

$$a = H_+^{-1}\left(\frac{1}{f_{\max}\alpha}\right), \quad b = f_{\max}\alpha H_+^{-1}\left(\frac{1}{f_{\max}\alpha}\right).$$

Hence, with probability 1,

$$\lim_{n\to\infty}\left(\frac{\Delta_n}{\log n} \times \frac{\log n}{n\theta r_n^d}\right) = \frac{b}{\alpha} = f_{\max}H_+^{-1}\left(\frac{1}{f_{\max}\alpha}\right),$$

proving (6.47) for the case $\alpha < \infty$.

Next suppose $\alpha = \infty$. Let $\varepsilon > 0$. Set $k_n = \lfloor (1+2\varepsilon)n\theta r_n^d f_{\max}\rfloor$ and $j_n = \lfloor (1-2\varepsilon)n\theta r_n^d f_{\max}\rfloor$. Then $(k_n/\log n) \to \infty$ and $(j_n/\log n) \to \infty$. By Theorem 6.11, we have, with probability 1, that $n\theta S_{k_n}(\mathcal{X}_n)^d/k_n \to f_{\max}^{-1}$ and $n\theta S_{j_n}(\mathcal{X}_n)^d/j_n \to f_{\max}^{-1}$ as $n \to \infty$, and therefore for large enough n,

$$n\theta S_{k_n}(\mathcal{X}_n)^d \ge \left(\frac{1+\varepsilon}{1+2\varepsilon}\right)\left(\frac{k_n}{f_{\max}}\right) > n\theta r_n^d,$$
$$n\theta S_{j_n}(\mathcal{X}_n)^d \le \left(\frac{1-\varepsilon}{1-2\varepsilon}\right)\left(\frac{k_n}{f_{\max}}\right) < n\theta r_n^d,$$

so that for large enough n, $S_{j_n}(\mathcal{X}_n) < r_n < S_{k_n}(\mathcal{X}_n)$ and $j_n \le \Delta_n \le k_n$. Therefore, with probability 1,

$$1 - 2\varepsilon \le \liminf_{n\to\infty}\left(\frac{\Delta_n}{n\theta f_{\max} r_n^d}\right) \le \limsup_{n\to\infty}\left(\frac{\Delta_n}{n\theta f_{\max} r_n^d}\right) \le 1 + 2\varepsilon.$$

Since ε is arbitrarily small, this gives us (6.47) for the case $\alpha = \infty$. □

6.4 Laws of large numbers for clique number

This section contains a strong law of large numbers (Theorem 6.16) for the clique number in the connectivity regime where $(n/\log n)r_n^d$ tends to a positive finite constant. First we consider the threshold for the clique number to exceed a value k_n growing logarithmically in n. For any finite set $\mathcal{X} \subset \mathbb{R}^d$, and any positive integer k, let $\rho(\mathcal{X}; \mathcal{C} \ge k)$ denote the minimum r such that the clique number of $G(\mathcal{X}; r)$ is at least k (if there is no such r set $\rho(\mathcal{X}; \mathcal{C} \ge k) := \infty$).

Define the function $H(a), a > 0$, and its inverse H_+^{-1}, as at (6.32). The strong convergence results in this section involve these functions in a similar manner to those given in the preceding section for the maximum degree. As before, we use the convention $1/\infty = 0$.

Theorem 6.15 *Suppose that f has compact support, and that $k_n/\log n \to b \in (0,\infty]$ as $n \to \infty$. Define $a \geq 1$ by $a/H(a) = b$ (so $a = 1$ if $b = \infty$). Then, with probability* 1,

$$\lim_{n\to\infty}\left(\frac{n\theta(\rho(\mathcal{X}_n;\mathcal{C} \geq k_n))^d}{k_n}\right) = \frac{2^d}{f_{\max}a}.$$

Proof As at (6.33), for $u > 0$ and $n > 0$ define

$$\rho_n(u) := (uk_n/(n\theta f_{\max}a))^{1/d}.$$

It suffices to prove that for any given $\alpha < 1$ and $\beta > 1$, with probability 1, for all large enough n,

$$2\rho_n(\alpha) \leq \rho(\mathcal{X}_n;\mathcal{C} \geq k_n) \leq 2\rho_n(\beta). \tag{6.49}$$

If $G(\mathcal{X};r)$ has a vertex X of degree k, and the vertices adjacent to X are denoted $Y_1,\ldots,Y_k$, then by the triangle inequality $\|Y_i - Y_j\| \leq 2r$ for all $i,j \leq k$, so that $G(\{Y_1,\ldots,Y_k\};2r)$ is a complete graph and hence $\mathcal{C}(G(\mathcal{X};2r)) \geq k$. Hence, $\rho(\mathcal{X};\mathcal{C} \geq k) \leq 2S_k(\mathcal{X})$, and the second inequality of (6.49) follows from Proposition 6.12, for any $\beta > 1$. It remains only to prove the first inequality of (6.49).

Given $\alpha < 1$, choose $\varepsilon_5 > 0$ satisfying

$$\alpha^{1/d} + \mathrm{diam}([0,\varepsilon_5^{1/d}] \times [0,\varepsilon_5^{1/d}] \times \cdots \times [0,\varepsilon_5^{1/d}]) < (1-\varepsilon_5)^{1/d}. \tag{6.50}$$

Choose $\varepsilon_6 \in (0,\varepsilon_5)$ such that

$$\left(\frac{1-\varepsilon_5}{a}\right) H\left(\frac{a(1-\varepsilon_6)}{1-\varepsilon_5}\right) > \frac{1+2\varepsilon_6}{b}. \tag{6.51}$$

Let Ω denote the support of f. For $a > 0$, let $a\mathbb{Z}^d := \{ax : x \in \mathbb{Z}^d\}$. Let $\tau_1^n,\ldots,\tau_{\kappa_n}^n$ be the collection of all finite subsets τ of $\rho_n(\varepsilon_5)\mathbb{Z}^d$ with the property that $\tau \oplus [0,\rho_n(\varepsilon_5))^d$ (a union of little cubes) has non-empty intersection with Ω and diameter at most $2\rho_n(1-\varepsilon_5)$. We assert that

$$\limsup_{n\to\infty}(\kappa_n(\log n)/n) < \infty. \tag{6.52}$$

This is because the number of choices for the first element of τ (according to the lexicographic ordering on square centres) is bounded by a constant times $n/\log n$, and given the first square, the number of choices for the remaining elements of τ is uniformly bounded.

Let F_i^n be the event

$$F_i^n := \{\mathcal{X}_n(\tau_i^n \oplus [0,\rho_n(\varepsilon_5))^d) \geq k_n(1-\varepsilon_6)\}.$$

By the Bieberbach isodiametric inequality (Corollary 5.12), each set $\tau_i^n \oplus [0,\rho_n(\varepsilon_5))^d$, $1 \leq i \leq \kappa_n$, has volume at most $\theta\rho_n(1-\varepsilon_5)^d$. Hence, for all large enough n and all i,

$$nF(\tau_i^n \oplus [0,\rho_n(\varepsilon_5))^d) \leq \frac{(1-\varepsilon_5)k_n}{a}. \tag{6.53}$$

Consider first the case with $b < \infty$. By Lemma 1.1, and (6.51) along with the assumption that $k_n \sim b \log n$, for large n and all i, we have

$$\begin{aligned} P[F_i^n] &\leq \exp\left(-\left(\frac{(1-\varepsilon_5)k_n}{a}\right) H\left(\frac{a(1-\varepsilon_6)}{1-\varepsilon_5}\right)\right) \\ &\leq \exp(-(1+\varepsilon_6)\log n) = n^{-(1+\varepsilon_6)}. \end{aligned}$$

Set $G_n := \cup_{i=1}^{\kappa_n} F_n^i$. By Boole's inequality and (6.52), for large enough n we have

$$P[G_n] \leq n^{-\varepsilon_6} \qquad (b < \infty). \tag{6.54}$$

For the case $b = \infty$ we have $a = 1$, so by Lemma 1.1 and inequalities (6.52) and (6.53), there is a constant $\gamma > 0$ such that

$$P[G_n] \leq n \exp(-\gamma k_n) \qquad (b = \infty). \tag{6.55}$$

which is summable in n since $k_n/\log n \to \infty$ in this case.

If $b < \infty$, pick a positive integer K such that $K\varepsilon_6 > 1$. If $b = \infty$, take $K := 1$. For $m = 1, 2, 3, \ldots$, let $\nu(m) := m^K$. By (6.54) or (6.55), we have in both cases that $\sum_{m=1}^{\infty} P[G_{\nu(m)}] < \infty$, and so by the Borel–Cantelli lemma, with probability 1, $G_{\nu(m)}$ occurs for only finitely many m.

Given n, take m such that $\nu(m-1) < n \leq \nu(m)$. Suppose $\rho(\mathcal{X}_n; \mathcal{C} \geq k_n) \leq 2\rho_n(\alpha)$. Then there is a subset $\mathcal{S}$ of $\mathcal{X}_n$ of cardinality at least k_n and diameter at most $2\rho_n(\alpha)$. Let $\tau(\mathcal{S})$ be the smallest set $\tau \subset \rho_{\nu(m)}(\varepsilon_5)\mathbb{Z}^d$ with the property that $\mathcal{S} \subseteq \tau \oplus [0, \rho_{\nu(m)}(\varepsilon_5))^d$. Then the diameter of the set $\tau \oplus [0, \rho_{\nu(m)}(\varepsilon_5))^d$ is bounded by

$$2\left(\frac{k_n \alpha}{n\theta f_{\max} a}\right)^{1/d} + 2\left(\frac{k_{\nu(m)}}{\nu(m)\theta f_{\max} a}\right)^{1/d} \operatorname{diam}([0, \varepsilon_5^{1/d}] \times \cdots \times [0, \varepsilon_5^{1/d}]).$$

By (6.50), for n large the above expression is at most $2\rho_{\nu(m)}(1-\varepsilon_5)$. Thus, $G_{\nu(m)}$ occurs. Hence, by the conclusion of the previous paragraph, the first inequality in (6.49) holds for all large enough n, almost surely. □

As a consequence of Theorem 6.15 we obtain a strong law of large numbers for the clique number $\mathcal{C}_n$ of of $G(\mathcal{X}_n; r_n)$, valid in the connectivity regime where $(n/\log n)r_n^d$ tends to a constant.

Theorem 6.16 *Suppose f has compact support. Suppose $\alpha \in (0, \infty]$ and $(r_n)_{n\geq 1}$ satisfies $n\theta r_n^d/\log n \to \alpha$ as $n \to \infty$. Then*

$$\lim_{n\to\infty} \left(\frac{\mathcal{C}_n}{n\theta r_n^d}\right) = \frac{f_{\max}}{2^d} H_+^{-1}\left(\frac{2^d}{f_{\max}\alpha}\right) \quad \text{a.s.}, \tag{6.56}$$

with the convention $1/\infty = 0$, so the limit is $f_{\max}/2^d$ if $\alpha = \infty$.

Proof First suppose $\alpha < \infty$. Given $b > 0$, define $a > 1$ by $a/H(a) = b$, and set $\psi(b) := 2^d/(f_{\max}H(a))$. If $(k_n)_{n\geq 1}$ is a sequence with $k_n/\log n \to b$ as $n \to \infty$, then by Theorem 6.15, with probability 1,

$$\lim_{n\to\infty}\left(\frac{n\theta\rho(\mathcal{X}_n;\mathcal{C}\geq k_n)^d}{\log n}\right) = \frac{2^d b}{f_{\max}a} = \psi(b). \tag{6.57}$$

Observe that $\psi(b)$ is a continuous, strictly increasing function of b. Choose $b < \psi^{-1}(\alpha) < b'$, and take integer-valued sequences $(k_n)_{n\geq 1}$ and $(k'_n)_{n\geq 1}$ with $k_n/\log n \to b$ and $k'_n/\log n \to b'$ as $n \to \infty$. By (6.57), for large n, $\rho(\mathcal{X}_n;\mathcal{C} \geq k_n) < r_n < \rho(\mathcal{X}_n;\mathcal{C}\geq k'_n)$, and hence $k_n \leq \mathcal{C}_n \leq k'_n$. It follows that with probability 1,

$$b \leq \liminf_{n\to\infty}(\mathcal{C}_n/\log n) \leq \limsup_{n\to\infty}(\mathcal{C}_n/\log n) \leq b',$$

and by taking $b \uparrow \psi^{-1}(\alpha)$ and $b' \downarrow \psi^{-1}(\alpha)$, we have $(\mathcal{C}_n/\log n) \to \psi^{-1}(\alpha)$ almost surely.

Now set $b = \psi^{-1}(\alpha)$, and choose $a > 1$ so that $a/H(a) = b$. Then by definition of the function ψ we have $H(a) = 2^d/(f_{\max}\alpha)$, and therefore

$$a = H_+^{-1}\left(\frac{2^d}{f_{\max}\alpha}\right), \quad b = 2^{-d}f_{\max}\alpha H_+^{-1}\left(\frac{2^d}{f_{\max}\alpha}\right),$$

whence

$$\lim_{n\to\infty}\left(\frac{\mathcal{C}_n}{\log n}\times\frac{\log n}{n\theta r_n^d}\right) = \frac{b}{\alpha} = 2^{-d}f_{\max}H_+^{-1}\left(\frac{2^d}{f_{\max}\alpha}\right),$$

proving (6.56) for the case $\alpha < \infty$.

Next suppose $\alpha = \infty$. Let $\varepsilon > 0$. Set $k_n := \lfloor(1+2\varepsilon)n\theta(r_n/2)^d f_{\max}\rfloor$, and set $j_n := \lfloor(1-2\varepsilon)n\theta(r_n/2)^d f_{\max}\rfloor$. Then $(k_n/\log n) \to \infty$ and $(j_n/\log n) \to \infty$, so by Theorem 6.15 we have, with probability 1, that $n\theta\rho(\mathcal{X}_n;\mathcal{C}\geq k_n)^d/k_n \to 2^d f_{\max}^{-1}$ as $n \to \infty$, and likewise for j_n. Hence, with probability 1, for large enough n,

$$n\theta\rho(\mathcal{X}_n;\mathcal{C}\geq k_n)^d \geq \left(\frac{1+\varepsilon}{1+2\varepsilon}\right)\left(\frac{2^d k_n}{f_{\max}}\right) > n\theta r_n^d,$$

$$n\theta\rho(\mathcal{X}_n;\mathcal{C}\geq j_n)^d \leq \left(\frac{1-\varepsilon}{1-2\varepsilon}\right)\left(\frac{2^d j_n}{f_{\max}}\right) < n\theta r_n^d,$$

so that $\rho(\mathcal{X}_n;\mathcal{C}\geq k_n) > r_n > \rho(\mathcal{X}_n;\mathcal{C}\geq j_n)$, and $j_n \leq \mathcal{C}_n \leq k_n$. Therefore,

$$(1-2\varepsilon)f_{\max} \leq \liminf_{n\to\infty}\left(\frac{2^d\mathcal{C}_n}{n\theta r_n^d}\right) \leq \limsup_{n\to\infty}\left(\frac{2^d\mathcal{C}_n}{n\theta r_n^d}\right) \leq (1+2\varepsilon)f_{\max} \quad \text{a.s.},$$

and making $\varepsilon \downarrow 0$ gives us (6.56) for the case with $\alpha = \infty$. □

6.5 The chromatic number

We write χ_n (not to be confused with $\mathcal{X}_n$) for the chromatic number of $G(\mathcal{X}_n; r_n)$, and $\chi(G)$ for the chromatic number of an arbitrary graph G. By standard (and easy) results in graph theory, we have the bounds

$$\mathcal{C}_n \leq \chi_n \leq \Delta_n + 1. \tag{6.58}$$

In the subconnective regime with $nr_n^d/\log n \to 0$ and $\log(1/(nr_n^d))/\log n \to 0$, Theorem 6.10, along with (6.58), shows that

$$\lim_{n\to\infty}\left(\frac{\chi_n \log\left((\log n)/(nr_n^d)\right)}{\log n}\right) = 1 \quad \text{in probability.}$$

In the connectivity regime with $nr_n^d/\log n \to$ const., Theorems 6.14 and 6.16, with (6.58), imply asymptotic upper and lower bounds for χ_n, and these upper and lower bounds are within a constant of each other.

This section contains sharper bounds for χ_n in the superconnectivity regime $nr_n^d/\log n \to \infty$. Our results require various notions of *packing* taken from Rogers (1964). Suppose B is a bounded convex set in $\mathbb{R}^d$, with $\mathbf{0} \in B$. For $x \in \mathbb{R}^d$, let the set $\{x\} \oplus B$ be called the *translate of B centred at x*. By a *B-packing* of $\mathbb{R}^d$ we mean a collection $\mathcal{K}$ of disjoint translates of B. Given such a packing, and given $L > 0$, the volume of the packing relative to $[-L/2, L/2]^d$, denoted $V_L(\mathcal{K})$, is the total volume of the set of translates of B in the packing that have non-empty intersection with $[-L/2, L/2]^d$, divided by L^d. The *upper density* of the packing $\mathcal{K}$ is given by $\limsup_{L\to\infty} V_L(\mathcal{K})$, and the *packing density* of B is the supremum of the upper densities of all B-packings, and is here denoted $\phi(B)$.

Suppose $\{v_1, \ldots, v_d\}$ is a linearly independent set of vectors in $\mathbb{R}^d$, and suppose that the collection of translates of B centred at points which are linear combinations of the vectors v_i with integer coefficients, is pairwise disjoint. Then this collection of sets forms a B-packing of $\mathbb{R}^d$, which we denote the *lattice B-packing* of $\mathbb{R}^d$ generated by $\{v_1, \ldots, v_d\}$. In the case of a lattice packing $\mathcal{K}$, the limit $\lim_{L\to\infty} V_L(\mathcal{K})$ exists. The *lattice packing density* of B is the supremum of all upper densities of lattice B-packings, and is here denoted $\phi_L(B)$.

We shall give lower and upper bounds for the clique number of geometric random graphs in terms of the packing density $\phi(B)$ and the lattice packing density $\phi_L(B)$, where B is the unit ball of the chosen norm. It is clear that $\phi_L(B) \leq \phi(B) \leq 1$ for any B, and if there is a periodic tessellation of $\mathbb{R}^d$ by translates of B (e.g. if B is the unit ball of the l_∞ norm), then $\phi_L(B) = \phi(B) = 1$.

If $d = 2$ and B is the Euclidean (l_2) unit ball, then it is known that $\phi_L(B) = \phi(B) = \pi/\sqrt{12} \approx 0.9069$, which is Thue's theorem; the optimal packing is by disks with centres at the points of a triangular lattice. For an exposition and short proof see Hales (2000); also Rogers (1964) and Pach and Agarwal (1995). More generally, it is known that the equality $\phi_L(B) = \phi(B)$ holds for *any* bounded convex set $B \subset \mathbb{R}^2$; see Rogers (1951), Rogers (1964), and Pach and Agarwal (1995).

In higher dimensions, determining when the equality $\phi_L(B) = \phi(B)$ holds has been a long-standing open problem. Hales and Ferguson recently proved that if B is the Euclidean unit ball in $\mathbb{R}^3$ then $\phi_L(B) = \phi(B) = \pi/\sqrt{18} \approx 0.7404$, by way of a lengthy computer-assisted proof in a series of preprints that are electronically available but not all published at the time of writing. For an overview see Hales (2000) or Oesterlé (2000).

Theorem 6.17 *Suppose that f has compact support, and that $(r_n)_{n\geq 1}$ is chosen so that $r_n \to 0$ and $nr_n^d/\log n \to \infty$ as $n \to \infty$. Let B be the interior of $B(\mathbf{0};1)$. Then*

$$\liminf_{n\to\infty} \left(\frac{\chi_n}{n\theta r_n^d}\right) \geq \frac{f_{\max}}{\phi(B)2^d} \qquad \text{a.s.} \tag{6.59}$$

Theorem 6.18 *Suppose that f has compact support, and that $(r_n)_{n\geq 1}$ is chosen so that $r_n \to 0$ and $nr_n^d/\log n \to \infty$. Let B be the interior of $B(\mathbf{0};1)$. Then*

$$\limsup_{n\to\infty} \left(\frac{\chi_n}{n\theta r_n^d}\right) \leq \frac{f_{\max}}{\phi_L(B)2^d} \qquad \text{a.s.} \tag{6.60}$$

The next result is immediate from Theorems 6.17, 6.18, and 6.16.

Corollary 6.19 *Suppose that f has compact support, and that $(r_n)_{n\geq 1}$ is chosen so that $r_n \to 0$ and $nr_n^d/\log n \to \infty$. Let B be the interior of $B(\mathbf{0};1)$, and suppose $\phi_L(B) = \phi(B)$ (true, for example, if $d = 2$). Then*

$$\lim_{n\to\infty} \left(\frac{\chi_n}{n\theta r_n^d}\right) = \frac{f_{\max}}{\phi_L(B)2^d} \qquad \text{a.s.}$$

If there is a periodic tessellation of $\mathbb{R}^d$ by translates of B, then $\lim_{n\to\infty}(\chi_n/\mathcal{C}_n) = 1$ a.s.

Proof of Theorem 6.17 In this proof, given $A \subset \mathbb{R}^d$ and $a \in [0,\infty)$, we write aA for the rescaled set $\{ax : x \in A\}$.

Let $\varepsilon \in (0,1)$. Choose $D > 0$ so that $B(\mathbf{0};1) \subset [-D/2, D/2]^d$. Choose $R > 0$ such that $((R+D)/R)^d \leq 1+\varepsilon$. Take a cube Q_n of side Rr_n, with $F(Q_n) \geq (1-\varepsilon)f_{\max}(Rr_n)^d$. Such a cube exists, for large enough n, by the Lebesgue density theorem. Then $E[\mathcal{X}_n(Q_n)] \geq n(1-\varepsilon)f_{\max}(Rr_n)^d$, and so by Lemma 1.1, there exists a constant $\gamma > 0$ such that, for all n,

$$P[\mathcal{X}_n(Q_n) \leq (1-2\varepsilon)nf_{\max}(Rr_n)^d] \leq \exp(-\gamma n r_n^d), \tag{6.61}$$

which is summable in n, by the assumption that $nr_n^d/\log n \to \infty$.

Given a finite graph G of order v, a *stable set* of vertices (also known as an *independent set*, in the graph-theoretic sense) is a set of vertices, no two of which are connected by an edge. The *stability number* (or *independence number*) is the maximum size of all stable sets of vertices, and is here denoted $\beta(G)$. Since for any admissible vertex colouring, the set of vertices assigned a given colour is stable, the stability number and the chromatic number always satisfy the relation

$$\beta(G)\chi(G) \geq v. \tag{6.62}$$

Suppose $\mathcal{Y}$ is an arbitrary stable set of vertices of the graph $G(\mathcal{X}_n \cap Q_n; r_n)$. Then the balls of radius $r_n/2$ centred at the points of $\mathcal{Y}$ are disjoint, by the triangle inequality and the definition of the geometric graph. Therefore, the balls of radius 1 centred at $\{(2/r_n)X : X \in \mathcal{Y}\}$ are disjoint translates of B, all centred in the cube $(2/r_n)Q_n$, which is a cube of side $2R$. Therefore, enlarging this cube slightly, we obtain a cube of side $2(R+D)$ which entirely contains all of these balls, which have total volume $\theta\text{card}(\mathcal{Y})$. Extending this set of ball centres periodically, we obtain a B-packing of $\mathbb{R}^d$ with upper density $\theta\text{card}(\mathcal{Y})/(2^d(R+D)^d)$. Since this upper density is at most $\phi(B)$, it follows that

$$\beta(G(\mathcal{X}_n \cap Q_n; r_n)) \leq \frac{\phi(B)2^d(R+D)^d}{\theta} \leq \frac{(1+\varepsilon)\phi(B)2^dR^d}{\theta},$$

so that, by (6.62), if $\mathcal{X}_n(Q_n) \geq (1-2\varepsilon)nf_{\max}(Rr_n)^d$, then

$$\chi_n \geq \chi(G(\mathcal{X}_n \cap Q_n; r_n)) \geq \frac{(1-2\varepsilon)nf_{\max}(Rr_n)^d\theta}{(1+\varepsilon)\phi(B)(2R)^d}.$$

By (6.61) and the Borel–Cantelli lemma, this occurs for all large enough n, almost surely. The result follows by taking $\varepsilon \downarrow 0$. □

Proof of Theorem 6.18 Let $\varepsilon > 0$, and choose $v_1, \ldots, v_d \in \mathbb{R}^d$ such that $\{v_1, \ldots, v_d\}$ generates a lattice B-packing of $\mathbb{R}^d$ with relative volume at least $(1-\varepsilon)\phi_L(B)$. Let $\mathcal{L}$ denote the collection of centres of this packing, that is, the set of all linear combinations of $v_1, \ldots, v_d$ with integer coefficients. Let V be the Voronoi cell of the origin in $\mathcal{L}$, that is, the set of points of $\mathbb{R}^d$ lying closer to the origin (using Euclidean distance) than to any other point of $\mathcal{L}$. The set of translates $\{V \oplus \{u\} : u \in \mathcal{L}\}$ forms a tessellation of $\mathbb{R}^d$. Note that for $u, u' \in \mathcal{L}$ with $u \neq u'$, we have $\|u-u'\| \geq 2$. Indeed, if this were untrue, then the midpoint $(u+u')/2$ would lie in the interiors of the balls $B(u;1)$ and $B(u';1)$, contrary to the definition of a packing.

For all large enough R, by the definition of the relative volume of the lattice packing,

$$\theta\text{card}(\mathcal{L} \cap [-R/2, R/2]^d) \geq (1-2\varepsilon)\phi_L(B)R^d,$$

and if $|V|$ denotes the Lebesgue measure of the Voronoi cell V, we have

$$|V|\text{card}(\mathcal{L} \cap [-R/2, R/2]^d) \leq (1+\varepsilon)R^d.$$

Combining these inequalities, we obtain

$$|V| \leq \left(\frac{1+\varepsilon}{1-2\varepsilon}\right)\frac{\theta}{\phi_L(B)}.$$

Let $\delta > 0$ and assume δ is so small that

$$\mathrm{diam}([0,\delta]^d) < \varepsilon. \tag{6.63}$$

Let $\{Q_1, \ldots, Q_m\}$ be the set of cubes of side δ, of the form $Q_j = \{\delta z\} \oplus [0,\delta]^d$, with $z \in \mathbb{Z}^d$, that have non-empty intersection with $(1+\varepsilon)V$. Assume δ is chosen to be small enough, so that the total volume of these covering cubes is at most $(1+2\varepsilon)^d|V|$, and therefore

$$m \leq (1+2\varepsilon)^d|V|/\delta^d \leq \frac{(1+2\varepsilon)^d(1+\varepsilon)\theta}{(1-2\varepsilon)\phi_L(B)\delta^d}. \tag{6.64}$$

Let $a_n := (1+\varepsilon)\delta^d n r_n^d f_{\max}/2^d$. For any cube of side $\delta r_n/2$, the number of points of $\mathcal{X}_n$ in such a cube has expectation at most $\delta^d n r_n^d f_{\max}/2^d$, and by Lemma 1.1 there exists $\gamma > 0$ such that the probability that this number exceeds a_n is at most $\exp(-\gamma n r_n^d)$, and so is less than n^{-3} (for large n) since $nr_n^d/\log n \to \infty$ by assumption.

Let $\{u_1, \ldots, u_{\kappa_n}\}$ be the set of $u \in \mathcal{L}$ such that the set $(1+\varepsilon)(r_n/2)(V \oplus \{u\})$ has non-empty intersection with the support of f. For $j = 1, 2, \ldots, \kappa_n$, let

$$V_j^n := (1+\varepsilon)(r_n/2)(V \oplus \{u_j\}).$$

The sets $V_1^n, \ldots, V_{\kappa_n}^n$ cover the support of f. Since f has bounded support and $nr_n^d \to \infty$, it follows that $\kappa_n \leq n$, for n large.

Each of the sets V_j^n is itself covered by the sets

$$Q_{ij}^n := (r_n/2)((Q_i \cap (1+\varepsilon)V) \oplus \{(1+\varepsilon)u_j\}), \qquad 1 \leq i \leq m.$$

The sets Q_{ij}^n are all cubes of side $\delta r_n/2$, except for those at the boundary of V_j^n which are subsets of such cubes; in the sequel we refer to all sets Q_{ij}^n as 'cubes' even if they lie at the boundary. Let F_n be the event that each of the cubes $Q_{ij}^n, 1 \leq i \leq m, 1 \leq j \leq \kappa_n$ contains at most a_n points of $\mathcal{X}_n$, that is,

$$F_n := \bigcap_{i=1}^{m} \bigcap_{j=1}^{\kappa_n} \{\mathcal{X}_n(Q_{ij}^n) \leq a_n\}.$$

By the Borel–Cantelli lemma, with probability 1, F_n occurs for all but finitely many n.

Assuming F_n occurs, let us adopt the following colouring of the points of X_n, using colours represented by integers $1, 2, \ldots, ma_n$. Let the points in Q_{ij}^n be assigned distinct colours in an arbitrary way from the set of colours $\{(i-1)a_n + 1, (i-1)a_n + 2, \ldots, ia_n\}$. This is possible because $\mathcal{X}_n(Q_{ij}^n) \leq a_n$. This colouring uses at most ma_n colours, and if two points X, Y have the same colour, then for some $i \leq m$ they must lie in cubes Q_{ij}^n and $Q_{ij'}^n$, for some $j \neq j'$. In this case,

$X-(1+\varepsilon)(r_n/2)u_j$ and $Y-(1+\varepsilon)(r_n/2)u_{j'}$ both lie in the cube $(r_n/2)Q_i$, and therefore by (6.63),

$$\|X - Y - (1+\varepsilon)(r_n/2)(u_j - u_{j'})\| < \varepsilon r_n/2.$$

Since $\|u_j - u_{j'}\| \geq 2$, it follows by the triangle inequality that

$$\begin{aligned}\|X - Y\| &\geq (1+\varepsilon)(r_n/2)\|u_j - u_{j'}\| \\ &\quad - \|X - Y - (1+\varepsilon)(r_n/2)(u_j - u_{j'})\| > r_n,\end{aligned}$$

so that X and Y are not adjacent in $G(\mathcal{X}_n; r_n)$. This shows that the colouring adopted is admissible. Finally, by (6.64) and the definition of a_n, the number of colours used is bounded by the expression

$$ma_n \leq \frac{(1+2\varepsilon)^d(1+\varepsilon)^2\theta\delta^d n r_n^d f_{\max}}{(1-2\varepsilon)\phi_L(B)\delta^d 2^d} \leq \frac{(1+2\varepsilon)^{3d} n\theta r_n^d f_{\max}}{(1-2\varepsilon)\phi_L(B)2^d}$$

and, since ε is arbitrarily small, this gives us the result (6.60). □

6.6 Notes and open problems

Notes A topic related to those discussed in this chapter is the *range* of the sample $\mathcal{X}_n$, that is, the value of $\max\{\|X - Y\| : X \in \mathcal{X}_n, Y \in \mathcal{X}_n\}$, which is also the threshold value of r above which $\mathcal{C}(G(\mathcal{X}_n; r)) = n$. For a class of cases where the underlying density f is spherically symmetric, asymptotic results for the range have been obtained by Henze and Klein (1996); see also Appel *et al.* (2002).

Section 6.1. The results in Section 6.1 are new, although a result for the scan statistic, along the lines of Theorem 6.1, is given by Månsson (1999). For Erdös–Rényi random graphs, there are analogous results in Chapter III, Section 2, of Bollobás (1985). Focusing results for the clique number and scan statistic in the thermodynamic limit, analogous to the one given for the maximum degree in Theorem 6.6, are given in Penrose (2002).

Section 6.2. The idea of the proof of Theorem 6.10 is partly due to McDiarmid (2003). Detailed strong laws for the scan statistic also appear in Auer *et al.* (1991) and Auer and Hornik (1994).

Sections 6.3 and 6.4. Some strong laws for maximum degree and for cliques, in the case of uniformly distributed points on the unit cube using the l_∞ norm, are given in Appel and Russo (1997*a*); we take these considerably further. Deheuvels *et al.* (1988) give some detailed strong laws for the clique number in the uniform case.

Section 6.5. The results here on chromatic number use ideas of McDiarmid and Reed (1999), who prove asymptotic results for the chromatic number of geometric graphs on a certain class of infinite deterministic sets in $\mathbb{R}^2$, using the Euclidean norm. In the important special case of the Euclidean norm with $d = 2$, Corollary 6.19 is in McDiarmid (2003).

Open problems We conjecture that in the intermediate regime with $nr_n^d \to \infty$, and $nr_n^d/\log n \to 0$, the clique number is asymptotically focused on just two values. If true, this would extend the result in Penrose (2002) which is, in effect, concerned only with cases where nr_n^d remains bounded.

It is natural to look to extend the weak laws of large numbers in Section 6.2 to strong laws with almost sure convergence. For some sequences $(r_n)_{n\geq 1}$ it is possible to strengthen this to almost sure convergence by methods similar to those used in proving Theorem 6.11, but not for all sequences (r_n) in the range considered in Section 6.2.

Regarding the chromatic number, one may ask whether any focusing phenomena hold for the chromatic number, analogous to those seen in Section 6.1 and in Penrose (2002) for the maximum degree and clique number. Another question concerns the *connectivity regime* with $\lim_{n\to\infty} nr_n^d/\log n = \alpha \in (0, \infty)$; can any strong laws for the chromatic number be established in this limiting regime, to go with those seen in Section 6.5 for the subconnective and superconnective regimes? More modestly, in this regime one might hope to improve on the asymptotic upper bound on the ratio $\chi_n/\mathcal{C}_n$, provided by Theorem 6.14, along with Theorem 6.16 and eqn (6.58). In fact, an improvement can be effected by deriving an analogous result to Theorem 6.14 for the maximum 'left-degree' of $G(\mathcal{X}_n; r_n)$ (i.e. the maximum, over $X \in \mathcal{X}_n$, of the number of points in $\mathcal{X}_n$ adjacent to X and to its left), and observing that this provides an upper bound for $\chi_n - 1$ since one may assign colours (i.e. positive integers, using the lowest available integer each time) to points in order from left to right. This argument should yield a limiting upper bound, for the ratio $\chi_n/\mathcal{C}_n$, of

$$\frac{2^{d-1} H_+^{-1}(2/(f_{\max}\alpha))}{H_+^{-1}(2^d/(f_{\max}\alpha))}.$$

Any further improvement on this upper bound would be of interest.

We have not considered in detail the limit theory for the stability number of $G(\mathcal{X}_n; r)$ (the stability number $\beta(G)$ arose in the proof of Theorem 6.17, and is also of interest in its own right). Consider the thermodynamic limit, taking $r_n := (\lambda/n)^{1/d}$. At least in the case of the uniform distribution $f := f_{\mathrm{U}}$, and the Poisson process $\mathcal{P}_n$, it should be possible by subadditive methods (see the Akcoglu–Krengel ergodic theorem, in Theorem 4.9 of Yukich (1998)) to show that $n^{-1}\beta(G(\mathcal{P}_n; r_n))$ converges in probability to a finite constant. In the subcritical case where $\lambda < \lambda_{\mathrm{c}}$, with the percolation threshold λ_{c} defined formally in Section 9.6 below, the so-called 'objective method' (see Steele (1997, Chapter 5)) can be used instead; a result of Penrose and Yukich (2003) can be applied in this subcritical case to show convergence in mean-square to a constant of either $n^{-1}\beta(G(\mathcal{X}_n; r_n))$ or $n^{-1}\beta(G(\mathcal{P}_n; r_n))$, for any bounded density f. The result of Penrose and Yukich also gives a formula for the value of the limiting constant in this subcritical case.

7

MINIMUM DEGREE: LAWS OF LARGE NUMBERS

Given a sequence $(r_n)_{n \geq 1}$, let δ_n be the minimum degree of $G(\mathcal{X}_n; r_n)$. This chapter contains laws of large numbers for δ_n. It is sometimes convenient to reformulate results on δ_n in terms of the threshold radius for the minimum degree to exceed a certain value. Given a finite set $\mathcal{X} \subset \mathbb{R}^d$, and given $k \in \mathbb{N}$, let $M_k(\mathcal{X})$ denote the *largest k-nearest-neighbour link*, that is, the smallest value of r for which $G(\mathcal{X}; r)$ has minimum degree at least k. The largest k-nearest-neighbour link is of considerable importance in combinatorial optimization; see Steele and Tierney (1986). In notation from Section 1.4, $M_k(\mathcal{X})$ is the threshold $\rho(\mathcal{X}; \delta \geq k)$, where δ denotes the minimum vertex degree of any given graph.

The first two sections are devoted to strong laws of large numbers for $M_k(\mathcal{X}_n)$, both for k fixed and k growing with n. In Section 7.3, we deduce strong laws for δ_n from these. The method of proof for the strong laws is similar to that used already for the maximum degree, and described at the start of Section 6.3, but extra complications arise, except in the especially simple case of points uniformly distributed in the torus, because the effective volume of balls near the boundary of the support of F is less than that in its interior, and also the number of balls of small radius that can be fitted along the boundary grows in a different way from the number of points that can be fitted in the region as a whole. The results demonstrate the interplay between different types of boundary effect, and their dependence on the underlying density f.

In this chapter, as in Section 5.2, Ω denotes the support of F and $\partial\Omega$ is the topological boundary of Ω. Let $f|_\Omega$ be the restriction of f to Ω, let f_0 denote the essential infimum of the function $f|_\Omega$, and let $f_1 := \inf_{\partial\Omega} f$. Let $\mathrm{Leb}(\cdot)$ denote Lebesgue measure (volume) of Borel sets in $\mathbb{R}^d$; as in previous chapters, set $\theta := \mathrm{Leb}(B(\mathbf{0}; 1))$. If A and B are non-empty compact sets in $\mathbb{R}^d$, then the Brunn–Minkowski inequality (Theorem 5.11) implies that

$$\mathrm{Leb}(A \oplus B) \geq \mathrm{Leb}(A) + \mathrm{Leb}(B), \tag{7.1}$$

which will be useful on more than one occasion.

7.1 Thresholds in smoothly bounded regions

This section and the next are concerned with strong laws of large numbers for the threshold $M_{k_n}(\mathcal{X}_n)$ for some given sequence $(k_n)_{n \geq 1}$. We assume either that $k_n / \log n \to \infty$, or k_n grows like a constant times $\log n$. The constant might be zero, so k_n fixed, and in particular $k_n = 1$ for all n, are included in the argument. The function $H(a), a > 0$, and its inverses H_+^{-1} and H_-^{-1}, are as defined at (6.32).

First, consider the case where the underlying distribution F is supported by the unit cube $[0,1]^d$, and the measure of distance between points is *toroidal*; that is $\mathrm{dist}(x,y) = \max_{z \in \mathbb{Z}^d} \|x - y + z\|$. In this case we say the points X_i are *distributed on the torus.*

Theorem 7.1 *Suppose that $d \geq 1$ and the points X_i are distributed on the torus, with $f_0 > 0$. Suppose $(k_n)_{n \geq 1}$ is a sequence of positive integers with $k_n / \log n \to b \in [0, \infty]$, and $k_n/n \to 0$ as $n \to \infty$. In the case $b < \infty$, assume also that the sequence $(k_n)_{n \geq 1}$ is nondecreasing, and define $a \in [0,1)$ by $a/H(a) = b$. Then if $b = \infty$,*

$$\lim_{n\to\infty} \left(\frac{n\theta M_{k_n}(\mathcal{X}_n)^d}{k_n} \right) = \frac{1}{f_0}, \qquad \text{a.s.}$$

If $b < \infty$,

$$\lim_{n\to\infty} \left(\frac{n\theta M_{k_n}(\mathcal{X}_n)^d}{\log n} \right) = \frac{1}{f_0 H(a)}, \qquad \text{a.s.}$$

The main subject of the present section is the more complicated case where $d \geq 2$ and Ω has a smooth boundary (the case $d = 1$ amounts to points distributed in an interval, and is covered by the study of points in the cube in Section 7.2). The notion of a $(d-1)$-dimensional C^2 submanifold in $\mathbb{R}^d$ was described in Section 5.2.

Theorem 7.2 *Suppose that $d \geq 2$, and that $\partial\Omega$ is a compact $(d-1)$-dimensional C^2 submanifold of $\mathbb{R}^d$. Suppose also that $f_0 > 0$, and $f|_\Omega$ is continuous at x for all $x \in \partial\Omega$. Let $(k_n)_{n\geq 1}$ be a sequence of positive integers, with $\lim_{n\to\infty}(k_n/n) = 0$ and $\lim_{n\to\infty}(k_n/\log n) = b \in [0,\infty]$. In the case $b < \infty$, assume also that the sequence $(k_n)_{n\geq 1}$ is nondecreasing, and define numbers a_0 and a_1 in $[0,1)$ by*

$$\frac{a_0}{H(a_0)} = b, \qquad \frac{a_1}{H(a_1)} = \frac{bd}{d-1}. \tag{7.2}$$

If $b = \infty$, then with probability 1,

$$\lim_{n\to\infty} \left(\frac{n\theta M_{k_n}(\mathcal{X}_n)^d}{k_n} \right) = \max\left(\frac{1}{f_0}, \frac{2}{f_1} \right), \tag{7.3}$$

while if $b < \infty$, then with probability 1,

$$\lim_{n\to\infty} \left(\frac{n\theta M_{k_n}(\mathcal{X}_n)^d}{\log n} \right) = \max\left(\frac{1}{f_0 H(a_0)}, \frac{2(d-1)}{d f_1 H(a_1)} \right).$$

It is clear that the toroidal setting is the same as that for Theorem 7.2, only without boundary effects. Therefore Theorem 7.1 is proved by a similar (easier) argument to the proof of Theorem 7.2. Details of the modifications required to prove Theorem 7.1 are left to the reader.

The rest of this section is devoted to proving Theorem 7.2, and *we assume throughout the rest of this section that $\partial\Omega$ is a compact $(d-1)$-dimensional*

C^2 submanifold of $\mathbb{R}^d$. The proof uses Poissonization; enlarging the probability space, assume that for each n there exist Poisson variables $N^-(n)$ and $M(n)$ with mean $n - n^{3/4}$ and $2n^{3/4}$, respectively, independent of each other and of $(X_1, X_2, \ldots)$. Define point processes

$$\mathcal{P}_n^- = \{X_1, X_2, \ldots, X_{N^-(n)}\}, \quad \mathcal{P}_n^+ = \{X_1, X_2, \ldots, X_{N^-(n)+M(n)}\}.$$

Then $\mathcal{P}_n^-$ (respectively, $\mathcal{P}_n^+$) is a Poisson process on $\mathbb{R}^d$ with intensity function $(n - n^{3/4})f(\cdot)$ (respectively, $(n + n^{3/4})f(\cdot)$). The point processes $\mathcal{P}_n^-$, $\mathcal{P}_n^+$, and $\mathcal{X}_n$ are coupled in such a way that $\mathcal{P}_n^- \subseteq \mathcal{P}_n^+$, and by Lemma 1.4, defining the event $H_n := \{\mathcal{P}_n^- \subseteq \mathcal{X}_n \subseteq \mathcal{P}_n^+\}$, we have for all large enough n that

$$P[H_n^{\mathrm{c}}] \leq 2\exp(-n^{1/2}/9). \tag{7.4}$$

With b, a_0, and a_1 as given in the statement of Theorem 7.2, in the proof it is useful to define the function $(\rho_n^j(u), u > 0)$ for $j = 0$ or $j = 1$ by

$$\rho_n^j(u) := \left(\frac{2^j u k_n}{n\theta f_j}\right)^{1/d} \quad \text{if } b = \infty,$$

$$\rho_n^j(u) := \left(\frac{2^j(d-j)u\log n}{nd\theta f_j H(a_j)}\right)^{1/d} \quad \text{if } b < \infty.$$

Lemma 7.3 *Suppose $j = 0$ or $j = 1$. Suppose $(k_n)_{n\geq 1}$, $b \in [0, \infty]$, and (if $b < \infty$) $a_j \in [0, 1)$ are as in the statement of Theorem 7.2. Suppose $0 < \beta < 1$. Then with probability 1, $M_{k_n}(\mathcal{X}_n) > \rho_n^j(\beta)$ for all large enough n.*

Proof Pick $\varepsilon_1 > 0$ satisfying

$$\varepsilon_1^{1/d} + \beta^{1/d} < (1 - 4\varepsilon_1)^{1/d}. \tag{7.5}$$

Recall that $B(x; r)$ denotes the r-ball centred at x. For $x \in \mathbb{R}^d$, define the event $E_n(x)$ by

$$E_n(x) := \{\mathcal{P}_n^+[B(x; \rho_n^j(1 - 4\varepsilon_1))] \leq k_n\} \cap \{\mathcal{P}_n^-[B(x; \rho_n^j(\varepsilon_1))] \geq 1\}.$$

If $\|y - x\| \leq \rho_n^j(\varepsilon_1)$ and $\|z - y\| \leq \rho_n^j(\beta)$, then by (7.5) and the triangle inequality, $\|z - x\| \leq \rho_n^j(1 - 4\varepsilon_1)$. So, if events H_n (defined at (7.4) above) and $E_n(x)$ occur there is a point X of $\mathcal{X}_n$ in $B(x; \rho_n^j(\varepsilon_1))$ with at most $k_n - 1$ other points of $\mathcal{X}_n$ in $B(X; \rho_n^j(\beta))$, and hence $M_{k_n}(\mathcal{X}_n) > \rho_n^j(\beta)$. Therefore

$$\{M_{k_n}(\mathcal{X}_n) \leq \rho_n^j(\beta)\} \subseteq H_n^{\mathrm{c}} \cup (\cup_{x\in\mathbb{R}^d} E_n(x))^{\mathrm{c}}. \tag{7.6}$$

First suppose $j = 0$, $b = \infty$. Let x_0 be a Lebesgue point of f with $f_0 \leq f(x_0) < f_0(1 - 3\varepsilon_1)/(1 - 4\varepsilon_1)$. For all large enough n, we have $F(B(x_0; \rho_n^0(1 - 4\varepsilon_1))) < (1 - 2\varepsilon_1)k_n/n$, so the expected number of points of $\mathcal{P}_n^+$ in $B(x_0; \rho_n^0(1 - 4\varepsilon_1))$ is at most $k_n(1 - \varepsilon_1)$. By Lemma 1.2, the probability that this number of points exceeds k_n decays exponentially in k_n. Also the probability that there is no point of $\mathcal{P}_n^-$ in $B(x_0; \rho_n^0(\varepsilon_1))$ decays exponentially in k_n. By these estimates, the fact that $k_n/\log n$ tends to infinity, and (7.4), we find that $1-P[E_n(x_0)\cap H_n]$ is summable in n. Therefore by the Borel–Cantelli lemma, with probability 1, $E_n(x_0) \cap H_n$ occurs for all but finitely many n, and the result follows, for this case, by (7.6).

Next consider the case with $j = 1$, $b = \infty$. Let $x_1 \in \partial\Omega$ with $f(x_1) = f_1$. By Lemma 5.6, for all large enough n, the volume of $B(x_1; \rho_n^1(1-4\varepsilon_1)) \cap \Omega$ is at most $k_n(1-3\varepsilon_1)/(nf_1)$, and $F(B(x_1; \rho_n^1(1-4\varepsilon_1))) < (1-2\varepsilon_1)(k_n/n)$, so that the expected number of points of $\mathcal{P}_n^+$ in $B(x_1; \rho_n^1(1-4\varepsilon_1))$ is at most $k_n(1-\varepsilon_1)$. By Lemma 1.2, the probability that this number of points exceeds k_n decays exponentially in k_n. Also the probability that there is no point of $\mathcal{P}_n^-$ in B'_n decays exponentially in k_n. By these estimates, the assumption that $k_n/\log n$ tends to infinity, and (7.4), we find that $1 - P[E_n(x_1) \cap H_n]$ is summable in n. Therefore, with probability 1, $E_n(x_1) \cap H_n$ occurs for all but finitely many n, and the result follows, for this case, by (7.6).

Next, suppose $b < \infty$, with j fixed satisfying either $j = 0$ or $j = 1$. Without loss of generality, assume in addition to (7.5) that $2\varepsilon_1 < 1 - a_j$. Set $\psi = f_j(1-3\varepsilon_1)/(1-4\varepsilon_1)$. Pick a collection, of maximal cardinality, of points $x_1^n, \ldots, x_{\sigma_n}^n$ in $\mathbb{R}^d$ such that the balls $B(x_i^n; \rho_n^0(1-4\varepsilon_1))$, $1 \le i \le \sigma_n$, are disjoint and satisfy

$$F(B(x_i^n; \rho_n^j(1-4\varepsilon_1))) \le 2^{-j}\psi\theta(\rho_n^j(1-4\varepsilon_1))^d = \frac{(1-3\varepsilon_1)(d-j)\log n}{ndH(a_j)},$$

and also

$$F(B(x_i^n; \rho_n^j(\varepsilon_1))) \ge f_j\theta\rho_n^j(\varepsilon_1)^d/8, \qquad 1 \le i \le \sigma_n. \tag{7.7}$$

Then by Lemma 5.2 in the case $j = 0$, or Lemma 5.8 in the case $j = 1$,

$$\liminf(((\log n)/n)^{(d-j)/d}\sigma_n) > 0. \tag{7.8}$$

For all n exceeding some n_0, and all $i \le \sigma_n$, we have

$$(n+n^{3/4})F(B(x_i^n; \rho_n^j(1-4\varepsilon_1))) \le \frac{(1-2\varepsilon_1)(d-j)\log n}{dH(a_j)} := \mu_n.$$

By the definition of a_j,

$$\lim_{n\to\infty}(k_n/\mu_n) = \frac{bdH(a_j)}{(d-j)(1-2\varepsilon_1)} = \frac{a_j}{1-2\varepsilon_1} \in (a_j, 1).$$

Therefore by Lemma 1.3, for large n, and $1 \le i \le \sigma_n$,

$$\begin{aligned} P[\mathcal{P}_n^+[B(x_i^n; \rho_n^j(1-4\varepsilon_1))] \le k_n] &\ge P[\mathrm{Po}(\mu_n) = k_n] \\ &\ge \left(\frac{\mathrm{e}^{-1/(12k_n)}}{\sqrt{2\pi k_n}}\right)\exp(-\mu_n H(k_n/\mu_n)) \\ &\ge \left(\frac{\mathrm{e}^{-1/(12k_n)}}{\sqrt{2\pi k_n}}\right)\exp\left(-\mu_n H\left(a_j\right)\right) \ge n^{(\varepsilon_1-1)(d-j)/d}. \end{aligned}$$

Given the number of points of $\mathcal{P}_n^+$ in $B(x_i^n; \rho_n^j(1-4\varepsilon_1))$, the conditional distribution of the number of points of $\mathcal{P}_n^-$ in $B(x_i^n; \rho_n^j(\varepsilon_1))$ is binomial with parameter

$$\frac{(n-n^{3/4})F(B(x_i^n; \rho_n^j(\varepsilon_1)))}{(n+n^{3/4})F(B(x_i^n; \rho_n^j(1-4\varepsilon_1)))},$$

which remains bounded away from zero by (7.7). Since also the mean number of points of $\mathcal{P}_n^+$ in $B(x_i^n; \rho_n^j(1-4\varepsilon_1))$ tends to infinity, there exists $\eta > 0$ such that for all large enough n,

$$\inf_{1 \le i \le \sigma_n} P[E_n(x_i^n) | \mathcal{P}_n^+[B(x_i^n; \rho_n^j(1-4\varepsilon_1))] \le k_n] \ge \eta.$$

Hence, for all large enough n,

$$\inf_{1 \le i \le \sigma_n} P[E_n(x_i^n)] \ge \eta n^{(\varepsilon_1 - 1)(d-j)/d}. \tag{7.9}$$

The events $E_n(x_i^n), 1 \le i \le \sigma_n$, are independent, so by (7.9), and the estimate $1 - t \le \mathrm{e}^{-t}$, for large enough n we have

$$P\left[\left(\bigcup_{i=1}^{\sigma_n} E_n(x_i^n)\right)^{\mathrm{c}}\right] \le \exp(-\eta \sigma_n n^{(\varepsilon_1-1)(d-j)/d}),$$

which is summable in n by (7.8). The result follows, for this case, by (7.4) and the Borel–Cantelli lemma. □

To complete the proof of Theorem 7.2, we need to find upper bounds on $M_{k_n}(\mathcal{X}_n)$. With $b = \lim_{n\to\infty}(k_n/\log n)$ as assumed in the statement of Theorem 7.2, define the constants ρ_n by

$$\rho_n := (k_n/n)^{1/d} \text{ if } b = \infty, \quad \rho_n := ((\log n)/n)^{1/d} \text{ if } b < \infty. \tag{7.10}$$

Given a graph G with vertex set V, let a subset U of the vertex set be denoted *k-separated* set if (i) U is non-empty, and (ii) at most k vertices in $V \setminus U$ lie adjacent to U. Recalling from Section 5.3 the definition of Minkowski addition $\oplus$ of sets in $\mathbb{R}^d$, observe that a non-empty subset U of a finite set $\mathcal{X} \subset \mathbb{R}^d$ is k-separated for $G(\mathcal{X}; r)$, if and only if $\mathcal{X}[U \oplus B(\mathbf{0}; r) \setminus U] \le k_n$.

Suppose a sequence $(k_n)_{n\ge1}$ is given. For $K > 0, t > 0$, define the event $E'_n(K; t)$ by

$$\begin{aligned} E'_n(K;t) := \{&\text{there exists a } k_n - \text{separated set } U \text{ for } G(\mathcal{X}_n; t\rho_n) \\ &\text{with } \mathrm{diam}(U) \le K\rho_n\}. \end{aligned} \tag{7.11}$$

If the minimum degree of a graph is at most k, then it has a k-separated set consisting of a single vertex. Hence, if $M_{k_n+1}(\mathcal{X}_n) > t\rho_n$, then $E'_n(K; t)$ occurs. Therefore Proposition 7.4 below provides the upper bound needed to complete the proof of Theorem 7.2.

Proposition 7.4 is stated in greater generality than required for the proof of Theorem 7.2, allowing as it does for k_n-separated sets which are not singletons. It is stated in this generality for use later on in proving results about connectivity.

Proposition 7.4 *Let $(k_n)_{n\geq 1}$, $b \in [0,\infty)$, a_0 and a_1 be as in the statement of Theorem 7.2, and assume the other hypotheses of that result hold. Let $K > 0$. In the case $b = \infty$, fix t satisfying*

$$\theta t^d > \max\left(\frac{1}{f_0}, \frac{2}{f_1}\right); \tag{7.12}$$

in the case $b < \infty$, fix t satisfying

$$\theta t^d > \max\left(\frac{1}{f_0 H(a_0)}, \frac{2(d-1)}{d f_1 H(a_1)}\right). \tag{7.13}$$

Then with probability 1, the event $E'_n(K;t)$ occurs for only finitely many n, and hence $M_{k_n+1}(\mathcal{X}_n) < t\rho_n$ for all but finitely many n.

To prove Proposition 7.4, define the constant c_1 by

$$c_1 = \mathrm{diam}([0,1]^d), \tag{7.14}$$

and recall the definition of $B^*(x;r,\eta,e)$ given at (5.9). With t fixed satisfying (7.12) or (7.13), pick $\varepsilon_2 > 0$ such that

$$f_0\theta(t - 5c_1\varepsilon_2)^d > 1 + \varepsilon_2 \quad \text{if } b = \infty, \tag{7.15}$$

$$f_0 H(a_0)\theta(t - 5c_1\varepsilon_2)^d > 1 + \varepsilon_2 \quad \text{if } b < \infty, \tag{7.16}$$

and also such that for any l_2 unit vector $e \in S^{d-1}$,

$$f_1 \mathrm{Leb}(B^*(\mathbf{0}; t - 3c_1\varepsilon_2, \varepsilon_2, e)) \geq 1 + 2\varepsilon_2 \quad \text{if } b = \infty, \tag{7.17}$$

$$d f_1 H(a_1) \mathrm{Leb}(B^*(\mathbf{0}; t - 3c_1\varepsilon_2, \varepsilon_2, e)) \geq (1 + 2\varepsilon_2)(d-1) \quad \text{if } b < \infty. \tag{7.18}$$

For $r > 0$, let $r\mathbb{Z}^d$ denote the set of points of the form $y = rz$ with $z \in \mathbb{Z}^d$, regarded as a subset of $\mathbb{R}^d$. For $y \in \varepsilon_2\rho_n\mathbb{Z}^d$, let $C_n(y) := \{y\} \oplus [-\varepsilon_2\rho_n/2, \varepsilon_2\rho_n/2]^d$, the rectilinear hypercube of side $\varepsilon_2\rho_n$ centred at y.

The proof of Proposition 7.4 proceeds by a discretization argument. With ρ_n defined at (7.10), instead of the precise configuration $\mathcal{X}_n$, one considers the set of $z \in \varepsilon_2\rho_n\mathbb{Z}^d$ for which $\mathcal{X}_n(C_n(z)) > 0$, and applies counting arguments to those possibilities for this set which are compatible with the existence of separated sets.

We shall use the subsequence trick; in the case $b < \infty$, choose a positive integer J such that

$$J\varepsilon_2(d-1)/d > 1, \tag{7.19}$$

but if $b = \infty$ set $J = 1$. For $m = 1, 2, 3, \ldots$, let $\nu(m) = m^J$. For $K > 0$, define $\mathcal{T}_m(K)$ (a collection of subsets of $\varepsilon_2\rho_{\nu(m)}\mathbb{Z}^d$) by

$$\mathcal{T}_m(K) := \{\tau \subset \varepsilon_2\rho_{\nu(m)}\mathbb{Z}^d : \mathrm{diam}(\tau) \leq 2K\rho_{\nu(m)}, C_{\nu(m)}(z) \cap \Omega \neq \emptyset \ \forall z \in \tau\}.$$

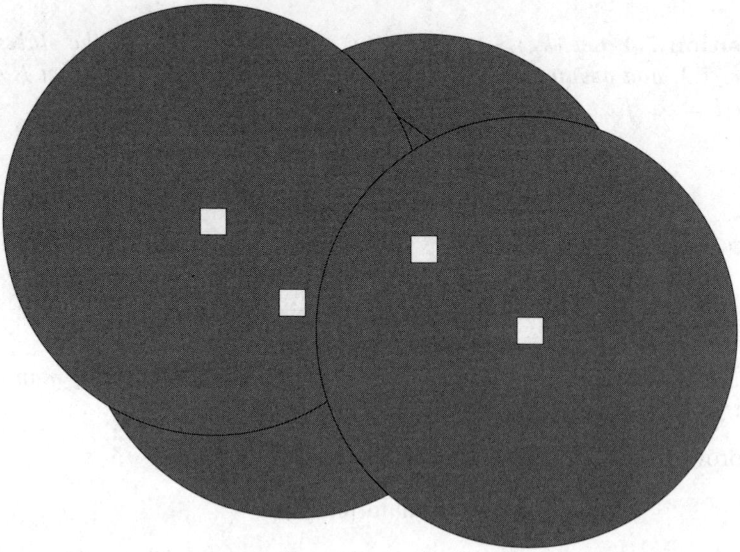

FIG. 7.1. The set $A_m(\tau)$ is shaded for a set τ with four elements.

Given $\tau \in \mathcal{T}_m(K)$, and $t > 0$, define the 'annulus-like' set $A_m(\tau)$ (see Fig. 7.1 for an example) by

$$A_m(\tau) := \left(\tau \oplus B(\mathbf{0}; (t - 2c_1\varepsilon_2)\rho_{\nu(m)})\right) \setminus \left(\tau \oplus C_{\nu(m)}(\mathbf{0})\right). \tag{7.20}$$

Define the event

$$F_m(\tau) := \{\mathcal{X}_m [A_m(\tau)] \leq k_m\} \quad \text{if } b = \infty, \tag{7.21}$$
$$F_m(\tau) := \{\mathcal{X}_{\nu(m)} [A_m(\tau)] \leq k_{\nu(m+1)}\} \quad \text{if } b < \infty. \tag{7.22}$$

The purpose of these definitions is demonstrated by the next result.

Lemma 7.5 *Let $K > \max(1, t)$. Then there exists m_0 such that if $m \geq m_0$ and $\nu(m) \leq n < \nu(m+1)$, then the event $E'_n(K;t)$ defined at (7.11) is contained in the union of the events $F_m(\tau), \tau \in \mathcal{T}_m(K)$.*

Proof First suppose $b < \infty$. Choose m_0 so that if $m \geq m_0$, then

$$c_1\varepsilon_2\rho_{\nu(m)} + (t - 2c_1\varepsilon_2)\rho_{\nu(m)} \leq t\rho_{\nu(m+1)}. \tag{7.23}$$

Suppose $m \geq m_0$ and $\nu(m) \leq n < \nu(m+1)$. Given $U \subseteq \mathcal{X}_n$, let $\tau(U)$ denote the discretization of U in $\varepsilon_2\rho_{\nu(m)}\mathbb{Z}^d$, that is, the set of $z \in \varepsilon_2\rho_{\nu(m)}\mathbb{Z}^d$ for which $U \cap C_{\nu(m)}(z) \neq \emptyset$. If $\operatorname{diam}(U) \leq K\rho_n$, then $\operatorname{diam}(\tau(U)) \leq 2K\rho_{\nu(m)}$, so that $\tau(U) \in \mathcal{T}_m(K)$. Also, by (7.23) and the triangle inequality,

$$A_m(\tau(U)) \subseteq U \oplus B(\mathbf{0}; t\rho_{\nu(m+1)}) \setminus U.$$

If also U is a k_n-separated set for $G(\mathcal{X}_n; t\rho_n)$, then $\mathcal{X}_n[U \oplus B(\mathbf{0}; t\rho_n) \setminus U] \leq k_{\nu(m+1)}$, and hence $\mathcal{X}_{\nu(m)}[A_m(\tau(U))] \leq k_{\nu(m+1)}$. This completes the proof in the case $b < \infty$.

When $b = \infty$ the argument is similar; replace $\rho_{\nu(m+1)}$ by ρ_m in the right hand side of (7.23). □

Let $\mathcal{T}_m^0(K)$ be the collection of all $\tau \in \mathcal{T}_m(K)$ such that $A_m(\tau) \subseteq \Omega$ and let $\mathcal{T}_m^1(K) = \mathcal{T}_m(K) \setminus \mathcal{T}_m^0(K)$. Let $|\mathcal{T}_m^0(K)|$ and $|\mathcal{T}_m^1(K)|$ be the cardinalities of $\mathcal{T}_m^0(K)$ and $\mathcal{T}_m^1(K)$, respectively.

Lemma 7.6 *Let $K > \max(1, t)$. Then either for $j = 0$ or for $j = 1$,*

$$\limsup_{m\to\infty} \left(\rho_{\nu(m)}^{d-j} |\mathcal{T}_m^j(K)|\right) < \infty. \tag{7.24}$$

Proof Given $\tau \in \mathcal{T}_m(K)$, let $y(\tau)$ be the first element of τ according to the lexicographic ordering on $\varepsilon_2\rho_n\mathbb{Z}^d$, and let

$$\tau' := \{(\varepsilon_2\rho_{\nu(m)})^{-1}(y - y(\tau)) : y \in \tau\}.$$

Then $y(\tau)$ and τ' together determine τ. Also, τ' is a subset of $\mathbb{Z}^d \cap B(\mathbf{0}; 2K/\varepsilon_2)$, and the number of such subsets is a constant independent of m. Therefore $|\mathcal{T}_m^j(K)|$ is bounded by a constant times the number of possibilities for $y(\tau)$ consistent with $\tau \in |\mathcal{T}_m^j(K)|$.

Since $y(\tau) \in \varepsilon_2\rho_{\nu(m)}\mathbb{Z}^d$ with $C_{\nu(m)}(y_\tau) \cap \Omega \neq \emptyset$, and Ω is bounded, the number of possibilities for $y(\tau)$ consistent with $\tau \in \mathcal{T}_m^0(K)$ is at most a constant times $\rho_{\nu(m)}^{-d}$, which gives us (7.24) for the case $j = 0$.

If $\tau \in \mathcal{T}_m^1(K)$, then $\operatorname{dist}(y(\tau), \partial\Omega) \leq 3K\rho_{\nu(m)}$. By Lemma 5.4, the number of balls of radius $3K\rho_{\nu(m)}$ centred at points of $\partial\Omega$, required to cover $\partial\Omega$, is $O(\rho_{\nu(m)}^{1-d})$. The number of points of $\varepsilon_2\rho_{\nu(m)}\mathbb{Z}^d$ lying in any ball of radius $3K\rho_{\nu(m)}$ is bounded by a constant independent of m, and it follows that for $\tau \in \mathcal{T}_m^1(K)$, the number of possibilities for $y(\tau)$ is $O(\rho_{\nu(m)}^{1-d})$. This proves (7.24) for the case $j = 1$. □

Lemma 7.7 *Suppose $j = 0$ or $j = 1$, and suppose $(k_n)_{n\geq 1}$, b, a_0, a_1, K and t satisfy the assumptions of Proposition 7.4. Then with probability 1, the event $\cup_{\tau\in\mathcal{T}_m^j(K)} F_m(\tau)$ occurs for only finitely many m.*

Proof By (7.14), $C_n(\mathbf{0}) \subseteq B(\mathbf{0}; c_1\varepsilon_2\rho_n)$, so that the triangle inequality gives us

$$C_n(\mathbf{0}) \oplus B(\mathbf{0}; (t - 3c_1\varepsilon_2)\rho_n) \subseteq B(\mathbf{0}; (t - 2c_1\varepsilon_2)\rho_n),$$

and therefore

$$A_m(\tau) \supseteq \left(\tau \oplus C_{\nu(m)}(\mathbf{0})\right) \oplus B(\mathbf{0}; (t - 3c_1\varepsilon_2)\rho_{\nu(m)}) \setminus \left(\tau \oplus C_{\nu(m)}(\mathbf{0})\right). \tag{7.25}$$

Hence by the Brunn–Minkowski inequality (7.1),

$$\operatorname{Leb}(A_m(\tau)) \geq \theta((t - 3c_1\varepsilon_2)\rho_{\nu(m)})^d.$$

If $\tau \in \mathcal{T}_m^0(K)$, then $A_m(\tau) \subseteq \Omega$, and hence,

$$E\mathcal{X}_{\nu(m)}[A_m(\tau)] \geq f_0\theta(t - 3c_1\varepsilon_2)^d \nu(m)\rho_{\nu(m)}^d := \mu_{m,0}.$$

By conditions (7.15) and (7.16) on ε_2,

$$\mu_{m,0} \geq (1 + \varepsilon_2)k_m \quad (b = \infty), \tag{7.26}$$

$$H(a_0)\mu_{m,0} \geq (1 + \varepsilon_2)\log \nu(m) \quad (b < \infty). \tag{7.27}$$

By Lemma 5.5 and a compactness argument, there exists a finite collection of triples (ζ_i, δ_i, e_i), $1 \leq i \leq \mu'$, with $\zeta_i \in \partial\Omega, \delta_i > 0$ and e_i a unit vector for each i, such that for all $x \in B(\zeta_i; \delta_i) \cap \Omega$ and $h < \delta_i$, we have $B^*(x; h, c_1\varepsilon_2, e_i) \subseteq \Omega$, and such that $\partial\Omega \subseteq \cup_{i=1}^{\mu'} B(\zeta_i; \delta_i/2)$.

Let $\psi := f_1(1 + \varepsilon_2)/(1 + 2\varepsilon_2)$. Suppose $\tau \in \mathcal{T}_m^1(K)$. Then, provided m is large enough, $f(x) \geq \psi$ for $x \in A_m(\tau) \cap \Omega$, and also $A_m(\tau) \subset B(\zeta_i; \delta_i)$ for some $i = i(\tau) \leq \mu'$. Then, by (7.25), we have

$$A_m(\tau) \cap \Omega \supseteq \left(\left(\bigcup_{y \in \tau} C_{\nu(m)}(y) \cap \Omega \right) \oplus B^*(\mathbf{0}; (t - 3c_1\varepsilon_2)\rho_{\nu(m)}, \varepsilon_2, e_i) \right) \setminus \left(\bigcup_{y \in \tau} C_{\nu(m)}(y) \cap \Omega \right).$$

Therefore, by the Brunn–Minkowski inequality (7.1),

$$\mathrm{Leb}(A_m(\tau) \cap \Omega) \geq \mathrm{Leb} B^*(\mathbf{0}; (t - 3c_1\varepsilon_2)\rho_{\nu(m)}, \varepsilon_2, e_i)$$

so that

$$E[\mathcal{X}_{\nu(m)}(A_m(\tau))] \geq \nu(m)\rho_{\nu(m)}^d \psi \mathrm{Leb} B^*(\mathbf{0}; t - 3c_1\varepsilon_2, \varepsilon_2, e_i) := \mu_{m,1}.$$

By (7.17) or (7.18), and (7.10),

$$\mu_{m,1} \geq (1 + \varepsilon_2)k_m \quad \text{if } b = \infty, \tag{7.28}$$

$$dH(a_1)\mu_{m,1} \geq (d - 1)(1 + \varepsilon_2)\log \nu(m) \quad \text{if } b < \infty. \tag{7.29}$$

First suppose $b = \infty$ and $j = 0$ or $j = 1$. By (7.26) for $j = 0$ or (7.28) for $j = 1$, and by Lemma 1.1, there exists $\delta > 0$ such that for all large enough m, and all $\tau \in \mathcal{T}_m(K)$, we have $P[F_m(\tau] \leq \exp(-\delta k_n)$. Since $k_m/\log m \to \infty$ by assumption, by (7.24) and Boole's inequality we have for large m that

$$P\left[\bigcup_{\tau \in \mathcal{T}_m^j(K)} F_m(\tau) \right] \leq m^{-2},$$

which is summable in m. The result follows by the Borel–Cantelli lemma, for the case $b = \infty$.

Suppose $b < \infty$, and $j = 0$ or $j = 1$. By (7.27) for $j = 0$ or (7.29) for $j = 1$, and by the fact that $k_{\nu(m+1)}/\log \nu(m) \to b$ by assumption, for m large we have

$$\frac{k_{\nu(m+1)}}{\mu_{m,j}} \le \frac{bdH(a_j)}{d-j} = a_j,$$

where the last inequality is from (7.2). Therefore, by (7.22) and Lemma 1.1, and by (7.27) or (7.29), for large enough m, if $\tau \in \mathcal{T}_m^j(K)$ then

$$P[F_m(\tau)] \le \exp\left(-H\left(a_j\right)\mu_{m,j}\right) \le m^{(1+\varepsilon_2)J(j-d)/d}.$$

By Boole's inequality and (7.24), for large enough m,

$$P\left[\bigcup_{\tau \in \mathcal{T}_m^j(K)} F_m(\tau)\right] \le m^{\varepsilon_2 J(j-d)/d},$$

which is summable in m by the choice of J at (7.19). The case $b < \infty$ of the result follows by the Borel–Cantelli lemma. □

Proof of Proposition 7.4 Immediate from Lemmas 7.5 and 7.7. □

Proof of Theorem 7.2 Immediate from Lemma 7.3 and Proposition 7.4. □

7.2 Strong laws for thresholds in the cube

This section contains analogous results to those in the preceding one, in a case where the support Ω of the underlying density f has 'corners'. Throughout this section we assume that *the norm* $\|\cdot\|$ *is one of the* l_p *norms,* $1 \le p \le \infty$. We also assume that the support Ω of f *is a product of finite closed intervals*, that is, that Ω is of the form $\prod_{i=1}^d [\alpha_i, \beta_i]$, for example, the unit cube. For $1 \le j \le d$, let ∂_j denote the union of all $(d-j)$-dimensional 'edges' (intersections of j hyperplanes bounding Ω), and let f_j denote the infimum of f over ∂_j. The results in this section are valid in any dimension, including $d = 1$.

Theorem 7.8 *Suppose that* $d \ge 1$, *that* Ω *is a product of finite closed intervals, that* $f_0 > 0$, *and that* $f|_\Omega$ *is continuous at* x *for all* $x \in \partial\Omega$. *Let* $(k_n)_{n\ge1}$ *be a sequence of integers satisfying* $\lim_{n\to\infty}(k_n/n) = 0$, *and* $\lim_{n\to\infty}(k_n/\log n) = b \in [0,\infty]$. *In the case* $b < \infty$, *assume also that the sequence* $(k_n)_{n\ge1}$ *is nondecreasing, and define* $a_0, \ldots, a_{d-1}$ *in* $[0,1)$ *by*

$$\frac{a_j}{H(a_j)} = \frac{bd}{d-j}. \tag{7.30}$$

If $b = \infty$, *then with probability 1,*

$$\lim_{n\to\infty}\left(n\theta M_{k_n}(\mathcal{X}_n)^d/k_n\right) = \max\{2^j/f_j : 0 \le j \le d\},$$

while if $b < \infty$ *then, with probability 1,*

$$\lim_{n\to\infty}\left(\frac{n\theta(M_{k_n}(\mathcal{X}_n))^d}{\log n}\right) = \max\left(\max_{0\le j\le d-1}\left(\frac{2^j(d-j)}{df_jH(a_j)}\right), \frac{2^d b}{f_d}\right). \tag{7.31}$$

The proof of Theorem 7.8 uses the same Poissonization device as the earlier proof of Theorem 7.2, and the coupled Poisson processes $\mathcal{P}_n^-, \mathcal{P}_n^+$ are as defined in the earlier proof. As before, let H_n denote the event that $\mathcal{P}_n^- \subseteq \mathcal{X}_n \subseteq \mathcal{P}_n^+$. For $u \geq 0$, and integer j, we define

$$\rho_n^j(u) := \left(\frac{2^j u k_n}{n\theta f_j}\right)^{1/d}, \quad 0 \leq j \leq d, \quad \text{if } b = \infty, \tag{7.32}$$

$$\rho_n^j(u) := \left(\frac{2^j (d-j) u \log n}{n\theta d f_j H(a_j)}\right)^{1/d}, \quad 0 \leq j \leq d-1, \quad \text{if } b < \infty, \tag{7.33}$$

$$\rho_n^d(u) = \left(\frac{2^d b u \log n}{n\theta f_d}\right)^{1/d} \quad \text{if } b < \infty. \tag{7.34}$$

Lemma 7.9 *Suppose $j \in \{0, 1, 2, \ldots, d\}$. Suppose f, $(k_n)_{n \geq 1}$, $b \in [0, \infty]$, and $a_j \in [0, 1)$ are as in the statement of Theorem 7.8. Suppose $0 < \beta < 1$. Then with probability 1, $M_{k_n}(\mathcal{X}_n) > \rho_n^j(\beta)$ for all large enough n.*

Proof When $j = 0$, the argument in the proof of Lemma 7.3 carries over to the present case, so it remains only to prove the result in the case $j > 0$. Assume from now on that $j > 0$. Choose $\varepsilon_3 > 0$ such that

$$\varepsilon_3^{1/d} + \beta^{1/d} < (1 - 4\varepsilon_3)^{1/d}. \tag{7.35}$$

For each $x \in \mathbb{R}^d$, define the event

$$E_n(x) := \{\mathcal{P}_n^+[B(x; \rho_n^j(1 - 4\varepsilon_3))] \leq k_n\} \cap \{\mathcal{P}_n^-[B(x; \rho_n^j(\varepsilon_3))] \geq 1\}.$$

By the triangle inequality, and by (7.35), if $E_n(x) \cap H_n$ occurs then there is a point of $\mathcal{X}_n$ whose k_n-nearest neighbour is at a distance at least $\rho_n^j(\beta)$. Hence we have the event inclusion

$$\{M_{k_n}(\mathcal{X}_n) \leq \rho_n^j(\beta)\} \subseteq H_n^c \cup (\cup_{x \in \mathbb{R}^d} E_n(x))^c. \tag{7.36}$$

Consider first the case $b = \infty$. Let $x_j \in \partial_j$ with $f(x_j) < f_j(1 - 3\varepsilon_3)/(1 - 4\varepsilon_3)$. For all large enough n, we have $F(B(x_j; \rho_n^j(1 - 4\varepsilon_3))) \leq (1 - 3\varepsilon_3) k_n / n$, and the expected number of points of $\mathcal{P}_n^+$ in $B(x_j; \rho_n^j(1 - 4\varepsilon_3))$ is at most $k_n(1 - \varepsilon_3)$. By Lemma 1.2, the probability that this number of points exceeds k_n decays exponentially in k_n. Also, the probability that there is no point of $\mathcal{P}_n^-$ in $B(x_j; \rho_n^j(\varepsilon_3))$ decays exponentially in k_n. Since $k_n/n \to \infty$ by assumption, these estimates together with (7.4) imply that $1 - P[E_n(x_j) \cap H_n]$ is summable in n. Therefore by the Borel–Cantelli lemma, with probability 1, $E_n(x_j) \cap H_n$ occurs all but finitely many n, and by (7.36) the result follows for the case $b = \infty$.

Now suppose $b < \infty$. Assume in addition to (7.35) that $\varepsilon_3 < 1 - a_j$. Take $x_1 \in \partial_j$ and $\varepsilon_4 > 0$ such that $f(x) < f_j(1 - 3\varepsilon_3)/(1 - 4\varepsilon_3)$ for $x \in B(x_1; 2\varepsilon_4) \cap \Omega$, and such that $B(x_1; 2\varepsilon_4)$ intersects only j of the hyperplanes bounding Ω. Set

$B_1 = B(x_1; \varepsilon_4)$. Recall the definition of the packing number $\sigma(U;r)$ in Section 5.2. If $j < d$, then since ∂_j is $(d-j)$-dimensional,

$$\liminf_{n\to\infty} \left(((\log n)/n)^{(d-j)/d} \sigma(B_1 \cap \partial_j; \rho_n^j(1-4\varepsilon_3)) \right) > 0. \tag{7.37}$$

Suppose $0 < j < d$, and $x \in B_1 \cap \partial_j$, and $r < \varepsilon_4$. Then the Lebesgue measure of $B(x;r) \cap \Omega$ is $2^{-j}\theta r^d$, so that for n big, by (7.33),

$$(n + n^{3/4}) F(B(x; \rho_n^j(1-4\varepsilon_3))) \leq \frac{(1-2\varepsilon_3)(d-j)\log n}{dH(a_j)} := \mu_{n,j}.$$

Suppose $j < d$. Since $k_n/\log n \to b$, using (7.30) we have

$$\lim_{n\to\infty} (k_n/\mu_{n,j}) = \frac{bdH(a_j)}{(d-j)(1-\varepsilon_3)} = \frac{a_j}{1-\varepsilon_3} \in (a_j, 1),$$

so that for large enough n, by Lemma 1.3,

$$\begin{aligned} P[\mathcal{P}_n^+[B(x;\rho_n^j(1-4\varepsilon_3))] \leq k_n] &\geq P[\mathrm{Po}(\mu_{n,j}) \geq k_n] \\ &\geq \left(\frac{\mathrm{e}^{-1/(12k_n)}}{\sqrt{2\pi k_n}} \right) \exp\left(-\mu_n H(a_j)\right) \\ &\geq n^{(\varepsilon_3-1)(d-j)/d}. \end{aligned}$$

Therefore by the same argument as for (7.9), there is a constant $\eta > 0$ such that

$$P[E_n(x)] \geq \eta n^{(\varepsilon_3-1)(d-j)/d}. \tag{7.38}$$

The rest of the proof for $0 < j < d$ (and $b < \infty$) proceeds as for Lemma 7.3, using (7.37) and (7.38) instead of (7.8) and (7.9), respectively.

Next suppose $j = d$ (and $b < \infty$). If $b = 0$, then $\rho_n^d(\beta) = 0$ and there is nothing to prove, so assume $0 < b < \infty$. Choose a corner point y of B_n with $f(y) = f_d$. Then there exists $\varepsilon_5 > 0$ such that, for large enough n, $nF[B(y;\rho_n^d(\varepsilon_3))] > \varepsilon_5 \log n$, and hence $P[\mathcal{X}_n[B(y;\rho_n^d(\varepsilon_3))] = 0] \leq n^{-\varepsilon_5}$.

Set $\varepsilon_6 = H(1/(1-\varepsilon_3))(1-2\varepsilon_3)b$. Choose integer $J > \max(\varepsilon_6^{-1}, \varepsilon_5^{-1})$, and set $\nu(m) = m^J$. For large enough m we have

$$\nu(m+1)F(B(y;\rho_n^d(1-4\varepsilon_3))) \leq (1-2\varepsilon_3)b\log\nu(m).$$

By Lemma 1.1, since $k_n \sim b\log n$, for large enough m we have the estimate

$$P[\mathcal{X}_{\nu(m+1)}[B(y;\rho_{\nu(m)}^d(1-4\varepsilon_3))] \geq k_{\nu(m)}] \leq m^{-\varepsilon_6 J},$$

and therefore, by the Borel–Cantelli lemma, with probability 1, the event

$$\{\mathcal{X}_{\nu(m+1)}[B(y;\rho_{\nu(m)}^d(1-4\varepsilon_3))] \leq k_{\nu(m)}\} \cap \{\mathcal{X}_{\nu(m)}[B(y;\rho_{\nu(m)}^d(\varepsilon_3))] \geq 1\}$$

occurs for all but finitely many m. By an application of the triangle inequality, and (7.35), the above event implies that for all n between $\nu(m)$ and $\nu(m+1)$ we have $M_{k_n}(\mathcal{X}_n) > \rho_n^d(\beta)$. □

It remains to prove upper bounds. As at (7.10), for each $n > 0$, define

$$\rho_n := (k_n/n)^{1/d} \text{ if } b = \infty, \qquad \rho_n := ((\log n)/n)^{1/d} \text{ if } b < \infty. \tag{7.39}$$

As at (7.11), let $E'_n(K;t)$ be the event that there exists a k_n-separated set U for $G(\mathcal{X}_n; t\rho_n)$ with $\mathrm{diam}(U) \leq K\rho_n$. As at Proposition 7.4, we prove a stronger result than is needed here, for later use in proving results about connectivity.

Proposition 7.10 *Suppose the hypotheses of Theorem 7.8 hold. Let $K > 0$. If $b = \infty$, then let t satisfy*

$$\theta t^d > \max_{0\leq j\leq d} (2^j/f_j). \tag{7.40}$$

If $b < \infty$, then let t satisfy

$$\theta t^d > \max\left(\max_{0\leq j\leq d-1}\left(\frac{2^j(d-j)}{df_jH(a_j)}\right), \frac{2^d b}{f_d}\right). \tag{7.41}$$

Then with probability 1, event $E'_n(K;t)$ occurs for only finitely many n, and hence $M_{k_n}(\mathcal{X}_n) \leq t\rho_n$ for all but finitely many n.

The proof of Proposition 7.10 is fairly similar to that of Proposition 7.4. Fix t satisfying (7.40) if $b = \infty$ or (7.41) if $b < \infty$. As at (7.14), let c_1 denote the diameter of the unit cube. Pick $\varepsilon_7 \in (0,1)$, in the case $b = \infty$, such that

$$\frac{\theta(t-4c_1\varepsilon_7)^d}{(1+\varepsilon_7)} > \max_{0\leq j\leq d}\left(\frac{2^j}{f_j}\right), \tag{7.42}$$

or in the case $b < \infty$, such that

$$\frac{\theta(t-4c_1\varepsilon_7)^d}{(1+\varepsilon_7)} > \max\left(\max_{0\leq j\leq d-1}\left(\frac{2^j(d-j)}{df_jH(a_j)}\right), \frac{2^d(b+2\varepsilon_7)}{f_d}\right). \tag{7.43}$$

We shall be using the subsequence trick. In the case $b < \infty$, choose an integer J so that

$$J\varepsilon_7/d > 1 \text{ and } J\varepsilon_7 H((b+\varepsilon_7)/(b+2\varepsilon_7)) > 2. \tag{7.44}$$

In the case $b = \infty$, take $J = 1$. For $m = 1, 2, 3, \ldots$, let $\nu(m) = m^J$.

Define the lattice $\varepsilon_7\rho_n\mathbb{Z}^d := \{\varepsilon_7\rho_n z : z \in \mathbb{Z}^d\}$. For $y \in \varepsilon_7\rho_n\mathbb{Z}^d$, let $C_n(y) := \{y\} \oplus [-\varepsilon_7\rho_n/2, \varepsilon_7\rho_n/2]^d$ the cube of side $\varepsilon_7\rho_n$ centred at y. Define the finite lattice $\mathcal{L}_n$ by $\mathcal{L}_n := \{y \in \varepsilon_7\rho_n\mathbb{Z}^d : C_n(y) \cap \Omega \neq \emptyset\}$. For $K > 0$, let $\mathcal{T}_m(K)$ (a collection of subsets of $\mathcal{L}_{\nu(m)}$) be given by

$$\mathcal{T}_m(K) := \{\tau \subset \mathcal{L}_{\nu(m)} : \mathrm{diam}(\tau) \leq 2K\rho_{\nu(m)}\}.$$

Given $\tau \in \mathcal{T}_m(K)$, define the 'annulus-like' set $A_m(\tau)$, as at (7.20), by

$$A_m(\tau) := \left(\tau \oplus B(\mathbf{0}; (t - 2c_1\varepsilon_7)\rho_{\nu(m)})\right) \setminus \left(\tau \oplus C_{\nu(m)}(\mathbf{0})\right).$$

As at (7.21) and (7.22), define the event $F_m(\tau)$ by

$$F_m(\tau) := \{\mathcal{X}_m [A_m(\tau)] \le k_m\} \quad \text{if } b = \infty, \tag{7.45}$$

$$F_m(\tau) := \{\mathcal{X}_{\nu(m)} [A_m(\tau)] \le k_{\nu(m+1)}\} \quad \text{if } b < \infty. \tag{7.46}$$

The next result is identical to Lemma 7.5.

Lemma 7.11 *Let $K > \max(1, t)$. Then there exists m_0 such that if $m \ge m_0$ and $\nu(m) \le n < \nu(m+1)$, then the event $E'_n(K; t)$, that there is a k_n-separated set U for $G(\mathcal{X}_n; t\rho_n)$ with* $\mathrm{diam}(U) \le K\rho_n$, *is contained in the union of the events $F_m(\tau), \tau \in \mathcal{T}_m(K)$.*

For $j = 0, 1, \dots, d$, let $\mathcal{T}_m^j(K)$ be the collection of all $\tau \in \mathcal{T}_m(K)$ such that $\tau \oplus B(\mathbf{0}; t\rho_{\nu(m)})$ intersects with precisely j of the hyperplanes bounding Ω. Let $|\mathcal{T}_m^j(K)|$ be the cardinality of $\mathcal{T}_m^j(K)$.

Lemma 7.12 *Let $j \in \{0, 1, \dots, d\}$ and let $K > \max(1, t)$. Then*

$$\limsup_{m \to \infty} |\mathcal{T}_m^j(K)| \rho_{\nu(m)}^{d-j} < \infty. \tag{7.47}$$

Proof Given $\tau \in \mathcal{T}_m(K)$, let $y(\tau)$ be the first element of τ according to the lexicographic ordering on $\mathcal{L}_{\nu(m)}$, and let

$$\tau' := \{(\varepsilon_7 \rho_{\nu(m)})^{-1}(y - y(\tau)) : y \in \tau\}.$$

Then $y(\tau)$ and τ' together determine τ. Also, τ' is a subset of $\mathbb{Z}^d \cap B(\mathbf{0}; 3K/\varepsilon_7)$, and the number of such subsets is a constant independent of m. Therefore $|\mathcal{T}_m^j(K)|$ is bounded by a constant times the number of possibilities for $y(\tau)$ consistent with $\tau \in \mathcal{T}_m^j(K)$.

First suppose $j = 0$. The number of possibilities for $y(\tau)$ is bounded by the cardinality of $\mathcal{L}_n$, and hence by a constant times $\rho_{\nu(m)}^{-d}$, and (7.47) follows for this case.

Next suppose $j > 0$. If $\tau \in \mathcal{T}_m^j(K)$, then $y(\tau)$ is at an l_∞ distance at most $3K\rho_{\nu(m)}$ from ∂_j, the $(d-j)$-dimensional part of the boundary of Ω. Therefore we can choose $y_0(\tau) \in \mathcal{L}_n$ satisfying (i) $\|y_0(\tau) - y(\tau)\|_\infty \le 4K\rho_{\nu(m)}$, and (ii) $C_n(y_0) \cap \partial_j \ne \emptyset$.

The number of possibilities for $y_0(\tau)$ satisfying condition (ii) is bounded by a constant times $\rho_n^{-(d-j)}$. Given $y_0(\tau)$, the number of possibilities for $y(\tau)$ is bounded by a constant because of condition (i). Therefore we obtain (7.47) for $j > 0$. □

Lemma 7.13 *Let $0 \le j \le d$. With probability 1, the event $\cup_{\tau \in \mathcal{T}_m^j(K)} F_m(\tau)$ occurs for only finitely many m.*

Proof Suppose $\tau \in \mathcal{T}_m^j(K)$. In the case $j > 0$, assume also that Ω is of the form $\prod_{i=1}^d [0, \beta_i]$, and that the j bounding hyperplanes of Ω that are intersected by $\tau \oplus B(\mathbf{0}; t\rho_{\nu(m)})$ are $\pi_1^{-1}(\{0\}), \pi_2^{-1}(\{0\}), \ldots, \pi_j^{-1}(\{0\})$, where $\pi_i : \mathbb{R}^d \to \mathbb{R}$ denotes projection onto the ith coordinate (other cases are treated similarly). For $r > 0$ let

$$B^+(\mathbf{0}; r) := \{x = (x_1, \ldots, x_d) \in B(\mathbf{0}; r) : x_1 \geq 0, \ldots, x_j \geq 0\}.$$

(In the case $j = 0$, take $B^+(\mathbf{0}; r) := B(\mathbf{0}; r)$.) Also let $C'_n(y) = C_n(y) \cap \Omega$. If for some $y \in \tau$ we have $x \in C'_{\nu(m)}(y)$, and $w \in B^+(\mathbf{0}; (t - 3c_1\varepsilon_7)\rho_{\nu(m)})$, then $x + w \in \Omega$, and also $x + w \in B(y; t - 2c_1\varepsilon_7\rho_{\nu(m)})$ by the triangle inequality. Hence,

$$A_m(\tau) \cap \Omega \supseteq \left(\bigcup_{y \in \tau} C'_{\nu(m)}(y) \right) \oplus B^+(\mathbf{0}; (t - 3c_1\varepsilon_7)\rho_{\nu(m)}) \setminus \left(\bigcup_{y \in \tau} C'_{\nu(m)}(y) \right).$$

Therefore, by the Brunn–Minkowski inequality (7.1),

$$\mathrm{Leb}(A_m(\tau) \cap \Omega) \geq 2^{-j}\theta((t - 3c_1\varepsilon_7)\rho_{\nu(m)})^d,$$

and therefore, for $\tau \in \mathcal{T}_m^j(K)$ and for m large enough,

$$E[\mathcal{X}_{\nu(m)}(A_m(\tau))] \geq 2^{-j} f_j \theta(t - 4c_1\varepsilon_7)^d \nu(m)\rho_{\nu(m)}^d := \mu_m. \tag{7.48}$$

First suppose $b = \infty$ (so $\nu(m)\rho_{\nu(m)}^d = k_m$). By (7.42), for all large enough m, $\mu_m \geq (1 + \varepsilon_7)k_m$, so by (7.45) and Lemma 1.1, there is a constant $\delta > 0$ such that for large enough m, if $\tau \in \mathcal{T}_m^j(K)$ then

$$P[F_m(\tau)] \leq \exp(-\delta k_m).$$

Hence by Boole's inequality,

$$P\left[\bigcup_{\tau \in \mathcal{T}_m^j(K)} F_m(\tau) \right] \leq |\mathcal{T}_m^j(K)| \exp(-\delta k_m),$$

which is summable in m by (7.47) and the assumption that $k_m / \log m \to \infty$. The result follows by the Borel–Cantelli lemma, in the case $b = \infty$.

Now suppose $b < \infty$ (so $\nu(m)\rho_{\nu(m)}^d = \log \nu(m)$). First suppose $0 \leq j < d$. By (7.43),

$$dH(a_j)\mu_m \geq (d - j)(1 + \varepsilon_7) \log \nu(m). \tag{7.49}$$

Since $k_{\nu(m+1)} / \log \nu(m) \to b$ by assumption, and by (7.30), for large m we have

$$\frac{k_{\nu(m+1)}}{\mu_m} \leq \frac{bdH(a_j)}{d - j} = a_j.$$

Therefore by (7.46), Lemma 1.1, and (7.49), for large enough m, if $\tau \in \mathcal{T}_m^j(K)$ then

$$P[F_m(\tau)] \leq \exp\left(-\mu_m H\left(a_j\right)\right) \leq m^{(1+\varepsilon_7)J(j-d)/d}.$$

By (7.47) and (7.39), for large enough m we have $|\mathcal{T}_m^j(K)| \leq m^{J(d-j)/d}$, and hence by Boole's inequality,

$$P\left[\bigcup_{\tau \in \mathcal{T}_m^j(K)} F_m(\tau)\right] \leq |\mathcal{T}_m^j(K)| m^{(1+\varepsilon_7)J(j-d)/d} \leq m^{J\varepsilon_7(j-d)/d},$$

which is summable in m by the choice of J at (7.44). The result follows by the Borel–Cantelli lemma, in the case where $j < d$ (and $b < \infty$).

Next suppose $j = d$ (and $b < \infty$). Then $\mu_m \geq (b + 2\varepsilon_7) \log \nu(m)$ by (7.43). Therefore by Lemma 1.1, since $k_{\nu(m+1)}/\log \nu(m) \to b$, for large enough m we have

$$P[F_m(\tau)] \leq \exp\left(-H\left(\frac{b+\varepsilon_7}{b+2\varepsilon_7}\right) \varepsilon_7 \log \nu(m)\right),$$

so by (7.47) and Boole's inequality, there is a constant c such that

$$P\left[\bigcup_{\tau \in \mathcal{T}_m^d(K)} F_m(\tau)\right] \leq c m^{-J\varepsilon_7 H((b+\varepsilon_7)/(b+2\varepsilon_7))},$$

which is summable in m by (7.44). Thus the result follows by the Borel–Cantelli lemma, in this case too. □

Proof of Proposition 7.10 Immediate from Lemmas 7.11 and 7.13. □

Proof of Theorem 7.8 Immediate from Lemma 7.9 and Proposition 7.10. □

7.3 Strong laws for the minimum degree

Recall that δ_n denotes the minimum degree of $G(\mathcal{X}_n; r_n)$. In this section we re-interpret the preceding results in terms of the minimum degree, thereby describing a.s. asymptotic behaviour of δ_n for a large class of sequences $(r_n)_{n \geq 1}$. We consider together the three possibilities for the support Ω of f that we have considered in the preceding sections. These are:

- *Case I*: $d \geq 1$ and Ω is the d-dimensional unit torus.
- *Case II*: $d \geq 2$, Ω is bounded in $\mathbb{R}^d$, and $\partial\Omega$ is a compact $(d-1)$-dimensional C^2 submanifold of $\mathbb{R}^d$.
- *Case III*: $d \geq 1$, the norm $\|\cdot\|$ is one of the l_p norms, $1 \leq p \leq \infty$, and Ω is a product of finite closed intervals.

Define the finite set $\mathcal{J}$ by $\mathcal{J} := \{0\}$ in Case I, $\mathcal{J} := \{0,1\}$ in Case II, and $\mathcal{J} := \{0,1,2,\ldots,d\}$ in Case III.

Keeping notation from the preceding sections, let f_0 denote the essential infimum of the function $f|_\Omega$, and in Case II let $f_1 := \inf_{\partial\Omega} f$. In Case III, for $0 \neq j \in \mathcal{J}$ let ∂_j denote the union of all $(d-j)$-dimensional 'edges' of Ω, and let $f_j := \inf_{x\in\partial_j} f(x)$. Assume $f_0 > 0$ (and hence $f_j > 0$ for all $j \in \mathcal{J}$).

The functions $H(\cdot)$, $H_+^{-1}(\cdot)$, and $H_-^{-1}(\cdot)$ were defined at (6.32). It is instructive to compare Case I of the following result with Corollary 6.14, and to note the similarities between the limiting behaviour of the maximum and minimum degree. In particular, in the case of uniformly distributed points in the torus, the right-hand side of (7.50) below is simply $H_-^{-1}(\alpha^{-1})$, while the right-hand side of (6.47) comes to $H_+^{-1}(\alpha^{-1})$.

Theorem 7.14 *Suppose that the conditions of Case I, Case II or Case III hold. Suppose also that $f|_\Omega$ is continuous at x for all $x \in \partial\Omega$. Suppose $r_n \to 0$ and $n\theta r_n^d/\log n \to \alpha \in [0,\infty]$ as $n \to \infty$. Then:*

If $\alpha < \max_{j\in\mathcal{J}\setminus\{d\}}\{2^j(d-j)/(df_j)\}$, then $\delta_n \to 0$, almost surely.

If $\max_{j\in\mathcal{J}\setminus\{d\}}\{2^j(d-j)/(df_j)\} \leq \alpha \leq \infty$, then in Case I or Case II, with probability 1,

$$\lim_{n\to\infty}\left(\frac{\delta_n}{n\theta r_n^d}\right) = \min_{j\in\mathcal{J}}\left(\frac{f_j}{2^j}H_-^{-1}\left(\frac{2^j(d-j)}{df_j\alpha}\right)\right), \tag{7.50}$$

while in Case III, with probability 1,

$$\lim_{n\to\infty}\left(\frac{\delta_n}{n\theta r_n^d}\right) = \min\left(\min_{0\leq j\leq d-1}\left(\frac{f_j}{2^j}H_-^{-1}\left(\frac{2^j(d-j)}{df_j\alpha}\right)\right), \frac{f_d}{2^d}\right), \tag{7.51}$$

with the interpretation (in all cases) $1/\infty = 0$, so if $\alpha = \infty$ the limit is $\min\{f_j/2^j : j \in \mathcal{J}\}$.

Proof First suppose $\alpha < \max\{2^j(d-j)/(df_j) : j \in \mathcal{J}\setminus\{d\}\}$. By the case $b = 0$ of Theorem 7.1 in Case I, of Theorem 7.2 in Case II, or of Theorem 7.8 in Case III, with probability 1, we have

$$\lim(n\theta M_1(\mathcal{X}_n)^d/\log n) = \max\{2^j(d-j)/(df_j) : j \in \mathcal{J}\setminus\{d\}\},$$

and hence $M_1(\mathcal{X}_n) > r_n$ for large enough n, so that $\delta_n = 0$ for large enough n, which proves the result for this case.

Next, suppose $\max\{2^j(d-j)/(df_j) : j \in \mathcal{J}\setminus\{d\}\} \leq \alpha < \infty$. Given $b > 0$, for $j \in \mathcal{J}\setminus\{d\}$ define $a_j \in (0,1)$ by $a_j/H(a_j) = bd/(d-j)$ as before, and define $\psi_j(b)$ by

$$\psi_j(b) := \frac{2^j(d-j)}{df_jH(a_j)}.$$

Also, in Case III, define $\psi_d(b) := 2^d b/f_d$. Set $\psi(b) := \min_{j\in\mathcal{J}}\psi_j(b)$.

Suppose that $(k_n)_{n\geq 1}$ is a nondecreasing sequence with $k_n/\log n \to b$. Then by Theorem 7.1 in Case I, Theorem 7.2 in Case II, or Theorem 7.8 in Case III, with probability 1 we have

$$\liminf_{n\to\infty}\left(\frac{n\theta(M_{k_n}(\mathcal{X}_n))^d}{\log n}\right) = \max_{j\in\mathcal{J}} \psi_j(b) = \psi(b). \tag{7.52}$$

Observe that for $j \in \mathcal{J}$, $\psi_j(\cdot)$ is a continuous, strictly increasing function; hence $\psi(\cdot)$ is also a continuous, strictly increasing function. Let $b < \psi^{-1}(\alpha) < b'$, and choose nondecreasing sequences $(k_n)_{n\geq 1}$ and $(k'_n)_{n\geq 1}$, such that $k_n/\log n \to b$ and $k'_n/\log n \to b'$ as $n \to \infty$. By (7.52), for large enough n we have $M_{k_n}(\mathcal{X}_n) < r_n < M_{k'_n}(\mathcal{X}_n)$, and hence $k_n \leq \delta_n \leq k'_n$. It follows that with probability 1,

$$b \leq \liminf(\delta_n/\log n) \leq \limsup(\delta_n/\log n) \leq b',$$

and hence, by taking $b \uparrow \psi^{-1}(\alpha)$ and $b' \downarrow \psi^{-1}(\alpha)$, we obtain

$$\lim_{n\to\infty}(\delta_n/(n\theta r_n^d)) = \alpha^{-1}\psi^{-1}(\alpha) = \alpha^{-1}\min\{\psi_j^{-1}(\alpha) : j \in \mathcal{J}\}, \qquad \text{a.s.} \tag{7.53}$$

For $j \in \mathcal{J}\setminus\{d\}$, if we set $b = \psi_j^{-1}(\alpha)$, and $a_j < 1$ is chosen so that $a_j/H(a_j) = bd/(d-j)$, then by definition of the function ψ_j we have $H(a_j) = 2^j(d-j)/(df_j\alpha)$, and therefore

$$a_j = H_-^{-1}\left(\frac{2^j(d-j)}{df_j\alpha}\right), \quad b = \frac{f_j\alpha}{2^j}H_-^{-1}\left(\frac{2^j(d-j)}{df_j\alpha}\right),$$

so that $\psi_j^{-1}(\alpha) = 2^{-j}f_j\alpha H_-^{-1}(2^j(d-j)/(df_j\alpha))$. Also, in Case III, $\psi_d^{-1}(\alpha) = 2^{-d}f_d\alpha$. The results (7.50) and (7.51), in the case $\alpha < \infty$, follow from these facts and (7.53).

Finally, suppose $\alpha = \infty$. If $(k_n/\log n) \to \infty$ and $k_n/n \to 0$, then by the case $b = \infty$ of Theorem 7.1 in Case I, of Theorem 7.2 in Case II, or of Theorem 7.8 in Case III, we have, with probability 1, that as $n \to \infty$,

$$\frac{n\theta M_{k_n}(\mathcal{X}_n)^d}{k_n}\min_{j\in\mathcal{J}}\left(\frac{f_j}{2^j}\right) \to \max_{j\in\mathcal{J}}\left(\frac{2^j}{f_j}\right)\min_{j\in\mathcal{J}}\left(\frac{f_j}{2^j}\right) = 1. \tag{7.54}$$

Let $\varepsilon > 0$. Then set $k_n := \lfloor(1-\varepsilon)n\theta r_n^d \min_{j\in\mathcal{J}}(2^{-j}f_j)\rfloor$, and also set $k'_n := \lfloor(1+\varepsilon)n\theta r_n^d\min_{j\in\mathcal{J}}(2^{-j}f_j)\rfloor$. Then $(k_n/\log n) \to \infty$ and $(k_n/n) \to 0$ as $n \to \infty$, so by (7.54), with probability 1, we have

$$\frac{n\theta M_{k_n}(\mathcal{X}_n)^d \min_{j\in\mathcal{J}}(2^{-j}f_j)}{(1-\varepsilon)n\theta r_n^d \min_{j\in\mathcal{J}}(2^{-j}f_j)} \to 1,$$

so that $M_{k_n}(\mathcal{X}_n) < r_n$ for n large, and similarly $M_{k'_n} > r_n$ for n large. Thus, with probability 1, $k_n \leq \delta_n \leq k'_n$ for all but finitely many n. Hence by taking $\varepsilon \downarrow 0$, we obtain $\delta_n/(n\theta r_n^d) \to \min_{j\in\mathcal{J}}(2^{-j}f_j)$, which is the required result for the case $\alpha = \infty$. □

7.4 Notes

Section 7.1. Theorem 7.2 generalizes a result in Penrose (1999*a*), where only the case $k_n = \text{const.}$ was considered. The case $k_n = \text{const.}$ of Theorem 7.1 (points on the torus) is a special case of a result given in Penrose (1999*a*) for points distributed on a general manifold.

Section 7.2. Theorem 7.8 considerably extends results of Appel and Russo (1997*b*) who considered only the case of uniformly distributed points on the unit cube using the l_∞ norm.

8

MINIMUM DEGREE: CONVERGENCE IN DISTRIBUTION

This chapter contains convergence in distribution results for the largest k-nearest-neighbour link $M_k(\mathcal{X}_n)$, with k fixed. These are achieved via convergence in distribution for the number of vertices of degree k in $G(\mathcal{X}_n; r_n)$, for a sequence of parameters r_n chosen in such a way as to give an honest limiting distribution. Let $W_{k,n}(r)$ (respectively, $W'_{k,n}(r)$) be the number of vertices of degree k in $G(\mathcal{X}_n; r)$ (respectively, $G(\mathcal{P}_n; r)$), and observe that

$$M_{k+1}(\mathcal{X}_n) \leq r \quad \text{if and only if} \quad \sum_{j=0}^{k} W_{j,n}(r) = 0. \tag{8.1}$$

Set

$$\psi_{n,r}^k(x) := (nF(B(x;r))^k \exp(-nF(B(x;r)))/k!. \tag{8.2}$$

By Palm theory (Theorem 1.6),

$$EW'_{k,n}(r) = n \int_{\mathbb{R}^d} \psi_{n,r}^k(x) f(x) \mathrm{d}x. \tag{8.3}$$

Given any underlying density f, and given any $k \in \mathbb{N}$, observe that for any fixed $K > 1$ it is the case that $\inf_{1 \leq s \leq K} E[W'_{k,n}(sn^{-1/d})]$ tends to infinity as $n \to \infty$; observe also that for each n, the function $r \mapsto EW'_{k,n}(r)$ is continuous in r and tends to zero as $r \to \infty$. Therefore, for any $\beta > 0$, we can always find a sequence $(r_n, n \geq 1)$ satisfying the conditions

$$\lim_{n \to \infty} (nr_n^d) = \infty, \qquad \lim_{n \to \infty} EW'_{k,n}(r_n) = \beta. \tag{8.4}$$

Indeed, by the intermediate value theorem and the above properties of $EW'_{k,n}(\cdot)$, it is possible to choose such a sequence with $EW'_{k,n}(r_n) = \beta$ for all n.

Given a sequence $(r_n)_{n \geq 1}$ satisfying (8.4), for any non-negative integer $j < k$, we have $\psi_{n,r_n}^j(x)/\psi_{n,r_n}^k(x) \to 0$ for each Lebesgue point x with $f(x) > 0$. Hence, by (8.3), (8.4), and the dominated convergence theorem, we have

$$EW'_{j,n}(r_n) \to 0, \qquad 0 \leq j < k. \tag{8.5}$$

Given a sequence $(r_n)_{n \geq 1}$ satisfying (8.4), it is not unreasonable to conjecture that $W'_{k,n} \xrightarrow{\mathcal{D}} \mathrm{Po}(\beta)$, and together with (8.5) and (8.1), this would give us a limit for $P[M'_k(\mathcal{P}_n) \leq r_n]$. If we can then de-Poissonize, this gives us a convergence in distribution result for $M_k(\mathcal{X}_n)$, suitably transformed. Theorem 6.7 gives us a tool for proving the Poisson convergence, but its application requires some effort. Moreover, finding an explicit sequence $(r_n)_{n\geq 1}$ satisfying (8.4) is not easy in general.

In this chapter we carry out the above programme for two specific underlying density functions, the uniform on the unit cube (or torus), and the standard multivariate normal (in the latter case, only for $k = 0$).

8.1 Uniformly distributed points I

Assume throughout this section that $f = f_{\mathrm{U}}$, so that F *is the uniform distribution on the unit cube* $[-\frac{1}{2}, \frac{1}{2}]^d$, which we denote C in this section. Assume also that *the metric on* C *is given either by an* l_p *norm with* $1 < p \leq \infty$, *or by a toroidal metric based on an arbitrary norm.*

Theorem 8.1 *Let* $d \geq 1$, *and* $k \in \mathbb{N} \cup \{0\}$. *Let* $\beta > 0$, *and suppose the sequence* $(r_n = r_n(\beta, k), n \geq 1)$ *satisfies* (8.4). *Then as* $n \to \infty$,

$$W_{k,n}(r_n) \xrightarrow{\mathcal{D}} \mathrm{Po}(\beta). \tag{8.6}$$

Also, for any non-negative integer $j < k$,

$$\lim_{n\to\infty} P[W_{j,n}(r_n) \neq 0] = 0. \tag{8.7}$$

Hence,

$$\lim_{n\to\infty} P[M_{k+1}(\mathcal{X}_n) \leq r_n] = \mathrm{e}^{-\beta}. \tag{8.8}$$

Explicit formulae for r_n satisfying (8.4) are deferred to Section 8.2. For now, we merely derive properties of r_n required to prove Theorem 8.1. Note first that with θ denoting the volume of the unit ball as usual, there is a constant $\delta_1 > 0$ such that for all $r < \delta_1$, and all $x \in C$,

$$2^{-d}\theta r^d \leq F(B(x;r)) \leq \theta r^d, \tag{8.9}$$

so that $\psi^k_{n,r}(x)$ given by (8.2) satisfies

$$2^{-dk}\theta^k n(nr^d)^k \exp(-n\theta r^d)/k! \leq n\psi^k_{n,r}(x) \tag{8.10}$$

and

$$\theta^k n(nr^d)^k \exp(-2^{-d}\theta n r^d)/k! \geq n\psi^k_{n,r}(x). \tag{8.11}$$

Taking $r = r_n(\alpha, k)$ in (8.10), integrating over $x \in C$, using (8.4), and taking logarithms, we have

$$\limsup_{n\to\infty}(\log n + k\log(nr_n^d) - \theta nr_n^d) \le \log k! + \log \beta - k\log(2^{-d}\theta).$$

Since $\log nr_n^d \ge 0$ by the assumption $nr_n^d \to \infty$, we thus have $\liminf_{n\to\infty}((nr_n^d\theta) - \log n) > -\infty$, and hence,

$$nr_n^d \ge (2\theta)^{-1}\log n, \qquad n \text{ large.} \tag{8.12}$$

By a similar argument using (8.11),

$$nr_n^d \le (2^{d+1}/\theta)\log n, \qquad n \text{ large.} \tag{8.13}$$

In particular, $r_n \to 0$ as $n \to \infty$.

We now prove a Poissonized version of Theorem 8.1.

Theorem 8.2 *Under the hypotheses of Theorem* 8.1,

$$W'_{k,n}(r_n) \xrightarrow{\mathcal{D}} \mathrm{Po}(\beta). \tag{8.14}$$

Proof For typographical reasons we shall sometimes write $r(n)$ for r_n in the proof. Given n, for $x, y \in C$, define

$$v_x := F(B(x;r_n)), \quad v_{x,y} := F(B(x;r_n)\cap B(y;r_n)), \quad v_{x\backslash y} := v_x - v_{x,y}.$$

Define the integrals $I_i = I_i(n)$ $(i = 1, 2, 3)$ by

$$I_1 := n^2\int_C \mathrm{d}x \int_{C\cap B(x;3r(n))} \mathrm{d}y\psi^k_{n,r(n)}(y)\psi^k_{n,r(n)}(x), \tag{8.15}$$

$$I_2 := n^2\int_C \mathrm{d}x \int_{C\cap B(x;3r(n))} \mathrm{d}yP[Z_1+Z_2 = Z_1+Z_3 = k-1], \tag{8.16}$$

$$I_3 := n^2\int_C \mathrm{d}x \int_{C\cap B(x;3r(n))\backslash B(x;r(n))} \mathrm{d}yP[Z_1+Z_2 = Z_1+Z_3 = k], \tag{8.17}$$

where Z_1, Z_2, and Z_3 denote independent Poisson variables with means $nv_{x,y}$, $nv_{x\backslash y}$ and $nv_{y\backslash x}$ respectively. By Theorem 6.7,

$$d_{\mathrm{TV}}(W'_{k,n}, \mathrm{Po}(EW'_{k,n})) \le 3(I_1 + I_2 + I_3),$$

so to prove (8.14) it suffices to prove that I_1, I_2 and I_3 tend to zero as $n \to \infty$.

First consider I_1. Let $\pi_1 : \mathbb{R}^d \to \mathbb{R}$ denote projection onto the first coordinate. Set $x_0 = (\frac{1}{2}, \frac{1}{2}, \ldots, \frac{1}{2})$, one of the corners of the cube C. Define the sets

$$C_0 := \{x \in C : \|x - x_0\|_\infty \le 4r_n\}, \quad C_1 := \{x \in C : |\pi_1(x)| \le \tfrac{1}{2} - 4r_n\}.$$

Thus C_0 is a region near the corner of C and C_1 is the set of $x \in C$ at a distance at least $4r_n$ from the left face and from the right face of C. By invariance of the norm under permutation of the coordinates, the integral over $x \in C$ in (8.15) is bounded by 2^d times the contribution from $x \in C_0$ plus d times the contribution from $x \in C_1$.

The volume of C_0 is $(4r_n)^d$, so by (8.11), there is a constant $c > 0$ such that

$$n^2 \int_{C_0} \mathrm{d}x \int_{B(x;3r(n))} \mathrm{d}y \psi^k_{n,r(n)}(y)\psi^k_{n,r(n)}(x) \leq c\left(nr_n^d(nr_n^d)^k \exp(-2^{-d}\theta nr_n^d)\right)^2,$$

which tends to zero since $nr_n^d \to \infty$. Also,

$$\begin{aligned} n^2 &\int_{C_1} \mathrm{d}x \int_{B(x;3r(n))\cap C} \mathrm{d}y \psi^k_{n,r(n)}(y)\psi^k_{n,r(n)}(x) \\ &\leq n^2 \int_{C_1} \mathrm{d}x \int_{y\in C:|\pi_1(y)-\pi_1(x)|\leq 3r(n)} \mathrm{d}y \psi^k_{n,r(n)}(y)\psi^k_{n,r(n)}(x) \\ &= \left(n\int_{C_1} \mathrm{d}x \psi^k_{n,r(n)}(x)\right)\left(n\int_{y\in C:|\pi_1(y)|\leq 3r(n)} \mathrm{d}y \psi^k_{n,r(n)}(y)\right), \end{aligned} \tag{8.18}$$

where the last line follows because the value of the integral of $\psi^k_{n,r(n)}(y)$ over $\{y \in C : |\pi_1(y) - t| \leq 3r_n\}$ is the same for all real t with $|t| \leq \frac{1}{2} - 4r_n$. Moreover, it is possible to find $\lfloor(1 - 8r_n)/(6r_n)\rfloor$ disjoint slabs in C of the form $\{y \in C : |\pi_1(y) - t| < 3r_n\}$ with $|t| \leq \frac{1}{2} - 4r_n$. Therefore, for n large,

$$n\int_{\{y\in C:|\pi_1(y)|\leq 3r(n)\}} \psi^k_{n,r(n)}(y)\mathrm{d}y \leq (7r_n)n\int_C \psi^k_{n,r(n)}(y)\mathrm{d}y,$$

so (8.18) tends to zero by (8.4) and the fact that $r_n \to 0$. Hence $I_1 \to 0$.

In the toroidal case, the proof that $I_1 \to 0$ is similar, but simpler, and is omitted.

Next consider I_3. Given $x \in C$, let D_x denote the set of points in C that are at least as close to the centre of C as x is, in the l_1 norm (the D stands for 'diamond'):

$$D_x := \{y \in C : \|y\|_1 \leq \|x\|_1\}. \tag{8.19}$$

The integrand in (8.17) is symmetric in x and y. Writing simply r for r_n, and recalling the definitions of v_x, $v_{x,y}$, and $v_{x\backslash y}$, we have $I_3 = 2\sum_{j=0}^k \nu_j$, where we set

$$\nu_j := n^2 \int_C \mathrm{d}x \int_{y\in D_x:r<\|y-x\|\leq 3r} \mathrm{d}y \exp(-n(v_x+v_{y\backslash x}))(nv_{x,y})^{k-j}(nv_{x\backslash y})^j(nv_{y\backslash x})^j.$$

By (8.9), $v_{x,y} \leq \theta r^d$. Also, there is a constant c such that

$$\max(v_{y\backslash x}, v_{x\backslash y}) \leq cr^{d-1}\|y - x\|. \tag{8.20}$$

Also, by Proposition 5.16 and some easy scaling, there is a constant $\eta_2 > 0$ such that for $y \in D_x$ with $\|y - x\| \leq 3r$,

$$v_{y\backslash x} \geq \eta_2 r^{d-1} \|y - x\|. \tag{8.21}$$

In the case of the toroidal metric, (8.20) and (8.21) remain true, if we slightly abuse notation and let $\|y - x\|$ denote the toroidal distance between x and y.

Combining these bounds, we find that for some constant, also denoted c,

$$\nu_j \leq cn^2 (nr^d)^{k+j} \int_C \mathrm{d}x \int_{B(x;3r)\backslash B(x;r)} \exp(-nv_x - \eta_2 nr^{d-1} \|y - x\|) \times \left(r^{-1}\|y - x\|\right)^{2j} \mathrm{d}y.$$

Changing variable to $w = nr^{d-1}(y - x)$, so that $\mathrm{d}w = (nr^{d-1})^d \mathrm{d}y$, we have

$$\nu_j \leq cn(nr^d)^{k+1-j-d} \int_C \mathrm{d}x \exp(-nv_x) \int_{\mathbb{R}^d \backslash B(\mathbf{0};nr^d)} \exp(-\eta_2\|w\|)\|w\|^{2j}\mathrm{d}w.$$

Using (8.2) and (8.10), we find that ν_j is bounded by a constant times

$$(nr_n^d)^{1-j-d} \left(\int_C n\psi^k_{n,r(n)}(x)\mathrm{d}x \right) \left(\int_{\mathbb{R}^d \backslash B(\mathbf{0};nr^d)} \exp(-\eta_2\|w\|)\|w\|^{2j}\mathrm{d}w \right). \tag{8.22}$$

The final factor in (8.22) tends to zero since $nr_n^d \to \infty$, while the second factor is bounded by (8.4), and the first factor is bounded, so that $\nu_j \to 0$ as $n \to \infty$, for each $j \in \{0, 1, 2, \ldots, k\}$, and $I_3 \to 0$.

A similar calculation to that yielding (8.22) shows that the jth term in the sum for I_2 is bounded by a constant times

$$(nr_n^d)^{-j-d} \left(\int_C n\psi^k_{n,r(n)}(x)\mathrm{d}x \right) \left(\int_{\mathbb{R}^d} \exp(-\eta_2\|w\|)\|w\|^{2j}\mathrm{d}w \right).$$

This time the first factor $(nr_n^d)^{-j-d}$ tends to zero, while the other two factors are bounded, so that $I_2 \to 0$, completing the proof. □

Proof of Theorem 8.1 We need to de-Poissonize Theorem 8.2. For each positive integer n, set $m(n) := \lfloor n - n^{3/4} \rfloor$. Then $(r_n)_{n \geq 1}$ satisfying (8.4) also satisfies

$$\lim_{n \to \infty} EW'_{k,m(n)}(r_n) = \beta, \tag{8.23}$$

owing to the fact that $m(n)/n \to 1$, while $n^{3/4} r_n^d \to 0$ by (8.13). Hence by Theorem 8.2,

$$W'_{k,m(n)}(r_n) \xrightarrow{\mathcal{D}} \mathrm{Po}(\beta) \qquad \text{as } n \to \infty.$$

As described in Section 1.7, let $N_{m(n)}$ be the number of points of $\mathcal{P}_{m(n)}$, and assume $\mathcal{P}_{m(n)}$ and $\mathcal{X}_n$ are coupled by setting $\mathcal{P}_{m(n)} = \{X_1, \ldots, X_{N_{m(n)}}\}$ and $\mathcal{X}_n = \{X_1, \ldots, X_n\}$. Also, set $\mathcal{Y}_n = \{X_1, X_2, \ldots, X_{N_{m(n)}+2\lfloor n^{3/4} \rfloor}\}$. Define events F_n, A_n, and B_n by

- $F_n := \{\mathcal{P}_{m(n)} \subseteq \mathcal{X}_n \subseteq \mathcal{Y}_n\}$.
- A_n is the event that there exists a point $Y \in \mathcal{Y}_n \setminus \mathcal{P}_{m(n)}$ such that Y has degree at most k in $G(\mathcal{P}_{m(n)} \cup \{Y\}; r_n)$.
- B_n is the event that at least one point of $\mathcal{Y}_n \setminus \mathcal{P}_{m(n)}$ lies within distance r_n of a point X of $\mathcal{P}_{m(n)}$ with degree at most k in $G(\mathcal{P}_{m(n)}; r_n)$.

Then

$$\{W_{k,n}(r_n) \neq W'_{k,n}(r_n)\} \subseteq A_n \cup B_n \cup F_n^{\mathrm{c}}.$$

By Chebyshev's inequality (or by Lemma 1.4), $P(n-2\lfloor n^{3/4} \rfloor \leq N_{m(n)} \leq n)$ tends to 1, so that $P[F_n]$ tends to 1. Also, by (8.3), the probability that an inserted point Y has degree j in $G(\mathcal{P}_{m(n)} \cup \{Y\}; r_n)$ is equal to $m(n)^{-1} EW'_{j,m(n)}(r_n)$; hence,

$$P[A_n] \leq 2n^{3/4} m(n)^{-1} \sum_{j=0}^{k} EW'_{j,m(n)}(r_n),$$

so that $P[A_n] \to 0$. Also, by Boole's inequality

$$P[B_n] \leq 2n^{3/4} \theta r_n^d \sum_{j=0}^{k} EW'_{j,m(n)}(r_n),$$

so that $P[B_n] \to 0$ by (8.23), (8.13), and (8.5), completing the proof of (8.6).

Finally, for any non-negative integer $j < k$, we have $EW'_{j,n}(r_n) \to 0$ by (8.5), so that $P[W'_{j,n}(r_n) \neq 0] \to 0$ by Markov's inequality, and $P[W'_{j,n}(r_n) \neq W_{j,n}(r_n)] \to 0$ by the argument above, so that (8.7) follows. Finally, (8.8) follows from (8.6), (8.7), and (8.1). □

8.2 Uniformly distributed points II

In this section, a weak convergence result for a suitable transformation of $M_{k+1}(\mathcal{X}_n)$ is derived from Theorem 8.1. To obtain this we need to find a sequence $(r_n)_{n \geq 1}$ satisfying (8.4). Let Z denote a random variable with the double exponential extreme-value distribution $P[Z \leq \alpha] = \exp(-e^{-\alpha})$ for all $\alpha \in \mathbb{R}$.

Theorem 8.3 *Suppose $f = f_{\mathrm{U}}$, let $\|\cdot\|$ be an arbitrary norm on $\mathbb{R}^d$, and suppose the chosen metric on C (with opposite faces identified) is the toroidal metric $\mathrm{dist}(x,y) = \min_{z \in \mathbb{Z}^d} \|x+z-y\|$. Suppose $k \in \mathbb{N} \cup \{0\}$. Then*

$$n\theta M_{k+1}(\mathcal{X}_n)^d - \log n - k \log\log n + \log(k!) \xrightarrow{\mathcal{D}} Z. \tag{8.24}$$

Proof By spatial homogeneity of the torus, the condition (8.4) says that as $n \to \infty$,

$$n((n\theta r_n^d)^k / k!) \exp(-n\theta r_n^d) \to \exp(-\alpha),$$

which is equivalent to

$$-n\theta r_n^d + \log n + k \log(n\theta r_n^d) - \log(k!) \to -\alpha,$$

which is satisfied if we define r_n by

$$n\theta r_n^d = \log n + k\log(\log n) - \log(k!) + \alpha.$$

Therefore with this choice of r_n, by Theorem 8.1 we have $P[M_{k+1}(\mathcal{X}_n) \le r_n] \to \exp(-e^{-\alpha})$, which implies (8.24). □

The case of uniform points in the unit d-cube, with the l_p norm, is much more complicated because of boundary effects. The result goes as follows.

Theorem 8.4 *Suppose that $f = f_U$, and $\|\cdot\| = \|\cdot\|_p$ with $1 < p \le \infty$. If $d \ge 2$, let θ_{d-1} be the Lebesgue measure of the unit radius l_p ball in $\mathbb{R}^{d-1}$ (e.g. $\theta_{d-1} = 2^{d-1}$ if $p = \infty$), and set $\theta_0 = 1$. Let $k \in \mathbb{N} \cup \{0\}$. Then if $1 \le k+1 < d$,*

$$\left(\frac{n\theta M_{k+1}(\mathcal{X}_n)^d}{2^{d-2}} - \frac{2}{d}\log n - (k+3-d-2d^{-1})\log(\log n)\right.$$
$$\left. + \log\left(k!\binom{d}{2}^{-1} 2^{1-k-(2/d)} (\theta d)^{3-d-(2/d)} d^k \theta_{d-1}^{d-2}\right)\right) \xrightarrow{\mathcal{D}} Z. \quad (8.25)$$

If $1 \le d < k+1$, or if $d = 1$ and $k = 0$, then

$$\left(\frac{n\theta M_{k+1}(\mathcal{X}_n)^d}{2^{d-1}} - \frac{\log n}{d} - (k+2-d-d^{-1})\log(\log n)\right.$$
$$\left. + \log\left(k! 2^{2-d-(1/d)} d^{k-1} (d\theta)^{2-d-(1/d)} \theta_{d-1}^{d-1}\right)\right) \xrightarrow{\mathcal{D}} Z. \quad (8.26)$$

If $k+1 = d \ge 2$, then if we set $T_n = n\theta 2^{1-d} M_{k+1}(\mathcal{X}_n)^d - d^{-1}\log n - (1 - d^{-1})\log\log n$, we have

$$T_n - \log\left(e^{-T_n} + (2d\theta)^{1-(1/d)}\theta_{d-1}^{-1}/(d-1)\right)$$
$$+ \log\left((d-2)!(2\theta)^{3-d-(2/d)}\theta_{d-1}^{d-2} d^{1-(2/d)}\right) \xrightarrow{\mathcal{D}} Z. \quad (8.27)$$

In some special cases, the constants in Theorem 8.4 simplify considerably. If $k = 0$ and $d = 1$ the result is simply $2nM_1(\mathcal{X}_n) - \log n \xrightarrow{\mathcal{D}} Z$, while if $k = 0$ and $d = 2$, the result is simply $n\theta M_1(\mathcal{X}_n)^2 - \log n \xrightarrow{\mathcal{D}} Z$. If $d = 1$, the result (8.26) reduces to (8.24), that is, the toroidal boundary conditions in Theorem 8.3 do not affect the asymptotics for the case $d = 1$.

To prove Theorem 8.4 we must find $(r_n)_{n\ge1}$ satisfying (8.4). The difficulty lies in determining whether the interior of the unit cube C or its boundary makes the dominant contribution to the integral at (8.3), and in the latter case which part of the boundary is dominant. A clue is provided by Theorem 7.8, in the special case under consideration here, where k_n takes a fixed value k for all n, and $f = f_U$. In the notation of that result we have $b = 0$ so that $a_j = 0$ and $H(a_j) = 1$ for all j, while $f_j = 1$ for all j. Therefore, the maximum on the right-hand side of (7.31) is $\max_{0\le j\le d-1}(2^j(d-j)/d)$, and this maximum is achieved at

$j = d-2$ and $j = d-1$. This suggests that the dominant contribution will come either from points near ∂_{d-2} or near ∂_{d-1}, the two-dimensional and the one-dimensional part of the boundary of C. It turns out that this is indeed the case (Lemma 8.6 below), and that the question of which of these two contributions dominates is determined by whether $k+1 < d$, $k+1 = d$ or $k+1 > d$; see (8.30) below.

For $j = 0, 1, \ldots, d$, let $C_{n,j}$ be the set of points $x \in C$ such that $B(x; r_n)$ intersects precisely j of the hyperplanes bounding C. Given a sequence $(r_n)_{n\geq 1}$, define $I_{n,j}$ by

$$I_{n,j} := n \int_{C_{n,j}} dx (nF(B(x; r_n)))^k \exp(-nF(B(x; r_n)))/k!.$$

Then by (8.3), $EW'_{k,n}(r_n) = \sum_{j=0}^{d} I_{n,j}$.

Set $J_n := I_{n,d-1}$ and $I_n := I_{n,d-2}$ (in the case $d = 1$, I_n is not defined). As mentioned above, I_n and J_n are of special interest because they provide the dominant contributions to $EW'_{k,n}(r_n)$.

Lemma 8.5 *Suppose $d \geq 1$ and $\|\cdot\| = \|\cdot\|_p$ with $1 \leq p \leq \infty$. Suppose $r_n \to 0$ and $nr_n^d \to \infty$ as $n \to \infty$. Then as $n \to \infty$,*

$$J_n \sim \gamma_1(d,k) n^{1/d} (nr_n^d)^{k+2-d-(1/d)} \exp\left(-\frac{n\theta r_n^d}{2^{d-1}}\right), \tag{8.28}$$

where we set $\gamma_1(d,k) := k!^{-1} d\theta^k \theta_{d-1}^{1-d} 2^{(d-1)(d-1-k)}$. If $d \geq 2$, then

$$I_n \sim \gamma_2(d,k) n^{2/d} (nr_n^d)^{k+3-d-(2/d)} \exp\left(-\frac{n\theta r_n^d}{2^{d-2}}\right), \tag{8.29}$$

where we set $\gamma_2(d,k) := k!^{-1} \binom{d}{2} \theta^k \theta_{d-1}^{2-d} 2^{(d-2)(d-2-k)}$. As a consequence,

$$I_n \sim (\gamma_2(d,k)/\gamma_1(d,k)^2) J_n^2 (nr_n^d)^{d-k-1}. \tag{8.30}$$

Proof First assume $d \geq 2$ and consider J_n, the contribution to $EW'_{k,n}(r_n)$ from points near one-dimensional edges of C formed by the intersection of $d-1$ bounding hyperplanes. The number of such one-dimensional edges is $d2^{d-1}$.

Let $\mathcal{O}_d$ be the orthant $[0,\infty)^d$. For $\boldsymbol{t} = (t_1 \ldots, t_{d-1}) \in [0,1]^{d-1}$, let $(\boldsymbol{t}, 2) = (t_1, t_2, \ldots, t_{d-1}, 2) \in \mathbb{R}^d$, and with Leb($\cdot$) denoting Lebesgue measure, set

$$g(\boldsymbol{t}) := \mathrm{Leb}(B((\boldsymbol{t}, 2); 1) \cap \mathcal{O}_d) - \theta/2^{d-1}.$$

This is the volume of the set of points in $B((\boldsymbol{t}, 2); 1) \cap \mathcal{O}_d$ having at least one of their first $d-1$ coordinates less than the corresponding coordinate of $\boldsymbol{t}$. Then

$$
\begin{aligned}
J_n &\sim \left(\frac{d2^{d-1}}{k!}\right) n \int_{[0,r_n]^{d-1}} \exp\left(-nr_n^d\left(\frac{\theta}{2^{d-1}} + g\left(r_n^{-1}\boldsymbol{t}\right)\right)\right) \\
&\quad \times \left(nr_n^d\left(\frac{\theta}{2^{d-1}} + g\left(r_n^{-1}\boldsymbol{t}\right)\right)\right)^k \mathrm{d}\boldsymbol{t} \\
&= \left(\frac{d2^{d-1}}{k!}\right) nr_n^{d-1} \exp\left(-\frac{n\theta r_n^d}{2^{d-1}}\right)(nr_n^d)^k \\
&\quad \times \int_{[0,1]^{d-1}} \left(\frac{\theta}{2^{d-1}} + g(\boldsymbol{u})\right)^k \exp(-nr_n^d g(\boldsymbol{u}))\mathrm{d}\boldsymbol{u}. \qquad (8.31)
\end{aligned}
$$

We need to determine the asymptotic behaviour of the last integral. We assert that as $\xi \to \infty$,

$$
\int_{[0,1]^{d-1}} (2^{1-d}\theta + g(\boldsymbol{u}))^k \exp(-g(\boldsymbol{u})\xi)\mathrm{d}\boldsymbol{u} \sim 2^{(d-1)(d-2-k)}\theta^k\theta_{d-1}^{1-d}\xi^{1-d}. \qquad (8.32)
$$

We sketch a proof of (8.32). Let $\varepsilon > 0$ with ε small. There exists $\delta > 0$ such that if $\|\boldsymbol{u}\|_\infty > \varepsilon$, then $g(\boldsymbol{u}) > \delta$, so that the contribution to the above integral from such values of $\boldsymbol{u}$ decays exponentially in ξ. On the other hand, if $\|\boldsymbol{u}\|_\infty < \varepsilon$ then $g(\boldsymbol{u})$ is approximately the volume of the union of $d-1$ disjoint slabs, the jth slab being (approximately) the product of an interval $[0, u_j]$ (in the jth coordinates) with a $(d-1)$-dimensional l_p unit ball (for all the other coordinates) with all but one of its coordinates restricted to values which exceed the value at the ball's centre, and therefore the jth slab has approximate volume $u_j 2^{2-d}\theta_{d-1}$. Hence $g(\boldsymbol{u}) \approx 2^{2-d}\theta_{d-1}(u_1 + \cdots + u_{d-1})$, with an error term which is $o(\varepsilon)$. Also, $(2^{1-d}\theta + g(\boldsymbol{u}))^k \approx (2^{1-d}\theta)^k$, for $\|\boldsymbol{u}\|_\infty < \varepsilon$. As $\xi \to \infty$,

$$
\begin{aligned}
\int_0^\varepsilon \cdots \int_0^\varepsilon & \exp(-2^{2-d}\theta_{d-1}(u_1 + \cdots + u_{d-1})\xi)\mathrm{d}u_{d-1}\cdots \mathrm{d}u_1 \\
&\sim (2^{2-d}\theta_{d-1}\xi)^{1-d}, \qquad (8.33)
\end{aligned}
$$

and we can deduce (8.32) by routine analysis.

Using (8.32) and the fact that we assume $nr_n^d \to \infty$, we obtain

$$
J_n \sim \left(\frac{d}{k!}\right) 2^{(d-1)(d-1-k)}\theta^k\theta_{d-1}^{1-d} nr_n^{d-1} \exp\left(-\frac{n\theta r_n^d}{2^{d-1}}\right)(nr_n^d)^{k+1-d},
$$

and (8.28) follows.

In the case $d = 1$, J_n is the contribution to the integral for $EW'_{k,n}(r_n)$ from all points of the interval C except those within distance r_n of the boundary, and so $J_n \sim n(2nr_n)^k \exp(-2nr_n)/k!$, which is consistent with (8.28) (recall that we set $\theta_0 = 1$).

We seek a similar analysis for I_n. First suppose $d \geq 3$. The number of two-dimensional edges of C is $\binom{d}{2}2^{d-2}$. For $\boldsymbol{u} = (u_1, \ldots, u_{d-2}) \in [0,1]^{d-2}$, let $(\boldsymbol{u}, 2, 2) = (u_1, \ldots, u_{d-2}, 2, 2) \in \mathbb{R}^d$, and set

$$h(\boldsymbol{u}) := \mathrm{Leb}(B((\boldsymbol{u},2,2);1)\cap\mathcal{O}_d) - \theta/2^{d-2}.$$

This is the volume of the set of points in $B((\boldsymbol{u},2,2);1)\cap\mathcal{O}_d$ having at least one of their first $d-2$ coordinates less than the corresponding coordinate of $\boldsymbol{u}$.

A similar argument to the derivation of (8.31) yields

$$I_n \sim \binom{d}{2}\left(\frac{2^{d-2}}{k!}\right) nr_n^{d-2}\exp\left(-\frac{n\theta r_n^d}{2^{d-2}}\right)(nr_n^d)^k$$
$$\times \int_{[0,1]^{d-2}}\left(\frac{\theta}{2^{d-2}}+h(\boldsymbol{u})\right)^k \exp(-nr_n^d h(\boldsymbol{u}))\mathrm{d}\boldsymbol{u}.$$

We assert that as $\xi\to\infty$,

$$\int_{[0,1]^{d-2}}(2^{2-d}\theta+h(\boldsymbol{u}))^k\exp(-\xi h(\boldsymbol{u}))\mathrm{d}\boldsymbol{u}\sim(2^{2-d}\theta)^k(2^{3-d}\theta_{d-1}\xi)^{2-d}. \tag{8.34}$$

The proof of (8.34) is similar to that of (8.32). The factor of $(2^{2-d}\theta_{d-1}\xi)^{1-d}$ coming from (8.33) in the derivation of (8.32) is replaced by $(2^{3-d}\theta_{d-1}\xi)^{2-d}$, because there are now two 'free coordinates', so for $\|\boldsymbol{u}\|_\infty$ small, each of the $d-2$ slabs contributing to $h(\boldsymbol{u})$ is the product of an interval $[0,u_i]$ (for the ith coordinate) with a $(d-1)$-dimensional ball (for the other coordinates) with $d-3$ of its coordinates restricted to exceed the coordinate of the ball's centre. By (8.34) and the fact that we assume $nr_n^d\to\infty$,

$$I_n\sim k!^{-1}\binom{d}{2}\theta^k\theta_{d-1}^{2-d}2^{(d-2)(d-k-2)}nr_n^{d-2}(nr_n^d)^{k+2-d}\exp\left(-\frac{n\theta r_n^d}{2^{d-2}}\right),$$

and (8.29) follows.

In the case $d=2$, I_n is the contribution to $EW'_{k,n}(r_n)$ from all points of the square C except for those within r_n of the boundary, and so $I_n\sim n(n\theta r_n^2)^k\exp(-n\theta r_n^2)/k!$, which is consistent with (8.29). □

Lemma 8.6 *Suppose the sequence $(r_n)_{n\geq1}$ is such that* $\liminf(nr_n^d/\log n)>0$ *and* $\limsup(nr_n^d/\log n)<\infty$, *and I_n and J_n remain bounded as $n\to\infty$. Then $I_{n,j}\to0$ as $n\to\infty$, for $j\in\{0,1,\dots,d\}\setminus\{d-1,d-2\}$.*

Proof The volume of $C_{n,j}$ is bounded by a constant times r_n^j. Also, for $x\in C_{n,j}$, the value of $F(B(x;r_n))$ is at least $\theta r_n^d/2^j$, and at most θr_n^d. Therefore, if we set

$$I_{n,j}^* := n^{(d-j)/d}\exp(-n\theta r_n^d/2^j),$$

there is a constant c such that

$$I_{n,j}\leq cnr_n^j(nr_n^d)^k\exp(-n\theta r_n^d/2^j) = cI_{n,j}^*(nr_n^d)^{k+(j/d)}. \tag{8.35}$$

By the earlier estimates (8.28) and (8.29) on I_n and J_n, that for appropriate choices of β_{d-1} and β_{d-2}, $I_{n,j}$ is asymptotic to a constant times $I_{n,j}^*(nr_n^d)^{\beta_j}$, both for $j=d-1$ and $j=d-2$.

First consider the case $j = d$. We have $I^*_{n,d} = (I^*_{n,d-1})^{1/2} n^{-1/(2d)}$, and therefore for suitable positive constants c, c', γ, using (8.35) we have

$$I_{n,d} \leq c I^*_{n,d} (nr_n^d)^{k+1} \leq c' I^{1/2}_{n,d-1} (nr_n^d)^{\gamma} n^{-1/(2d)},$$

which tends to zero by the assumptions on r_n and J_n. Next consider $j < d-2$. We have

$$I^*_{n,j} = (I^*_{n,j+1})^2 n^{(j+2-d)/d},$$

so if $I^*_{n,j+1}$ is bounded by $(nr_n^d)^\gamma$ for some γ, then $I^*_{n,j}$ tends to zero and $I_{n,j}$ tends to zero. Hence by considering $j = d-3, d-4, \ldots, 0$ in turn, we can deduce that $I_{n,j} \to 0$ for each of these values of j. □

Proof of Theorem 8.4 First suppose $k+1 < d$. Let $\alpha \in \mathbb{R}$. We look for a sequence $(r_n)_{n \geq 1}$ such that $I_n \to \mathrm{e}^{-\alpha}$. By (8.29), this is equivalent to

$$-\frac{n\theta r_n^d}{2^{d-2}} + \frac{2}{d} \log n + (k+3-d-\tfrac{2}{d}) \log(nr_n^d) + \log(\gamma_2(d,k)) \to -\alpha.$$

This convergence holds if we take

$$\frac{n\theta r_n^d}{2^{d-2}} = \frac{2}{d} \log n + (k+3-d-\tfrac{2}{d}) \log\left(\frac{2^{d-1}}{\theta d} \log n\right) + \log(\gamma_2(d,k)) + \alpha. \quad (8.36)$$

With r_n defined in this way, we have $I_n \to \mathrm{e}^{-\alpha}$. Since $nr_n^d \to \infty$ and $k+1 < d$, it follows by (8.30) and Lemma 8.6 that J_n and all other contributions to $EW'_{k,n}(r_n)$ tend to zero, so that (8.4) holds with $\beta = \mathrm{e}^{-\alpha}$. Therefore by Theorem 8.1, if r_n is defined by (8.36), then $P[M_{k+1}(\mathcal{X}_n) \leq r_n] \to \exp(-\mathrm{e}^{-\alpha})$; thus,

$$\lim_{n\to\infty} P\left[\frac{n\theta M_{k+1}(\mathcal{X}_n)^d}{2^{d-2}} - \frac{2}{d}\log n - (k+3-d-\tfrac{2}{d})\log(\log n) \right.$$
$$\left. + (k+3-d-\tfrac{2}{d}) \log\left(\frac{\theta d}{2^{d-1}}\right) - \log(\gamma_2(d,k)) \leq \alpha\right] = \exp(-\mathrm{e}^{-\alpha}),$$

and rearranging terms in the constant completes the proof for the case $k+1 < d$.

Next, suppose $k+1 > d$. Let $\alpha \in \mathbb{R}$. This time we seek r_n giving us $J_n \to \mathrm{e}^{-\alpha}$. By (8.28), this is equivalent to

$$-\frac{n\theta r_n^d}{2^{d-1}} + \frac{\log n}{d} + (k+2-d-\tfrac{1}{d}) \log(nr_n^d) + \log \gamma_1(d,k) \to -\alpha.$$

This holds if we define r_n by

$$\frac{n\theta r_n^d}{2^{d-1}} = \frac{\log n}{d} + (k+2-d-\tfrac{1}{d}) \log\left(\frac{2^{d-1}\log n}{d\theta}\right) + \log \gamma_1(d,k) + \alpha. \quad (8.37)$$

With r_n defined by (8.37), we have $J_n \to \mathrm{e}^{-\alpha}$. Since $nr_n^d \to \infty$ and $k+1 > d$, (8.30) and Lemma 8.6 imply that other contributions to $EW'_{k,n}(r_n)$ vanish, so that (8.4) holds with $\beta = \mathrm{e}^{-\alpha}$, and so $P[M_{k+1}(\mathcal{X}_n) \leq r_n] \to \exp(-\mathrm{e}^{-\alpha})$ by Theorem 8.1. Therefore,

$$\lim_{n\to\infty} P\left[\frac{n\theta M_{k+1}(\mathcal{X}_n)^d}{2^{d-1}} - \frac{\log n}{d} - (k+2-d-\tfrac{1}{d})\log(\log n) \right.$$
$$\left. +(k+2-d-\tfrac{1}{d})\log\left(\frac{d\theta}{2^{d-1}}\right) - \log\gamma_1(d,k) \leq \alpha\right] = \exp(-\mathrm{e}^{-\alpha}),$$

and rearranging terms in the constant completes the proof of (8.26).

Next, suppose $k+1 = d = 1$. In this case, I_n is undefined and by the same analysis as in the preceding case, defining r_n by (8.37) gives us $J_n \to \mathrm{e}^{-\alpha}$ and hence $P[M_{k+1}(\mathcal{X}_n) \leq r_n] \to \exp(-\mathrm{e}^{-\alpha})$, so that again (8.26) holds.

Finally, suppose $k+1 = d \geq 2$. We write simply γ_1 for $\gamma_1(d, d-1)$ and γ_2 for $\gamma_2(d, d-1)$. In this case, (8.30) gives us $\lim_{n\to\infty}(I_n/J_n^2) = \gamma_2/\gamma_1^2$. Therefore, if we can find $(r_n)_{n\geq 1}$ such that

$$\lim_{n\to\infty}(J_n + (\gamma_2/\gamma_1^2)J_n^2) = \mathrm{e}^{-\alpha}, \tag{8.38}$$

then by Lemma 8.6, we will have (8.4) with $\beta = \mathrm{e}^{-\alpha}$. Since $J_n \geq 0$, (8.38) is equivalent to

$$\lim_{n\to\infty}(J_n) = \frac{\gamma_1^2}{2\gamma_2}\left(\sqrt{4(\gamma_2/\gamma_1^2)\mathrm{e}^{-\alpha}+1} - 1\right),$$

and by (8.28) and the assumption $k+1 = d$, this is equivalent to

$$-\frac{n\theta r_n^d}{2^{d-1}} + \frac{\log n}{d} + (1-\tfrac{1}{d})\log(nr_n^d) + \log\gamma_1$$
$$\to \log\left(\frac{\gamma_1^2}{2\gamma_2}\left(\sqrt{4(\gamma_2/\gamma_1^2)\mathrm{e}^{-\alpha}+1} - 1\right)\right),$$

which is satisfied if we define r_n by

$$\frac{n\theta r_n^d}{2^{d-1}} = \frac{\log n}{d} + \left(1 - \frac{1}{d}\right)\log\left(\frac{2^{d-1}\log n}{d\theta}\right)$$
$$-\log\left(\frac{\gamma_1}{2\gamma_2}\left(\sqrt{4(\gamma_2/\gamma_1^2)\mathrm{e}^{-\alpha}+1} - 1\right)\right). \tag{8.39}$$

With r_n defined by (8.39), we have (8.38), and hence (8.4) with $\beta = \mathrm{e}^{-\alpha}$, so that $P[M_{k+1}(\mathcal{X}_n) \leq r_n] \to \exp(-\mathrm{e}^{-\alpha})$ by Theorem 8.1. Therefore, if we set $T_n = n\theta 2^{1-d}M_{k+1}(\mathcal{X}_n)^d - d^{-1}\log n - (1-d^{-1})\log\log n$, we obtain

$$P\left[T_n \leq -\log\left(\left(\frac{d\theta}{2^{d-1}}\right)^{1-(1/d)}\frac{\gamma_1}{2\gamma_2}\left(\sqrt{4(\gamma_2/\gamma_1^2)\mathrm{e}^{-\alpha}+1} - 1\right)\right)\right]$$
$$\to \exp(-\mathrm{e}^{-\alpha}). \tag{8.40}$$

Define the constant η by

$$\eta := (2^{1-d}d\theta)^{1-(1/d)}\gamma_1/\gamma_2 = (2d\theta)^{1-(1/d)}\theta_{d-1}^{-1}/(d-1),$$

where the second equality comes from the definitions of γ_1 and γ_2, and some routine manipulation. By (8.40) we have

$$\begin{aligned}\exp(-\mathrm{e}^{-\alpha}) &= \lim_{n\to\infty} P\left[\frac{2\mathrm{e}^{-T_n}}{\eta} + 1 \geq \sqrt{4(\gamma_2/\gamma_1^2)\mathrm{e}^{-\alpha}+1}\right] \\ &= \lim_{n\to\infty} P\left[\eta^{-2}\mathrm{e}^{-2T_n} + \eta^{-1}\mathrm{e}^{-T_n} \geq (\gamma_2/\gamma_1^2)\mathrm{e}^{-\alpha}\right] \\ &= \lim_{n\to\infty} P\left[\mathrm{e}^{-T_n}(\mathrm{e}^{-T_n}+\eta)\gamma_1^2/(\gamma_2\eta^2) \geq \mathrm{e}^{-\alpha}\right] \\ &= \lim_{n\to\infty} P\left[T_n - \log(\mathrm{e}^{-T_n}+\eta) + \log(\gamma_2\eta^2/\gamma_1^2) \leq \alpha\right],\end{aligned}$$

and since

$$\gamma_2\eta^2/\gamma_1^2 = \gamma_2^{-1}\left(2^{1-d}d\theta\right)^{2-(2/d)} = (d-2)!(2\theta)^{3-d-(2/d)}\theta_{d-1}^{d-2}d^{1-(2/d)},$$

this gives us (8.27). □

8.3 Normally distributed points I

Now consider the largest nearest-neighbour link for points having a multivariate standard normal distribution. We assume throughout this section and the next that $d \geq 2$ and $\|\cdot\|$ *is the Euclidean* (l_2) *norm.* The standard multivariate normal density function is given by

$$\phi(x) := (2\pi)^{-d/2}\exp(-\|x\|^2/2).$$

Assume throughout this section and the next that $f = \phi$. The main distinction from the uniform cases previously considered is that the distribution of points has unbounded support.

Once again, let Z denote a random variable with the double exponential distribution $P[Z \leq \alpha] = \exp(-\mathrm{e}^{-\alpha})$ for all $\alpha \in \mathbb{R}$. Recall that the gamma function $(\Gamma(t), t \geq 0)$ is given by $\Gamma(t) := \int_0^\infty x^{t-1}\mathrm{e}^{-x}\mathrm{d}x$.

Theorem 8.7 *Suppose $f = \phi$. Then as $n \to \infty$,*

$$M_1(\mathcal{X}_n)\sqrt{2\log n} - (d-1)\log_2 n + \left(\frac{d-1}{2}\right)\log_3 n + \log\kappa_d \xrightarrow{\mathcal{D}} Z, \qquad (8.41)$$

where $\kappa_d := 2^{-d/2}(2\pi)^{-1/2}\Gamma(d/2)(d-1)^{(d-1)/2}$, and

$$\log_2 n := \log(\log n), \qquad \log_3 n := \log(\log_2 n). \qquad (8.42)$$

Theorem 8.7 shows that each percentage point of the distribution of $M_1(\mathcal{X}_n)$ behaves, to first order, like $(d-1)\log_2 n/\sqrt{2\log n}$, as $n \to \infty$. This contrasts with

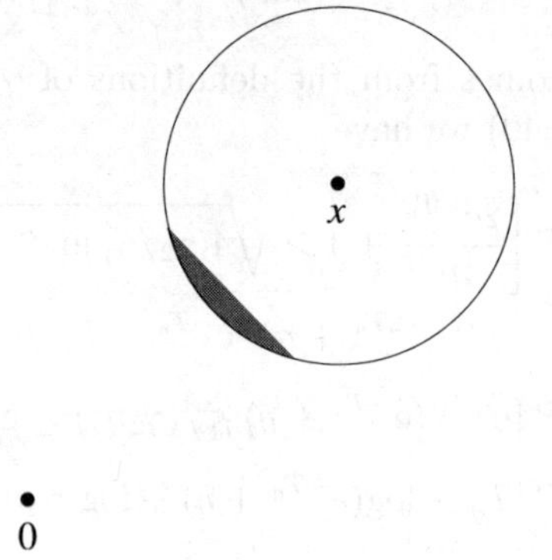

FIG. 8.1. The segment $B_\delta(x;r)$ is shaded.

the case of points uniformly distributed on the cube, where each percentage point of $M_1(\mathcal{X}_n)$ decays like a constant times $((\log n)/n)^{1/d}$, as seen in the preceding sections. Thus the asymptotics are completely different, and more delicate, in the normal case. The decay of the percentage points is slower because there are regions of $\mathbb{R}^d$ where the density function ϕ is very small, but not zero.

As before, let $W'_{k,n}(r)$ denote the number of vertices of degree k in $G(\mathcal{P}_n, r)$. The proof of Theorem 8.7 follows the same scheme as that for Theorem 8.4, but this time we proceed in a different order. First we find $(r_n)_{n\geq 1}$ such that $E[W'_{0,n}(r_n)]$ tends to a limit, and then we deduce a Poisson limit analogous to Theorem 8.1.

For $x \in \mathbb{R}^d, r > 0$, and $\delta \in (0, 2]$, define $B_\delta(x;r)$ to be the segment of $B(x;r)$ of thickness δr that is closest to the origin, that is,

$$B_\delta(x;r) := \{y \in B(x;r) : \|x\|^{-1} x \cdot y \leq \|x\| - (1-\delta)r\}, \tag{8.43}$$

where $x \cdot y$ is the Euclidean inner product (see Fig. 8.1). In terms of earlier notation from (5.9), $B_\delta(x;r)$ is the closure of $B^*(x;r,1-\delta,-e_x)$, where $e_x := \|x\|^{-1}x$, the unit vector in the direction of x. Define the d-dimensional integrals

$$I(x;r) = \int_{B(x;r)} \phi(y)\mathrm{d}y, \qquad I_\delta(x;r) = \int_{B_\delta(x;r)} \phi(y)\mathrm{d}y. \tag{8.44}$$

Thus $I(x;r) = I_2(x;r)$. Also, by (8.3),

$$E[W'_{0,n}(r_n)] = n \int_{\mathbb{R}^d} \exp(-nI(x;r_n))\phi(x)\mathrm{d}x. \tag{8.45}$$

For $\rho > 0$ define $I(\rho;r) := I(\rho e;r)$, where e is the d-dimensional unit vector $(1,0,0,\ldots,0)$. Define $I_\delta(\rho;r)$ similarly.

We start with large-ρ asymptotics for $I(\rho; r)$. Set $\theta_d := \pi^{d/2}/\Gamma((d/2)+1)$, the volume of the unit ball in $\mathbb{R}^d$ (see, e.g., Huang (1987, p. 139)).

Lemma 8.8 *Let $(\rho_n)_{n\geq 1}$ and $(r_n)_{n\geq 1}$ be sequences of positive numbers with $r_n \to 0$ and $r_n\rho_n \to \infty$ as $n \to \infty$. Let $\delta \in (0, 2]$. Then*

$$I_\delta(\rho_n; r_n) \sim (2\pi)^{-1/2} r_n^d \exp(\rho_n r_n - \rho_n^2/2)(\rho_n r_n)^{-(d+1)/2} \quad \text{as } n \to \infty, \tag{8.46}$$

where $\sim$ means that the ratio of the two sides tends to 1. Also,

$$\inf\left\{\left(\frac{I_\delta(\rho; r)}{\exp(\rho r - \rho^2/2) r^d (\rho r)^{-(d+1)/2}}\right) : r \leq \tfrac{1}{2}, \rho \geq 1/r\right\} > 0 \tag{8.47}$$

and

$$\inf\left\{\left(\frac{I_\delta(\rho; r)}{\exp(-\rho^2/2) r^d}\right) : r \leq \tfrac{1}{2}, \rho \geq 0\right\} > 0. \tag{8.48}$$

Proof In the definition of $I_\delta(\rho_n e; r_n)$ write $y = (\rho_n + r_n t, r_n \boldsymbol{s})$ with $\boldsymbol{s} \in \mathbb{R}^{d-1}$ to obtain

$$\begin{aligned}
I_\delta(\rho_n; r_n) &= \int_{t=-1}^{(\delta-1)} \int_{\boldsymbol{s}\in\mathbb{R}^{d-1}:\|\boldsymbol{s}\|^2<1-t^2} (2\pi)^{-d/2} \\
&\quad \times \exp(-((\rho_n + r_n t)^2 + (r_n\|\boldsymbol{s}\|)^2)/2) r_n^d \mathrm{d}\boldsymbol{s}\mathrm{d}t \\
&\sim (2\pi)^{-d/2} r_n^d \exp(-\rho_n^2/2) \int_{t=-1}^{\delta-1} \exp(-\rho_n r_n t)\theta_{d-1}(1-t^2)^{(d-1)/2}\mathrm{d}t \\
&= \theta_{d-1}(2\pi)^{-d/2} r_n^d \exp(-\rho_n^2/2) \\
&\quad \times \int_{u=0}^{\delta} \exp(-\rho_n r_n(u-1))(u(2-u))^{(d-1)/2}\mathrm{d}u \\
&= \theta_{d-1}(2\pi)^{-d/2} r_n^d \exp(\rho_n r_n - \rho_n^2/2) \\
&\quad \times \int_{v=0}^{\delta\rho_n r_n} \mathrm{e}^{-v}\left(2 - \frac{v}{\rho_n r_n}\right)^{(d-1)/2} v^{(d-1)/2}(\rho_n r_n)^{-(d+1)/2}\mathrm{d}v \\
&\sim 2^{(d-1)/2}\theta_{d-1}(2\pi)^{-d/2} r_n^d \exp(\rho_n r_n - \rho_n^2/2)(\rho_n r_n)^{-(d+1)/2} \\
&\quad \times \int_0^\infty \mathrm{e}^{-v} v^{(d-1)/2}\mathrm{d}v.
\end{aligned}$$

Since the constant $2^{(d-1)/2}\pi_{d-1}(2\pi)^{-d/2}\Gamma((d+1)/2)$ simplifies to $(2\pi)^{-1/2}$, this gives us (8.46). Also, at each stage where the symbol $\sim$ occurs in the above, it can be replaced by the symbol $\geq c$ for some suitable positive constant c, uniformly over those (ρ, r) for which $r \leq \frac{1}{2}$ and $\rho r \geq 1$, and (8.47) follows.

The final inequality (8.48) is elementary, following from the fact that for a suitable choice of $\delta > 0$ we have $\|y\| \leq \|x\|$ for all $y \in B_\delta(x, r)$, for all x with $\|x\| \geq 1$ and $r \leq \frac{1}{2}$, while $I_\delta(x; r)/r^d$ is bounded away from zero on $\|x\| \leq 1, r \leq \frac{1}{2}$. ∎

If $R := \|X_1\|$, then R^2 has a chi-square distribution and $R^2/2$ has a gamma distribution with density function $f_d(t) := t^{(d/2)-1}\mathrm{e}^{-t}/\Gamma(d/2)$, $t > 0$. By (8.45),

$$E[W'_{0,n}(r)] = n\int_0^\infty \exp(-nI(\sqrt{2s}; r))f_d(s)\mathrm{d}s. \tag{8.49}$$

Set

$$a_n := \log n + ((d/2) - 1)\log_2 n - \log(\Gamma(d/2)). \tag{8.50}$$

Then $nP[(R^2/2) - a_n \in \mathrm{d}t] = g_n(t)\mathrm{d}t$, where we set

$$\begin{aligned} g_n(t) &:= nf_d(t + a_n) = n(t + a_n)^{(d/2)-1}\mathrm{e}^{-t-a_n}/\Gamma(d/2) \\ &= (t + a_n)^{(d/2)-1}\mathrm{e}^{-t}(\log n)^{1-(d/2)}, \qquad t > -a_n. \end{aligned} \tag{8.51}$$

Then, for all $t \in \mathbb{R}$,

$$\lim_{n\to\infty} g_n(t) = \mathrm{e}^{-t}. \tag{8.52}$$

Also, for n so big that $a_n < 2\log n$, we have

$$g_n(t) \le \mathrm{e}^{-t}\left(\frac{3\max(t, \log n)}{\log n}\right)^{(d/2)-1} = 3^{(d/2)-1}\mathrm{e}^{-t}\max\left(1, \frac{t}{\log n}\right)^{(d/2)-1}. \tag{8.53}$$

By (8.49), setting $\rho_n(t) := (2(t + a_n))^{1/2}$, we have

$$E[W'_{0,n}(r)] = \int_{-a_n}^\infty \exp(-nI(\rho_n(t); r))g_n(t)\mathrm{d}t. \tag{8.54}$$

By (8.52), the second factor in the integrand is pointwise convergent, and we now show the same for the first factor, with a suitable choice of $r = r_n$.

Lemma 8.9 *Let $\alpha \in \mathbb{R}$, and suppose $(r_n)_{n\ge1}$ satisfies*

$$r_n\sqrt{2\log n} - (d-1)\log_2 n + \left(\frac{d-1}{2}\right)\log_3 n \to \alpha, \tag{8.55}$$

as $n \to \infty$. Let $t \in \mathbb{R}$, and set $\rho_n(t) := (2(t + a_n))^{1/2}$. Then for $0 < \delta \le 2$,

$$\lim_{n\to\infty} nI_\delta(\rho_n(t); r_n(\alpha)) = \kappa_d \mathrm{e}^{\alpha - t}, \tag{8.56}$$

with $\kappa_d = (2\pi)^{-1/2}\Gamma(d/2)(d-1)^{(d-1)/2}2^{-d/2}$.

Proof We use Lemma 8.8. Set $\rho_n := \rho_n(t)$ and assume (8.55) holds. Then

$$\exp(-\rho_n^2/2) = \mathrm{e}^{-t}\mathrm{e}^{-a_n} = \Gamma(d/2)\mathrm{e}^{-t}n^{-1}(\log n)^{1-(d/2)} \tag{8.57}$$

and

$$r_n^d \sim ((d-1)\log_2 n)^d (2\log n)^{-d/2}. \tag{8.58}$$

Also,

$$\begin{aligned} r_n\rho_n &= \left((d-1)\log_2 n + \left(\frac{1-d}{2}\right)\log_3 n + \alpha + o(1)\right) \\ &\quad \times \left(1 + \frac{t}{\log n} + \frac{((d/2)-1)\log_2 n - \log\Gamma(d/2)}{\log n}\right)^{1/2} \\ &= (d-1)\log_2 n + \left(\frac{1-d}{2}\right)\log_3(n) + \alpha + O\left(\frac{t\log_2 n}{\log n}\right) + o(1), \end{aligned} \tag{8.59}$$

where the $o(1)$ term does not depend on t. Hence,

$$\exp(r_n\rho_n) \sim (\log n)^{d-1}(\log_2 n)^{(1-d)/2}\mathrm{e}^{\alpha} \tag{8.60}$$

and

$$(r_n\rho_n)^{-(d+1)/2} \sim ((d-1)\log_2 n)^{-(d+1)/2}. \tag{8.61}$$

Combining (8.57), (8.58), (8.60) and (8.61), we have

$$n\exp(\rho_n r_n - \rho_n^2/2)r_n^d(\rho_n r_n)^{-(d+1)/2} \sim \Gamma(d/2)(d-1)^{(d-1)/2}2^{-d/2}\mathrm{e}^{\alpha-t},$$

and (8.56) now follows from (8.46). □

Proposition 8.10 *Let $\alpha \in \mathbb{R}$, let $(r_n)_{n\geq 1}$ satisfy (8.55), and let $\rho_n(t) := (2(a_n + t))^{1/2}$ as in the preceding lemma. Then, for $0 < \delta \leq 2$,*

$$\lim_{n\to\infty}\int_{-a_n}^{\infty}\exp(-nI_\delta(\rho_n(t);r_n))g_n(t)\mathrm{d}t = \mathrm{e}^{-\alpha}/\kappa_d. \tag{8.62}$$

In particular, $\lim_{n\to\infty} E[W'_{0,n}(r_n)] = \mathrm{e}^{-\alpha}/\kappa_d$.

Proof By (8.52) and (8.56), the integrand satisfies

$$\lim_{n\to\infty}\exp(-nI_\delta(\rho_n(t);r_n))g_n(t) = \exp(-\kappa_d\mathrm{e}^{\alpha-t})\mathrm{e}^{-t}. \tag{8.63}$$

To prove the result by dominated convergence, we need upper bounds holding for all large enough n.

First consider $t \geq 0$. By the bound (8.53) on g_n we have, for some c, that

$$\exp(-nI_\delta(\rho_n(t);r_n))g_n(t) \leq c\mathrm{e}^{-t}\max(t^{(d/2)-1}, 1), \qquad t \geq 0, \tag{8.64}$$

and this upper bound is integrable over $(0,\infty)$.

Next consider t with $-(\log n/\log_2 n) \leq t \leq 0$. By (8.59), there is a constant c such that for all such t,

$$r_n\rho_n(t) \geq (d-1)\log_2 n + ((1-d)/2)\log_3 n - c \tag{8.65}$$

and hence

$$\exp(r_n\rho_n(t)) \geq \mathrm{e}^{-c}(\log n)^{d-1}(\log_2 n)^{(1-d)/2}. \tag{8.66}$$

Also, there is a constant c' such that for t in this interval, $r_n\rho_n(t) \leq c'\log_2 n$; combining this with (8.57), (8.58), and (8.66), we obtain for some c that

$$n\exp(\rho_n r_n - \rho_n^2/2)r_n^d(\rho_n r_n)^{-(d+1)/2} \geq c\mathrm{e}^{-t}, \quad -(\log n/\log_2 n) \leq t \leq 0.$$

The lower bound in (8.65) exceeds 1 for n big, that is, $\rho_n(t) \geq 1/r_n(t)$ for n big, and so by (8.47), for n sufficiently large we have $nI_\delta(\rho_n(t); r_n) \geq c\mathrm{e}^{-t}$. Hence, by (8.53), for some $c' > 0$, we have

$$\exp(-nI_\delta(\rho_n(t); r_n))g_n(t) \leq c'\exp(-c\mathrm{e}^{-t})\mathrm{e}^{-t},$$
$$\text{if} \quad -\left(\frac{\log n}{\log_2 n}\right) \leq t \leq 0. \tag{8.67}$$

This upper bound is integrable over $t \in (-\infty, 0)$. By (8.63), (8.64), (8.67), and the dominated convergence theorem we have

$$\begin{aligned}\lim_{n\to\infty}\int_{-\log n/\log_2 n}^{\infty} & \exp(-nI_\delta(\rho_n(t); r_n))g_n(t)\mathrm{d}t \\ &= \int_{-\infty}^{\infty}\exp(-\kappa_d \mathrm{e}^{\alpha-t})\mathrm{e}^{-t}\mathrm{d}t \\ &= \int_0^{\infty}\exp(-\kappa_d\mathrm{e}^{\alpha}u)\mathrm{d}u = \mathrm{e}^{-\alpha}/\kappa_d. \end{aligned} \tag{8.68}$$

Now consider $t \leq -\log n/\log_2 n$. By (8.48), (8.57), and (8.58), there exist c and c' such that

$$nI_\delta(\rho_n(t); r_n) \geq c'nr_n^d\exp(-\rho_n^2/2) \geq h_n\mathrm{e}^{-t}, \tag{8.69}$$

with

$$h_n := c(\log n)^{1-d}(\log_2 n)^d. \tag{8.70}$$

Hence by (8.53), setting $c'' = 3^{(d/2)-1}$, we have

$$\int_{-a_n}^{-\log n/\log_2 n}\exp(-nI_\delta(\rho_n(t); r_n))g_n(t)\mathrm{d}t \leq c''\int_{-\infty}^{-\log n/\log_2 n}\exp(-h_n\mathrm{e}^{-t})\mathrm{e}^{-t}\mathrm{d}t$$
$$= c''\int_{\exp(\log n/\log_2 n)}^{\infty}\exp(-h_n u)\mathrm{d}u = c''h_n^{-1}\exp(-h_n\exp(\log n/\log_2 n)),$$

which converges to zero as can be seen by taking logarithms. Combining this with (8.68), we have (8.62). The limit for $EW'_{0,n}(r_n)$ then follows at once, by (8.54). □

Lemma 8.11 *Let* $\alpha \in \mathbb{R}$ *and* $r_n = r_n(\alpha)$ *as in* (8.55). *Then for all* $\delta \in (0, 2]$,

$$\limsup_{n\to\infty} \sup_{\rho \geq 0} n e^{-\rho^2/2} \exp(-nI_\delta(\rho; r_n)) < \infty. \tag{8.71}$$

Proof Given ρ, set $t := (\rho^2/2) - a_n$ so that $\rho = (2(t + a_n))^{1/2} = \rho_n(t)$. Then by (8.57),

$$n e^{-\rho^2/2} \exp(-nI_\delta(\rho; r_n)) \leq \Gamma(d/2) e^{-t} \exp(-nI_\delta(\rho_n(t); r_n)).$$

Set $u_n(t) := \exp(-t - nI_\delta(\rho_n(t); r_n))$. Then $u_n(t) \leq 1$ for $t \geq 0$, and by the proof of (8.67),

$$u_n(t) \leq c' \exp(-t - c e^{-t}), \qquad -(\log n/ \log_2 n) \leq t \leq 0.$$

On $t \leq (-\log n / \log_2 n)$, by (8.69) we have $u_n(t) \leq \exp(-t - h_n e^{-t})$. The function $-t - h_n e^{-t}$ has its maximum at $t = \log h_n$, and so is maximized over $t \in (-\infty, -\log n/\log_2 n]$ by its value at the right-hand end of this interval; hence

$$\log u_n(t) \leq \frac{\log n}{\log_2 n} - h_n \exp\left(\frac{\log n}{\log_2 n}\right), \qquad t \leq -\log n/ \log_2 n,$$

and by taking logs again we see that this bound is negative for large n. Combining these bounds for $u_n(t)$, we see that $u_n(t)$ is bounded uniformly in t and n, as required. □

8.4 Normally distributed points II

In this section the proof of Theorem 8.7 is completed; we make the same assumptions about the norm and the density function f as in the preceding section. First, consider the Poisson process $\mathcal{P}_n$. As in the statement of Theorem 8.7, set $\kappa_d := 2^{-d/2}(2\pi)^{-1/2}\Gamma(d/2)(d-1)^{(d-1)/2}$. We first give a Poisson limit for the number of isolated vertices.

Theorem 8.12 *Let* $\alpha \in \mathbb{R}$, *and suppose* $(r_n)_{n\geq 1}$ *satisfies* (8.55). *Then* $W'_{0,n}(r_n) \xrightarrow{\mathcal{D}} \mathrm{Po}(e^{-\alpha}/\kappa_d)$.

Proof By Theorem 6.7, $d_{\mathrm{TV}}(W'_{0,n}(r_n), \mathrm{Po}(E[W'_{0,n}(r_n)]))$ is bounded by $3(J_1(n) + J_2(n))$, with $J_1(n)$ and $J_2(n)$ defined as follows. Setting $I^{(2)}(x, y; r) := \int_{B(x;r)\cup B(y;r)} \phi(z) dz$, define

$$J_1(n) := n^2 \int_{\mathbb{R}^d} dx \phi(x) \exp(-nI(x; r_n)) \int_{B(x;r_n)} dy \phi(y) \exp(-nI(y; r_n)),$$

$$J_2(n) := n^2 \int_{\mathbb{R}^d} dx \phi(x) \int_{r_n \leq \|y-x\| \leq 3r_n} dy \phi(y) \exp(-nI^{(2)}(x, y; r_n)).$$

By the uniform bound of Lemma 8.11, there is a constant c such that for all large enough n,

$$\begin{aligned} J_1(n) &\leq c \left(n \int_{\mathbb{R}^d} \mathrm{d}x \phi(x) \exp(-nI(x;r_n)) \right) r_n^d \\ &= c r_n^d E[W'_{0,n}(r_n)], \end{aligned}$$

which converges to zero as $n \to \infty$, by Proposition 8.10 and the fact that $r_n \to 0$.

We can (and do) pick $\delta > 0$ such that for any $r \leq \frac{1}{2}$, any $x \in \mathbb{R}^d$ with $\|x\| \geq 1$, and any y with $\|y - x\| \geq r$, the regions $B_\delta(x;r)$ and $B_\delta(y;r)$ are disjoint, so that $I^{(2)}(x,y;r_n) \geq I_\delta(x;r_n) + I_\delta(y;r_n)$, and

$$\begin{aligned} J_2(n) \leq\; & n^2 \int_{\|x\| \leq 1} \mathrm{d}x \phi(x) \int_{r_n \leq \|y-x\| \leq 3r_n} \mathrm{d}y \phi(y) \exp(-nI(x;r_n)) \\ & + n^2 \int_{\mathbb{R}^d} \mathrm{d}x \phi(x) \exp(-nI_\delta(x;r_n)) \\ & \times \int_{r_n \leq \|y-x\| \leq 3r_n} \mathrm{d}y \phi(y) \exp(-nI_\delta(y;r_n)). \end{aligned} \tag{8.72}$$

The first term on the right-hand side of (8.72) is bounded by $n^2 \exp(-cnr_n^d)$ for some $c > 0$, and this tends to zero as $n \to \infty$. The second term tends to zero as $n \to \infty$ by the same argument as for $J_1(n)$.

By the above estimates, along with Theorem 6.7 and Proposition 8.10, $W'_{0,n}(r_n)$ converges in distribution to $\mathrm{Po}(\mathrm{e}^{-\alpha}/\kappa_d)$. □

We now de-Poissonize Theorem 8.12.

Theorem 8.13 *Let $\alpha \in \mathbb{R}$. If $(r_n)_{n\geq 1}$ satisfies* (8.55), *then*

$$W_{0,n}(r_n) \xrightarrow{\mathcal{D}} \mathrm{Po}(\mathrm{e}^{-\alpha}/\kappa_d). \tag{8.73}$$

Proof For each positive integer n, set $m(n) := \lfloor n - n^{3/4} \rfloor$. Let $\alpha \in \mathbb{R}$. Then $(r_n)_{n\geq 1}$ satisfying (8.55) also satisfies

$$r_n \sqrt{2 \log m(n)} - (d-1) \log_2 m(n) + \left(\frac{d-1}{2} \right) \log_3 m(n) \to \alpha,$$

and hence by Theorem 8.12,

$$W'_{0,m(n)}(r_n) \xrightarrow{\mathcal{D}} \mathrm{Po}(\exp(-\mathrm{e}^{-\alpha}/\kappa_d)). \tag{8.74}$$

Assume $\mathcal{P}_{m(n)}$ and $\mathcal{X}_n$ are coupled as described in Section 1.7; let $N_{m(n)}$ be the number of points of $\mathcal{P}_{m(n)}$. Also, set $\mathcal{Y}_n = \{X_1, X_2, \ldots, X_{N_{m(n)}+2\lfloor n^{3/4} \rfloor}\}$. Let B_n be the event that one or more point of $\mathcal{Y}_n \setminus \mathcal{P}_{m(n)}$ lies within distance r_n of a point X of $\mathcal{P}_{m(n)}$ with degree 0 in $G(\mathcal{P}_{m(n)}; r_n)$. It suffices to prove that $\lim_{n\to\infty} P[B_n] = 0$, since the rest of the proof follows the de-Poissonization argument used to prove Theorem 8.1 at the end of Section 8.1.

Let X' denote a standard normal random d-vector, independent of $\{X_1, X_2, \ldots\}$. By Boole's inequality, $P[B_n]$ is bounded by $2n^{3/4}$ times the probability that there is an isolated point of $G(\mathcal{P}_{m(n)}; r_n)$ in $B(X'; r_n)$, and hence by $2n^{3/4}$ times the mean number of such points. Therefore by Palm theory (Theorem 1.6),

$$P[B_n] \leq 2n^{7/4} \int_{\mathbb{R}^d} \phi(y)\mathrm{d}y \int_{B(y;r_n)} \exp(-m(n)I(x;r_n))\phi(x)\mathrm{d}x$$

and by interchanging the order of integration, we obtain

$$\begin{aligned} P[B_n] &\leq 2n^{7/4} \int_{\mathbb{R}^d} I(x;r_n)\exp(-m(n)I(x;r_n))\phi(x)\mathrm{d}x \\ &\leq 2\int_{-a_n}^{\infty} n^{3/4} I(\rho_n(t);r_n)\exp(-m(n)I(\rho_n(t);r_n))g_n(t)\mathrm{d}t, \qquad (8.75) \end{aligned}$$

with a_n defined at (8.50) and $\rho_n(t) := (2(t+a_n))^{1/2}$. Since $nI(\rho_n(t);r_n)$ converges to a finite limit for each t, the integrand in (8.75) tends to zero pointwise.

For $t \geq 0$, $nI(\rho_n(t);r_n) \leq nI(\rho_n(0);r_n)$ which is bounded by (8.56), so for some $c > 0$,

$$\int_0^{\infty} n^{3/4} I(\rho_n(t);r_n)\exp(-m(n)I(\rho_n(t);r_n))g_n(t)\mathrm{d}t \leq c\int_0^{\infty} n^{-1/4} g_n(t)\mathrm{d}t,$$

which tends to zero by (8.53).

Since $\mathrm{e}^{x/2} \geq x$ for all $x \in \mathbb{R}$, we have $nI(\rho_n(t);r_n) \leq \exp(nI(\rho_n(t);r_n)/2)$, and since $m(n) > 3n/4$, we have

$$nI(\rho_n(t);r_n)\exp(-m(n)I(\rho_n(t);r_n)) \leq \exp(-(n/4)I(\rho_n(t);r_n)).$$

Therefore by the same argument as for $t \leq -\log n/\log_2 n$ in the proof of Proposition 8.10, the contribution from t in this range to the integral in (8.75) tends to zero.

For $-(\log n/\log_2 n) \leq t \leq 0$, by (8.59) there is a constant c such that

$$\exp(r_n\rho_n(t)) \leq c(\log n)^{d-1}(\log_2 n)^{(1-d)/2},$$

and also $(r_n\rho_n(t))^{-(d+1)/2} \leq c(\log_2 n)^{-(d+1)/2}$. Also the proof of Lemma 8.8 shows that the right-hand side of (8.46) is also an upper bound. Hence

$$I(\rho_n(t);r_n) \leq cn^{-1}\mathrm{e}^{-t}, \qquad -(\log n/\log_2 n) \leq t \leq 0. \qquad (8.76)$$

By the proof of (8.67), additionally $I(\rho_n(t);r_n) \geq 2c'n^{-1}\mathrm{e}^{-t}$, for some c'. Hence, for n large enough we have $m(n)I(\rho_n(t);r_n) \geq c'\mathrm{e}^{-t}$ for $-(\log n/\log_2 n) \leq t \leq 0$. Applying (8.76) again to the first integrand in (8.75), and using (8.53), we have

$$\int_{-\log n/\log_2 n}^{0} n^{3/4} I(\rho_n(t); r_n) \exp(-m(n) I(\rho_n(t); r_n)) g_n(t) \mathrm{d}t$$
$$\leq c \int_{-\log n/\log_2 n}^{0} n^{-1/4} \mathrm{e}^{-t} \exp(-c' \mathrm{e}^{-t}) \mathrm{e}^{-t} \mathrm{d}t,$$

which converges to zero. Thus $P[B_n] \to 0$. □

Proof of Theorem 8.7 By (8.1) and Theorem 8.13, if $(r_n)_{n \geq 1}$ satisfies (8.55), then

$$\lim_{n\to\infty} P[M_1(\mathcal{X}_n) \leq r_n] = \lim_{n\to\infty} P[W_{0,n}(r_n) = 0] = \exp(-\mathrm{e}^{-\alpha}/\kappa_d). \qquad (8.77)$$

Hence,

$$P\left((2\log n)^{1/2} M_1(\mathcal{X}_n) - (d-1)\log_2 n + \left(\frac{d-1}{2}\right)\log_3 n \leq \alpha - \log \kappa_d\right)$$
$$\to \exp\left(-\mathrm{e}^{-(\alpha - \log \kappa_d)}/\kappa_d\right) = P[Z \leq \alpha],$$

as required. □

8.5 Notes and open problems

Notes *Sections 8.1 and 8.2.* The proof of weak convergence is adapted from Penrose (1997, 1999*c*). Previous work on results of this type followed work by Henze (1982, 1983) in which convergence of probabilities is proved using Bonferroni bounds. In particular, Steele and Tierney (1986) consider the case $k = 0$ of Theorem 8.3, while Dette and Henze (1989, 1990) give Theorem 8.4 in the special cases of the l_∞ norm (for all d, k), the l_2 norm (for $d = 3, k = 0$) and for all l_p norms (for $d = 2, k = 0$).

Sections 8.3 and 8.4. Theorem 8.7 is taken from Penrose (1998). The statements of Theorems 8.12 and 8.13 are new, but most of the work for proving them is done in Penrose (1998). Recently, Hsing and Rootzén (2002) have extended Theorem 8.7 to a general class of two-dimensional distributions having densities with unbounded support and with logarithm satisfying certain regularity conditions, including a form of regular variation. In particular, elliptically contoured densities such as the bivariate normal are included in their result.

Open problems An extension of the results of this chapter would be to consider density functions other than the uniform and standard normal cases considered here. For example, a uniform distribution on a polyhedral domain should not present any problems; a smooth density function on such a domain, bounded away from zero and infinity, might also be feasible. It might also be possible to generalize the case of the standard normal distribution to a class of spherically symmetric density function. For example, Henze and Klein (1996) considered such a general class of density functions in their analysis of the range of $\mathcal{X}_n$. As mentioned above, Hsing and Rootzén (2002) have recently addressed the two-dimensional case of this problem.

9

PERCOLATIVE INGREDIENTS

This chapter contains topological and probabilistic preliminaries which will be useful in proving results about global connectivity and large components of random geometric graphs.

9.1 Unicoherence

If (Ω, τ) is a topological space and $x, y \in \Omega$, a *path* in Ω from x to y is a continuous function π from $[0, 1]$ to Ω with $\pi(0) = x$ and $\pi(y) = 1$. If $x = y$, such a path is called a *loop*. A topological space (Ω, τ) is said to be *unicoherent* if for any two closed connected sets $A_1 \subseteq \Omega, A_2 \subseteq \Omega$ with union $A_1 \cup A_2 = \Omega$, the intersection $A_1 \cap A_2$ is connected. It is said to be *simply-connected* if any two elements can be connected by a path, and every loop can be deformed continuously to a single point.

An example of a non-unicoherent space is provided by the unit circle $\{z \in \mathbb{R}^2 : \|z\|_2 = 1\}$. This example is, of course, not simply-connected either. The following result shows that this example is typical of all such counterexamples.

Lemma 9.1 *If Ω is simply-connected, then it is unicoherent.*

Proof See Dugundji (1966). □

A topological space (Ω, τ) is *bicoherent* if for any two closed connected sets $A_1 \subseteq \Omega, A_2 \subseteq \Omega$ with union $A_1 \cup A_2 = \Omega$, the intersection $A_1 \cap A_2$ has at most two components. Although the d-dimensional torus is not unicoherent, we have the following result.

Lemma 9.2 *Let $d \in \mathbb{N}$. Then the d-dimensional torus is bicoherent.*

It is not hard to see that this result holds for $d = 1$, and the case of general d follows from a result of Eilenberg (1936, §1.4, Theorem 5) on multicoherence of Cartesian products. See also Illanes Mejia (1985).

9.2 Connectivity and Peierls arguments

When proving results on connectedness properties of random geometric graphs, one useful technique is the discretization of the continuum into blocks; the study of analogous connectivity properties on random subgraphs of a discrete lattice is lattice *percolation theory*. One of the technical uses of such a discretization lies in the availability of combinatorial arguments for enumerating the sets in $\mathbb{Z}^d$ with certain connectedness properties. These are the subject of the current section.

We shall require a variety of notions of connectivity for sets in the integer lattice $\mathbb{Z}^d$. A set $A \subseteq \mathbb{Z}^d$ is said to be *symmetric* if $-x \in A$ for all $x \in A$. Given a finite symmetric set $A \subseteq \mathbb{Z}^d$, let $\sim_A$ denote the relation on $\mathbb{Z}^d$ whereby $x \sim_A y$ if and only if $y - x \in A$. Let $(\mathbb{Z}^d, \sim_A)$ denote the graph with vertex set $\mathbb{Z}^d$ and adjacency relation $\sim_A$. Let us say that a subset S of $\mathbb{Z}^d$ is *A-connected* if it induces a connected subgraph of the graph $(\mathbb{Z}^d, \sim_A)$, that is, if the maximal subgraph of $(\mathbb{Z}^d, \sim_A)$ with vertex set S is connected.

The main examples of use here are as follows. In the case where $A = \{z \in \mathbb{Z}^d : \|z\|_1 = 1\}$, the graph $(\mathbb{Z}^d, \sim_A)$ is the 'usual' nearest-neighbour integer lattice, and in this case we shall refer to A-connected subsets of $\mathbb{Z}^d$ as simply being *connected.* In the case where $A = \{z \in \mathbb{Z}^d : \|z\|_\infty = 1\}$, we shall refer to A-connected subsets of $\mathbb{Z}^d$ as being *$*$-connected.* Finally, in the case where $A = \{z \in \mathbb{Z}^d : 0 < \|z\| \leq r\}$ for some constant r and some norm $\|\cdot\|$ (the same norm as in the definition of the geometric graph), we write $\sim_r$ for the adjacency relationship $\sim_A$, we refer to A-connected subsets of $\mathbb{Z}^d$ as being *r-connected.* The last type of connectivity will be of use in making lattice approximations to connected regions of continuous space made up of a union of balls in the chosen norm.

The following lemma says that the number of A-connected subsets of $\mathbb{Z}^d$ of size n containing the origin grows at most exponentially in n. This fact was used by Peierls (1936) in his work on the Ising model, and the result is usually named after him.

Lemma 9.3 (Peierls argument) *Let A be a finite symmetric subset of $\mathbb{Z}^d$ with $|A|$ elements. The number of A-connected subsets of $\mathbb{Z}^d$ containing the origin, of cardinality n, is at most $2^{|A|n}$.*

Proof Let S be an A-connected subset of $\mathbb{Z}^d$ with n elements. We shall construct a nondecreasing sequence of lists $L_1, L_2, \ldots, L_t$, where a 'list' means an ordered sequence of distinct elements of S. For each (ordered) list L_j, write S_j for the (unordered) set of its elements. Let $L_1 = (z_1)$ with z_1 denoting the origin of $\mathbb{Z}^d$. Let $L_2 = (z_1, \ldots, z_{k_2})$, where $z_2, \ldots z_{k_2}$ are the elements of $S \setminus S_1$ lying adjacent to z_1, taken in lexicographic order.

If the jth list is $L_j = (z_1, \ldots, z_{k_j})$, then to obtain L_{j+1}, take the list L_j and add to the end of it all elements of S which are adjacent to z_j but are not already included in the list L_j (possibly an empty set of added elements), putting the added vertices in lexicographic order, to get the list $L_{j+1} = (z_1, \ldots, z_{k_{j+1}})$ with $k_{j+1} \geq k_j$. Continue in this way until at some termination time t, the list L_t is of length t and the tth element of the list z_t has no neighbour in S that is not part of the list L_t.

The termination will always take place at time $t = n$ leaving us with a list of the entire set S. For if the algorithm terminated earlier, then S_t would have fewer than n elements, and there would be no element of $S \setminus S_t$ lying adjacent to any element of S_t, contradicting the A-connectivity of S.

At each step of the algorithm, the number of possibilities for the added set of elements is bounded by the number of subsets of the set of elements of $\mathbb{Z}^d$ lying adjacent to z_j, so is bounded by $2^{|A|}$. Therefore, the result follows. □

For positive integers $m_1, \ldots, m_d$, define the lattice rectangle $B_{\mathbb{Z}}(m_1, \ldots, m_d)$ by

$$B_{\mathbb{Z}}(m_1, \ldots, m_d) := \prod_{i=1}^{d} ([1, m_i] \cap \mathbb{Z}), \tag{9.1}$$

and for integer $m > 0$ let $B_{\mathbb{Z}}(m)$ be the lattice box $B_{\mathbb{Z}}(m, m, \ldots, m)$.

Corollary 9.4 *Let A be a finite symmetric subset of $\mathbb{Z}^d$ with $|A|$ elements. Then for all positive integers $n, m_1, \ldots, m_d$, the number of A-connected subsets of the lattice box $B_{\mathbb{Z}}(m)$ of cardinality n is at most $2^{|A|n} \prod_{i=1}^{d} m_i$.*

Proof By Lemma 9.3, for each $z \in \mathbb{Z}^d$, the number of A-connected subsets of $B_{\mathbb{Z}}(m)$ of cardinality n containing z is at most $2^{|A|n}$. Since the number of $z \in B_{\mathbb{Z}}(m)$ is $\prod_{i=1}^{d} m_i^d$, the result follows by the combinatorial version of Boole's inequality. □

We shall also be concerned with connected subsets of the lattice torus. Let us say that a subset S of $B_{\mathbb{Z}}(m)$ is *toroidally $*$-connected* if it is a connected subgraph of the graph $(B_{\mathbb{Z}}(m), \sim)$ with the adjacency $x \sim y$ if and only if $\|x - y + mz\|_\infty = 1$ for some $z \in \mathbb{Z}^d$.

Lemma 9.5 *For all positive integers m, n, the number of subsets of the lattice box $B_{\mathbb{Z}}(m)$ of cardinality n having at most two toroidally $*$-connected components is at most $nm^{2d}2^{3^d n}$.*

Proof Given $x \in B_{\mathbb{Z}}(m)$, the number of toroidally $*$-connected subsets of $B_{\mathbb{Z}}(m)$ of cardinality j containing x is at most $2^{3^d j}$, by the proof of Lemma 9.3 (adapted to the torus). For any $n \geq 2$, and any $S \subset B_{\mathbb{Z}}(m)$ of cardinality n having at most two toroidally $*$-connected components, we can find $j \in \{1, 2, \ldots, n-1\}$ and $x, y \in B_{\mathbb{Z}}(m)$ such that S is the union of a toroidally $*$-connected set of cardinality j containing x, and a toroidally $*$-connected set of cardinality $n - j$ containing y. The number of choices of (j, x, y) is at most nm^{2d}, and given (j, x, y) the number of ways to choose a toroidally $*$-connected set of cardinality j containing x and a toroidally $*$-connected set of cardinality $n - j$ containing y is at most $2^{3^d j} 2^{3^d (n-j)} = 2^{3^d n}$. The result follows. □

The next result is a lattice version of the unicoherence property of $\mathbb{R}^d$ or of the unit cube.

Lemma 9.6 *Let W be either the set $\mathbb{Z}^d$ or the set $B_{\mathbb{Z}}(m)$ for some m. Suppose $A \subset W$ is such that both A and $W \setminus A$ are connected. Let $\partial_W^- A$ denote the internal vertex-boundary of A, that is, the set of lattice sites $z \in A$ such that $\{y \in W \setminus A : \|z - y\|_1 = 1\}$ is non-empty. Then $\partial_W^- A$ is $*$-connected.*

Proof For $S \subseteq \mathbb{Z}^d$ set $S^* := S \oplus [-\frac{1}{2}, \frac{1}{2}]^d$, the union of the closed rectilinear unit cubes centred at points of S. Since both A and $W \setminus A$ are connected in the lattice, both A^* and $(W \setminus A)^*$ are connected subsets of W^*, and by unicoherence of W^* (Lemma 9.1), their intersection is connected. Hence, $\partial_W^- A$ is $*$-connected. □

9.3 Bernoulli percolation

Motivated mainly by the study of random physical media, percolation theory is the study of connectivity properties of random sets in space. Lattice percolation in particular has been much studied (see Grimmett (1999) for a thorough treatment of the subject; also Kesten (1982) and Stauffer and Aharony (1994)). In the present context, its importance arises from various dicretizations of continuum processes, and the most relevant lattice percolation models are concerned with properties of geometric graphs on subsets of the integer lattice $\mathbb{Z}^d$, embedded in the continuous space $\mathbb{R}^d$.

For our purposes, the most relevant lattice percolation model is *site percolation* on $\mathbb{Z}^d$, defined as follows. Given $p \in [0,1]$, let $Z^p = (Z_x^p, x \in \mathbb{Z}^d)$ be a family of mutually independent Bernoulli(p) random variables. The sites $x \in \mathbb{Z}^d$ for which $Z_x^p = 1$ are denoted *open* and the sites $x \in \mathbb{Z}^d$ for which $Z_x^p = 0$ are denoted *closed*. Let $\mathcal{B}_p$ denote the (random) set of open sites; here $\mathcal{B}$ stands for 'Bernoulli' and we shall sometimes refer either to Z^p or to $\mathcal{B}_p$ as a *Bernoulli process*.

Embedding $\mathbb{Z}^d$ in $\mathbb{R}^d$, we may view any (typically random) set $\mathcal{B} \subseteq \mathbb{Z}^d$, as a subset of $\mathbb{R}^d$, on which geometric graphs are defined as for sets in the continuum. Let $G_{\mathbb{Z}}(\mathcal{B})$ denote the graph $G(\mathcal{B};1)$ using the l_1 norm, and for $r > 0$ let $G_{\mathbb{Z}}(\mathcal{B};r)$ denote the graph $G(\mathcal{B};r)$ using an arbitrary norm (the norm of choice for defining continuum geometric graphs).

The components of the graph $G_{\mathbb{Z}}(\mathcal{B})$ (i.e. the maximal connected subsets of $\mathcal{B}$) are denoted the *open clusters* (or just *clusters*) in $\mathcal{B}$. The components of the graph $G_{\mathbb{Z}}(\mathcal{B}, r)$ (i.e. the maximal r-connected subsets of $\mathcal{B}$, in notation from Section 9.2) are the *open r-clusters* (or just *r-clusters*) in $\mathcal{B}$. We shall avoid using the term 'cluster' for random geometric graphs in the continuum.

The *cluster at the origin* for $\mathcal{B}_p$ is the open cluster in $\mathcal{B}_p$ containing the origin $\mathbf{0}$ (or the empty set if $\mathbf{0}$ is closed). Let $\theta_{\mathbb{Z}}(p)$ denote the probability that this cluster is infinite. Then $\theta_{\mathbb{Z}}(p)$ is nondecreasing in p, so there is a critical value p_c of p such that if $p < p_c$ then $\theta_{\mathbb{Z}}(p) = 0$ and if $p > p_c$ then $\theta_{\mathbb{Z}}(p) > 0$. If $p < p_c$ then the Bernoulli process $\mathcal{B}_p$ is *subcritical*, while if $p > p_c$ the Bernoulli process $\mathcal{B}_p$ is *supercritical*. It is well known that $p_c \in (0,1)$, and in fact

$$\theta_{\mathbb{Z}}(p) \to 1 \quad \text{as } p \to 1. \tag{9.2}$$

See Grimmett (1999, p. 18) for a proof of this for bond percolation in two dimensions, based on a Peierls argument, which can be adapted to site percolation in two or more dimensions. Alternatively, see Grimmett (1999, Theorem 8.8).

More generally, for any norm $\|\cdot\|$ and any $r > 0$, define the *r-cluster at the origin* for $\mathcal{B}_p$ to be the open r-cluster in $\mathcal{B}_p$ containing the origin $\mathbf{0}$ (or the empty set if $\mathbf{0}$ is closed); let $\theta_{\mathbb{Z}}(p; r)$ denote the probability that this cluster is infinite. There is a critical value $p_c(r)$ of p such that if $p < p_c(r)$ then $\theta_{\mathbb{Z}}(p; r) = 0$ and if $p > p_c(r)$ then $\theta_{\mathbb{Z}}(p; r) > 0$. If $p < p_c(r)$ then the Bernoulli process $\mathcal{B}_p$ is subcritical with regard to the adjacency relationship $\sim_r$, while if $p > p_c(r)$ the Bernoulli process $\mathcal{B}_p$ is supercritical with regard to $\sim_r$.

One significant result from lattice percolation theory that we shall have occasion to use says that for subcritical Bernoulli percolation, the distribution of the order of the cluster at the origin has an exponentially decaying tail.

Theorem 9.7 *Suppose $r > 0$ and $p < p_c(r)$. Then if C_0 denotes the r-cluster at the origin for $\mathcal{B}_p$, and $|C_0|$ denotes its order,*

$$\limsup_{n\to\infty} \left(n^{-1} \log P[|C_0| \geq n]\right) < 0.$$

Proof For bond percolation on the usual nearest-neighbour lattice, the result is given by Grimmett (1999, Theorem 6.75). The argument is adapted easily enough to site percolation on $(\mathbb{Z}^d, \sim_r)$. □

Of particular interest to us is lattice percolation restricted to a box. For $n \in \mathbb{N}$, define the lattice box $B_{\mathbb{Z}}(n)$, as at (9.1), by $B_{\mathbb{Z}}(n) := ([1, n] \cap \mathbb{Z})^d$. Given an integer $n > 0$, let $\mathcal{B}_{p,n} := \mathcal{B}_p \cap B_{\mathbb{Z}}(n)$, the set of open sites in the lattice box $B_{\mathbb{Z}}(n)$. Certain renormalization techniques mean that we shall be particularly interested in percolation on the lattice box at *high densities*. In particular, the following result on lattice percolation will be important later on.

For any finite graph G, and $j \in \mathbb{N}$, let $L_j(G)$ denote the order of the jth largest component of G, that is, the jth largest of the orders of the components of G (let $L_j(G) = 0$ if G has fewer than j components). The next result is concerned with the probability of large deviations for $L_1(G_{\mathbb{Z}}(\mathcal{B}_{p,n}))$ as $n \to \infty$, with p fixed but close to 1.

Theorem 9.8 (Deuschel and Pisztora 1996). *Suppose $d \geq 2$. Let $\varepsilon \in (0, \frac{1}{3})$. Then there exists $p_0 = p_0(\varepsilon) \in (0, 1)$ such that for $p_0 \leq p < 1$,*

$$\limsup_{n\to\infty} n^{-(d-1)} \log(P[L_1(G_{\mathbb{Z}}(\mathcal{B}_{p,n})) \leq (1 - \varepsilon)n^d]) < 0.$$

To prove this, we need to define certain notions of boundaries of sets in $B_{\mathbb{Z}}(n)$. Let $\mathbb{Z}^d$ be endowed with the usual graph structure in which $x, y \in \mathbb{Z}^d$ are deemed adjacent if and only if $\|x - y\|_1 = 1$. If $A \subset B_{\mathbb{Z}}(n)$, its *edge-boundary in $B_{\mathbb{Z}}(n)$* is the set of pairs $\{x, y\}$ satisfying $x \in A$, $y \in B_{\mathbb{Z}}(n) \setminus A$ and x adjacent to y. The *internal vertex-boundary in $B_{\mathbb{Z}}(n)$* of A is the set of $x \in A$ which lie adjacent to some $y \in B_{\mathbb{Z}}(n) \setminus A$, and the *external vertex-boundary in $B_{\mathbb{Z}}(n)$* of A is the set of $y \in B_{\mathbb{Z}}(n) \setminus A$, which lie adjacent to some $x \in A$. Let $\partial_{B(n)} A$ (respectively, $\partial^-_{B(n)} A$, $\partial^+_{B(n)} A$) denote the edge-boundary (respectively, internal vertex-boundary, external vertex-boundary) of A. Finally, let $|A|$ denote the number of elements of A.

First we require a version of the isoperimetric inequality.

Lemma 9.9 *If A is a subset of $B_{\mathbb{Z}}(n)$ (not necessarily connected), with $|A| \leq 2n^d/3$, then*

$$\min(|\partial^-_{B(n)}A|, |\partial^+_{B(n)}A|) \geq (2d)^{-1}(1-(\tfrac{2}{3})^{1/d})|A|^{(d-1)/d}. \tag{9.3}$$

Proof Since the size of the edge-boundary is at most $2d$ times the size of the internal vertex-boundary, and is also at most $2d$ times the size of the external vertex-boundary, it suffices to prove that

$$|\partial_{B(n)}A| \geq (1-(\tfrac{2}{3})^{1/d})|A|^{(d-1)/d}. \tag{9.4}$$

A *down-set* in $B_{\mathbb{Z}}(n)$ is a set $D \subset B_{\mathbb{Z}}(n)$ such that if $x = (x_1, \ldots, x_d) \in D$ and $y = (y_1, \ldots, y_d) \in B_{\mathbb{Z}}(n)$ with $y_i \leq x_i$ for all i, then $y \in D$. The first step is to show that we can assume A is a down-set with no loss of generality, by a discrete version of the argument used at the start of the proof of Proposition 5.13.

Let $A \subset B_{\mathbb{Z}}(n)$ with $|A| \leq \frac{2}{3}n^d$. For $i = 1, 2, \ldots, d$, let Π_i denote the ith coordinate hyperplane, that is, the set of $x = (x_1, \ldots, x_d) \in \mathbb{Z}^d$ such that $x_i = 0$. For $x \in \Pi_i$, let $A_i(x)$ be the x-section of A, that is, the set of $z \in \mathbb{Z}$ such that $x + ze_i \in A$, where e_i denotes the unit vector in the direction of the ith coordinate. Define the *i-compression* of A to be the set $C_i(A) \subseteq B_{\mathbb{Z}}(n)$ with x-section, for each $x \in \Pi_i$, given by

$$\begin{aligned} (C_i(A))_i(x) &= \emptyset && \text{if } A_i(x) = \emptyset, \\ (C_i(A))_i(x) &= \{1, 2, \ldots, |A_i(x)|\} && \text{otherwise.} \end{aligned}$$

Loosely speaking, $C_i(A)$ is obtained by squashing each linear section of A in the i-direction down as far as possible towards the lower i-face of the cube $B_{\mathbb{Z}}(n)$. It is not hard to see that $|\partial_{B(n)}C_i(A)| \leq |\partial_{B(n)}A|$; moreover, by successively taking the 1-compression, then the 2-compression, and so on up to the d-compression, one ends up with a set $A' := C_d \circ C_{d-1} \circ \cdots \circ C_1(A)$, which is a down-set; for details see Bollobás and Leader (1991, Lemma 1). Therefore, there exists a down-set A' with the same cardinality as A and with $|\partial_{B(n)}A'| \leq |\partial_{B(n)}A|$, and from now on, we may assume that A itself is a down-set.

For $h > 0$ set $B_h := [-(h/2), (h/2)]^d$, the rectilinear cube of side h centred at the origin; let $A^* := A \oplus B(1)$. Then by the Brunn–Minkowski inequality (Theorem 5.11), with $\mathrm{Leb}(\cdot)$ denoting Lebesgue measure,

$$\mathrm{Leb}(A^* \oplus B(h)) \geq (\mathrm{Leb}(A^*)^{1/d} + h)^d,$$

so that

$$\mathrm{Leb}((A^* \oplus B(h)) \setminus A) \geq d\mathrm{Leb}(A^*)^{(d-1)/d}h. \tag{9.5}$$

For $1 \leq i \leq d$, let ψ_i denote projection onto the hyperplane Π_i, and let $S_i = \psi_i(A)$. Choose $j \in \{1, \ldots, d\}$ satisfying $|S_j| = \max_{1 \leq i \leq d}|S_i|$. Taking the limit

$h \to 0$ in (9.5), and using the fact that A is assumed to be a down-set and, therefore, that the size of its edge-boundary in $\mathbb{Z}^d$ (*not* in $B_{\mathbb{Z}}(n)$) is $2\sum_{i=1}^d |S_i|$, we obtain $\sum_{i=1}^d |S_i| \geq d|A|^{(d-1)/d}$, so that by the choice of j,

$$|S_j| \geq |A|^{(d-1)/d}. \tag{9.6}$$

Let F be the set of $x \in S_j$ such that $A_j(x) = \{1, 2, \dots, n\}$. Then $n|F| \leq |A|$ so by (9.6), and the assumption that $|A| \leq \frac{2}{3}n^d$,

$$\frac{|F|}{|S_j|} \leq \frac{|A|/n}{|A|^{(d-1)/d}} = \frac{|A|^{1/d}}{n} \leq \left(\frac{2}{3}\right)^{1/d}.$$

Using (9.6) once more, we obtain

$$|S_j \setminus F| \geq \left(1 - (\tfrac{2}{3})^{1/d}\right)|S_j| \geq \left(1 - (\tfrac{2}{3})^{1/d}\right)|A|^{(d-1)/d}.$$

For each $x \in S_j \setminus F$, there exists $r \in \{1, 2, \dots, n-1\}$ such that $x + re_j \in A$ and $x + (r+1)e_j \notin A$. Hence, $|\partial_{B(n)} A| \geq |S_j \setminus F|$, and (9.4) follows. □

Lemma 9.10 *Suppose $n \geq 1$ is an integer, and suppose Λ, Λ' are disjoint subsets of $B_{\mathbb{Z}}(n)$ with no edge of the lattice $\mathbb{Z}^d$ connecting Λ to Λ', and with $|\Lambda| > n^d/3$. If the $*$-connected components of $B_{\mathbb{Z}}(n) \setminus (\Lambda \cup \Lambda')$ are denoted $C_1, \dots, C_l$, then*

$$\sum_{i=1}^{l} |C_i|^{d/(d-1)} \geq \delta_1(n^d - |\Lambda|),$$

where we set $\delta_1 = ((1 - (\frac{2}{3})^{1/d})/(2d))^{d/(d-1)}$.

Proof Let $F_1, \dots, F_k$ be the connected components of $B_{\mathbb{Z}}(n) \setminus \Lambda$. Since $|\Lambda| > n^d/3$ by assumption, we have $|F_i| < 2n^d/3$ for each i and so by Lemma 9.9,

$$\sum_{i=1}^{k} |\partial^-_{B(n)} F_i|^{d/(d-1)} \geq \delta_1 \sum_{i=1}^{k} |F_i| \geq \delta_1(n^d - |\Lambda|).$$

For each $i \in \{1, 2, \dots, k\}$, both F_i and $B_{\mathbb{Z}}(n) \setminus F_i$ are connected in the lattice, so that by unicoherence (Lemma 9.6), the set $\partial^-_{B(n)} F_i$ is $*$-connected. Moreover, each set $\partial^-_{B(n)} F_i$ is disjoint from Λ' because of the assumption that Λ is disconnected from Λ'. Therefore, if the $*$-connected components of $B(n) \setminus (\Lambda \cup \Lambda')$ are denoted $C_1, \dots, C_l$, each of the sets $\partial^-_{B(n)} F_i$ is contained entirely within one of the sets $C_1, \dots, C_l$. By Minkowski's inequality (see, e.g., Rudin (1987)), for any finite sequence of non-negative numbers (a_n) and any $\alpha > 1$, we have $(\sum_n a_n^\alpha)^{1/\alpha} \leq \sum_n a_n$, and so we obtain

$$\sum_{m=1}^{l} |C_m|^{d/(d-1)} \geq \sum_{m=1}^{l} \left(\sum_{\{i:\partial^-_{B(n)} F_i \subseteq C_m\}} |\partial^-_{B(n)} F_i| \right)^{d/(d-1)}$$
$$\geq \sum_{m=1}^{l} \left(\sum_{\{i:\partial^-_{B(n)} F_i \subseteq C_m\}} |\partial^-_{B(n)} F_i|^{d/(d-1)} \right)$$
$$= \sum_{i=1}^{k} |\partial^-_{B(n)} F_i|^{d/(d-1)} \geq \delta_1 (n^d - |\Lambda|),$$

as asserted. □

The last lemma required is a classical result on large deviations for the sample mean of random variables with sub-exponentially decaying tails.

Lemma 9.11 *Suppose that a, b, y_0 are positive constants and $r \in (0,1)$ is a constant. Let Y be a random variable satisfying*

$$P[Y > y] \leq a \exp(-by^r), \qquad y \geq y_0. \tag{9.7}$$

Suppose $Y_1, Y_2, Y_3, \ldots$ are independent copies of Y, and set $S_n = \sum_{i=1}^n Y_i$. Then, for $s > E[Y]$,

$$\limsup_{n \to \infty} n^{-r} \log P[S_n \geq ns] < 0. \tag{9.8}$$

Proof Take s_1, s_2 satisfying $E[Y] < s_1 < s_2 < s$. Set $Y'_{i,n} := Y_i \mathbf{1}_{\{Y_i \leq n\}}$, and $S'_n := \sum_{i=1}^n Y'_{i,n}$. Then, by Boole's inequality,

$$P[S_n \geq ns] \leq P[S'_n \geq ns] + nP[Y > n],$$

and on the right-hand side, by (9.7) the second probability decays exponentially in n^r. Therefore, it suffices to show that the first probability decays exponentially in n^r. Put $Y'_n := Y \mathbf{1}_{\{Y \leq n\}}$ and set $t = t(n) := bn^{-q}/2$, with $q := 1 - r$. We have

$$P[S'_n \geq ns] = P[\mathrm{e}^{tS'_n} \geq \mathrm{e}^{tsn}]$$
$$\leq \mathrm{e}^{-tsn} (E[\mathrm{e}^{tY'_n}])^n = \exp\left\{ -nt \left(s - \frac{\log E[\mathrm{e}^{tY'_n}]}{t} \right) \right\}. \tag{9.9}$$

Let F_n (respectively, F) be the cumulative distribution function of Y'_n (respectively, Y). By the inequality $\log x \leq x - 1$ ($x > 0$), and Fubini's theorem,

$$t^{-1} \log E\mathrm{e}^{tY'_n} \leq t^{-1}(E\mathrm{e}^{tY'_n} - 1)$$
$$= \int_0^\infty \left(\int_0^x \mathrm{e}^{ty} \mathrm{d}y \right) \mathrm{d}F_n(x) = \int_0^\infty \mathrm{e}^{ty} (1 - F_n(y)) \mathrm{d}y.$$

For $y_1 > y_0$, we have

$$\int_{y_1}^{\infty} e^{t(n)y}(1 - F_n(y))dy \le a \int_{y_1}^{n} \exp\left(\frac{by}{2n^q} - by^r\right) dy$$
$$\le a \int_{y_1}^{n} \exp(-by^r/2)dy,$$

and, provided y_1 is sufficiently big, this is less than $s_1 - E[Y]$. On the other hand, for any (fixed) y_1 we have

$$\lim_{n\to\infty} \int_0^{y_1} e^{t(n)y}(1 - F_n(y))dy = \int_0^{y_1} (1 - F(y))dy \le E[Y],$$

by the integration by parts formula for expectation. Combining these, we have for large enough n that $t^{-1} \log Ee^{tY_n'} \le s_2$, and therefore by (9.9) and the definition of $t(n)$ we obtain the desired sub-exponentially decaying bound. □

Proof of Theorem 9.8 By Lemma 1.1, provided $p > \frac{1}{2}$, the probability $P[|\mathcal{B}_{p,n}| < n^d/2]$ decays exponentially in n^d. Therefore, it suffices to find $p > \frac{1}{2}$ such that the probability of the event F_n decays exponentially in n^{d-1}, where we set

$$F_n := \{|\mathcal{B}_{p,n}| \ge n^d/2\} \cap \{L_1(G_{\mathbb{Z}}(\mathcal{B}_{p,n})) \le (1-\varepsilon)n^d\}.$$

Suppose there are M open clusters in $\mathcal{B}_{p,n}$. For each $j \le M$ set $\xi_j := \sum_{i=1}^{j} L_i(G_{\mathbb{Z}}(\mathcal{B}_{p,n}))$, and let $J = \min\{j : \xi_j > n^d/3\}$. If F_n occurs, then $\xi_M \ge n^d/2$, so J exists. Moreover, if F_n occurs, then either $n^d/3 < \xi_1 \le (1-\varepsilon)n^d$, or $\xi_1 \le n^d/3$ and $\xi_{i+1} - \xi_i \le n^d/3$ for all $i < M$. In either case, we see that $n^d/3 < \xi_J \le (1-\varepsilon)n^d/3$ whenever F_n occurs. Therefore, by Lemma 9.10, with Λ taken to be the union of the J largest open clusters in $\mathcal{B}_{p,n}$ and Λ' to be the set $\mathcal{B}_{p,n} \setminus \Lambda$, the event F_n is contained in the event A_n defined by

$$A_n := \left\{ \sum_i |W_i|^{d/(d-1)} \ge \delta_1 \varepsilon n^d \right\},$$

where $\{W_i\}$ are the so-called *dual clusters*, that is, the $*$-connected components of the set of *closed* sites in $B_{\mathbb{Z}}(n)$.

For $x \in B_{\mathbb{Z}}(n)$, let C_x be the $*$-connected component of $\mathbb{Z}^d \setminus \mathcal{B}_p$ containing x (or the empty set if $x \in \mathcal{B}_p$). Let $(C_x^*, x \in \mathbb{Z}^d)$ be the so-called *pre-clusters* at x, that is, let $(C_x^*, x \in \mathbb{Z}^d)$ be independent random subsets of $\mathbb{Z}^d$ with each C_x^* having the same distribution as C_x.

The pre-clusters can be used to generate a realization of the Bernoulli process $\mathcal{B}_{p,n}$ as follows. List the elements of $B_{\mathbb{Z}}(n)$ in lexicographic order. Let x_1 be the first element of the list, let C_1 be the component containing x_1 of $C_{x_1}^* \cap B_{\mathbb{Z}}(n)$, and let C_1' be the union of C_1 and $\partial_{B(n)}^+ C_1$ (if C_1 is non-empty) or the set $\{x_1\}$ (if C_1 is empty). Let all elements of C_1 be denoted closed, and all elements of $C_1' \setminus C_1$ be denoted open.

Inductively, suppose subsets $C_1, \ldots, C_m$ of $B_{\mathbb{Z}}(n)$ have been defined. Let x_{m+1} be the first element of $B_{\mathbb{Z}}(n)$ (in the lexicographic order) not lying in

$\cup_{i=1}^m C_i'$; if no such element exists, the process terminates. Let C_{m+1} be the component containing x_{m+1} of $(C^*_{x_{m+1}} \cap B_{\mathbb{Z}}(n)) \setminus \cup_{i=1}^m (C_i')$, and let C'_{m+1} be the set $C_{m+1} \cup (\partial^+_{B(n)} C_{m+1}) \setminus \cup_{i=1}^m C_i'$ (if C_{m+1} is non-empty) or the set $\{x_{m+1}\}$ (if C_{m+1} is empty). Let all elements of C_{m+1} be denoted closed and let all elements of $C'_{m+1} \setminus C_{m+1}$ be denoted open.

In this procedure, each site is open with probability p, independent of all other sites. This is because the procedure amounts to the successive examination of state of the sites in $\mathcal{B}_{n,p}$, in some order, where the choice of the next site to be examined is determined by the states of the sites already examined, where once the status of a site in $B_{\mathbb{Z}}(n)$ has been determined, it is not subsequently changed, and where, on examination, a site always has probability p of being declared open.

In this construction, every dual cluster W_i arises as a subset of one of the pre-clusters $C_x^*, x \in B$, and none of these pre-clusters is used more than once; therefore,

$$P[A_n] \leq P\left[\sum_{i=1}^{n^d} V_i^{d/(d-1)} \geq \delta_1 \varepsilon n^d\right], \tag{9.10}$$

where $V_1, V_2, \ldots$ are independent copies of a variable V given by the order of the pre-cluster including the origin.

By a Peierls argument (Lemma 9.3), there is a constant $\gamma > 0$ such that

$$P[V \geq m] \leq \gamma^m (1-p)^m,$$

and so, provided p satisfies $(1-p)\gamma < 1$, we have exponential decay of the tail of V.

Set $Y = V^{d/(d-1)}$ and let $Y_1, Y_2, \ldots$ denote independent copies of Y. Further use of the Peierls argument shows that if p is sufficiently close to 1, then $E[Y] < \delta_1 \varepsilon / 2$. Therefore, by Lemma 9.11,

$$\limsup_{m \to \infty} m^{-(d-1)/d} \log P[Y_1 + \cdots + Y_m > \delta_1 \varepsilon m] < 0.$$

Hence, by (9.10) and the fact that event F_n is contained in A_n,

$$\limsup_{n \to \infty} n^{-(d-1)} \log P[F_n] < 0,$$

which completes the proof. □

9.4 k-Dependent percolation

Suppose S is a finite or countable set, and that for $i = 0, 1$, $Y^{(i)} = (Y_z^{(i)}, z \in S)$ is an S-indexed family of Bernoulli random variables (a random field). We say $Y^{(1)}$ stochastically dominates $Y^{(0)}$, and write $Y^{(1)} \geq_{\text{st}} Y^{(0)}$, if $E[f(Y^{(1)})] \geq$

$E[f(Y^{(0)})]$ for all bounded, increasing, measurable functions $f : \{0,1\}^S \to \mathbb{R}$. (A function $f : \{0,1\}^S \to \mathbb{R}$ is denoted increasing if $f(\boldsymbol{x}) \geq f(\boldsymbol{y})$ whenever $\boldsymbol{x} = (x_z, z \in S) \in \{0,1\}^S$ and $\boldsymbol{y} = (y_z, z \in S) \in \{0,1\}^S$ satisfy $x_z \geq y_z$ for all $z \in S$.)

Given $k \in \{0,1,2,\ldots\}$, we say the $\mathbb{Z}^d$-indexed random field $(Y_z, z \in \mathbb{Z}^d)$ is *k-dependent* if, for any two sets $A \subset \mathbb{Z}^d$ and $B \subset \mathbb{Z}^d$ with $\|a - b\|_1 > k$ for all $a \in A, b \in B$, the family of variables $(Y_z, z \in A)$ is independent of the family of variables $(Y_z, z \in B)$.

Recall from Section 9.3 that Z^p denotes a $\mathbb{Z}^d$-indexed family of independent Bernoulli(p) variables. We quote Grimmett (1999, Theorem 7.65) without giving a proof.

Theorem 9.12 *Let $d, k \geq 1$. There exists a non-decreasing function $\pi : [0,1] \to [0,1]$ satisfying $\pi(\delta) \to 1$ as $\delta \to 1$ such that the following holds. If $Y = (Y_z : z \in \mathbb{Z}^d)$ is a k-dependent family of Bernoulli random variables satisfying*

$$P[Y_z = 1] \geq \delta, \qquad z \in \mathbb{Z}^d,$$

then $Y \geq_{\mathrm{st}} Z^{\pi(\delta)}$.

9.5 Ergodic theory

Some of the results in Chapter 10 make use of a rather primitive version of the multidimensional ergodic theorem, involving only L^1 rather than almost sure convergence, which can be deduced quite easily from the classical one-dimensional ergodic theorem. To make this presentation more self-contained, we give the result and a sketch of its proof here.

Theorem 9.13 *Suppose $\xi = (\xi(z), z \in \mathbb{Z}^d)$ is a collection of independent identically distributed $\mathcal{S}$-valued random variables, where $\mathcal{S}$ is some measurable space. For $x \in \mathbb{Z}^d$ let $S_x\xi(z) := \xi(z - x)$, so that S_x is a shift operator $\mathcal{S}^{\mathbb{Z}^d}$, and $S_x\xi$ is a shifted version of the family of random variables ξ.*

Suppose h is a measurable function from $\mathcal{S}^{\mathbb{Z}^d}$ to $\mathbb{R}$, set $Y_x = h(S_x\xi)$ for each $x \in \mathbb{Z}^d$, and assume $E[|Y_{\mathbf{0}}|] < \infty$. Then $n^{-d} \sum_{x \in B_{\mathbb{Z}}(n)} Y_x \xrightarrow{L^1} EY_{\mathbf{0}}$ as $n \to \infty$.

Proof Let $e_1 := (1, 0, \ldots, 0) \in \mathbb{Z}^d$. The variables $Y_{ne_1}, n \geq 1$, form an ergodic sequence because they take the form $f(T^n(V))$, where T is a shift operator on an independent identically distributed sequence $V = (V_z, z \in \mathbb{Z})$. By the one-dimensional ergodic theorem (see, e.g., Durrett (1991, Chapter 6)),

$$n^{-1}(Y_{e_1} + Y_{2e_1} + \cdots + Y_{ne_1}) \xrightarrow{L^1} E[Y_{\mathbf{0}}].$$

Since we can partition $B_{\mathbb{Z}}(n)$ into n^{d-1} translates of the set $\{e_1, 2e_1, \ldots, ne_1\}$, and since for any $x \in \mathbb{Z}^d$ the joint distribution of $(Y_{x+e_1}, \ldots, Y_{x+e_n})$ is the same as that of $(Y_{e_1}, \ldots, Y_{e_n})$, the result follows. □

9.6 Continuum percolation: fundamentals

Let $\mathcal{H}_\lambda$ denote a homogeneous Poisson process of intensity λ on $\mathbb{R}^d$, that is, a Poisson process with constant intensity function $g(x) = \lambda$ for all $x \in \mathbb{R}^d$ (see Section 1.7). In its simplest form, continuum percolation can loosely be characterized as the study of large components of the infinite graph $G(\mathcal{H}_\lambda; 1)$. Equivalently, one may study the connected components of the union of balls of radius $\frac{1}{2}$ centred at the points of $\mathcal{H}_\lambda$. Continuum percolation is of interest in its own right; for example, the balls centred at the points of $\mathcal{H}_\lambda$ could represent pores in a piece of rock, or regions accessible to radio transmitters. The principal mathematical reference is Meester and Roy (1996) (see also Grimmett (1999), Stauffer and Aharony (1994), and Torquato (2002)), but we shall develop here some results on percolation that are not treated fully there or in other texts. The basic continuum percolation model readily lends itself to generalizations such as balls of random radius, but we shall concentrate here on the basic model.

Strictly speaking, Meester and Roy (1996) restrict attention to the case where $\|\cdot\|$ is the Euclidean norm, but usually their arguments can be adapted to other norms. Some of the basic results on continuum percolation are given in Penrose (1991) using a formulation that allows for arbitrary norms.

For $s > 0$, define $B(s)$ to be the (continuum) box of side s centred at the origin, and let $\mathcal{H}_{\lambda,s}$ be the restriction of the homogeneous Poisson process $\mathcal{H}_\lambda$ to the box $B(s)$. In other words, define

$$B(s) := [-s/2, s/2]^d, \quad \mathcal{H}_{\lambda,s} := \mathcal{H}_\lambda \cap B(s). \tag{9.11}$$

The random geometric graphs which are the subject of this book, and also most physical systems that one might model by continuum percolation, are on large but finite vertex sets, and therefore the large-s behaviour of the graph $G(\mathcal{H}_{\lambda,s}; 1)$ is of interest. To see the relevance to random geometric graphs as described in earlier chapters, consider the case where the underlying density f of points is the uniform density f_{U} on the unit cube, and suppose $r_n = (\lambda/n)^{1/d}$. Then, re-scaling space by a factor of r_n^{-1}, it can be seen (cf. Theorem 9.17 below) that the random geometric graph $G(\mathcal{P}_n; r_n)$ (with the Poisson process $\mathcal{P}_n$ defined in Section 1.7) is isomorphic to a copy of the graph $G(\mathcal{H}_{nr_n^d, r_n^{-1}}; 1)$.

We introduce further notation concerned with percolation. It is useful to have a continuum analogue for the cluster at the origin, and with this in mind let $\mathcal{H}_{\lambda,0}$ denote the point process $\mathcal{H}_\lambda \cup \{\mathbf{0}\}$, where $\mathbf{0}$ is the origin in $\mathbb{R}^d$. For $k \in \mathbb{N}$, let $p_k(\lambda)$ denote the probability that the component of $G(\mathcal{H}_{\lambda,0}; 1)$ containing the origin is of order k; see (9.15) below for a formula for $p_k(\lambda)$. The *percolation probability* $p_\infty(\lambda)$ is the probability that $\mathbf{0}$ lies in an infinite component of the graph $G(\mathcal{H}_{\lambda,0}; 1)$, and is defined by

$$p_\infty(\lambda) = 1 - \sum_{k=1}^{\infty} p_k(\lambda).$$

The *critical value* (continuum percolation threshold) λ_c is defined by

$$\lambda_c = \inf\{\lambda > 0 : p_\infty(\lambda) > 0\}. \tag{9.12}$$

The value of λ_c depends on the dimension d and the choice of norm. The fundamental result of continuum percolation says that $0 < \lambda_c < \infty$, provided $d \geq 2$; see Meester and Roy (1996), Grimmett (1999) or Penrose (1991).

Exact values for λ_c or for $p_\infty(\lambda)$ are not known. For $d = 2$, with the Euclidean (l_2) norm, simulation studies, such as Quintanilla *et al.* (2000), indicate that $1 - e^{-\lambda_c \pi/4} \approx 0.676$ so that $\lambda_c \approx 1.44$, while rigorous bounds $0.696 < \lambda_c < 3.372$ are given in Meester and Roy (1996, Chapter 3.9). For $d = 3$ (again with the l_2 norm), simulation studies by Rintoul and Torquato (1997) indicate that $1 - e^{-(4\pi/3)\lambda_c/8} \approx 0.290$. For an overview of simulation methods, see Torquato (2002).

An upper bound for $p_\infty(\lambda)$ is provided by the survival probability of a Galton–Watson branching process with a $\text{Po}(\lambda\theta)$ offspring distribution, and hence a lower bound for λ_c is $1/\theta$. At least in the case of the Euclidean norm, this lower bound becomes sharp as $d \to \infty$; see Penrose (1996).

It is widely believed that, for all $d \geq 2$, $p_\infty(\lambda_c) = 0$. This is actually known to be true for $d = 2$ (Meester and Roy 1996, Theorem 4.5), and also known to be true for all but at most finitely many d (Tanemura 1996).

To conclude this section, we state various basic results about percolation and Poisson processes.

Theorem 9.14 (Superposition theorem) *Suppose $\mathcal{P}$ is a Poisson process on $\mathbb{R}^d$ with intensity function $g(\cdot)$ and $\mathcal{P}'$ is a Poisson process on $\mathbb{R}^d$ with intensity function $g'(\cdot)$, independent of $\mathcal{P}$. Then $\mathcal{P} \cup \mathcal{P}'$ is a Poisson process on $\mathbb{R}^d$ with intensity function $g(\cdot) + g'(\cdot)$.*

Proof See, for example, Kingman (1993). □

Theorem 9.15 (Thinning theorem) *Suppose $\mathcal{P}$ is a Poisson process on $\mathbb{R}^d$ with intensity function $g(\cdot)$ and suppose $p : \mathbb{R}^d \to [0,1]$ is a measurable function. For each point X of $\mathcal{P}$, let X be accepted with probability $p(X)$ and rejected if not accepted, independently of all other points; let $\mathcal{P}'$ be the point process of accepted points. Then $\mathcal{P}'$ is a Poisson process on $\mathbb{R}^d$ with intensity function $p(\cdot)g(\cdot)$.*

Proof Immediate from the marking theorem (with mark space $\{0,1\}$) and the restriction theorem in Kingman (1993). □

For point processes $\mathcal{Y}_1$ and $\mathcal{Y}_2$ in $\mathbb{R}^d$, we shall say $\mathcal{Y}_2$ *dominates* $\mathcal{Y}_1$ if there exist coupled point processes $\mathcal{Y}_1'$ and $\mathcal{Y}_2'$, such that $\mathcal{Y}_i'$ and $\mathcal{Y}_i$ have the same distribution for $i = 1, 2$, and such that $\mathcal{Y}_1' \subseteq \mathcal{Y}_2'$ almost surely.

Corollary 9.16 *Suppose that for $i = 1, 2$ the point process $\mathcal{Y}_i$ is a Poisson process in $\mathbb{R}^d$ with intensity function g_i, and $g_1(x) \leq g_2(x)$ for all $x \in \mathbb{R}^d$. Then $\mathcal{Y}_2$ dominates $\mathcal{Y}_1$.*

Proof Immediate from either Theorem 9.15 or Theorem 9.14. □

For Borel $A \subseteq \mathbb{R}^d$, and $\lambda > 0$, a homogeneous Poisson process of intensity λ on A is a Poisson process on $\mathbb{R}^d$ with intensity function $\lambda \mathbf{1}_A$.

Theorem 9.17 (Scaling theorem) *Suppose $\mathcal{H}$ is a homogeneous Poisson process on a region $A \subseteq \mathbb{R}^d$ of intensity λ. Let $a > 0$ and let $a\mathcal{H}$ (respectively, aA) be the image of $\mathcal{H}$ (respectively, A) under the mapping $x \mapsto ax$. Then $a\mathcal{H}$ is a homogeneous Poisson process on aA with intensity $a^{-d}\lambda$.*

Proof This is a special case of the mapping theorem in Kingman (1993). □

Corollary 9.18 *Let $\lambda > 0$ and $r > 0$. Then the (possibly defective) probability distribution of the order of the component containing the origin of $G(\mathcal{H}_{\lambda,0}; 1)$ is the same as that of the component containing the origin of $G(\mathcal{H}_{r^{-d}\lambda,0}; r)$.*

Proof Clearly, the order of the component containing the origin of $G(\mathcal{H}_{\lambda,0}; 1)$ is the same as that of the component containing the origin of $G(r\mathcal{H}_{\lambda,0}; r)$, and the result then follows from Theorem 9.17. □

Theorem 9.19 *Suppose $p_\infty(\lambda) > 0$. Then, with probability 1, the graph $G(\mathcal{H}_\lambda; 1)$ has precisely one infinite component.*

Proof Let N be the number of infinite components of $G(\mathcal{H}_\lambda; 1)$. If $p_\infty(\lambda) > 0$, then $P[N \geq 1] > 0$.

For Borel $A \subseteq \mathbb{R}^d$, let $\mathcal{F}_A$ be the σ-field generated by the Poisson configuration in A, that is, the smallest σ-field with respect to which all variables of the form $\mathcal{H}_\lambda(B)$, with B a Borel subset of A, are measurable. Let A_1 be the box $B(1)$ and for $n \geq 2$ let A_n be the annulus $B(n) \setminus B(n-1)$. Then the event $\{N \geq 1\}$ lies in the tail σ-field of the independent σ-fields $\mathcal{F}_{A_1}, \mathcal{F}_{A_2}, \mathcal{F}_{A_3}, \ldots$, and so by the Kolmogorov zero–one law $P[N \geq 1] = 1$. In fact, the Kolmogorov zero–one law stated in texts such as those mentioned in Section 1.6 refers to the tail σ-field of a sequence of independent *random variables*, but the proof carries through to the tail σ-field of a sequence of independent *σ-fields*.

The fact that $P[N = 1] = 1$ (uniqueness of the infinite component) is much deeper; see Meester and Roy (1996, Theorem 3.6) for a proof. □

Theorem 9.20 *As a function of λ, the percolation probability $p_\infty(\lambda)$ is monotonically nondecreasing, is continuous at λ for all $\lambda \neq \lambda_c$, and is right continuous at $\lambda = \lambda_c$.*

Proof The monotonicity follows easily from Corollary 9.16. See Meester and Roy (1996, Theorem 3.9) for a proof of continuity. The proof there of *right* continuity carries over to the case $\lambda = \lambda_c$. □

The next result may be viewed as a generalization of continuity of $p_\infty(\lambda)$ to the case $\lambda = \infty$.

Proposition 9.21 *It is the case that $p_\infty(\lambda) \to 1$ as $\lambda \to \infty$.*

Proof Divide $\mathbb{R}^d$ into boxes of side ε, centred at points of the form $\varepsilon z, z \in \mathbb{Z}^d$, with $\varepsilon > 0$ chosen so that $\|x-y\| < 1$ for any two points x, y lying in neighbouring boxes. Let each lattice site $z \in \mathbb{Z}^d$ be denoted *open* if the corresponding box contains at least one Poisson point. Then each lattice site is open with probability

$p(\lambda) = 1 - \exp(-\lambda\varepsilon^d)$. Then the origin will be part of an infinite component of $G(\mathcal{H}_{\lambda,0}; 1)$ if there is a path of open sites starting at the origin. Since $p(\lambda) \to 1$ as $\lambda \to \infty$, the result follows by (9.2). □

The next preliminary result adds to the earlier result on the Palm theory of finite Poisson processes (Theorem 1.6) and says that the infinite Poisson process $\mathcal{H}_\lambda$ is also its own Palm point process.

Theorem 9.22 (Palm theory for infinite Poisson process) *Suppose $h(x; \mathcal{X})$ is a bounded measurable real-valued function defined on all pairs of the form $(x, \mathcal{X})$ with $\mathcal{X}$ a locally finite subset of $\mathbb{R}^d$ and x an element of $\mathcal{X}$. Assume that h is translation-invariant, meaning that $h(x; \mathcal{X}) = h(\mathbf{0}; \mathcal{X} \oplus \{-x\})$ for any $(x, \mathcal{X})$. Then*

$$E \sum_{X \in \mathcal{H}_\lambda \cap B(s)} h(X; \mathcal{H}_\lambda) = \lambda s^d E\left[h(\mathbf{0}; \mathcal{H}_{\lambda,0})\right]. \tag{9.13}$$

Proof Consider $\mathcal{H}_\lambda$ as the union of two independent Poisson processes, namely, $\mathcal{H}_{\lambda,s}$ (a homogeneous Poisson process of intensity λ on $B(s)$) and $\tilde{\mathcal{H}}_{\lambda,s}$ (a homogeneous Poisson process of intensity λ on $\mathbb{R}^d \setminus B(s)$). Then, by Theorem 1.6,

$$E\left[\sum_{X \in \mathcal{H}_\lambda \cap B(s)} h(X; \mathcal{H}_\lambda) | \tilde{\mathcal{H}}_{\lambda,s}\right] = \lambda \int_{B(s)} E[h(x; \{x\} \cup \mathcal{H}_{\lambda,s} \cup \tilde{\mathcal{H}}_{\lambda,s}) | \tilde{\mathcal{H}}_{\lambda,s}] \mathrm{d}x,$$

and taking the expectation of both sides, we obtain (9.13). □

Next, we give a formula for $p_k(\lambda)$, in a form somewhat different from that seen, for example, in Meester and Roy (1996, Proposition 6.2).

Theorem 9.23 (Formula for $p_k(\lambda)$) *Given $x_0, x_1, \ldots, x_k \in \mathbb{R}^d$, let the function $h(x_0, x_1, \ldots, x_k)$ take the value 1 if $G(\{x_0, x_1, \ldots, x_k\}; 1)$ is connected* and *$x_0, \ldots, x_k$ are in left-to-right order, that is, $\pi_1(x_0) < \pi_1(x_2) < \cdots < \pi_1(x_k)$, where π_1 denotes projection onto the first coordinate. Otherwise, set $h(x_0, x_1, \ldots, x_k) = 0$. Also, set*

$$A(x_0, x_1, \ldots, x_k) := \mathrm{Leb}\left(\bigcup_{i=0}^{k} B(x_i; 1)\right), \tag{9.14}$$

the volume (area) of the union of balls of radius 1 centred at $x_0, x_1, \ldots, x_k$. Then, for $k \in \mathbb{N} \cup \{0\}$,

$$p_{k+1}(\lambda) = (k+1)\lambda^k \int_{\mathbb{R}^d} \cdots \int_{\mathbb{R}^d} h(\mathbf{0}, x_1, \ldots, x_k) \times \exp(-\lambda A(\mathbf{0}, x_1, \ldots, x_k)) \mathrm{d}x_1 \cdots \mathrm{d}x_k. \tag{9.15}$$

Proof Let $\tilde{p}_k(\lambda)$ be the probability that (i) the component $\tilde{C}_0$ containing the origin of $G(\mathcal{H}_{\lambda,0}; 1)$ is of order k, and (ii) the origin is the left-most vertex of $\tilde{C}_0$, that is, the projection on to the first coordinate is less for the origin than for any other vertex of $\tilde{C}_0$.

Let M_s be the number of points of $\mathcal{H}_\lambda$ lying in $B(s)$ which are in components of $G(\mathcal{H}_\lambda; 1)$ of order k and let $\tilde{M}_s$ be the number of points X of $\mathcal{H}_\lambda$ lying in $B(s)$ for which (i) X is in a component of $G(\mathcal{H}_\lambda; 1)$ of order k, and (ii) X is the left-most vertex of that component. By (9.13),

$$E[M_s] = \lambda s^d p_k(\lambda), \qquad E[\tilde{M}_s] = \lambda s^d \tilde{p}_k(\lambda).$$

Since $|M_s - k\tilde{M}_s|$ is bounded by the number of points of $\mathcal{H}_\lambda$ lying within a distance k of the boundary of $B(s)$, it is the case that $s^{-d}E[|M_s - k\tilde{M}_s|] \to 0$ as $s \to \infty$, and therefore

$$p_k(\lambda) = \lambda^{-1} \lim_{s\to\infty} s^{-d} E[M_s] = \lambda^{-1} \lim_{s\to\infty} s^{-d} k E[\tilde{M}_s] = k\tilde{p}_k(\lambda). \tag{9.16}$$

Let B be the ball $B(\mathbf{0}; k+3)$, and let $|B|$ denote the Lebesgue measure of B. For finite point sets $\mathcal{Y} \subseteq \mathcal{X}$ in $\mathbb{R}^d$, let $g(\mathcal{Y}, \mathcal{X})$ be the indicator of the event that $G(\mathcal{Y} \cup \{\mathbf{0}\}; 1)$ is a component of $G(\mathcal{X} \cup \{\mathbf{0}\}; 1)$, of order $k+1$ with $\mathbf{0}$ as its left-most vertex. Then

$$\tilde{p}_{k+1}(\lambda) = E \sum_{\mathcal{Y} \subseteq (\mathcal{H}_\lambda \cap B)} g(\mathcal{Y}, \mathcal{H}_\lambda \cap B).$$

Now regard $\mathcal{H}_\lambda \cap B$ as a finite Poisson process whose total number of points has a $\mathrm{Po}(\lambda|B|)$ distribution, each point being uniformly distributed over B. Let $\mathcal{U}_k$ be a point process consisting of k points uniformly distributed over B, independently of each other and of $\mathcal{H}_\lambda$. Then, by Theorem 1.6,

$$\tilde{p}_{k+1}(\lambda) = \frac{(\lambda|B|)^k}{k!} E g(\mathcal{U}_k; \mathcal{U}_k \cup (\mathcal{H}_\lambda \cap B)),$$

so that if $\tilde{h}(x_1, \ldots, x_k)$ denotes the indicator of the event that $G(\{\mathbf{0}, x_1, \ldots, x_k\}; 1)$ is connected with $\mathbf{0}$ as its left-most vertex, then

$$\tilde{p}_{k+1}(\lambda) = \frac{\lambda^k |B|^k}{k!} \int_B \frac{dx_1}{|B|} \cdots \int_B \frac{dx_k}{|B|} \tilde{h}(x_1, \ldots, x_k) \exp(-\lambda A(\mathbf{0}, x_1, \ldots, x_k)),$$

and since the integrand is symmetric in its arguments $x_1, \ldots, x_k$, the multiple integral is equal to $k!$ times its restriction to $x_1, \ldots, x_k$ in left-to-right order, so that

$$\tilde{p}_{k+1}(\lambda) = \lambda^k \int_B dx_1 \cdots \int_B dx_k h(\mathbf{0}, x_1, \ldots, x_k) \exp(-\lambda A(\mathbf{0}, x_1, \ldots, x_k)).$$

Combining this with (9.16) yields (9.15). □

Finally in this section, we give a continuum version of Grimmett (1999, Theorem 2.45), which will be used in Chapter 12. If A is a measurable collection of finite subsets of the box $B(s)$, we shall say that A is *increasing* (or A is an *up-set*) if $\mathcal{X} \cup \{x\} \in A$ for all $\mathcal{X} \in A$ and $x \in B(s)$. For $k \in \mathbb{N}$ let $I_k(A)$, the *k-interior* of A, be the set of $\mathcal{X} \in A$ such that $\mathcal{X} \setminus \mathcal{Y} \in A$ for all $\mathcal{Y} \subseteq \mathcal{X}$ with at most k elements, that is, the set of configurations which remain in A even after the removal of up to k points.

Theorem 9.24 *Suppose $s > 0$ and $0 < \mu < \lambda$. Suppose A is a measurable increasing collection of finite subsets of $B(s)$. Then*

$$P[\mathcal{H}_{\lambda,s} \notin I_k(A)] \leq \left(\frac{\lambda}{\lambda - \mu}\right)^k P[\mathcal{H}_{\mu,s} \notin A].$$

Proof By the thinning theorem (Theorem 9.15), a realization of $\mathcal{H}_{\mu,s}$ can be obtained by retaining each point of $\mathcal{H}_{\lambda,s}$ with probability (μ/λ) and discarding with probability $(\lambda - \mu)/\lambda$. If $\mathcal{H}_{\lambda,s} \notin I_k(A)$, then pick a set $\mathcal{S}$ of at most k points of $\mathcal{H}_{\lambda,s}$ such that $\mathcal{H}_{\lambda,s} \setminus \mathcal{S} \notin A$; given the configuration of $\mathcal{H}_{\lambda,s}$, the probability that each point in $\mathcal{S}$ is discarded is at least $((\lambda-\mu)/\lambda)^k$, and since A is increasing, this is a lower bound for the conditional probability that $\mathcal{H}_\mu \notin A$, given this configuration for $\mathcal{H}_\lambda$. Hence

$$P[\mathcal{H}_{\mu,s} \notin A | \mathcal{H}_{\lambda,s} \notin I_k(A)] \geq \left(\frac{\lambda - \mu}{\lambda}\right)^k$$

so that

$$P[\mathcal{H}_{\mu,s} \notin A] \geq \left(\frac{\lambda - \mu}{\lambda}\right)^k P[\mathcal{H}_{\lambda,s} \notin I_k(A)],$$

and the result follows. □

10

PERCOLATION AND THE LARGEST COMPONENT

In Chapter 3 we considered components in $G(\mathcal{X}_n; r_n)$ or $G(\mathcal{P}_n; r_n)$ of fixed size. In this chapter we begin an investigation of 'large' components. Throughout this chapter we assume that *the norm* $\|\cdot\|$ *of choice is one of the* l_p *norms,* $1 \leq p \leq \infty$.

For any graph G, let $L_j(G)$ denote the order of its jth-largest component, that is, the jth-largest of the orders of its components, or zero if it has fewer than j components. A fundamental result in the theory of the independent Erdös–Rényi random graph $G(n,p)$ (see, e.g., Janson *et al.* (2000, Theorem 5.4)) states that if $\lambda > 0$ then, as $n \to \infty$,

$$n^{-1} L_1(G(n, \lambda/n)) \xrightarrow{P} \phi(\lambda), \tag{10.1}$$

whereas $L_2(G(n, \lambda/n)) \xrightarrow{P} 0$, where the function $\phi(\cdot)$ satisfies $\phi(\lambda) = 0$ for $\lambda \leq 1$ and $\phi(\lambda) > 0$ for $\lambda > 1$. In other words, if the mean vertex degree is fixed at a value exceeding a critical value of 1, then a giant component emerges containing a non-vanishing proportion of vertices. As we shall see, a similar phenomenon occurs for random geometric graphs, when we take the *thermodynamic limit* in which nr_n^d, and therefore the mean vertex degree, tends to a finite limit. In this case the critical value of $\lambda = \lim_{n\to\infty} nr_n^d$ is at $\lambda = \lambda_c$, the continuum percolation threshold defined at (9.12).

Recall from (9.11) that $B(s)$ denotes a box of side s centred at the origin and for $s > 0$, $\mathcal{H}_{\lambda,s}$ is the restriction of the homogeneous Poisson process $\mathcal{H}_\lambda$ to that box. The basic result on the largest component for the geometric random graph $G(\mathcal{H}_{\lambda,s}; 1)$, providing an analogue to the fundamental result (10.1) on Erdös–Rényi random graphs, is that if $\lambda \neq \lambda_c$ then

$$s^{-d} L_1(G(\mathcal{H}_{\lambda,s}; 1)) \xrightarrow{P} \lambda p_\infty(\lambda) \quad \text{as } s \to \infty, \tag{10.2}$$

and

$$s^{-d} L_2(G(\mathcal{H}_{\lambda,s}; 1)) \xrightarrow{P} 0 \quad \text{as } s \to \infty. \tag{10.3}$$

In all cases where $p_\infty(\lambda_c) = 0$, it can be shown by a routine continuity argument using the case $\lambda > \lambda_c$ of (10.2) and the right continuity of $p_\infty(\cdot)$ (Theorem 9.20) that (10.2) and (10.3) are true for $\lambda = \lambda_c$ as well.

This chapter contains a proof of (10.2) and (10.3), along with various refinements. These include results on the growth rate of $L_1(G(\mathcal{H}_{\lambda,s}; 1))$ in the subcritical case and of $L_2(G(\mathcal{H}_{\lambda,s}; 1))$ in the supercritical case. In the supercritical case, we also give results on the rate of sub-exponential decay of the probability of large deviations of $L_1(G(\mathcal{H}_{\lambda,s};1))$, and a central limit theorem for $L_1(G(\mathcal{H}_{\lambda,s};1))$.

Recall from Section 9.6 that $\mathcal{H}_{\lambda,0}$ denotes a homogeneous Poisson process of intensity λ on $\mathbb{R}^d$ with a point added at the origin, and that $(p_n(\lambda), n \in \mathbb{Z}^+)$ is the probability mass function of the order of the component containing the origin of $G(\mathcal{H}_{\lambda,0};1)$. In Sections 10.1 and 10.4, we shall establish new results on the large-n behaviour of the sequence $(p_n(\lambda), n \in \mathbb{N})$, adding to analogous known results for lattice percolation. These are needed for our investigation of geometric graphs.

10.1 The subcritical regime

Given $\lambda > 0$, it is clear that $\sum_{k \geq n} p_k(\lambda)$ decays to zero as $n \to \infty$. It is of interest to characterize the rate of decay, both for its own sake as a feature of continuum percolation, and also as an aid to the understanding the asymptotic behaviour of the size of the large clusters of the random geometric graph $G(\mathcal{X}_n;r_n)$ in the thermodynamic limit. In the present section we consider the subcritical case $\lambda < \lambda_c$; in this case the sum $\sum_{k \geq n} p_k(\lambda)$ is the tail of the distribution of V, where V denotes the order of the component containing the origin of $G(\mathcal{H}_{\lambda,0};1)$. We show that, loosely speaking, the tail behaviour of the distribution of V approximates to that of a geometric random variable.

Theorem 10.1 *Suppose $\lambda > 0$. Then the limit*

$$\zeta(\lambda) := \lim_{n \to \infty} \left(-n^{-1} \log p_n(\lambda)\right) \tag{10.4}$$

$$= \lim_{n \to \infty} \left(-\frac{1}{n} \log \sum_{k=n}^{\infty} p_k(\lambda) \right) \tag{10.5}$$

exists. Also, $\zeta(\lambda)$ is a continuous and monotone nonincreasing function of λ, and $\zeta(\lambda) \to \infty$ as $\lambda \to 0$ from above. If $\lambda < \lambda_c$ then $\zeta(\lambda) > 0$; if $\lambda \geq \lambda_c$ then $\zeta(\lambda) = 0$.

The case $\lambda \geq \lambda_c$ is included in the statement of this result for the sake of completeness, but the main interest in the present section is in the case $\lambda < \lambda_c$. The first step in the proof is to show exponential decay for $p_n(\lambda)$ for $\lambda < \lambda_c$.

Lemma 10.2 *Suppose $0 < \lambda < \lambda_c$. Then*

$$\limsup_{n \to \infty} \left(n^{-1} \log p_n(\lambda)\right) < 0 \tag{10.6}$$

and

$$\lim_{\lambda \to 0} \left(\limsup_{n \to \infty} \left(n^{-1} \log p_n(\lambda)\right) \right) = -\infty. \tag{10.7}$$

Proof Take $\lambda' > \lambda$ and $\varepsilon > 0$ such that $\lambda'(1+4\varepsilon d)^d < \lambda_c$. Then by scaling (Corollary 9.18), the component containing the origin of $G(\mathcal{H}_{\lambda',0};1+4\varepsilon d)$ is almost surely finite.

Set $l := (1 + 2\varepsilon d)/\varepsilon$, $p := 1 - \exp(-\lambda\varepsilon^d)$, and $p' := 1 - \exp(-\lambda'\varepsilon^d)$. For $z \in \mathbb{Z}^d$, set $B_z := B(\varepsilon) \oplus \{\varepsilon z\}$, the box of side ε centred at εz. Then $\mathcal{H}_{\lambda'}$ induces a realization of the Bernoulli site percolation process $\mathcal{B}_{p'}$ on $\mathbb{Z}^d$, by setting each site $z \in \mathbb{Z}^d$ to be open if $\mathcal{H}_{\lambda'}(B_z) > 0$ and closed otherwise, and $\mathcal{H}_\lambda$ induces a realization of $\mathcal{B}_p$ in an analogous manner. If $z, z' \in \mathbb{Z}^d$ with $\|z - z'\| \leq l$, then any two points $X \in B_z$, $Y \in B_{z'}$ will satisfy $\|X - Y\| \leq 1 + 4\varepsilon d$. Since the component containing the origin of $G(\mathcal{H}_{\lambda',0}; 1 + 4\varepsilon d)$ is almost surely finite, the open l-cluster containing the origin in the induced realization of $\mathcal{B}_{p'} \cup \{\mathbf{0}\}$ is almost surely finite.

It follows that if $p_c(l)$ denotes the critical parameter for Bernoulli site percolation on the graph $(\mathbb{Z}^d, \sim_l)$ (see Sections 9.2 and 9.3), then $p' \leq p_c(l)$ and $p < p_c(l)$ (the strict inequality here was the purpose of introducing λ'). Therefore, if C_0 denotes the l-cluster containing the origin in the induced realization of $\mathcal{B}_p \cup \{\mathbf{0}\}$, by Theorem 9.7, there exist constants $\mu > 0$, $n_0 > 0$ such that

$$P[|C_0| \geq n] \leq \mathrm{e}^{-\mu n}, \qquad n \geq n_0. \tag{10.8}$$

If $x, x' \in \mathbb{R}^d$ with $\|x - x'\| \leq 1$, and if $z, z' \in \mathbb{Z}^d$ with $x \in B_z, x' \in B_{z'}$, then $\|z - z'\| \leq l$. Let V denote the order of the component of $G(\mathcal{H}_{\lambda,0}; 1)$ containing the origin. By a Peierls argument (Lemma 9.3), there is a constant $\gamma = \gamma(\varepsilon)$ such that, for all n, the number of l-connected subsets of $\mathbb{Z}^d$ of cardinality n containing the origin is at most γ^n. Let $K \geq \mathrm{e}^2\varepsilon^d\lambda$. If $|C_0| < n$ and $V \geq Kn+1$, then for at least one of these subsets of $\mathbb{Z}^d$ the union of the associated cubes B_z contains at least Kn points of $\mathcal{H}_\lambda$. Therefore, by (1.12), we have

$$\begin{aligned} P[\{|C_0| < n\} \cap \{V \geq Kn + 1\}] &\leq \gamma^n P[\mathrm{Po}(n\varepsilon^d\lambda) \geq Kn] \\ &\leq \gamma^n \exp\left(-\left(\frac{Kn}{2}\right)\log\left(\frac{K}{\varepsilon^d\lambda}\right)\right). \end{aligned} \tag{10.9}$$

If we take K sufficiently large, we see from (10.8) and (10.9) that $P[V \geq Kn+1]$ decays exponentially in n, so that (10.6) follows.

To obtain (10.7), take $\varepsilon = 1$ and $\gamma = \gamma(1)$ in the argument above. Then for λ small enough, a Peierls argument yields

$$P[|C_0| \geq n] \leq \gamma^n(1 - \mathrm{e}^{-\lambda})^n \leq (\lambda\gamma)^n. \tag{10.10}$$

By taking $K = 1$ in (10.9) we obtain for $\lambda \leq \mathrm{e}^{-2}$ that

$$P[\{|C_0| < n\} \cap \{V \geq n + 1\}] \leq \exp\left\{n\left(\tfrac{1}{2}\log\lambda + \log\gamma\right)\right\},$$

which, combined with (10.10), shows that, for all sufficiently small λ,

$$\limsup_{n\to\infty} n^{-1}\log(P[V \geq n]) \leq \tfrac{1}{4}\log\lambda,$$

which implies (10.7). □

Proof of Theorem 10.1 We show existence of a limit in (10.4) by showing a form of supermultiplicativity for $p_k(\lambda)$. As in Theorem 9.23, let $h(x_0, \ldots, x_k)$ denote the indicator of the event that $G(\{x_0, \ldots, x_k\}; 1)$ is connected and $x_0, \ldots, x_k$ are in left-to-right order, and let $A(x_0, \ldots, x_k)$ denote the volume of $\cup_{i=0}^k B(x_i; 1)$. By (9.15), with $\tilde{p}_k(\lambda) := p_k(\lambda)/k$, we have

$$\begin{aligned}
&\tilde{p}_{k+1}(\lambda)\tilde{p}_{j+1}(\lambda) \\
&\quad = \lambda^{k+j} \int_{\mathbb{R}^d} \cdots \int_{\mathbb{R}^d} h(\mathbf{0}, x_1, \ldots, x_k) h(\mathbf{0}, y_1, \ldots, y_j) \\
&\qquad \times \exp(-\lambda(A(\mathbf{0}, x_1, \ldots, x_k) + A(\mathbf{0}, y_1, \ldots, y_j))) \mathrm{d}x_1 \cdots \mathrm{d}x_k \mathrm{d}y_1 \cdots \mathrm{d}y_j \\
&\quad = \lambda^{k+j} \int_{\mathbb{R}^d} \cdots \int_{\mathbb{R}^d} h(\mathbf{0}, x_1, \ldots, x_k) h(x_k, \ldots x_{k+j}) \\
&\qquad \times \exp(-\lambda(A(\mathbf{0}, x_1, \ldots, x_k) + A(x_k, x_{k+1}, \ldots, x_{k+j}))) \mathrm{d}x_1 \cdots \mathrm{d}x_{k+j}. \quad (10.11)
\end{aligned}$$

By the subadditivity of measure, we have the inequality

$$A(\mathbf{0}, x_1, \ldots, x_k) + A(x_k, x_{k+1}, \ldots, x_{k+j}) \geq A(\mathbf{0}, x_1, \ldots, x_{k+j}),$$

and since the union of two connected geometric graphs having a vertex in common is connected, we also have for all $x_1, \ldots, x_{k+j}$ that

$$h(\mathbf{0}, x_1, \ldots, x_k) h(x_k, \ldots x_{k+j}) \leq h(\mathbf{0}, x_1, \ldots, x_{k+j}).$$

Putting these inequalities into (10.11), we obtain

$$\begin{aligned}
\tilde{p}_{k+1}(\lambda)\tilde{p}_{j+1}(\lambda) &\leq \lambda^{k+j} \int_{\mathbb{R}^d} \cdots \int_{\mathbb{R}^d} h(\mathbf{0}, x_1, \ldots, x_{k+j}) \\
&\qquad \times \exp(-\lambda A(\mathbf{0}, x_1, \ldots, x_{k+j})) \mathrm{d}x_1 \cdots \mathrm{d}x_{k+j} \\
&= \tilde{p}_{k+j+1}(\lambda),
\end{aligned}$$

and hence, for all $k, j \geq 1$,

$$\tilde{p}_{k+j-1}(\lambda) \geq \tilde{p}_k(\lambda)\tilde{p}_j(\lambda). \quad (10.12)$$

It is well known how to deduce existence of a limit using a supermultiplicative property such as (10.12). Let $q_k := -\log \tilde{p}_k(\lambda)$; for all k, j, by (10.12) we have

$$q_{k+j-1} \leq q_k + q_j. \quad (10.13)$$

Set $\zeta(\lambda) := \inf_{k\geq 2}(q_k/(k-1))$. Then $\zeta(\lambda) \in [0, \infty)$. Given $\varepsilon > 0$, choose $m \geq 2$ such that $q_m/(m-1) < \zeta(\lambda) + \varepsilon$. By (10.13) and induction on j, we have for $r, j \in \mathbb{N}$ that

$$q_{r+j(m-1)} \leq q_r + jq_m. \quad (10.14)$$

Given n, choose integers k, r with $r \in \{1, 2, \ldots, m-1\}$ such that $n = k(m-1)+r$. By (10.14) we have $q_n \leq q_r + kq_m$, so that

$$\frac{q_n}{n-1} \le \frac{\max\{q_1, \ldots, q_{m-1}\}}{n-1} + \frac{kq_m}{k(m-1)}.$$

Taking $n \to \infty$ we have $\limsup(q_n/(n-1)) \le \zeta(\lambda) + \varepsilon$, and since $\varepsilon > 0$ is arbitrary we have $q_n/(n-1) \to \zeta(\lambda)$ and $q_n/n \to \zeta(\lambda)$ as $n \to \infty$. Therefore, since $p_n(\lambda) = n\tilde{p}_n(\lambda)$,

$$-\frac{\log p_n(\lambda)}{n} = \frac{q_n}{n} - \frac{\log n}{n} \to \zeta(\lambda),$$

proving (10.4). It is straightforward to deduce (10.5) from (10.4). Also, it follows from Lemma 10.2 that the limit $\zeta(\lambda)$ is strictly positive for $\lambda < \lambda_{\mathrm{c}}$, and tends to ∞ as $\lambda \to 0$.

It remains to prove that limiting exponent $\zeta(\lambda)$ defined in (10.4) is a continuous nonincreasing function of λ. To this end, set $\rho(\lambda) := \mathrm{e}^{-\zeta(\lambda)}$ and $p_n^+(\lambda) := \sum_{k \ge n} p_k(\lambda)$. For $\lambda < \lambda' < \lambda_{\mathrm{c}}$, by the superposition theorem (Theorem 9.14), the union of independent homogeneous Poisson processes of intensity λ and $\lambda' - \lambda$ is a homogeneous Poisson process of intensity λ', so that $p_n^+(\lambda) \le p_n^+(\lambda')$; hence $p_n^+(\lambda)$ is nondecreasing in λ in the range $(0, \lambda_{\mathrm{c}})$. Since $\rho(\lambda) = \lim_{n\to\infty} p_n^+(\lambda)^{1/n}$ by (10.5), $\rho(\lambda)$ is also nondecreasing in λ and $\zeta(\lambda)$ is nonincreasing in λ, at least for λ in the range $(0, \lambda_{\mathrm{c}})$.

To show continuity, let $0 < \lambda < \mu$. By the thinning theorem (Theorem 9.15), we can obtain coupled realizations of the Poisson processes $\mathcal{H}_\lambda$ and $\mathcal{H}_\mu$ in which $\mathcal{H}_\lambda$ is obtained from $\mathcal{H}_\mu$ by retaining each point of $\mathcal{H}_\mu$ with probability λ/μ, discarding it otherwise, and taking $\mathcal{H}_\lambda$ to consist of all retained points. With this coupling, one way for the component containing the origin of $G(\mathcal{H}_{\lambda,0}; 1)$ to have n vertices is for the component containing the origin of $G(\mathcal{H}_{\mu,0}; 1)$ to have n vertices and all of these vertices to be retained. Therefore,

$$p_n(\lambda) \ge \left(\frac{\lambda}{\mu}\right)^n p_n(\mu),$$

so that

$$\rho(\lambda) = \lim_{n\to\infty} (p_n(\lambda)^{1/n}) \ge \left(\frac{\lambda}{\mu}\right) \rho(\mu), \qquad 0 < \lambda < \mu < \infty. \tag{10.15}$$

This inequality, together with monotonicity of $\rho(\cdot)$ in the range $(0, \lambda_{\mathrm{c}})$, ensures that $\rho(\lambda)$ is continuous in λ, and hence that $\zeta(\lambda)$ is also continuous in λ, at least for λ in the range $(0, \lambda_{\mathrm{c}})$.

For $\lambda > \lambda_{\mathrm{c}}$, it follows from Theorem 10.14 below that $\zeta(\lambda) = 0$ and $\rho(\lambda) = 1$; since $\rho(\lambda) \le 1$ for all λ, it follows from this and (10.15) that $\rho(\lambda_{\mathrm{c}}) = 1$ and $\rho(\cdot)$ is continuous at λ_{c}, so $\zeta(\lambda_{\mathrm{c}}) = 0$ and $\zeta(\cdot)$ is continuous at λ_{c}. □

We now apply Theorem 10.1 to describe the behaviour of the order of the largest component in the random geometric graph $G(\mathcal{H}_{\lambda,s}; 1)$.

Theorem 10.3 *Suppose* $0 < \lambda < \lambda_c$, *and let* $\zeta(\lambda) = -\log \lim_n (p_n(\lambda)^{1/n})$, *as described in Theorem* 10.1. *Then, as* $s \to \infty$,

$$\frac{L_1(G(\mathcal{H}_{\lambda,s};1))}{\log s} \xrightarrow{P} \frac{d}{\zeta(\lambda)}.$$

Proof Let $\alpha > d/\zeta(\lambda)$. Let $N(\alpha)$ be the number of vertices of $G(\mathcal{H}_{\lambda,s};1)$ lying in components of order at least $\alpha \log s$. By Markov's inequality and then by Theorem 1.6 (Palm theory),

$$\begin{aligned} P[L_1(G(\mathcal{H}_{\lambda,s};1)) \geq \alpha \log s] &\leq E[N(\alpha)] \\ &= \lambda \int_{B(s)} P[V_x \geq \alpha \log s] \mathrm{d}x, \end{aligned}$$

where V_x is the order of the component of $G(\mathcal{H}_{\lambda,s} \cup \{x\};1)$ containing x. Take $\zeta' \in (d/\alpha, \zeta(\lambda))$. By the definition (10.5) of $\zeta(\lambda)$, for large enough s, and all $x \in B(s)$,

$$P[V_x \geq \alpha \log s] \leq \sum_{n \geq \alpha \log s} p_n(\lambda) \leq \exp(-\zeta' \alpha \log s)$$

so that

$$P[L_1(G(\mathcal{H}_{\lambda,s};1)) \geq \alpha \log s] \leq \lambda s^{d-\zeta'\alpha} \to 0. \tag{10.16}$$

Conversely, let $\beta < d/\zeta(\lambda)$. Given s, let $\{B_{1,s}, B_{2,s}, \ldots, B_{m(s),s}\}$ be a collection of disjoint balls of radius $2\beta \log s$ contained in $B(s)$, of maximal cardinality. Then, clearly

$$\liminf_{s \to \infty} \left\{ m(s) \left(\frac{\log s}{s} \right)^d \right\} > 0. \tag{10.17}$$

Let $x_{i,s}$ denote the centre of the ball $B_{i,s}$. Take $\lambda' \in (0, \lambda)$ such that $\beta\zeta(\lambda') < d$. This is possible by continuity of $\zeta(\cdot)$ (Theorem 10.1). By the superposition theorem (Theorem 9.14), we may assume, without loss of generality, that $\mathcal{H}_\lambda$ is obtained as the union of two independent Poisson processes $\mathcal{H}_{\lambda'}$ and $\mathcal{H}_{\lambda-\lambda'}$. Take $\zeta'' > \zeta(\lambda')$ such that $\zeta''\beta < d$.

If $\mathcal{H}_{\lambda-\lambda'} \cap B(x_{i,s};1)$ consists of a single point, then let that point be denoted $X_{i,s}$, and let $V_{i,s}$ be the order of the component of $G((\mathcal{H}_{\lambda'} \cap B_{i,s}) \cup \{X_{i,s}\};1)$ that includes $X_{i,s}$. If $\mathcal{H}_{\lambda-\lambda'}(B(x_{i,s};1)) \neq 1$, then set $V_{i,s} = 0$. Then, by independence of $\mathcal{H}_{\lambda'}$ and $\mathcal{H}_{\lambda-\lambda'}$, for all large enough s and for $i = 1, 2, \ldots, m(s)$, recalling that θ is the volume of the unit ball, we have that

$$\begin{aligned} P[V_{i,s} \geq \beta \log s] &\geq \theta(\lambda - \lambda')e^{\theta(\lambda'-\lambda)} p_{\lceil \beta \log s \rceil}(\lambda') \\ &\geq \exp(-\zeta''\beta \log s) = s^{-\zeta''\beta}, \end{aligned}$$

where the last inequality comes from (10.4).

The variables $V_{1,s}, \ldots, V_{m(s),s}$ are independent, since they are determined by the Poisson configurations in disjoint balls, so that

$$P\left[\bigcap_{i=1}^{m(s)} \{V_{i,s} < \beta \log s\}\right] \leq \left(1 - s^{-\zeta''\beta}\right)^{m(s)}$$
$$\leq \exp(-s^{-\zeta''\beta} m(s)),$$

which tends to zero by (10.17) and the condition $\zeta''\beta < d$. But, if for some i we have $V_{i,s} \geq \beta \log s$, then $L_1(G(\mathcal{H}_{\lambda,s}; 1)) \geq \beta \log s$. Combined with (10.16) this gives us the result. □

10.2 Existence of a crossing component

We now turn our attention to the largest component of a random geometric graph in the supercritical phase $\lambda > \lambda_c$, with the goal of establishing the giant component phenomenon asserted in (10.2) and (10.3). It is convenient to define 'large' components of $G(\mathcal{H}_{\lambda,s}; 1)$ in terms of a *crossing* property, defined as follows.

Suppose $B \subset \mathbb{R}^d$ is a set of the form $B = \prod_{i=1}^d [a_i, b_i]$. For $k = 1, 2, \ldots, d$, let $\pi_k : \mathbb{R}^d \to \mathbb{R}$ denote projection onto the kth coordinate. If $G(\mathcal{X}; r)$ is a geometric graph with vertex set $\mathcal{X}$, we shall say that $G(\mathcal{X}; r)$ is *k-crossing* for B if there exist vertices $x^-, x^+ \in \mathcal{X}$, such that $|\pi_k(x^-) - a_k| \leq r/2$ and $|\pi_k(x^+) - b_k| \leq r/2$, and x^- and x^+ lie in the same component of $G(\mathcal{X}; r)$. If $\mathcal{X} \subset B$, this means that there is a continuum path between the opposite faces in the k-direction of B that stays inside the union of balls of radius $r/2$ centred at the points of $\mathcal{X}$ (see Fig. 10.1 for the case $d = 2$). We shall say $G(\mathcal{X}; r)$ is *crossing* for B if it is k-crossing for all $k \in \{1, 2, \ldots, d\}$.

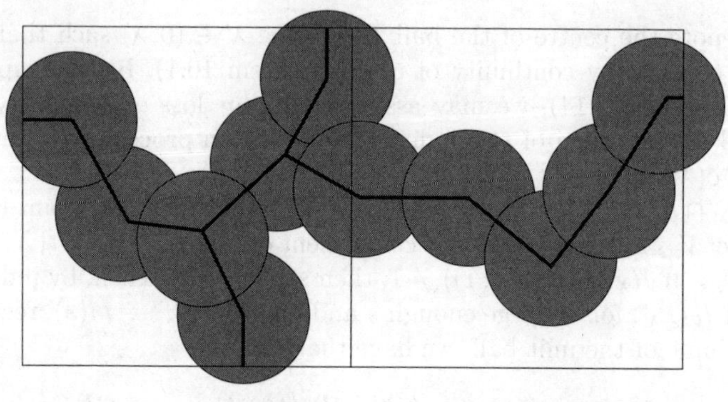

FIG. 10.1. If the disks are of radius $r/2$, and their centres are the points of $\mathcal{X}$, then $G(\mathcal{X}; r)$ is 1-crossing for the horizontal rectangle and is 2-crossing for the left-hand square, and these crossings must intersect.

In this section we show that the probability of non-existence of a component of $G(\mathcal{H}_{\lambda;s};1)$ that is 1-crossing for the box $B(s) := [-s/2, s/2]^d$ decays exponentially. We start with the case $d = 2$. In this case it is useful to work with rectangles; set

$$B(a,j) := [0, ja] \times [0, a], \qquad j = 1, 2, 4.$$

Let LR_a denote the event that there is a component of $G(\mathcal{H}_\lambda \cap B(a,2);1)$ that is 1-crossing for $B(a,2)$ (i.e. it crosses this rectangle the long way, from left to right). Let LLR_a denote the event that there is a component of $G(\mathcal{H}_\lambda \cap B(a,4);1)$ that is 1-crossing for $B(a,4)$, and let SLR_a denote the event that there is a component of $G(\mathcal{H}_\lambda \cap B(a,1);1)$ that is 1-crossing for $B(a,1)$.

Lemma 10.4 *Suppose $d = 2$ and $\lambda > \lambda_c$. Then $P[LR_a \to 1]$ and $P[SLR_a \to 1]$ as $a \to \infty$.*

Proof At first sight, this appears to follow directly from Meester and Roy (1996, Corollary 4.1). However, the 'occupied crossings' of a given rectangle described in Meester and Roy (1996) are continuum crossings in the intersection of the rectangle with region occupied by the union of balls of radius $\frac{1}{2}$ centred at *all* points in $\mathcal{H}_\lambda$, whereas the crossings the definition of events LR_a and SLR_a above correspond to continuum crossing paths in the union of balls centred at Poisson points *in the rectangle*; from our point of view, Poisson points outside the rectangle 'do not count'. This means we have some extra work to do. For similar reasons, it is not immediately clear how to prove the intuitively plausible monotonicity relation $P[SLR_a] \geq P[LR_a]$ (which would be obvious if we were using Meester and Roy's interpretation of crossing).

Let ν be a large fixed integer. Given a, divide $B(a,2)$ lengthwise into ν narrow strips $T_{1,a}, \ldots, T_{\nu,a}$ of dimensions $2a \times (a/\nu)$. For each $i \leq \nu$, let $T'_{i,a}$ be the $\frac{1}{2}$-interior of $T_{i,a}$, that is, let $T'_{i,a} := \{x : B(x,1/2) \subseteq T_{i,a}\}$. Then $T'_{i,a}$ is a slightly narrower strip of dimensions $(2a-1) \times ((a/\nu) - 1)$, contained in $T_{i,a}$.

Take $\lambda' \in (\lambda_c, \lambda)$. By the superposition theorem (Theorem 9.14), we may assume that $\mathcal{H}_\lambda$ is obtained as the union of two independent homogeneous Poisson processes $\mathcal{H}_{\lambda'}$ and $\mathcal{H}_{\lambda-\lambda'}$.

Let $F_{i,a}$ be the event that there is a continuum path in $T'_{i,a}$ from the left edge to the right edge that stays in the occupied region $\mathcal{H}_{\lambda'} \oplus B(\mathbf{0}; \frac{1}{2})$ (see Fig. 10.2). This is an occupied crossing of $T'_{i,a}$ in the sense of Meester and Roy (1996), and by Meester and Roy (1996, Corollary 4.1), which is also valid for all l_p norms, since $\lambda' > \lambda_c$ and the aspect ratio of the rectangles $T'_{i,a}$ is less than 3ν for all large enough a, we obtain

$$\lim_{a \to \infty} P[F_{1,a}] = 1. \tag{10.18}$$

Let $G_{i,a}$ be the event that in addition to event $F_{i,a}$ occurring, there is a continuum path in $(\mathcal{H}_\lambda \cap T_{i,a}) \oplus B(\mathbf{0}; \frac{1}{2})$ from the left edge to the right edge of $T_{i,a}$. We assert that there is a constant $\delta > 0$, independent of i or ν, such that

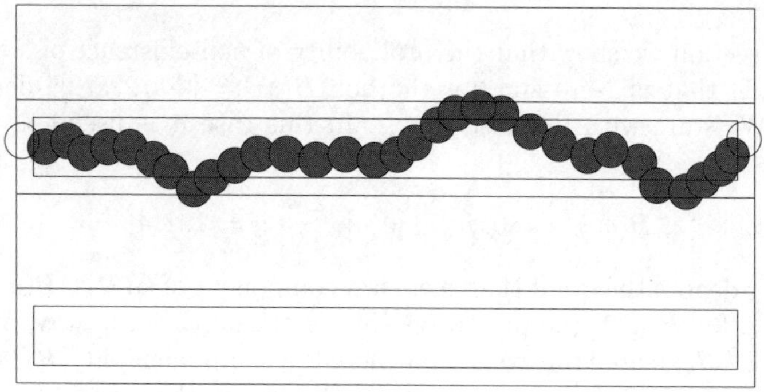

FIG. 10.2. The strips $T_{i,a}$ are shown for $\nu = 4$. Also two of the smaller strips $T'_{i,a}$ are shown, and event $F_{i,a}$ is illustrated for one of them.

$$P[G_{i,a}|F_{i,a}] \geq \delta. \tag{10.19}$$

The reason for this is as follows. If $F_{i,a}$ occurs then all the disks used in the crossing defined in the definition of that event are centred at points of $\mathcal{H}_{\lambda'}$ in $T_{i,a}$. We need at most one extra disk to connect the left edge of $T_{i,a}$ to that of $T'_{i,a}$ and at most one extra disk to connect the right edge of $T_{i,a}$ to that of $T'_{i,a}$ and the conditional probability that we have such a pair of disks centred at points of the independent Poisson process $\mathcal{H}_{\lambda-\lambda'}$ (the unshaded disks in Fig. 10.2), given the configuration of $\mathcal{H}_{\lambda'}$, is bounded away from zero.

By (10.18) and (10.19), we have for all large enough a and all $i \in \{1, 2, \ldots, \nu\}$ that $P[G_{i,a}] \geq \delta/2$, and since events $G_{1,a}, \ldots, G_{\nu,a}$ are independent, we obtain

$$\limsup_{a\to\infty} P[(LR_a)^{\text{c}}] \leq \limsup_{a\to\infty} \prod_{i=1}^{\nu} P[G^{\text{c}}_{i,a}] \leq (1 - \delta/2)^{\nu},$$

and, since ν is arbitrarily large and δ does not depend on ν, this shows that $P[LR_a] \to 1$. A similar argument shows that $P[SLR_a] \to 1$. □

Lemma 10.5 *Suppose $d = 2$ and $\lambda > \lambda_{\text{c}}$. Then there exist $c > 0$ and $a_1 > 0$ such that $1 - P[LR_a] \leq \exp(-ca)$ for all $a \geq a_1$.*

Proof By Lemma 10.4, $P[LR_a] \to 1$ and $P[SLR_a] \to 1$ as $a \to \infty$. Choose a_0 with $P[LR_a] > 49/50$ and $P[SLR_a] > 49/50$ for all $a > a_0$.

Suppose $b > 0$ with $P[LR_b] \leq 1 - \delta/25$ and $P[SLR_b] \leq 1 - \delta/25$ for some $\delta < 1$. If we set $H_i := [ib, (i+2)b] \times [0, b]$ $(i = 0, 1, 2)$ and $V_i = H_{i-1} \cap H_i$ $(i = 1, 2)$, then the occurrence of horizontal crossings of each of the horizontal rectangles H_0, H_1, H_2 and vertical crossings of each of the squares V_1, V_2 implies the occurrence of LLR_b, since horizontal and vertical crossings of a square must intersect (see Figs. 10.1 and 10.3). So, by Boole's inequality,

$$1 - P[LLR_b] \leq 5(\delta/25) = \delta/5,$$

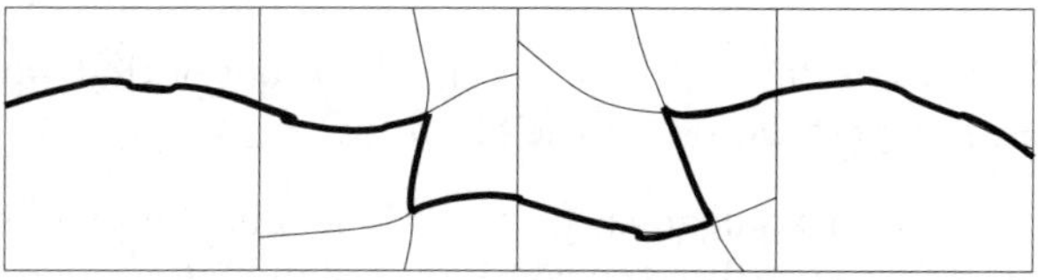

FIG. 10.3. horizontal crossings of the three horizontal $2b \times b$ rectangles shown, together with vertical crossings of the two squares formed by their intersections, imply a long-way crossing of a $4b \times b$ rectangle.

and, since $B(2b, 2)$ is the union of two disjoint translates of $B(b, 4)$, independence yields

$$1 - P[LR_{2b}] \leq (1 - P[LLR_b])^2 \leq \delta^2/25.$$

Likewise, since $B(2b, 1)$ is the union of two translates of $B(b, 2)$,

$$1 - P[SLR_{2b}] \leq (\delta/25)^2 \leq \delta^2/25.$$

Repeating this argument, we can deduce by induction that for every non-negative integer n, $1 - P[LR_{2^n b}] \leq \delta^{2^n}/25$ and $1 - P[SLR_{2^n b}] \leq \delta^{2^n}/25$. Then for any $a > a_0$, if we choose integer m so that $a_0 2^m \leq a < a_0 2^{m+1}$, and then set $b := 2^{-m}a$, by definition of a_0 we then have $P[LR_b] \leq 1 - \delta/25$ and $P[SLR_b] \leq 1 - \delta/25$, with $\delta = \frac{1}{2}$, so that

$$1 - P[LR_a] \leq (1/2)^{2^m}/25 \leq (1/2)^{a/(2a_0)}/25,$$

and the result follows. □

Different techniques are needed to prove an analogue to Lemma 10.5 in the case $d \geq 3$. The result goes as follows:

Proposition 10.6 *Suppose $d \geq 3$ and $\lambda > \lambda_c$. Then*

$$\limsup_{s \to \infty} s^{-1} \log P[\{\exists a \text{ 1-crossing component of } G(\mathcal{H}_{\lambda,s}; 1) \text{ for } B(s)\}^c] < 0.$$

The first step towards a proof of Proposition 10.6 is to consider Bernoulli site percolation on the lattice $(\mathbb{Z}^d, \sim_r)$ in which a pair of vertices $z, z' \in \mathbb{Z}^d$ are deemed adjacent if $\|z - z'\| \leq r$. As in Section 9.3, let $\theta_{\mathbb{Z}}(p; r)$ denote the probability that the open r-cluster at the origin for $\mathcal{B}_p$ is infinite, and let $p_c(r) := \inf\{p : \theta_{\mathbb{Z}}(p; r) > 0\}$. Also, define the lattice slab

$$S_{\mathbb{Z}}(K, n) := ([1, K] \cap \mathbb{Z}) \times ([1, n] \cap \mathbb{Z})^{d-1}.$$

Lemma 10.7 *Let $d \geq 3$, $r \geq 1$, and $p \in (p_c(r), 1)$. Then there exist an integer $K = K(p, r) > 0$ and $\delta \in (0, 1)$ such that, for any $n \geq 1$ and any z_1, z_2 in $S_{\mathbb{Z}}(K, n)$, the probability that z_1 and z_2 lie in the same open r-cluster in $\mathcal{B}_p \cap S_{\mathbb{Z}}(K, n)$ exceeds δ.*

Proof See Grimmett (1999, Lemma 7.78). The proof of that result is adapted easily enough to site percolation on the graph $(\mathbb{Z}^d, \sim_r)$. □

The next step is an analogous 'finite slab lemma' in the continuum. For $K > 0$, $a > 0$, let $S(K, a)$ denote the continuum slab $[0, K] \times [0, a]^{d-1}$. If $\mathcal{Y} \subset \mathbb{R}^d$ is locally finite, and $x \in \mathbb{R}^d$ (not necessarily in $\mathcal{Y}$), let $C(x; \mathcal{Y})$ denote the vertex set of the component of $G(\mathcal{Y} \cup \{x\}; 1)$ which contains x. For $A \subseteq \mathbb{R}^d$, set

$$C(A; \mathcal{Y}) := \bigcup_{x \in A} C(x; \mathcal{Y}). \tag{10.20}$$

Lemma 10.8 *Suppose $d \geq 3$, $\lambda > \lambda_c$. Then there exists $K = K(\lambda) \in (0, \infty)$ such that*

$$\liminf_{a \to \infty} \inf_{x,y \in S(K,a)} P[y \in C(x; (\mathcal{H}_\lambda \cap S(K, a)) \cup \{y\})] > 0 \tag{10.21}$$

and

$$\liminf_{a \to \infty} \inf_{x \in S(K,a)} P[C(x; \mathcal{H}_\lambda \cap S(K, a)) \text{ is 2-crossing for } S(K, a)] > 0. \tag{10.22}$$

Proof Choose $\varepsilon \in (0, 1/(4d))$ and $\lambda' < \lambda$ such that $\lambda'(1 - 4\varepsilon d)^d > \lambda_c$ and $\varepsilon^{-1} \in \mathbb{Z}$. Then by scaling (Corollary 9.18) the component containing the origin of $G(\mathcal{H}_{\lambda',0}; 1 - 4\varepsilon d)$ is infinite with non-zero probability.

For $z \in \mathbb{Z}^d$, set $B_z := \{\varepsilon z\} \oplus (-\varepsilon, 0]^d$, a cube (box) of side ε with one corner at εz. Let $\mathcal{H}_\lambda$ induce a realization of the Bernoulli process $\mathcal{B}_p$, with $p = 1 - \exp(-\varepsilon^d \lambda)$, by setting each site $z \in \mathbb{Z}^d$ to be open if $\mathcal{H}_\lambda(B_z) > 0$ and closed otherwise. Similarly, let $\mathcal{H}_{\lambda'}$ induce a Bernoulli process $\mathcal{B}_{p'}$ with $p' = 1 - \exp(-\varepsilon^d \lambda')$.

Set $l = (1 - 2\varepsilon d)/\varepsilon$. If $x, y \in \mathbb{Z}^d$, and if X and Y are points of $\mathcal{H}_{\lambda'}$ with $X \in B_x$, $Y \in B_y$, and $\|X - Y\| \leq 1 - 4\varepsilon d$, then $\|\varepsilon x - \varepsilon y\| \leq 1 - 2\varepsilon d$, so $x \sim_l y$. Therefore, the probability that $\mathcal{B}_{p'}$ has an infinite open l-cluster containing the origin is strictly positive. Hence, $p' \geq p_c(l)$ and $p > p_c(l)$.

Set $K_1 := \varepsilon K(p, l)$, with $K(p, l)$ given in Lemma 10.7. By that result, for $z, z' \in S_{\mathbb{Z}}(K_1/\varepsilon, n)$, the probability that z and z' lie in the same open l-cluster in $\mathcal{B}_p \cap S_{\mathbb{Z}}(K_1/\varepsilon, n)$ is bounded away from zero, uniformly in n, z, z'. Note that for any open $z, z' \in \mathcal{B}_p$, with $\|z - z'\| \leq l$, there exist points $X \in \mathcal{H}_\lambda \cap B_z$ and $X' \in \mathcal{H}_\lambda \cap B_{z'}$, and by the triangle inequality these satisfy $\|X - X'\| \leq 1$.

Given $a > 0$, let $n = \lfloor a/\varepsilon \rfloor$. Given $x, x' \in S(K_1, a)$, we can choose $z, z' \in S_{\mathbb{Z}}(K_1/\varepsilon, n)$ such that $\|\varepsilon z - x\| \leq 1 - d\varepsilon$ and $\|\varepsilon z' - x'\| \leq 1 - d\varepsilon$. If z and z' lie in the same open l-cluster in $\mathcal{B}_p \cap S_{\mathbb{Z}}(K_1/\varepsilon, n)$ then there is a path in the graph $G((\mathcal{H}_\lambda \cap S(K_1, a)) \cup \{x, x'\}; 1)$ from x to x', so that $x' \in C(x; (\mathcal{H}_\lambda \cap S(K, a)) \cup \{x'\})$. Thus, the probability that this occurs is bounded away from zero, uniformly in a, x, and x', and (10.21) follows, with $K(\lambda) = K_1$. The argument for (10.22) is similar. □

Proof of Proposition 10.6 Let $K = K(\lambda)$, as given by Lemma 10.8, and let π_2 denote projection onto the second coordinate. Divide $B(s)$ into slabs U_j of thickness K, by setting $U_j := B(s) \cap \pi_2^{-1}(((j-1)K, jK])$. Let H_j denote the event that there is no component of $G(\mathcal{H}_\lambda \cap U_j; 1)$ that is 1-crossing for $B(s)$. By Lemma 10.8, $P[H_j]$ is bounded below by some constant $\delta > 0$, independent of s. Therefore, by independence $P[\cap_{j=1}^{\lfloor s/(2K) \rfloor} H_j^c] \leq (1-\delta)^{\lfloor s/(2K) \rfloor}$, and the result follows. □

10.3 Uniqueness of the giant component

Let the *metric diameter* of a geometric graph $G(\mathcal{X}; r)$ be the value of $\mathrm{diam}(\mathcal{X})$ (see (1.2)). The word 'metric' is used here to distinguish this concept from the graph-theoretic notion of the diameter of a graph.

This section contains a proof of the fundamental result (10.2),(10.3) concerning the giant component in the supercritical phase. The result says that not only is there a unique giant component measured in terms of order, but also with respect to metric diameter.

Theorem 10.9 *Suppose $d \geq 2$ and $\lambda > \lambda_c$. Then*

$$s^{-d} L_1(G(\mathcal{H}_{\lambda,s}; 1)) \xrightarrow{P} \lambda p_\infty(\lambda) \quad \text{as } s \to \infty. \tag{10.23}$$

Also,

$$s^{-d} L_2(G(\mathcal{H}_{\lambda,s}; 1)) \xrightarrow{P} 0 \quad \text{as } s \to \infty. \tag{10.24}$$

Moreover, given $(\phi_s, s \geq 0)$ with $\phi_s \leq s$ for all s and $\phi_s / \log s \to \infty$ as $s \to \infty$, with probability approaching 1 as $s \to \infty$, the largest component of $G(\mathcal{H}_{\lambda,s}; 1)$ is crossing for $B(s)$, and no other component has metric diameter greater than ϕ_s.

This section also contains an exponentially decaying bound on the probability that there is not a unique cluster of metric diameter greater than ϕ_s (see Proposition 10.13 below).

Remark Suppose $\mathcal{S}$ is a measurable space, and suppose X, Y are independent $\mathcal{S}$-valued random variables. Suppose $g : \mathcal{S} \times \mathcal{S} \to \mathbb{R}$ is bounded and measurable with respect to the product σ-field on $\mathcal{S} \times \mathcal{S}$. Define the function $g_1 : \mathcal{S} \to \mathbb{R}$ by $g(x) = E[g(x, Y)]$. By the monotone-class theorem (see, e.g., Williams (1991)),

$$E[g(X, Y)|X] = g_1(X) \quad \text{a.s.} \tag{10.25}$$

We shall apply this fact in cases where $\mathcal{S}$ is a space of point configurations.

Recall the notation $C(x; \mathcal{Y})$ used in Lemma 10.8. Given a slab S, by making $\mathrm{dist}(x, S)$ very close to 1, the probability that $C(x; \mathcal{H}_\lambda \cap S) \neq \{x\}$ may be made arbitrarily small; however, given that $C(x; \mathcal{H}_\lambda \cap S)$ is not a singleton, the *conditional* probability that it is big can be bounded away from 0, as shown by our next lemma (the ratios in this lemma could be re-written as conditional probabilities, in the case $\lambda = \mu$). For the time being, only the case $\lambda = \mu$ of this result will be used; however, the case $\lambda < \mu$ will be used in a later chapter.

Lemma 10.10 *Suppose $d \geq 3$, and $\mu \geq \lambda > \lambda_c$. Then there exists $K' = K'(\lambda) > 0$ such that*

$$\liminf_{s\to\infty} \inf_{A} \frac{P[C(A;\mathcal{H}_\lambda \cap S(K',s)) \text{ is 2-crossing for } S(K',s)]}{P[C(A;\mathcal{H}_\mu \cap S(K',n)) \neq A]} > 0, \quad (10.26)$$

where the infimum is over all non-empty $A \subseteq (-1,0) \times [0,s]^{d-1}$; and also

$$\liminf_{s\to\infty} \inf_{A,B} \frac{P[C(A;\mathcal{H}_\lambda \cap S(K',s)) \cap C(B;\mathcal{H}_\lambda \cap S(K',s)) \neq \emptyset]}{P[C(A;\mathcal{H}_\mu \cap S(K',s)) \neq \emptyset; C(B;\mathcal{H}_\mu \cap S(K',s)) \neq \emptyset]} > 0, \quad (10.27)$$

where the infimum is over all non-empty $A, B \subseteq (-1,0) \times [0,s]^{d-1}$.

Proof Choose $\lambda' \in (\lambda_c, \lambda)$. Take $K' = K(\lambda')$ as given by Lemma 10.8. Set $U_A := S(K',s) \cap \cup_{x\in A} B(x;1)$, the 1-neighbourhood of A in $S(K',s)$. By the superposition theorem (Theorem 9.14) we can, and do, assume without loss of generality that $\mathcal{H}_\lambda$ is the union of two independent homogeneous Poisson processes $\mathcal{H}_{\lambda'}$ and $\mathcal{H}_{\lambda-\lambda'}$ with intensities indicated by the subscripts. Then by (10.25),

$$\begin{aligned} &P[C(A;\mathcal{H}_\lambda \cap S(K',s)) \text{ is 2-crossing for } S(K',s)] \\ &\geq P[\mathcal{H}_{\lambda-\lambda'}(U_A) > 0] \inf_{x\in S(K',s)} P[C(x;\mathcal{H}_{\lambda'} \cap S(K',s)) \text{ is 2-crossing for } S(K',s)] \\ &\geq \text{const.} \times P[\mathcal{H}_{\lambda-\lambda'}(U_A) > 0], \end{aligned} \quad (10.28)$$

where the last line follows from Lemma 10.8. The denominator of (10.26) is equal to $P[\mathcal{H}_\mu(U_A) > 0]$, and by assuming $\mathcal{H}_{\lambda-\lambda'}$ is obtained from $\mathcal{H}_\mu$ by thinning (see Theorem 9.15), with each point of $\mathcal{H}_\mu$ retained with probability $(\lambda - \lambda')/\mu$, we see that

$$\frac{P[\mathcal{H}_{\lambda-\lambda'}(U_A) > 0]}{P[\mathcal{H}_\mu(U_A) > 0]} \geq \frac{\lambda - \lambda'}{\mu},$$

so (10.26) follows. The proof of (10.27) is similar. □

If $\mathcal{Y} \subset \mathbb{R}^d$ is locally finite, and $x \in \mathbb{R}^d$ (not necessarily in $\mathcal{Y}$), at most one component of $G(\mathcal{Y};1)$ can include a vertex in $B(x;\frac{1}{2})$; define $\tilde{C}(x;\mathcal{Y})$ to be the vertex set of the component of $G(\mathcal{Y};1)$ that includes a vertex in $B(x;\frac{1}{2})$, if such a component exists, or to be the empty set if not.

Lemma 10.11 *Suppose $d \geq 3$ and $\lambda > \lambda_c$. Suppose $k_1, k_2 \in \{1, 2, \ldots, d\}$ with $k_1 \neq k_2$. Let G_s denote the event that $G(\mathcal{H}_{\lambda,s};1)$ has a component which is k_1-crossing but not k_2-crossing for $B(s)$. Then $\limsup_{s\to\infty} s^{-1} \log P[G_s] < 0$.*

Proof It suffices to prove the result for $k_1 = 1, k_2 = 2$. Set $\beta := 1/(2d)$. Then for every $y \in B(s)$ there exists $x \in B(\beta^{-1}s) \cap \mathbb{Z}^d$ such that $\|\beta x - y\| \leq \frac{1}{2}$. For $x \in \mathbb{Z}^d$, define $G_{x,s}$ to be the event that $\tilde{C}(\beta x;\mathcal{H}_{\lambda,s})$ is 1-crossing but not 2-crossing for $B(s)$. Then

$$P[G_s] \leq \sum_{x\in\mathbb{Z}^d\cap B(\beta^{-1}s):\pi_1(x)\leq 0} P[G_{x,s}]. \quad (10.29)$$

Fix $x \in B(\beta^{-1}s) \cap \mathbb{Z}^d$ with $\pi_1(x) \leq 0$. To estimate $P[G_{x,s}]$, divide $B(s)$ into slabs $S_j, j \in \mathbb{Z}$, of thickness K', with $K' = K'(\lambda)$ given in Lemma 10.10, by setting

$$T_j := \{w \in B(s) : \pi_1(w) \leq \pi_1(\beta x) + 1 + jK'\}, \qquad S_j := T_j \setminus T_{j-1}. \quad (10.30)$$

For $j \in \mathbb{Z}$, set $\mathcal{F}_j := \mathcal{F}_{T_j}$, the σ-field generated by the the configuration of $\mathcal{H}_\lambda \cap T_j$, that is, the σ-field generated by the collection of all random variables of the form $\mathcal{H}_\lambda \cap B$ with B a Borel subset of T_j.

Let A_j be the event that $\tilde{C}(\beta x; \mathcal{H}_\lambda \cap T_j) \cap S_j \neq \emptyset$. Let B_j be the event that $\tilde{C}(\beta x; \mathcal{H}_\lambda \cap T_j)$ does not contain any 2-crossing component of $G(\mathcal{H}_\lambda \cap S_j; 1)$. Then A_j and B_j are $\mathcal{F}_j$-measurable. By (10.26) and (10.25), there exists a constant $\gamma > 0$, independent of x or s, such that, for $1 \leq j+1 \leq ((s/2)-1)/K'$,

$$P[B^{\mathrm{c}}_{j+1} | \mathcal{F}_j] \geq \gamma P[A_{j+1} | \mathcal{F}_j] \qquad \text{a.s.},$$

which implies that a.s.

$$\begin{aligned} P[A_{j+1} \cap B_{j+1} | \mathcal{F}_j] &\leq \min(P[A_{j+1} | \mathcal{F}_j], P[B_{j+1} | \mathcal{F}_j]) \\ &\leq \min(P[A_{j+1} | \mathcal{F}_j], 1 - \gamma P[A_{j+1} | \mathcal{F}_j]) \\ &\leq (1+\gamma)^{-1} = \exp(-\delta), \quad \text{say}. \end{aligned} \quad (10.31)$$

Therefore, by repeated conditioning,

$$P[G_{x,s}] \leq P\left[\bigcap_{j=1}^{\lfloor ((s/2)-1)/K' \rfloor} (A_j \cap B_j)\right] \leq \exp(-\delta \lfloor ((s/2)-1)/K' \rfloor).$$

This upper bound is uniform on $\{x \in \mathbb{Z}^d \cap B(\beta^{-1}s) : \pi_1(x) \leq 0\}$, so the desired result follows by (10.29), since the number of terms in the summation there is $O(s^d)$. □

Lemma 10.12 *Suppose $d \geq 3$ and $\lambda > \lambda_{\mathrm{c}}$. Suppose $k \in \{1, 2, \ldots, d\}$. Suppose the function $(\phi_s, s \geq 0)$ satisfies $(\phi_s / \log s) \to \infty$ as $s \to \infty$ and $\phi_s \leq s$ for all s. Let G'_s denote the event that there exist distinct components C and C' of $G(\mathcal{H}_{\lambda,s}; 1)$, such that* (i) *$C$ is k-crossing for $B(s)$, and* (ii) *$\mathrm{diam}(\pi_k(C')) \geq \phi_s$. Then $\limsup_{s\to\infty} \phi_s^{-1} \log P[G'_s] < 0$.*

Proof It suffices to prove the result for $k = 1$. Let $\beta := 1/(2d)$. For $x, y \in \mathbb{Z}^d \cap B(\beta^{-1}s)$, define $C_x := \tilde{C}(\beta x; \mathcal{H}_{\lambda,s})$ and $C_y := \tilde{C}(\beta y; \mathcal{H}_{\lambda,s})$. Also, let $G'_{x,y,s}$ denote the event that C_x and C_y are distinct, and are both 1-crossing for $[\pi_1(\beta x), \pi_1(\beta x) + \phi_s] \times [-s/2, s/2]^{d-1}$. Then, if G'_s occurs, $G'_{x,y,s}$ must occur for some $x, y \in \mathbb{Z}^d \cap B(\beta^{-1}s)$ with $\pi_1(\beta x) = \pi_1(\beta y) \leq s/2 - \phi_s$. So

$$P[G'_s] \leq \sum_{x,y \in \mathbb{Z}^d \cap B(\beta^{-1}s) : \pi_1(\beta x) = \pi_1(\beta y) \leq s/2 - \phi_s} P[G'_{x,y,s}].$$

Fix distinct $x, y \in B(\beta^{-1}s) \cap \mathbb{Z}^d$ with $\pi_1(x) = \pi_1(y) \leq (s/2) - \phi_s$. Define slabs S_j and T_j by (10.30) and the σ-field $\mathcal{F}_j = \mathcal{F}_{T_j}$ as before. Writing $C_{j-1}(x)$ for the set $\tilde{C}(\beta x; \mathcal{H}_\lambda \cap T_{j-1})$ and $C_{j-1}(y)$ for the set $\tilde{C}(\beta y; \mathcal{H}_\lambda \cap T_{j-1})$, define events

$$A'_j := \{C(C_{j-1}(x); \mathcal{H}_\lambda \cap S_j) \neq C_{j-1}(x)\} \cap \{C(C_{j-1}(y); \mathcal{H}_\lambda \cap S_j) \neq C_{j-1}(y)\}$$

and

$$B'_j := \{C(C_{j-1}(x); \mathcal{H}_\lambda \cap S_j) \cap C(C_{j-1}(y); \mathcal{H}_\lambda \cap S_j) = \emptyset\}.$$

Then A'_j and B'_j are $\mathcal{F}_j$-measurable. By (10.27) and (10.25), there exists a constant $\gamma' > 0$, independent of x, y or s, such that for $1 \leq j+1 \leq (\phi_s - 1)/K'$,

$$P[(B'_{j+1})^c | \mathcal{F}_j] \geq \gamma' P[A'_{j+1} | \mathcal{F}_j] \quad \text{a.s.}$$

and the remainder of the proof is much the same as for Lemma 10.11, since

$$G'_{x,y,s} \subseteq \bigcap_{j=1}^{\lfloor (\phi_s - 1)/K' \rfloor} (A'_j \cap B'_j). \qquad \square$$

Proposition 10.13 *Suppose $d \geq 2$ and $\lambda > \lambda_c$. Suppose $(\phi_s, s \geq 0)$ is increasing with $(\phi_s / \log s) \to \infty$ as $s \to \infty$, and with $\phi_s \leq s$ for all s. Let E_s denote the event that* (i) *there is a unique component of $G(\mathcal{H}_{\lambda,s}; 1)$ that is crossing for $B(s)$, and* (ii) *no other component of $G(\mathcal{H}_{\lambda,s}; 1)$ has metric diameter greater than ϕ_s. Then $\limsup_{s\to\infty} \phi_s^{-1} \log P[E_s^c] < 0$.*

Proof First suppose $d \geq 3$. The existence (except on an event of exponentially decaying probability) of a crossing component follows from Proposition 10.6 and Lemma 10.11. Its uniqueness follows from Lemma 10.12, as does the non-existence of any other component of diameter greater than ϕ_s, since for any set C with $\mathrm{diam}(C) \geq \phi_s$, we have $\mathrm{diam}(\pi_k(C)) \geq \phi_s/d$ for some k.

Now suppose $d = 2$. Set $m_s := \lceil 4s/\phi_s \rceil$ and $\psi_s := s/m_s$. Define horizontal and vertical 'dominoes' (rectangles with aspect ratio 2) $H_{i,j}$ and $V_{i,j}$ by

$$H_{i,j} = \left[-\frac{s}{2} + (i-1)\psi_s, -\frac{s}{2} + (i+1)\psi_s\right] \times \left[-\frac{s}{2} + (j-1)\psi_s, -\frac{s}{2} + j\psi_s\right],$$

$$V_{i,j} = \left[-\frac{s}{2} + (i-1)\psi_s, -\frac{s}{2} + i\psi_s\right] \times \left[-\frac{s}{2} + (j-1)\psi_s, -\frac{s}{2} + (j+1)\psi_s\right].$$

Then, by Lemma 10.5, there exists a constant $c > 0$ such that, for large enough s,

$$P[G(\mathcal{H}_\lambda \cap H_{i,j}; 1) \text{ is 1-crossing for } H_{i,j}] > 1 - e^{-cs/m_s}, \tag{10.32}$$

$$P[G(\mathcal{H}_\lambda \cap V_{i,j}; 1) \text{ is 2-crossing for } V_{i,j}] > 1 - e^{-cs/m_s}. \tag{10.33}$$

Therefore, if I_s denotes the intersection over all $(i, j) \in B_{\mathbb{Z}}(m_s)$ (defined at (9.1)) of the events described in (10.32) and (10.33) above (see Fig. 10.4), $P[I_s]$

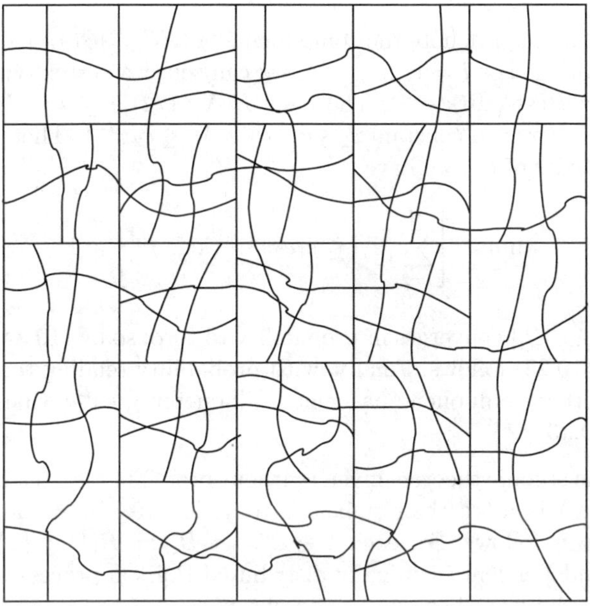

FIG. 10.4. Event I_s in the case where $m_s = 5$ to change.

exceeds $1-m_s^2 \exp(-cs/m_s)$, and therefore exceeds $1-\exp(-\text{const.}\times\phi_s)$, for large s. But, on the event I_s, the 1-crossing components of $G(\mathcal{H}_\lambda \cap H_{i,j}; 1)$ and the 2-crossing components of $G(\mathcal{H}_\lambda \cap V_{i,j}; 1)$ must all be part of the same big component of $G(\mathcal{H}_{\lambda,s}; 1)$ (since the long-way crossings for rectangles that intersect at right angles must overlap; see Fig. 10.1). Also on I_s, no other component can have metric diameter greater than ϕ_s without intersecting this big component. □

Proof of Theorem 10.9 By Proposition 10.13, with probability tending to 1, there is a unique big component of $G(\mathcal{H}_{\lambda,s}; 1)$ with metric diameter greater than $(\log s)^2$. Also, by scaling the clique number of $G(\mathcal{H}_{\lambda,s}; (\log s)^2)$ has the same distribution as that of $G(\mathcal{H}_{\lambda s^d,1}; s^{-1}(\log s)^2)$. Hence, by Theorem 6.16 and a simple Poissonization argument, the clique number of $G(\mathcal{H}_{\lambda,s}; (\log s)^2)$ is $O(\log s)^{2d}$ in probability, so that there is a constant c_1 such that, with probability tending to 1, all components of $G(\mathcal{H}_{\lambda,s}; 1)$, other than the biggest one, have order at most $c_1(\log s)^{2d}$. This shows that $s^{-d} L_2(G(\mathcal{H}_{\lambda,s}; 1))$ converges to zero in probability.

Since $\lambda > \lambda_c$, by uniqueness of the infinite component in continuum percolation (Theorem 9.19) the graph $G(\mathcal{H}_\lambda; 1)$ a.s. has precisely one infinite component. Let the vertex set of this infinite component be denoted $\mathcal{C}_\infty$, viewed as a point process.

By Palm theory (Theorem 9.22), $E[\mathcal{C}_\infty(B(s))] = \lambda p_\infty(\lambda) s^d$. By an ergodic theorem (see the remark below),

$$s^{-d}\mathcal{C}_\infty(B(s)) \xrightarrow{P} \lambda p_\infty(\lambda) \quad \text{as } s \to \infty. \tag{10.34}$$

Let $C_1, C_2, \ldots, C_M$ denote the components of $G(\mathcal{C}_\infty \cap B(s); 1)$, taken in order of decreasing order. For $i = 1, 2, \ldots, M$, we can select a vertex X_i from C_i lying in the annulus $B(s) \setminus B(s-2)$. The balls $B(X_i; 1/2), 1 \leq i \leq M$, are disjoint and, therefore, there is a constant c_2 such that $M \leq c_2 s^{d-1}$. Therefore, with $|C_i|$ denoting the order of C_i, we have

$$\lim_{s \to \infty} P\left[\sum_{i=2}^{M} |C_i| \leq c_1 c_2 s^{d-1} (\log s)^{2d}\right] = 1.$$

Hence, $s^{-d} \sum_{i=2}^{M} |C_i|$ converges in probability to zero, so by (10.34), $s^{-d}|C_1| \xrightarrow{P} \lambda p_\infty(\lambda)$, and (10.23) follows. Finally, with probability tending to 1, C_1 is crossing, and no other component has metric diameter greater than ϕ_s, both by Proposition 10.13. □

Remark The ergodic theorem given (without proof) in Meester and Roy (1996, Proposition 2.3) is more than sufficient to yield (10.34). A more elementary approach goes as follows. For each $z \in \mathbb{Z}^d$, set $B_z := B(1) \oplus \{z\}$. Since $\mathcal{H}_\lambda$ is the union of independent identically distributed Poisson processes on the cubes $B_z, z \in \mathbb{Z}^d$, we can use Theorem 9.13 to obtain

$$s^{-d} \sum_{z \in B(s-1) \cap \mathbb{Z}^d} \mathcal{C}_\infty(B_z) \xrightarrow{L^1} \lambda p_\infty(\lambda). \tag{10.35}$$

Also, $\mathcal{H}_\lambda(B(s)) - \mathcal{H}_\lambda\left(\cup_{z \in B(s-1) \cap \mathbb{Z}^d} B_z\right)$ is an upper bound for the non-negative random variable $\mathcal{C}_\infty(B(s)) - \sum_{z \in B(s-1) \cap \mathbb{Z}^d} \mathcal{C}_\infty(B_z)$, and simply taking expectations shows that

$$s^{-d}(\mathcal{H}_\lambda(B(s)) - \mathcal{H}_\lambda\left(\bigcup_{z \in B(s-1) \cap \mathbb{Z}^d} B_z\right) \xrightarrow{L^1} 0,$$

which, together with (10.35), yields (10.34).

10.4 Sub-exponential decay for supercritical percolation

We have already seen in Section 10.1 that for $\lambda < \lambda_c$, $p_n(\lambda)$ decays exponentially in n. The results in the present section show that if $\lambda > \lambda_c$, then $p_n(\lambda)$, and also the tail sum $\sum_{k=n}^{\infty} p_k(\lambda)$, decay exponentially in $n^{(d-1)/d}$ rather than in n as in the subcritical case. It is the continuum analogue of known results for lattice percolation, and will be applied to geometric graphs later on. The first result is a sub-exponentially decaying lower bound.

Theorem 10.14 *Suppose $d \geq 2$ and $\lambda > \lambda_c$. Then*

$$\liminf_{n \to \infty} \left(n^{-(d-1)/d} \log p_n(\lambda)\right) > -\infty. \tag{10.36}$$

The second result is an upper bound taking the same form.

Theorem 10.15 *Suppose $d \geq 2$ and $\lambda > \lambda_c$. Then*

$$\limsup_{n\to\infty} \left(n^{-(d-1)/d} \log \sum_{k\geq n} p_k(\lambda) \right) < 0. \tag{10.37}$$

Theorems 10.14 and 10.15 suggest the conjecture that

$$\lim_{n\to\infty} \left(n^{-(d-1)/d} \log p_n(\lambda) \right) \quad \text{exists.}$$

This remains an open problem, although analogous results for lattice percolation have been proved by Alexander *et al.* (1990) for $d = 2$ and by Cerf (2000) for $d = 3$. If the limit $\psi(\lambda) := \lim_{n\to\infty} \left(n^{-(d-1)/d} \log p_n(\lambda) \right)$ does exist, then by Alexander (1991, Theorem 2.5) and a diagonal argument, it must satisfy $\lim_{\lambda\to\infty} \lambda^{-1/d}\psi(\lambda) = -d\theta^{1/d}$ (K. S. Alexander, personal communication).

The proof of Theorem 10.15 uses a block construction which will reappear, in modified form, in Sections 10.5 and 10.6. The space $\mathbb{R}^d$ is divided into blocks of side M, where M is large but essentially 'fixed'. A block is deemed 'good' if the geometric graph in the block has a big component and if the geometric graph in the associated concentric box of side $5M$ satisfies a 'uniqueness of the big component' condition which guarantees that the big components in neighbouring good blocks are linked together. Results from Sections 10.3 and 9.4 then ensure that, for any $\varepsilon > 0$, if M is big enough the collection of good blocks dominates a Bernoulli process with parameter $1 - \varepsilon$. Then we can use Peierls-type arguments from Bernoulli lattice percolation theory to estimate probabilities for clusters of blocks. Finally, we typically need some extra Poisson estimate along the lines of (10.9) to deal with the possibility of an unusually large number of Poisson points associated with a moderate-sized block.

For later use we introduce an alternative formulation, without reference to a special inserted point at the origin, as follows. Let $p_n^*(\lambda)$ denote the probability that there exists a component of $G(\mathcal{H}_\lambda; 1)$ of order n having at least one vertex in the unit cube $B(1)$ centred at the origin. If M_n denotes the number of points of $\mathcal{H}_\lambda$ lying in the unit cube $B(1)$ and also in a component of $G(\mathcal{H}_\lambda; 1)$ of order n, then by Markov's inequality and by Palm theory for the Poisson process (Theorem 9.22),

$$p_n^*(\lambda) = P[M_n \geq 1] \leq E[M_n] = \lambda p_n(\lambda), \tag{10.38}$$

Let $p_\infty^*(\lambda)$ denote the probability that there exists an infinite component of $G(\mathcal{H}_\lambda; 1)$ containing at least one vertex in $B(1)$. By Theorem 9.22, the mean number of Poisson points in $B(1)$ lying in an infinite component of $G(\mathcal{H}_\lambda; 1)$ is $\lambda p_\infty(\lambda)$, and so is strictly positive if $\lambda > \lambda_c$. Therefore, $p_\infty^*(\lambda)$ is strictly positive if $\lambda > \lambda_c$.

Lemma 10.16 *Suppose $d \geq 2$ and $\lambda > \lambda_c$. Given $\varepsilon > 0$, there exists $\alpha > 0$ and $s_1 > 0$ such that, for all $s \geq s_1$, there exists an integer $k = k(s) \geq 5$ with*

$$(1-\varepsilon)\lambda p_\infty(\lambda)s^d \leq k \leq (1+\varepsilon)\lambda p_\infty(\lambda)s^d, \tag{10.39}$$

such that

$$p_k^*(\lambda) \geq k \exp(-\alpha k^{(d-1)/d}) \tag{10.40}$$

and

$$p_k(\lambda) \geq k \exp(-\alpha k^{(d-1)/d}). \tag{10.41}$$

Proof Let $F_s^{(1)}$ denote the event that, firstly,

$$(1-\varepsilon)\lambda p_\infty(\lambda)s^d < L_1(G(\mathcal{H}_{\lambda,s};1)) < (1+\varepsilon)\lambda p_\infty(\lambda)s^d,$$

and secondly, no component of $G(\mathcal{H}_{\lambda,s};1)$, other than the largest one, has metric diameter greater than $s^{1/2}$. By Theorem 10.9,

$$\lim_{s\to\infty} P[F_s^{(1)}] = 1. \tag{10.42}$$

Let $F_s^{(2)}$ be the event that there exists a component of $G(\mathcal{H}_{\lambda,s};1)$ having at least one vertex in $B(1)$ and having metric diameter greater than $s^{1/2}$. Then, for all $s > 16$,

$$P[F_s^{(2)}] \geq p_\infty^*(\lambda). \tag{10.43}$$

Finally, let $F_s^{(3)}$ denote the event that there are no points of $\mathcal{H}_\lambda$ in the annulus $B(s+2) \setminus B(s)$. Then, for all large enough s,

$$P[F_s^{(3)}] \geq \exp(-3\lambda d s^{d-1}). \tag{10.44}$$

Moreover, $F_s^{(3)}$ is independent of $F_s^{(1)} \cap F_s^{(2)}$, so by (10.42)–(10.44), there exists $\alpha_1 > 0$ such that for all large enough s we have

$$P[F_s^{(1)} \cap F_s^{(2)} \cap F_s^{(3)}] \geq \tfrac{1}{2} p_\infty^*(\lambda) \exp(-\alpha_1 s^{d-1}). \tag{10.45}$$

On event $F_s^{(1)} \cap F_s^{(2)}$, the largest component of $G(\mathcal{H}_{\lambda,s};1)$ includes at least one vertex in $B(1)$, and has order in the range $(1 \pm \varepsilon)\lambda p_\infty(\lambda)s^d$. On event $F_s^{(3)}$, no point of $\mathcal{H}_{\lambda,s}$ lies within unit distance of any point of $\mathcal{H}_\lambda$ lying outside $B(s)$, and therefore on the event $F_s^{(1)} \cap F_s^{(2)} \cap F_s^{(3)}$, there is a component of $G(\mathcal{H}_\lambda; 1)$ containing at least one vertex in $B(1)$, whose order lies in the range $(1 \pm \varepsilon)\lambda p_\infty(\lambda)s^d$. Therefore, by (10.45), there exists $s_0 > 0$ such that for all $s > s_0$ we have

$$\sum_{k=\lceil(1-\varepsilon)\lambda p_\infty(\lambda)s^d\rceil}^{\lfloor(1+\varepsilon)\lambda p_\infty(\lambda)s^d\rfloor} p_k^*(\lambda) \geq \exp(-\alpha_1 s^{d-1})p_\infty^*(\lambda)/2,$$

and therefore, for all $s \geq s_0$, there exists $k = k(s)$ satisfying (10.39) such that

$$p_k^*(\lambda) \geq \exp(-\alpha_1 s^{d-1})p_\infty^*(\lambda)/(4\varepsilon\lambda p_\infty(\lambda)s^d).$$

Hence, by (10.39) and (10.38), there are constants $s_1 \geq s_0$ and $\alpha > 0$ such that, for all $s \geq s_1$ and for each $k = k(s)$, (10.40) and (10.41) hold, and also $k(s) \geq 5$. □

Lemma 10.17 *Suppose $(r(n), n \geq 1)$ is a sequence of positive integers satisfying $r(1) \geq 4$ and $2r(i) \leq r(i+1) \leq 4r(i)$ for each i. For all $n = 1, 2, 3, \ldots$, if we set $I := \max\{i : r(i) \leq n\}$, there exist integers $w_0, w_1, w_2, \ldots, w_I$ such that*

$$n = w_0 + w_1 r(1) + \cdots + w_I r(I),$$

with $1 \leq w_0 \leq r(1)$ and $0 \leq w_i \leq 4$ for $i = 1, 2, \ldots, I$.

Proof The conclusion of the lemma is clearly true for $n \leq r(1)$. Suppose inductively that it is true for $n = 1, 2, \ldots, m$ with $m \geq r(1)$. Set $r(0) = 1$. Let $I := \max\{i : r(i) \leq m\}$, and using the inductive hypothesis, take $w_0, \ldots, w_I$ such that $m + 1 - r(I) = \sum_{i=0}^{I} w_i r(i)$, with $1 \leq w_0 \leq r(1)$ and $0 \leq w_i \leq 4$ for $i = 1, 2, \ldots, I$. Then

$$m + 1 = w_0 + w_1 r(1) + \cdots + w_{I-1} r(I-1) + (1 + w_I) r(I). \qquad (10.46)$$

It remains to prove that $1 + w_I \leq 4$. This holds because (10.46) implies

$$(1 + w_I) r(I) \leq m + 1 \leq r(I+1) \leq 4r(I).$$ □

Proof of Theorem 10.14 Let $\varepsilon_1 > 0$ be chosen sufficiently small so that $2(1+\varepsilon_1)^2 < 3(1-\varepsilon_1)^2$. Let s_1, α, and $k(s), s \geq s_1$ be as described in Lemma 10.16 (with $\varepsilon = \varepsilon_1$). Recursively, choose a sequence of integers $k_1, k_2, \ldots$ as follows. Set $k_1 = k(s_1)$. Given k_i with $k_i = k(s)$, say, choose t such that

$$t^d(1 - \varepsilon_1) = 2s^d(1 + \varepsilon_1)$$

and take $k_{i+1} = k(t)$. Then using (10.39) we have

$$k_{i+1} \geq (1 - \varepsilon_1)\lambda p_\infty(\lambda)t^d = 2(1 + \varepsilon_1)\lambda p_\infty(\lambda)s^d \geq 2k_i, \qquad (10.47)$$

and by the choice of ε_1, we have

$$k_{i+1} \leq (1 + \varepsilon_1)\lambda p_\infty(\lambda)t^d = 2\left(\frac{(1+\varepsilon_1)^2}{1 - \varepsilon_1}\right)\lambda p_\infty(\lambda)s^d \leq 3k_i. \qquad (10.48)$$

By assumption, $k_1 \geq 5$. Set $r(i) = k_i - 1$ for $i = 1, 2, 3, \ldots$. Then, by (10.47) and (10.48), we have for $i = 1, 2, 3, \ldots$ that $2r(i) \leq r(i+1) \leq 4r(i)$.

Take integer $n \geq 1$, and let $I := \max\{i : r(i) \leq n\}$. Using Lemma 10.17, take integers $w_0 \in [1, r(1)]$ and $w_j \in [0, 4], 1 \leq j \leq I$, such that

$$n = w_0 + \sum_{j=1}^{I} w_j(k_j - 1).$$

Set $\tilde{p}_k(\lambda) = p_k(\lambda)/k$. By (10.41), we have $\tilde{p}_{k_i}(\lambda) \geq \exp(-\alpha k_i^{(d-1)/d})$ for each $i = 1, 2, 3, \ldots$. So, by the supermultiplicative inequality (10.12), we have

$$\begin{aligned}\tilde{p}_n(\lambda) &\geq \tilde{p}_{w_0}(\lambda)\tilde{p}_{k_1}(\lambda)^{w_1}\tilde{p}_{k_2}(\lambda)^{w_2}\cdots\tilde{p}_{k_I}(\lambda)^{w_I} \\ &\geq \left(\min_{1\leq i\leq r(1)} \tilde{p}_i(\lambda)\right)\exp\left(-\alpha\sum_{j=1}^{I} w_j k_j^{(d-1)/d}\right).\end{aligned}$$

By the fact that $w_j \leq 4$ for each j and by (10.47), we have

$$\begin{aligned}\sum_{j=1}^{I} w_j k_j^{(d-1)/d} &\leq 4k_I^{(d-1)/d}\sum_{i=0}^{I-1}\left(\frac{k_{I-i}}{k_I}\right)^{(d-1)/d} \\ &\leq 4k_I^{(d-1)/d}\sum_{i=0}^{\infty}\left(\frac{1}{2}\right)^{i(d-1)/d},\end{aligned}$$

which is bounded by a constant times $n^{(d-1)/d}$, since by definition of I, $k_I \leq n+1$. Therefore, there is a constant $\gamma > 0$ such that, for all large enough n,

$$p_n(\lambda) \geq n\left(\min_{1\leq i\leq r(1)} \tilde{p}_i(\lambda)\right)\exp\left(-\gamma n^{(d-1)/d}\right)$$

and this gives us (10.36) as required. □

Proof of Theorem 10.15 We sketch a proof along the lines of that of the analogous result for lattice percolation; see Grimmett (1999, Theorem 8.65). In this argument, $|\cdot|$ denotes cardinality.

Let us say a finite set $\Gamma \subset \mathbb{Z}^d$ *disconnects the origin from infinity* if $\mathbf{0}$ does not lie in the infinite component of $\mathbb{Z}^d \setminus \Gamma$. Let $\mathcal{A}_n$ denote the collection of $*$-connected subsets of $\mathbb{Z}^d$ ('animals') of cardinality n that disconnect the origin from infinity. By a Peierls argument (Lemma 9.3), there exist combinatorial constants κ, γ, and $\beta > \gamma$ such that $\mathcal{A}_n$ has at most $\kappa n^d \gamma^n$ elements, and hence at most β^n elements for n large enough.

Given $M > 0$ (a constant), define variables $X_z, z \in \mathbb{Z}^d$ as follows. For $z \in \mathbb{Z}^d$, let B_z and B_z^+ be concentric boxes (cubes) of side M and $5M$, respectively, centred at Mz, that is, set

$$B_z := B(M) \oplus \{Mz\}, \quad B_z^+ := B(5M) \oplus \{Mz\}. \tag{10.49}$$

Set $X_z = 1$ if (i) there exists a component of $G(\mathcal{H}_\lambda \cap B_z; 1)$ that is crossing for B_z, and (ii) there is only one component of $G(\mathcal{H}_\lambda \cap B_z^+; 1)$ of metric diameter at least $M/3$; set $X_z = 0$ if either of (i) or (ii) fails.

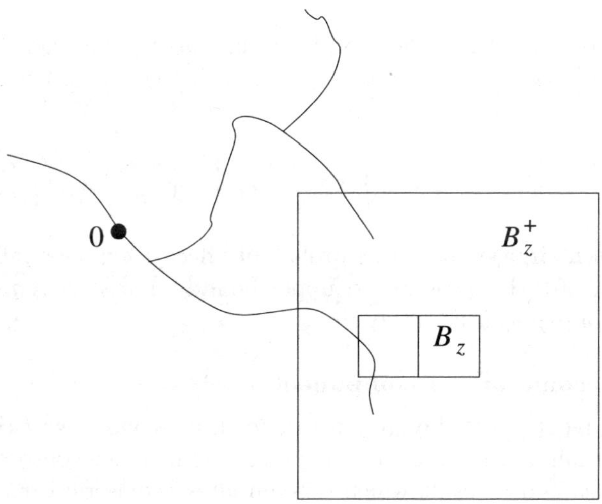

FIG. 10.5. Illustration for proof of Theorem 10.15.

There exists k, independent of M, such that $(X_z, z \in \mathbb{Z}^d)$ is a k-dependent random field. Also, by Proposition 10.13, given $\delta > 0$, we can choose M so that $P[X_z = 1] > 1 - \delta$ for all z. Therefore by Theorem 9.12 we can (and do) choose M so that the process $(X_z, z \in \mathbb{Z}^d)$ stochastically dominates the independent Bernoulli process $(Z_z^p, z \in \mathbb{Z}^d)$, with parameter $p = 1 - (2\beta)^{-1}$, where β is the combinatorial constant described above.

Let $\boldsymbol{C}_0$ be the set of $z \in \mathbb{Z}^d$ such that the cube B_z contains at least one point of the component containing the origin of $G(\mathcal{H}_{\lambda,0}; 1)$. Clearly, $\boldsymbol{C}_0$ is finite if this component is finite. If $\boldsymbol{C}_0$ is finite, then let D_0 be the exterior complement of $\boldsymbol{C}_0$, that is, the infinite connected component of $\mathbb{Z}^d \setminus \boldsymbol{C}_0$. By unicoherence (Lemma 9.6), the set $D_{\text{ext}}(\boldsymbol{C}_0)$ of vertices of D_0 lying adjacent to $\boldsymbol{C}_0$ is $*$-connected, and moreover by an isoperimetric inequality (Lemma 9.9; note that the lower bound in (9.3) does not depend on n), there is a constant $\eta > 0$ such that $|D_{\text{ext}}(\boldsymbol{C}_0)| \geq \eta |\boldsymbol{C}_0|^{(d-1)/d}$.

If $|\boldsymbol{C}_0| \geq 2^d + 1$, then $X_z = 0$ for every $z \in D_{\text{ext}}\boldsymbol{C}_0$. Indeed, if $z \in D_{\text{ext}}\boldsymbol{C}_0$, then there is a component of $G(\mathcal{H}_\lambda \cap B_z^+; 1)$ of metric diameter at least $M/3$, that does not intersect B_z at all (see Fig. 10.5).

Therefore, since $(X_z, z \in \mathbb{Z}^d)$ dominates the Bernoulli process $(Z_z^p, z \in \mathbb{Z}^d)$, with parameter $p = 1 - (2\beta)^{-1}$, we obtain

$$
\begin{aligned}
P[n \leq |\boldsymbol{C}_0| < \infty] &\leq \sum_{m \geq \eta n^{(d-1)/d}} |\mathcal{A}_m| (1-p)^m \\
&\leq \sum_{m \geq \eta n^{(d-1)/d}} \beta^m (1-p)^m \\
&\leq 2(2^{-\eta n^{(d-1)/d}}). \qquad (10.50)
\end{aligned}
$$

Also, if V is the order of the component containing the origin of $G(\mathcal{H}_{\lambda,0};1)$, then, by (10.9), there is a constant $\gamma_1 > 0$ such that, for all n and any $K \geq e^2 M^d \lambda$,

$$P[\{|C_0| < n\} \cap \{V > Kn+1\}] \leq \gamma_1^n \exp\left\{ -\left(\frac{Kn}{2}\right) \log\left(\frac{K}{M^d \lambda}\right)\right\},$$

so if K is chosen large enough, this probability decays exponentially in n. Combined with (10.50), this gives us an upper bound for $P[V > Kn + 1]$ with the required rate of exponential decay. □

10.5 The second-largest component

The following result gives the growth rate for the second-largest component for a geometric graph on the points of a supercritical homogeneous Poisson process on a cube. This is one result which differentiates geometric from Erdös–Rényi random graphs: for the Erdös–Rényi random graph $G(n, c/n)$ with $c > 1$, the order of the second-largest component grows as a constant times $\log n$ (see Janson *et al.* (2000, Theorem 5.4)), whereas for the geometric graph on a supercritical Poisson process the order of the second-largest component grows like a larger power of the logarithm of the number of points.

The proof, and also later arguments, use the following notation. For odd integer n, set

$$B'_{\mathbb{Z}}(n) := ([-(n-1)/2, (n-1)/2] \cap \mathbb{Z})^d, \tag{10.51}$$

a translate of the lattice cube $B_{\mathbb{Z}}(n)$ defined at (9.1).

Theorem 10.18 *Suppose $d \geq 2$ and $\lambda > \lambda_c$. Then there exist constants c_1, c_2 such that with probability tending to 1 as $s \to \infty$,*

$$c_1 < \frac{L_2(G(\mathcal{H}_{\lambda,s};1))}{(\log s)^{d/(d-1)}} < c_2. \tag{10.52}$$

Proof By Lemma 10.16, there are strictly positive constants α, s_1, and c_0 such that for all $t \geq s_1$ there exists $k = k(t) \in (c_0 t, 2c_0 t)$ satisfying

$$p^*_{k(t)}(\lambda) \geq \exp(-\alpha k(t)^{(d-1)/d}) \geq \exp(-\alpha(2c_0 t)^{(d-1)/d}). \tag{10.53}$$

Given s, let $\{B_{1,s}, B_{2,s}, \ldots, B_{m(s),s}\}$ be a collection of disjoint balls of radius $2(\alpha^{-1}\log s)^{d/(d-1)}$ contained in $B(s)$, of maximal cardinality. Then, clearly

$$\liminf_{s\to\infty} \left\{ m(s) \left(\frac{(\log s)^{d/(d-1)}}{s} \right)^d \right\} > 0. \tag{10.54}$$

Let $x_{i,s}$ denote the centre of the ball $B_{i,s}$. Let $A_{i,s}$ be the event that there exists a component of $G(\mathcal{H}_\lambda \cap B_{i,s};1)$ of order $k((2c_0)^{-1}(\alpha^{-1}\log s)^{d/(d-1)})$ having at least one vertex in the rectilinear unit cube centred at $x_{i,s}$. Then, for all large enough s and for $i = 1, 2, \ldots, m(s)$,

$$P[A_{i,s}] \geq \exp(-\log s) = s^{-1}.$$

Also, the events $A_{i,s}, i = 1, 2, \ldots, m(s)$, are independent, since they are determined by the Poisson configurations in disjoint balls, so that

$$P\left[\bigcap_{i=1}^{m(s)} A_{i,s}^c\right] \leq (1 - s^{-1})^{m(s)} \leq \exp(-s^{-1}m(s)),$$

which tends to zero by (10.54). But, if for any i the event $A_{i,s}$ occurs, and if also there is a component that is crossing for $B(s)$, then

$$L_2(G(\mathcal{H}_{\lambda,s}; 1)) \geq k((2c_0)^{-1}(\alpha^{-1}\log s)^{d/(d-1)}) \geq \tfrac{1}{2}(\alpha^{-1}\log s)^{d/(d-1)}.$$

This gives us the lower bound in (10.52).

For the upper bound, the proof follows a plan similar to the outline of the proof of Theorem 10.15 described at the start of Section 10.4. Let W_s be the number of points of $\mathcal{H}_{\lambda,s}$ which lie in a component of $G(\mathcal{H}_{\lambda,s}; 1)$ with more than $c_2(\log s)^{d/(d-1)}$ elements, and metric diameter less than $s^{1/2}$. By Theorem 10.9 it suffices to prove that $P[W_s \geq 1] \to 0$ as $s \to \infty$, so by Markov's inequality, it suffices to prove that $E[W_s] \to 0$ as $s \to \infty$. By Theorem 1.6,

$$E[W_s] = \int_{B(s)} P[\{|V_x| \geq c_2(\log s)^{d/(d-1)}\} \cap \{\mathrm{diam} V_x \leq s^{1/2}\}]\mathrm{d}x, \quad (10.55)$$

where V_x denotes the component containing x of $G(\mathcal{H}_{\lambda,s} \cup \{x\}; 1)$, with its order denoted $|V_x|$ and its metric diameter denoted $\mathrm{diam} V_x$.

With the lattice cube $B'_{\mathbb{Z}}(n)$ defined at (10.51), by Theorem 9.8 we can find $p_0 \in (0, 1)$ such that for Bernoulli site percolation on $B'_{\mathbb{Z}}(n)$ with parameter $p \geq p_0$, there is a big open cluster $\boldsymbol{C}_{\mathrm{b}}$ with at least $\frac{3}{4}n^d$ elements, except on an event with probability decaying exponentially in n^{d-1}. Take $p_1 \in [p_0, 1)$ such that $2^{3^d}(1 - p_1) \leq \frac{1}{2}$ and

$$(2d)^{-1}(1 - (\tfrac{2}{3})^{1/d}) \log(2^{3^d}(1 - p_1)) < -(2d + 2). \quad (10.56)$$

Given $M > 0$, let the random field $(X_z, z \in \mathbb{Z}^d)$ be defined as follows. Define concentric cubes B_z, B_z^+ centred at the point Mz, as at (10.49), by $B_z := B(M) \oplus \{Mz\}$ and $B_z^+ := B(5M) \oplus \{Mz\}$. Set $X_z = 1$ if (i) there exists a path in $G(\mathcal{H}_\lambda \cap B_z; 1)$ that is crossing for B_z, and (ii) for every $z' \in \mathbb{Z}^d$ with $\|z' - z\|_\infty \leq 2$, the component of $G(\mathcal{H}_\lambda \cap B_{z'}^+; 1)$ of metric diameter at least $M/3$ is unique; set $X_z = 0$ if either (i) or (ii) fails.

There exists k, independent of M, such that $(X_z, z \in \mathbb{Z}^d)$ is a k-dependent random field. Also, by Theorem 10.9, given $\delta > 0$, we can choose M_δ so that as

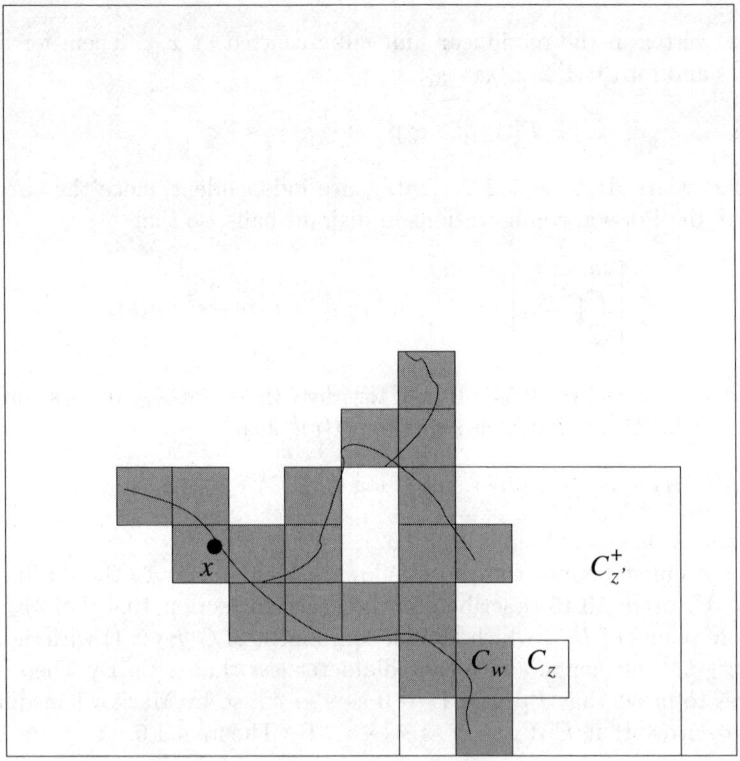

FIG. 10.6. If $z \in DC_x$ then $X_z = 0$. The centres of the shaded squares are at $\{My : y \in \boldsymbol{C}_x\}$.

long as $M \geq M_\delta$, $P[X_z = 1] > 1 - \delta$ for all z. Therefore, by Theorem 9.12 we can (and do) choose $M_0 \geq 1$ such that as long as $M \geq M_0$ we have the stochastic domination

$$(X_z, z \in \mathbb{Z}^d) \geq_{\text{st}} (Z_z^{p_1}, z \in \mathbb{Z}^d), \tag{10.57}$$

where $(Z_z^{p_1}, z \in \mathbb{Z}^d)$ is a family of independent variables taking the value 1 with probability p_1 and zero otherwise.

Recall the notation $G_{\mathbb{Z}}(\mathcal{B})$ from Section 9.3. For large enough s we take $M(s)$ so that $M_0 \leq M(s) \leq 2M_0$ and also $n(s) := s/M(s)$ is an odd integer. Then, except on an event with probability decaying exponentially in s^{d-1}, the graph $G_{\mathbb{Z}}(\{z \in B'_{\mathbb{Z}}(n(s)) : X_z = 1\})$ has a big component $\boldsymbol{C}_{\text{b}}$ of order more than $\frac{3}{4}n(s)^d$.

Given $x \in \mathbb{R}^d$, let $\boldsymbol{C}_x$ denote the set of $y \in B'_{\mathbb{Z}}(n(s))$ such that the cube C_y contains at least one vertex of the component V_x of $G(\mathcal{H}_{\lambda,s} \cup \{x\}; 1)$, corresponding to centres of shaded squares in Fig. 10.6. Then $\boldsymbol{C}_x$ is $*$-connected. Let DC_x denote the set of $z \in B'_{\mathbb{Z}}(n(s)) \setminus \boldsymbol{C}_x$ lying adjacent to $\boldsymbol{C}_x$.

Suppose that $|\boldsymbol{C}_x| > 3^d$. We assert that if $z \in D\boldsymbol{C}_x$ then $X_z = 0$. For if $X_z = 1$ there would be a component containing X of $G(\mathcal{H}_\lambda \cap B_z; 1)$ that was crossing for B_z and also a vertex $w \in \boldsymbol{C}_x$ with $\|w - z\|_\infty \leq 1$. But then there would exist z' with $\|z' - z\|_\infty \leq 2$ such that $B^+_{z'}$ contains $B_{w'}$ for all $w' \in B'_{\mathbb{Z}}(n(s))$ with $\|w' - z\|_\infty \leq 2$, but is itself contained in $B(s)$ (we can take $z' = z$, except when z lies at or adjacent to the boundary of $B'_{\mathbb{Z}}(n(s))$; see Fig. 10.6). Then the crossing component of $G(\mathcal{H}_\lambda \cap B_z; 1)$, and a part of V_x, would be part of disjoint components in $G(\mathcal{H}_\lambda \cap B^+_{z'}; 1)$, both of metric diameter at least $M/3$. which would contradict condition (ii) for $X_z = 1$. This justifies the assertion, from which it follows that each cluster in $\{z \in B'_{\mathbb{Z}}(n(s)) : X_z = 1\}$ is either contained in $\boldsymbol{C}_x$ or disjoint from $\boldsymbol{C}_x$.

Let $\Lambda_1, \ldots, \Lambda_l$ denote the connected components of $B'_{\mathbb{Z}}(n(s)) \setminus \boldsymbol{C}_x$. If the order $|\boldsymbol{C}_x|$ of $\boldsymbol{C}_x$ satisfies

$$(\log s)^{d/(d-1)} \leq |\boldsymbol{C}_x| \leq \tfrac{1}{2} n(s)^d;$$

then $\boldsymbol{C}_{\mathrm{b}}$ must be disjoint from $\boldsymbol{C}_x$ (since it is too big to be contained in $\boldsymbol{C}_x$), so that one of the components Λ_i, say Λ_1, contains $\boldsymbol{C}_{\mathrm{b}}$. In this case, the sets Λ_1 and $\boldsymbol{C}_x \cup \Lambda_2 \cup \cdots \cup \Lambda_l$ are disjoint complementary connected subsets of $B'_{\mathbb{Z}}(n(s))$, so by unicoherence (Lemma 9.6), the set $D_{\mathrm{ext}}\boldsymbol{C}_x$ of vertices of Λ_1 lying adjacent to $B'_{\mathbb{Z}}(n(s)) \setminus \Lambda_1$ is $*$-connected, and by the isoperimetric inequality (Lemma 9.9), its cardinality satisfies

$$|D_{\mathrm{ext}}\boldsymbol{C}_x| \geq (2d)^{-1}(1 - (\tfrac{2}{3})^{1/d})|\boldsymbol{C}_x|^{(d-1)/d}.$$

Let $\mathcal{A}_{m,s}$ denote the collection of $*$-connected subsets of $B'_{\mathbb{Z}}(n(s))$ of cardinality m. By the above, if $n(s)^d/2 \geq |\boldsymbol{C}_x| \geq (\log s)^{d/(d-1)}$, then there exists a set A in $\mathcal{A}_{m,s}$ such that $X_z = 0$ for all $z \in A$, for some $m \geq \beta \log s$, with $\beta := (2d)^{-1}(1 - (\frac{2}{3})^{1/d})$. Hence,

$$P\left[(\log s)^{d/(d-1)} \leq |\boldsymbol{C}_x| \leq n(s)^d/2\right] \leq P\left[\bigcup_{m \geq \beta \log s} \bigcup_{\sigma \in \mathcal{A}_{m,s}} \{X_z = 0, \forall\, z \in \sigma\}\right].$$

By a Peierls argument (Corollary 9.4), the cardinality $|\mathcal{A}_{m,s}|$ of $\mathcal{A}_{m,s}$ is bounded by $(n(s))^d\gamma^m$, with $\gamma := 2^{3^d}$. If $\mathrm{diam}(V_x) \leq s^{1/2}$ then $|\boldsymbol{C}_x| \leq n(s)^2/2$, so that by (10.57),

$$\begin{aligned} &P[\{\mathrm{diam}(V_x) \leq s^{1/2}\} \cap \{|\boldsymbol{C}_x| \geq (\log s)^{d/(d-1)}\}] \\ &\quad \leq n(s)^d \sum_{m \geq \beta \log s} \gamma^m (1 - p_1)^m \leq 2M_0^{-d} s^d (\gamma(1 - p_1))^{\beta \log s} \\ &\quad = 2M_0^{-d} s^d s^{\beta \log(\gamma(1-p_1))} \leq 2M_0^{-d} s^{-2-d}, \end{aligned} \tag{10.58}$$

the last inequality coming from (10.56).

By the same argument as at (10.9) (with $\varepsilon = M$), provided c_2 is chosen so that $c_2 \geq e^2(2M_0)^d\lambda$ and also $c_2 \log(c_2/((2M_0)^d\lambda)) > 4\log\gamma$, we have for some $\delta > 0$ that

$$P[\{|C_x| < (\log s)^{d/(d-1)}\} \cap \{|V_x| \geq c_2(\log n)^{d/(d-1)}\}] \\ \leq \exp(-\delta(\log s)^{d/(d-1)}), \tag{10.59}$$

which is $o(s^{-d})$. Combining (10.58), (10.59), and (10.55) gives us the result. □

10.6 Large deviations in the supercritical regime

Having given a law of large numbers for the order of the largest component of $G(\mathcal{H}_{\lambda,s};1)$ in Theorem 10.9, we now show that the probability of large deviations from its limiting value decays exponentially in s^{d-1}.

Theorem 10.19 *Suppose $d \geq 2$, and $\lambda > \lambda_c$. Suppose $0 < \varepsilon < \frac{1}{2}$. Let E_s be the event that* (i) $L_2(G(\mathcal{H}_{\lambda,s};1)) < \varepsilon\lambda p_\infty(\lambda)s^d$ *and* (ii)

$$(1-\varepsilon)\lambda p_\infty(\lambda) \leq s^{-d}L_1(G(\mathcal{H}_{\lambda,s};r_s)) \leq (1+\varepsilon)\lambda p_\infty(\lambda). \tag{10.60}$$

Then there exist constants $c_1 > 0$ and $s_0 > 0$, such that

$$P[E_s^c] \leq \exp(-c_1 s^{d-1}), \quad s \geq s_0. \tag{10.61}$$

Moreover, there is a lower *bound of the form* $\exp(-c_2 s^{d-1})$ *(with $c_2 > 0$) for the probability that property* (i) *fails.*

The next result characterizes the largest component in terms of metric diameter rather than order (with a weaker large deviations bound).

Theorem 10.20 *Suppose that $\lambda > \lambda_c, \varepsilon \in (0, \frac{1}{2})$, and that $(\phi_s, s \geq 0)$ satisfies $(\phi_s/\log s) \to \infty$ as $s \to \infty$, and $\phi_s \leq s/2$ for all s. Let G_s denote the event that there exists a unique component $C_b(B(s))$ of $G(\mathcal{H}_{\lambda,s};1)$ of metric diameter at least ϕ_s. Let E'_s be the event that G_s holds and additionally the order of $C_b(B(s))$ satisfies*

$$(1-\varepsilon)\lambda p_\infty(\lambda) \leq s^{-d}\text{order}(C_b(B(s))) \leq (1+\varepsilon)\lambda p_\infty(\lambda). \tag{10.62}$$

Then there exist constants $c_1 > 0, c_2 > 0, s_0 > 0$ such that

$$\exp(-c_1\phi_s) \leq P[G_s^c] \leq P[(E'_s)^c] \leq \exp(-c_2\phi_s), \quad s \geq s_0. \tag{10.63}$$

The proof of Theorem 10.19 uses a block construction similar to the ones used in previous two sections, as outlined at the start of Section 10.4. The 'extra Poisson estimate' needed in this case is more complex than in previous cases and is based on the following result.

Proposition 10.21 *For $\mu > 0$, let $Y, Y_1, Y_2, Y_3, \ldots$ be independent Poisson random variables with parameter μ; let $Y^n_{(1)}, Y^n_{(2)}, \ldots, Y^n_{(n)}$ denote the order statistics of $\{Y_1, \ldots, Y_n\}$ (in* decreasing *order). Suppose $0 < \delta < \frac{1}{2}$. Then there exists $\mu_0 = \mu_0(\delta) > 0$ such that, for any $\mu \geq \mu_0$,*

$$\limsup_{n\to\infty} n^{-1} \log P\left[\sum_{1\leq i\leq \delta n} Y^n_{(i)} > 4\mu\delta n\right] < 0.$$

Proof Choose μ_0 so that for $\mu > \mu_0$, we have $P[\mathrm{Po}(\mu) = k] \leq \delta/2$ for all $k \in \mathbb{Z}$. Now fix $\mu > \mu_0$. We can then choose u_μ with

$$2\delta < P[Y > u_\mu] < P[Y > u_\mu - 1] \leq 3\delta\,. \tag{10.64}$$

By Lemma 1.1, there exists $c_1 > 0$ such that, for large n,

$$P[Y^n_{(\lfloor n\delta \rfloor)} \leq u_\mu] = P\left[\sum_{i=1}^{n} \mathbf{1}_{\{Y_i > u_\mu\}} < \lfloor n\delta \rfloor\right] \leq \exp(-c_1 n). \tag{10.65}$$

This means that with high probability all the $\lfloor n\delta \rfloor$ largest values of $Y_1, \ldots, Y_n$ are larger than u_μ. We now show that the sum of all values exceeding u_μ is smaller than $4n\delta\mu$ up to large deviations of order n. This will complete the proof.

The random variable $Y\mathbf{1}_{\{Y > u_\mu\}}$ has a well-behaved logarithmic moment generating function, and by (10.64) its mean satisfies

$$E[Y\mathbf{1}_{\{Y>u_\mu\}}] = \mu P[u_\mu - 1 < Y] \leq 3\delta\mu, \tag{10.66}$$

where the equality can be verified by direct computation. By Cramér's large deviations theorem (see, e.g., (9.3) and (9.4) of Durrett (1991, Chapter 1)), there is a constant c_2 such that, for large n,

$$P\left[\sum_{i=1}^{n} Y_i \mathbf{1}_{\{Y_i > u_\mu\}} > 4n\delta\mu\right] \leq \exp(-c_2 n). \tag{10.67}$$

Therefore, by (10.65) we have

$$\begin{aligned}
&P\left[\sum_{1\leq i\leq n\delta} Y^n_{(i)} > 4\delta n\mu\right] \\
&\leq P\left[\left\{\sum_{1\leq i\leq n\delta} Y^n_{(i)} > 4\delta n\mu\right\} \cap \{Y^n_{(\lfloor n\delta\rfloor)} > u_\mu\}\right] + \exp(-c_1 n) \\
&\leq P\left[\sum_{1\leq k\leq n} Y_k \mathbf{1}_{\{Y_k > u_\mu\}} > 4\delta n\mu\right] + \exp(-c_1 n) \leq \exp(-c_2 n) + \exp(-c_1 n),
\end{aligned}$$

and the result follows. □

Proof of Theorem 10.19 Given $\varepsilon \in (0, \frac{1}{2})$, choose $\delta > 0$ with $(1-\delta)^2 > 1-\varepsilon$ and with $(2^{d+2}+2)\delta < \varepsilon p_\infty(\lambda)$. Also, let $\mu_0 = \mu_0(\delta)$ be given by Proposition 10.21.

Given $M > 0$, define blocks (i.e. translates of $B(M)$) B_z, $z \in \mathbb{Z}^d$, by $B_z := B(M) \oplus \{Mz\}$. Also, set $B_z^+ := B(3M) \oplus \{Mz\}$ and $B_z^- := B(M - 2M^{1/2}) \oplus \{Mz\}$. For $z \in \mathbb{Z}^d$, set $X_z = 1$ if (i) there is a unique component of $G(\mathcal{H}_\lambda \cap B_z; 1)$ that is crossing for B_z, denoted $C_{\rm b}(B_z)$; (ii) for each $y \in \mathbb{Z}^d$ with $\|y - z\|_\infty \leq 1$, at most one component of $G(\mathcal{H}_\lambda \cap B_y^+; 1)$ has metric diameter greater than $M^{1/2}$; (iii) the order of $C_{\rm b}(B_z)$ satisfies

$$(1-\delta)\lambda p_\infty(\lambda) < M^{-d}\mathrm{order}(C_{\rm b}(B_z)) < (1+\delta)\lambda p_\infty(\lambda), \qquad (10.68)$$

(iv) no other component has order greater than δM^d, and (v)

$$\mathcal{H}_\lambda(B_z \setminus B_z^-) \leq \delta \lambda M^d. \qquad (10.69)$$

Set $X_z = 0$ if any of conditions (i)–(v) fail.

Note that $P[X_z = 1]$ depends on M but not on z, and if this probability is denoted r_M then $r_M \to 1$ as $M \to \infty$, since the probability that condition (v) fails tends to zero by Markov's inequality, while the probability that any of conditions (i)–(iv) fail tends to zero by Theorem 10.9.

Recall from (10.51) that for odd $m \in \mathbb{N}$, the set $B'_{\mathbb{Z}}(m)$ is the lattice box of side m centred at the origin. Given $M \in \mathbb{R}$ and odd $m \in \mathbb{N}$, let $A_{M,m}$ denote the event that in the renormalized process $(X_z, z \in \mathbb{Z}^d)$ with block size M, there is a lattice cluster $\boldsymbol{C}$ in $\{z \in B'_{\mathbb{Z}}(m) : X_z = 1\}$, with cardinality $|\boldsymbol{C}| > (1-\delta)m^d$. By Theorem 9.8 and the fact that $r_M \to 1$, we can choose $c_3 > 0$, $M_0 > \max((2d/\delta)^2, (\mu_0/\lambda)^{1/d}/2)$, and $m_0 \in \mathbb{N}$, such that

$$P[A^{\rm c}_{M,m}] \leq \exp(-c_3 m^{d-1}), \qquad M \in [M_0, \infty), \;\; m \geq m_0. \qquad (10.70)$$

Given M with $M_0 \leq M \leq 2M_0$, set $Y_z := \mathcal{H}_\lambda(B_z)$ and denote by $Y_{(1)}, \ldots, Y_{(m^d)}$ the order statistics (in decreasing order) of the Poisson variables $(Y_z)_{z \in B'_{\mathbb{Z}}(m)}$. Define the event $H_{M,m}$ by

$$H_{M,m} := \left\{ \sum_{1 \leq i \leq \delta m^d} Y_{(i)} \leq 4\delta\lambda m^d (2M)^d \right\}. \qquad (10.71)$$

Let $(Z_z)_{z \in B'_{\mathbb{Z}}(m)}$ be independent $\mathrm{Po}(\lambda(2M_0)^d)$ variables, with order statistics $Z_{(1)}, \ldots, Z_{(m^d)}$. Then Z_z stochastically dominates Y_z. By Proposition 10.21, there exists $m_1 \in \mathbb{N}$ and $c_4 > 0$, such that

$$P[H^{\rm c}_{M,m}] \leq P\left[\sum_{1 \leq i \leq \delta m^d} Z_{(i)} > 4\delta\lambda m^d (2M_0)^d\right]$$
$$\leq \exp(-c_4 m^d), \qquad m \geq m_1, \;\; M \in [M_0, 2M_0]. \qquad (10.72)$$

Set $m_2 := \max(2m_0, 2m_1, 4)$. For $s \geq m_2 M_0$ (not necessarily an integer), choose an odd integer $m(s)$ so that $s/m(s) \in [M_0, 2M_0]$, and let $M(s) := s/m(s)$. Then

$s = m(s)M(s)$, and $\cup_{z \in B'_{\mathbb{Z}}(m)} B_z = B(s)$. Define the event $A'_{M,m} := A_{M,m} \cap H_{M,m}$. By (10.70) and (10.72),

$$P[(A'_{M,m(s)})^{\mathrm{c}}] \leq \exp(-c_3 m(s)^{d-1}) + \exp(-c_4 m(s)^d) \\ \leq 2\exp(-c_5 s^{d-1}), \tag{10.73}$$

for $c_5 := \min(c_3, c_4)/(2M_0)^{d-1}$. Therefore, to prove (10.61), it suffices to prove that with E_s as defined there,

$$A'_{M,m} \subseteq E_s. \tag{10.74}$$

If $X_z = 1$, then by (10.68), the graph $G(\mathcal{H}_\lambda \cap B_z; 1)$ contains a unique big component of approximately the expected size, denoted $C_{\mathrm{b}}(B_z)$ which we abbreviate to C_z. Condition (ii) in the definition of the event $\{X_z = 1\}$ ensures that if $z \in B'_{\mathbb{Z}}(m(s))$ and $y \in B'_{\mathbb{Z}}(m(s))$ are adjacent (i.e. $\|z - y\|_1 = 1$) and satisfy $X_z = X_y = 1$, then the components C_z and C_y are part of the same component of $G(\mathcal{H}_{\lambda,s}; 1)$. Therefore, if $A'_{M,m}$ occurs, then $\cup_{z \in \boldsymbol{C}} C_z$ is connected, and

$$(1-\delta)^2 \lambda p_\infty(\lambda) s^d \leq \sum_{z \in \boldsymbol{C}} \mathcal{H}_\lambda(C_z) \leq (1+\delta)\lambda p_\infty(\lambda) s^d. \tag{10.75}$$

Let C denote the vertex set of the component of $G(\mathcal{H}_{\lambda,s}; 1)$ which contains $\cup_{z \in \boldsymbol{C}} C_z$. We now estimate the size of the set $D := C \setminus \cup_{z \in \boldsymbol{C}} C_z$. Note that

$$D \subseteq \left(\bigcup_{z \in B'_{\mathbb{Z}}(m) \setminus \boldsymbol{C}} B_z \right) \cup \bigcup_{z \in \boldsymbol{C}} ((C \cap B_z) \setminus C_z). \tag{10.76}$$

By condition (ii) in the definition of event $\{X_z = 1\}$, for $z \in \boldsymbol{C}$, the set $(C \setminus C_z) \cap B_z$ is contained in $B_z \setminus B_z^-$, a set which has at most $\delta \lambda M^d$ points of $\mathcal{H}_\lambda$ in it by (10.69). It follows that

$$\mathrm{card}\left(\bigcup_{z \in \boldsymbol{C}} (C \cap B_z \setminus C_z) \right) \leq \delta \lambda s^d. \tag{10.77}$$

By (10.71), (10.76), and (10.77), if $A'_{M,m}$ occurs, then the total number of points of $\mathcal{H}_\lambda$ in D is bounded by $(2^{d+2} + 1)\lambda \delta s^d$. Thus, by (10.75),

$$(1-\delta)^2 s^d \lambda p_\infty(\lambda) \leq \mathrm{card}(C) \leq \lambda(p_\infty(\lambda) + (2^{d+2} + 2)\delta) s^d \ . \tag{10.78}$$

Hence, by the definition of δ, $\mathrm{card}(C)$ lies in the range $(1 \pm \varepsilon)\lambda p_\infty(\lambda)$.

We now check that all other components are small. Every component other than C is contained either in a single cube B_z, or in the union of $\cup_{z \notin \boldsymbol{C}} B_z$ and $\cup_{z \in \boldsymbol{C}} (B_z \setminus B_z^-)$. Therefore, if $A'_{M,m}$ occurs, then no component other than C has order more than $(2^{d+2} + 1)\delta \lambda s^d$, and hence, by the choice of δ, $L_2(G(\mathcal{H}_{\lambda,s}; 1)) < \varepsilon \lambda p_\infty(\lambda) s^d$. Then (10.74) follows, and hence, so does (10.61).

To prove the lower bound on the probability that condition (i) for the event E_s fails, take $\varepsilon = \frac{1}{2}$ in Lemma 10.16. Let $F_s^{(1)}, F_s^{(2)}$, and $F_s^{(3)}$ be as described in the proof of that result. Then $F_{s/4}^{(1)} \cap F_{s/4}^{(2)} \cap F_{s/4}^{(3)}$ is an event determined by the configuration of points of $\mathcal{H}_\lambda$ in the box $B(s/4+2)$, which guarantees the existence of a component of $G(\mathcal{H}_\lambda;1)$ that is contained in that box $B(s/4+2)$ and isolated from the complement of that box, of order at least $\frac{1}{2}\lambda p_\infty(\lambda)(s/4)^d$. By (10.45), for suitable $\alpha_2 > 0$ the probability of this event exceeds $\exp(-\alpha_2 s^{d-1})$.

Take a second box, also of side $s/4+2$, contained in $B(s)$ and disjoint from $B(s/4+2)$. Clearly, the probability that there exists a component of $G(\mathcal{H}_\lambda;1)$ that is of order greater than $\frac{1}{2}\lambda p_\infty(\lambda)(s/4)^d$ and is contained in the second box is also bounded below by $\exp(-\alpha_2 s^{d-1})$. Therefore, by independence the probability that there are disjoint components in both of these two boxes and both of order greater than $\frac{1}{2}\lambda p_\infty(\lambda)(s/4)^d$, exceeds $\exp(-2\alpha_2 s^{d-1})$. This gives us the lower bound. □

Proof of Theorem 10.20 The upper bound in (10.63) follows at once from Theorem 10.19 and Proposition 10.13. For the lower bound, take $\beta_1 := (5d)^{-1}$, so that $\mathrm{diam}(B(\beta_1)) < \frac{1}{4}$. For $z \in \mathbb{Z}^d$, let $B_z(\beta_1)$ denote the translate $B(\beta_1) \oplus \{\beta_1 z\}$ of $B(\beta_1)$.

Let $Q_s \subseteq B(s)$ be given by the union of an arbitrary row of $\lceil \phi_s/\beta_1 \rceil + 1$ neighbouring cubes of the form $B_z(\beta_1)$ in a straight line. Let Q'_s be defined similarly, with $\mathrm{dist}(Q_s, Q'_s) > 1$. Let E_{Q_s} denote the event that each cube in the row contains at least one Poisson point but that there are no points of $\mathcal{H}_{\lambda,s}$ in the 1-neighbourhood of Q_s, other than those in Q_s itself. Then there exists $c > 0$ such that

$$P[E_{Q_s} \cap E_{Q'_s}] \geq \exp(-c\phi_s).$$

Also, the occurrence of $E_{Q_s} \cap E_{Q'_s}$ implies that $G(\mathcal{H}_{\lambda,s} \cap Q_s;1)$ and $G(\mathcal{H}_{\lambda,s} \cap Q'_s;1)$ are disjoint components of $G(\mathcal{H}_{\lambda,s};1)$, each of metric diameter greater than ϕ_s. □

10.7 Fluctuations of the giant component

This section contains a central limit theorem for the order of the largest component $L_1(G(\mathcal{H}_{\lambda,s};1))$, $\lambda > \lambda_c$. Later on, we shall de-Poissonize this result to deduce a central limit theorem for $L_1(G(\mathcal{X}_n;r_n))$ in the case where the underlying distribution is uniform on the unit cube and $nr_n^d = \lambda > \lambda_c$, that is, in the supercritical thermodynamic limit. These central limit theorems are analogous to known central limit theorems for the giant component of the independent random graph $G(n,p)$, as discussed in Barraez *et al.* (2000).

Let H be the real-valued functional defined for all finite subsets of $\mathbb{R}^d$ by

$$H(\mathcal{X}) := L_1(G(\mathcal{X};1)). \tag{10.79}$$

Then H is translation-invariant, meaning that $H(\mathcal{X} \oplus \{y\}) = H(\mathcal{X})$ for all $\mathcal{X} \subset \mathbb{R}^d$ and all $y \in \mathbb{R}^d$. By scaling, the following central limit theorem for $H(\mathcal{H}_{\lambda,s})$

implies a central limit theorem for $L_1(G(\mathcal{P}_n; (\lambda/n)^{1/d}))$, when the underlying density function is $f = f_{\mathrm{U}}$.

Theorem 10.22 *Suppose* $d > 2$ *and* $\lambda > \lambda_c$. *There exists a constant* $\sigma^2 = \sigma^2(\lambda) \geq 0$ *such that, as* $s \to \infty$,

$$s^{-d}\mathrm{Var}(H(\mathcal{H}_{\lambda,s})) \to \sigma^2 \tag{10.80}$$

and

$$s^{-d/2}(H(\mathcal{H}_{\lambda,s}) - EH(\mathcal{H}_{\lambda,s})) \xrightarrow{\mathcal{D}} \mathcal{N}(0, \sigma^2). \tag{10.81}$$

Later on, in Section 11.5, we shall verify that σ^2 is strictly positive.

In the proof, we shall need to consider translates of the cubes $B(s)$. Let $\mathcal{B}$ be the collection of all regions $A \subset \mathbb{R}^d$ of the form $A = B(s) \oplus \{x\}$ with $x \in \mathbb{R}^d$, $s \geq 1$; we shall call such regions *boxes*. We assume for the remainder of this section that $d \geq 2$ and $\lambda > \lambda_c$.

The first step is a uniformly exponentially decaying bound on the probability that there are two large components meeting the unit cube.

Lemma 10.23 *For each box* $B \in \mathcal{B}$ *and* $r > 0$, *let* $E'(B; r)$ *denote the event that there are two distinct components in* $G(\mathcal{H}_\lambda \cap B \setminus B(1); 1)$ *which both have at least one vertex in* $B(3)$ *but both have metric diameter greater than* r. *Then*

$$\limsup_{r \to \infty}(r^{-1} \sup\{\log P[E'(B; r)] : B \in \mathcal{B}\}) < 0.$$

Proof Assume that B has side length greater than r/d (otherwise, trivially, $P[E'(B; r)] = 0$). Assume also that the centre of B lies in the closed positive orthant $[0, \infty)^d$ (other cases are treated similarly). Take a box of side r/d centred at the origin, and if it extends beyond the boundary of B, then translate it just enough so it does not, to obtain a box $B' \subset B$. In other words, if B is the product $\prod_{i=1}^d [b_i, b_i + s]$, with $s > r/d$, then let B' be the product of intervals $\prod_{i=1}^d [a_i, a_i + r/d]$, with $a_i = \max\{-r/(2d), b_i\}$ (see Fig. 10.7).

Let $E''(B; r)$ be the event that there exist two disjoint components of $G(\mathcal{H}_\lambda \cap B; 1)$, each of which has at least one vertex in $B(3)$ and also at least one vertex in $B \setminus B'$. Since any subset of B' has diameter at most r, if $E'(B; r)$ occurs and also $\mathcal{H}_\lambda(B(1)) = 0$ then $E''(B; r)$ occurs; hence

$$e^{-\lambda} P[E'(B; r)] \leq P[E''(B; r)]. \tag{10.82}$$

The distance from $B(3)$ to $B \setminus B'$ is at least $(r/2d) - \frac{3}{2}$, so that if $E''(B; r)$ occurs then there exist disjoint components of $G(\mathcal{H}_\lambda \cap B'; 1)$, both of which have metric diameter at least $(r/2d) - \frac{5}{2}$. The chance of this occurring decays exponentially in r by Proposition 10.13. Then the result follows by (10.82). □

For $z \in \mathbb{Z}^d$, set $Q_z := B(1) \oplus \{z\}$, the unit cube centred at z. The proof of Theorem 10.22 involves comparing the homogeneous Poisson process $\mathcal{H}_\lambda$ with a modification of $\mathcal{H}_\lambda$ created by replacing those Poisson points lying in a unit

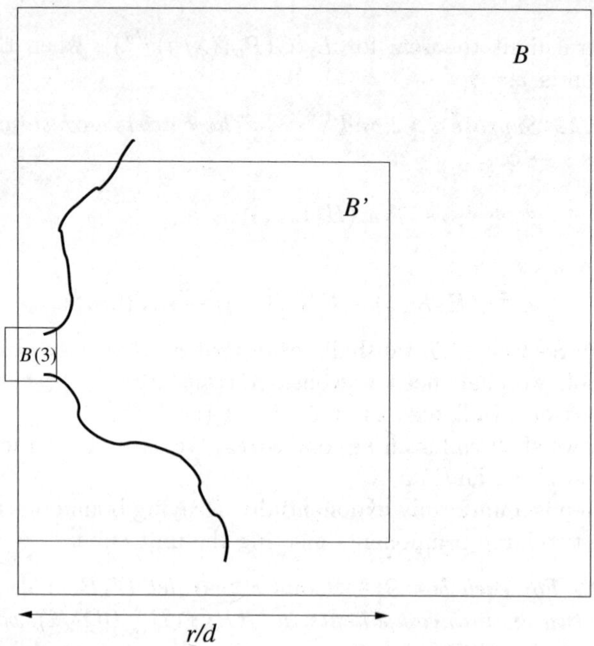

FIG. 10.7. Illustration for the proof of Lemma 10.23.

cube with an independent Poisson process on that unit cube, as follows. Let $\mathcal{H}'_\lambda$ be an independent copy of the Poisson process $\mathcal{H}_\lambda$. For $x \in \mathbb{Z}^d$, set

$$\mathcal{H}''_\lambda(x) := (\mathcal{H}_\lambda \setminus Q_x) \cup (\mathcal{H}'_\lambda \cap Q_x),$$

and for $A \in \mathcal{B}$, define $\Delta_x(A)$ (the effect on $H(\mathcal{H}_\lambda \cap A)$ of this modification) by

$$\Delta_x(A) := H(\mathcal{H}''_\lambda(x) \cap A) - H(\mathcal{H}_\lambda \cap A). \tag{10.83}$$

The next step is to check a *stabilization* condition, which says, loosely speaking, that the effect of changing $\mathcal{H}_\lambda$ to $\mathcal{H}''_\lambda(x)$ is local. Given $x \in \mathbb{Z}^d$, define the random variable $\Delta_x(\infty)$ as follows. Let C^x_∞ be the infinite component of $G(\mathcal{H}_\lambda \setminus Q_x; 1)$, which is almost surely unique, by Theorem 9.19 and the fact that $P[\mathcal{H}_\lambda(Q_x) = 0] > 0$. Let $\mathcal{T}_1(x)$ be the set of points of $\mathcal{H}_\lambda \setminus C^x_\infty$ connected to C^x_∞ by a path in $G(\mathcal{H}_\lambda; 1)$, and let $\mathcal{T}_2(x)$ be the number of points of $\mathcal{H}''_\lambda(x) \setminus C^x_\infty$ connected to C^x_∞ by a path in $G(\mathcal{H}''_\lambda(x); 1)$. Then $\mathcal{T}_1(x)$ and $\mathcal{T}_2(x)$ are almost surely finite, since they are both finite unions of finite components. With $|\cdot|$ denoting cardinality, define

$$\Delta_x(\infty) := |\mathcal{T}_1(x)| - |\mathcal{T}_2(x)|. \tag{10.84}$$

Definition 10.24 *A sequence of boxes* $(A_n)_{n\geq 1}$ *with* A_n *of side* s_n *is comparable if* (i) $\lim_{n\to\infty} s_n = \infty$, *and* (ii) *there exists* $\delta > 0$ *such that* $B(\mathbf{0}; \delta s_n) \subseteq A_n$ *for all but finitely many* n.

Lemma 10.25 *For any $x \in \mathbb{Z}^d$, and any comparable sequence of boxes $(A_n)_{n\geq 1}$, we have*

$$\Delta_x(A_n) \xrightarrow{P} \Delta_x(\infty) \quad \text{as } n \to \infty. \tag{10.85}$$

Proof It suffices to consider the case $x = \mathbf{0}$. Let $(A_n)_{n\geq 1}$ be a comparable sequence of boxes with each A_n of side a_n. Using comparability, choose $\delta > 0$ such that $B(\mathbf{0}; 2\delta a_n) \subseteq A_n$ for all large enough n. Let $\mathcal{E}_n$ be the event that $\mathcal{T}_1(\mathbf{0})$ and $\mathcal{T}_2(\mathbf{0})$ are contained in $B(\mathbf{0}; \delta a_n)$. Since $\mathcal{T}_1(\mathbf{0})$ and $\mathcal{T}_2(\mathbf{0})$ are almost surely finite, $P[\mathcal{E}_n] \to 1$. Let G_n be the event that at least one vertex of the infinite component of $G(\mathcal{H}_\lambda; 1)$ lies in $B(\mathbf{0}; \delta a_n)$. Then $\lim_{n\to\infty} P[G_n] = 1$.

Let F_n be the event that $G(\mathcal{H}_\lambda \cap A_n; 1)$ has a unique component that is crossing for A_n, and no other component of order greater than $a_n^{1/3}$. Let F_n'' be the event that $G(\mathcal{H}_\lambda''(\mathbf{0}) \cap A_n; 1)$ has a component that is crossing for A_n, and no other component of order greater than $a_n^{1/3}$. By Proposition 10.13 and Theorem 10.18, $P[F_n] \to 1$ and $P[F_n''] \to 1$ as $n \to \infty$.

If $\mathcal{E}_n \cap G_n \cap F_n \cap F_n''$ occurs, then the largest component of $G(\mathcal{H}_\lambda \cap A_n; 1)$ is part of the intersection of the infinite component of $G(\mathcal{H}_\lambda; 1)$ with A_n, and the change in this induced by changing the points in the unit cube $Q_\mathbf{0}$ from points of $\mathcal{H}_\lambda$ to points of $\mathcal{H}'_\lambda$ is precisely $\Delta_\mathbf{0}(\infty)$.

By the estimates above, $P[\mathcal{E}_n \cap G_n \cap F_n \cap F_n''] \to 1$ as $n \to \infty$. By the Borel–Cantelli lemma, for any increasing subsequence of the natural numbers we can take a sub-subsequence such that $\mathcal{E}_n \cap G_n \cap F_n \cap F_n''$ occurs for all but finitely many n in the sub-subsequence, almost surely. Therefore (see Williams (1991, A 13.2(e))) $\Delta_\mathbf{0}(A_n) \to \Delta_\mathbf{0}(\infty)$ in probability. □

Lemma 10.26 *The functional H satisfies the moments condition*

$$\sup_{A\in\mathcal{B},\ x\in\mathbb{Z}^d} \max(E[(\Delta_x(A))^4], E[(\Delta_x(\infty))^4]) < \infty. \tag{10.86}$$

Proof It suffices to consider the case $x = \mathbf{0}$. Suppose that event $E'(A; r)$ defined in Lemma 10.23 does not occur, and suppose also that $\mathcal{H}_\lambda(B(2r+3)) \leq \lambda(3r)^d$ and $\mathcal{H}_\lambda''(\mathbf{0})(B(2r+3)) \leq \lambda(3r)^d$. Then changes in $Q_\mathbf{0}$ do not change the order of the largest component by more than $\lambda(3r)^d$.

By Lemma 10.23 the probability of event $E'(A; r)$ decays exponentially in r, uniformly in A, and so does the probability of the event that there are more than $\lambda(3r)^d$ points of $\mathcal{H}_\lambda$ or $\mathcal{H}_\lambda''(\mathbf{0})$ in $B(2r+3)$. Hence, the change in the order of the largest component has a sub-exponentially decaying tail, uniformly in B, that is, there exists $\alpha > 0$ such that for large enough t, and all boxes $A \in \mathcal{B}$,

$$P[|\Delta_\mathbf{0}(A)| > t] \leq \exp(-\alpha t^{1/d}).$$

By the integration by parts formula for expectation, it follows that $E[\Delta_\mathbf{0}(A)^4]$ is bounded, uniformly in A. Also, by (10.85) we can choose a sequence of boxes $(A_n)_{n\geq 1}$ with $\Delta_\mathbf{0}(A_n) \to \Delta_\mathbf{0}(\infty)$ almost surely. Hence, by Fatou's lemma, $E[\Delta_\mathbf{0}(\infty)^4] < \infty$. □

Proof of Theorem 10.22 Let $(s_n)_{n \geq 1}$ be a sequence of numbers in $[1, \infty)$ satisfying $\lim_{n \to \infty}(s_n) = \infty$, and let $B_n := B(s_n)$ for each $n \geq 1$. For $x \in \mathbb{Z}^d$, let $\mathcal{F}_x$ denote the σ-field generated by the points of $\mathcal{H}_\lambda$ in $\cup_{y \preceq x} Q_y$, where $y \preceq x$ means $y \in \mathbb{Z}^d$ and y precedes or equals x in the lexicographic ordering on $\mathbb{Z}^d$. In other words, $\mathcal{F}_x$ is the smallest σ-field, with respect to which the number of Poisson points in any bounded Borel subset of $\cup_{y \preceq x} Q_y$ is measurable.

Let B'_n be the set of lattice points $x \in \mathbb{Z}^d$ such that $Q_x \cap B_n \neq \emptyset$. Label the elements of B'_n in lexicographic order as $x_1, \ldots, x_{k_n}$; then k_n / s_n^d tends to 1. Define the filtration $(\mathcal{G}_0, \mathcal{G}_1, \ldots, \mathcal{G}_{k_n})$ as follows: let $\mathcal{G}_0$ be the trivial σ-field, and let $\mathcal{G}_i = \mathcal{F}_{x_i}$ for $1 \leq i \leq k_n$. Then $H(\mathcal{H}_{\lambda,s}) - EH(\mathcal{H}_{\lambda,s}) = \sum_{i=1}^{k_n} D_i$ where we set

$$D_i = E[H(\mathcal{H}_{\lambda,s})|\mathcal{G}_i] - E[H(\mathcal{H}_{\lambda,s})|\mathcal{G}_{i-1}] = E[\Delta_{x_i}(B_n)|\mathcal{F}_{x_i}], \tag{10.87}$$

with $\Delta_{x_i}(B_n)$ defined by (10.83). By orthogonality of martingale differences, $\mathrm{Var}[H(\mathcal{H}_{\lambda,s})] = E\sum_{i=1}^{k_n} D_i^2$. By this fact, along with the central limit theorem for martingale differences (Theorem 2.10), it suffices to prove the conditions

$$\sup_{n \geq 1} E\left[\max_{1 \leq i \leq k_n} \left(k_n^{-1/2}|D_i|\right)^2\right] < \infty, \tag{10.88}$$

$$k_n^{-1/2} \max_{1 \leq i \leq k_n} |D_i| \xrightarrow{P} 0, \tag{10.89}$$

and, for some $\sigma^2 \geq 0$,

$$k_n^{-1} \sum_{i=1}^{k_n} D_i^2 \xrightarrow{L^1} \sigma^2. \tag{10.90}$$

Using the representation $D_i = E[\Delta_{x_i}(B_n)|\mathcal{F}_{x_i}]$, we may easily check conditions (10.88) and (10.89). Indeed, by the conditional Jensen inequality (see Section 1.6), we have

$$k_n^{-1} E[\max_{i \leq k_n} D_i^2] \leq k_n^{-1} \sum_{i=1}^{k_n} E[D_i^2] \leq k_n^{-1} \sum_{i=1}^{k_n} E[\Delta_{x_i}(B_n)^2],$$

which is uniformly bounded by the moments condition (10.86).

For the second condition (10.89), let $\varepsilon > 0$ and use Boole's and Markov's inequalities to obtain

$$P\left[\max_{1 \leq i \leq k_n} |D_i| \geq k_n^{1/2}\varepsilon\right] \leq \sum_{i=1}^{k_n} \frac{E[D_i^4]}{k_n^2 \varepsilon^4},$$

which tends to zero, again by (10.86).

We now prove (10.90). For each $x \in \mathbb{Z}^d$ let $\Delta_x(\infty)$ be given by (10.84). For $x \in \mathbb{Z}^d$ and $A \in \mathcal{B}$, let

$$W_x(A) := E[\Delta_x(A)|\mathcal{F}_x], \qquad W_x := E[\Delta_x(\infty)|\mathcal{F}_x].$$

Then $W_{x_i}(B_n) = D_i$ for each $i \leq k_n$. Also, $E[W_{\mathbf{0}}^2] < \infty$ by (10.86) and the conditional Jensen inequality. Also, $(W_x, x \in \mathbb{Z}^d)$ is a stationary family of random variables. In fact, W_x is of the form $h(S_x(\xi))$ where, as in Section 9.5, $\xi = (\xi_x, x \in \mathbb{Z}^d)$ is an independent identically distributed set of $\mathcal{S}$-valued random variables and S_x is a shift operator. In the present case, $\mathcal{S}$ is the space of point configurations on $B(1)$, and for each $x \in \mathbb{Z}^d$, ξ_x is the image of restriction of the point process $\mathcal{H}_\lambda$ to Q_x, under the translation $X \mapsto X - x$ (and hence ξ_x is a homogeneous Poisson process on $B(1)$). It follows by an application of Theorem 9.13 (the ergodic theorem) that, setting $\sigma^2 := E[W_0^2]$, we have

$$k_n^{-1} \sum_{x \in B'_n} W_x^2 \xrightarrow{L^1} \sigma^2. \tag{10.91}$$

We need to show that $W_x(B_n)^2$ approximates to W_x^2. We consider x at the origin $\mathbf{0}$. For any $A \in \mathcal{B}$, by the Cauchy–Schwarz inequality,

$$E[|W_\mathbf{0}(A)^2 - W_\mathbf{0}^2|] \le (E[(W_\mathbf{0}(A) + W_\mathbf{0})^2])^{1/2}(E[(W_\mathbf{0}(A) - W_\mathbf{0})^2])^{1/2}. \tag{10.92}$$

By the definition of $W_\mathbf{0}$ and the conditional Jensen inequality,

$$\begin{aligned} E[(W_\mathbf{0}(A) + W_\mathbf{0})^2] &= E[(E[\Delta_\mathbf{0}(A) + \Delta_\mathbf{0}(\infty)|\mathcal{F}_\mathbf{0}])^2] \\ &\le E[E[(\Delta_\mathbf{0}(A) + \Delta_\mathbf{0}(\infty))^2|\mathcal{F}_\mathbf{0}]] \\ &= E[(\Delta_\mathbf{0}(A) + \Delta_\mathbf{0}(\infty))^2], \end{aligned}$$

which is uniformly bounded by the moments condition (10.86). Similarly,

$$E[(W_\mathbf{0}(A) - W_\mathbf{0})^2] \le E[(\Delta_\mathbf{0}(A) - \Delta_\mathbf{0}(\infty))^2]. \tag{10.93}$$

By (10.86) this is also uniformly bounded. For any comparable $\mathcal{B}$-valued sequence $(A_n)_{n \ge 1}$, the sequence $(\Delta_0(A_n) - \Delta_\mathbf{0}(\infty))^2$ tends to zero in probability by (10.85), and is uniformly integrable by (10.86), and therefore (see Section 1.6) the expression (10.93) tends to zero so that, by (10.92), $E[|W_\mathbf{0}(A_n)^2 - W_\mathbf{0}^2|] \to 0$.

Returning to the given sequence (B_n), let $\varepsilon > 0$. It follows from the conclusion of the previous paragraph and translation-invariance that

$$\lim_{n \to \infty} \sup_{x \in B((1-\varepsilon)s_n) \cap \mathbb{Z}^d} E[|W_x(B_n)^2 - W_x^2|] = 0. \tag{10.94}$$

Using (10.94), the uniform boundedness of $E[|W_x(B_n)^2 - W_x^2|]$, and the fact that ε can be taken arbitrarily small, it is routine to deduce that

$$k_n^{-1} \sum_{x \in B_n} (W_x(B_n)^2 - W_x^2) \xrightarrow{L^1} 0,$$

and therefore (10.91) remains true with W_x replaced by $W_x(B_n)$; that is, (10.90) holds and the proof of Theorem 10.22 is complete. □

10.8 Notes and open problems

Notes *Section 10.1.* Theorem 10.1 is new, but is adapted from the analogous lattice result in Grimmett (1999). In fact, Grimmett (1999, p. 373) asserts that a result along the lines of Theorem 10.1 is 'not difficult' to show, but does not provide a proof. Theorem 10.3 is new.

Sections 10.2 and 10.3. Tanemura (1993) gave the first finite slab result in the continuum along the lines of Lemma 10.8. Theorem 10.9 appears in Penrose and Pisztora (1996) but is proved there only for $d \geq 3$. The proof of Proposition 10.13 in the case $d = 2$ uses ideas from Roy and Sarkar (1992).

Sections 10.4 and 10.5. Theorems 10.14 and 10.15 are new, but are adapted from results for lattice percolation found in Grimmett (1999). Theorem 10.18 is new.

Sections 10.6 and 10.7. Theorems 10.19 and 10.20 are from Penrose and Pisztora (1996). The central limit theorem in Theorem 10.22 is new. A similar approach is used in Penrose (2001), Penrose and Yukich (2001) to prove a variety of central limit theorems in spatial probability. In particular, a lattice version of Theorem 10.22 appears in Penrose (2001).

Open problems It is an open problem to investigate the growth of $L_1(G(\mathcal{H}_{\lambda_c,s};1))$ or of $L_1(G(\mathcal{H}_{\lambda(s),s};1))$ when $\lambda(s)$ is a function approaching λ_c as $s \to \infty$. A lattice version of this problem is considered by Borgs *et al.* (2001).

As mentioned just after Theorem 10.15, it is an open problem to show that, when $\lambda > \lambda_c$, the limit of $n^{-(d-1)/d} \log p_n(\lambda)$ exists.

Theorem 10.18 suggests the conjecture that $(\log s)^{-d/(d-1)} L_2(G(\mathcal{H}_{\lambda,s};1))$ should converge in probability to a positive finite constant, as $s \to \infty$.

11

THE LARGEST COMPONENT FOR A BINOMIAL PROCESS

The results in the preceding chapter describe many aspects of the asymptotic behaviour of the largest component order $L_1(G(\mathcal{H}_{\lambda,s};1))$, and hence by scaling, that of $L_1(G(\mathcal{P}_n;r_n))$ in the case with $f = f_U$ and $nr_n^d = \lambda$ (the thermodynamic limit for points uniformly distributed on the unit cube). In the present chapter, we de-Poissonize some of these results to describe aspects of the asymptotic behaviour of $L_1(G(\mathcal{X}_n;r_n))$, and related quantities, in the thermodynamic limit $nr_n^d \to \rho \in (0,\infty)$. The lack of spatial independence for the binomial point process $\mathcal{X}_n$ is overcome, with some effort, by coupling $\mathcal{X}_n$ with certain Poisson processes.

When proving laws of large numbers in this chapter, we do *not* restrict attention to the uniform density f_U. This enables us to discuss some interesting statistical applications, establishing consistency results for certain statistical tests based on geometric graphs. These are described in Sections 11.3 and 11.4. In the case of the central limit theorem for the order of the largest component, on the other hand, we restrict attention to the uniform case $f = f_U$ (Section 11.5).

We assume throughout this chapter that *the norm* $\|\cdot\|$ *is one of the* l_p *norms,* $1 \le p \le \infty$. Recall that θ denotes the volume of the unit ball in the chosen norm. Recall from Section 1.7 that $\mathcal{P}_\lambda$ is the coupled Poisson process $\{X_1,\ldots,X_{N_\lambda}\}$, where N_λ is a Po(λ) variable independent of $(X_1, X_2, \ldots)$. Recall also from Section 9.6 that $\mathcal{H}_{\lambda,s}$ is the restriction of the homogeneous Poisson process $\mathcal{H}_\lambda$ to the box $B(s) := [-s/2, s/2]^d$ $(s > 0)$, while $\mathcal{H}_{\lambda,0} := \mathcal{H}_\lambda \cup \{\mathbf{0}\}$. Recall from Section 1.5 that $f_{\max}$ denotes the essential supremum of f, always assumed finite.

11.1 The subcritical case

This section is concerned with the graph $G(\mathcal{X}_n \cap \Gamma; r_n)$ on the restriction to some specified Borel set $\Gamma \subseteq \mathbb{R}^d$ of a random sample $\mathcal{X}_n$ of size n from a probability density function f on $\mathbb{R}^d$. Giving the result in this generality (rather than just considering $G(\mathcal{X}_n;r_n)$) will be useful later on. We take the thermodynamic limit with $r_n = \Theta(n^{-1/d})$, below the percolation threshold. Let $f\mathbf{1}_\Gamma$ be the function on $\mathbb{R}^d$ which takes the value $f(x)$ for $x \in \Gamma$ and 0 for $x \in \mathbb{R}^d \setminus \Gamma$. Let $(f\mathbf{1}_\Gamma)_{\max}$ denote the essential supremum of the function $f\mathbf{1}_\Gamma$.

Recall from Theorem 10.1 that the limit $\zeta(\lambda) := -\log \lim_n p_n(\lambda)^{1/n}$ exists, and is continuous in λ, where $(p_n(\lambda), n \in \mathbb{N})$ is the probability mass function of the order of the component of $G(\mathcal{H}_{\lambda,0};1)$ containing the origin, and that $\zeta(\cdot)$ is continuous.

Theorem 11.1 *Suppose f and Γ are such that $f\mathbf{1}_\Gamma$ is almost everywhere continuous, and suppose $nr_n^d(f\mathbf{1}_\Gamma)_{\max} \to \lambda$ as $n \to \infty$, with $0 < \lambda < \lambda_c$. Then*

$$\frac{L_1(G(\mathcal{X}_n \cap \Gamma; r_n))}{\log n} \xrightarrow{P} \frac{1}{\zeta(\lambda)} \quad \text{as } n \to \infty.$$

The following non-asymptotic bound will be used in proving Theorem 11.1, and again later on.

Proposition 11.2 *Let $\lambda \in (0, \lambda_c)$, and let $\zeta' \in (0, \zeta(\lambda))$, with $\zeta(\lambda)$ defined in Theorem 10.1. Then there is a constant $\eta > 0$ and an integer m_0 such that whenever f, Γ, n, r satisfy $nr^d(f\mathbf{1}_\Gamma)_{\max} \leq \lambda$, we have for all $m \geq m_0$ that*

$$P[L_1(G(\mathcal{X}_n \cap \Gamma; r)) \geq m] \leq n(\exp(-\zeta' m) + \exp(-\eta n)). \tag{11.1}$$

Proof By Boole's inequality,

$$P[L_1(G(\mathcal{X}_n; r)) \geq m] \leq n \int_\Gamma P[A(n, m, x)] f(x) \mathrm{d}x, \tag{11.2}$$

where $A(n, m, x)$ denotes the event that the component containing x of $G((\mathcal{X}_{n-1} \cup \{x\}) \cap \Gamma; r)$ has at least m vertices.

Using the continuity of $\zeta(\cdot)$, choose $\mu > \lambda$ such that $\zeta' < \zeta(\mu)$. Suppose f, Γ, n, r satisfy $nr^d(f\mathbf{1}_\Gamma)_{\max} \leq \lambda$. If $E(n, m, x)$ denotes the event that the component containing x of $G(\mathcal{P}_{n\mu/\lambda} \cup \{x\}; r)$ has at least m vertices, we have for all $x \in \mathbb{R}^d$ that

$$P[A(n, m, x)] \leq P[E(n, m, x)] + P[N_{n\mu/\lambda} < n - 1]. \tag{11.3}$$

By assumption, $(n\mu/\lambda)(f\mathbf{1}_\Gamma)_{\max} \leq \mu r^{-d}$. Since $\mathcal{P}_{n\mu/\lambda} \cap \Gamma$ is a Poisson process on $\mathbb{R}^d$ with intensity function $(n\mu/\lambda)f\mathbf{1}_\Gamma$, by Corollary 9.16 it is dominated by the homogeneous Poisson process $\mathcal{H}_{\mu r^{-d}}$. Therefore, if $F(n, m)$ denotes the event that the component containing $\mathbf{0}$ of $G(\mathcal{H}_{\mu r^{-d}, 0}; r)$ has at least m vertices, we have for all $x \in \Gamma$ that

$$P[E(n, m, x)] \leq P[F(n, m)].$$

By scaling (Corollary 9.18), $P[F(n, m)]$ equals the probability that the component containing $\mathbf{0}$ of $G(\mathcal{H}_{\mu, 0}; 1)$ has at least m vertices. Therefore by (10.5), since $\zeta' < \zeta(\mu)$, we have for all large enough m and all $x \in \mathbb{R}^d$ that

$$P[E(n, m, x)] \leq \sum_{k \geq m} p_k(\mu) \leq \exp(-\zeta' m),$$

so by (11.2) and (11.3),

$$P[L_1(G(\mathcal{X}_n; r_n)) \geq m] \leq n(\mathrm{e}^{-\zeta' m} + P[N_{n\mu/\lambda} < n - 1]).$$

By Lemma 1.2, the expression $nP[N_{n\mu/\lambda} < n - 1]$ decays exponentially in n, to give us (11.1). $\square$

Proof of Theorem 11.1 Suppose $\alpha > 1/\zeta(\lambda)$. Choose $\zeta' \in (1/\alpha, \zeta(\lambda))$; then by Proposition 11.2 there exists $\eta > 0$ such that, for all large enough n,

$$P[L_1(G(\mathcal{X}_n \cap \Gamma; r_n)) > \alpha \log n] \leq n(n^{-\alpha\zeta'} + e^{-\eta n}) \to 0. \tag{11.4}$$

Conversely, let $\beta \in (0, 1/\zeta(\lambda))$. Using the continuity of $\zeta(\cdot)$, choose $\varepsilon > 0$ such that $\zeta(\lambda(1-3\varepsilon)) < 1/\beta$. Using the almost everywhere continuity of $f\mathbf{1}_\Gamma$, choose $x_0 \in \mathbb{R}^d$ such that $f(x_0) > (1-\varepsilon)(f\mathbf{1}_\Gamma)_{\max}$ and $f\mathbf{1}_\Gamma$ is continuous at x_0. Choose $\delta > 0$ such that $f\mathbf{1}_\Gamma(x) > (1-\varepsilon)(f\mathbf{1}_\Gamma)_{\max}$ for all x in the cube $C(x_0;\delta) := B(\delta) \oplus \{x_0\}$ of side δ centred at x_0. Then

$$\begin{aligned} P[L_1(G(\mathcal{X}_n \cap \Gamma; r_n)) < \beta \log n] \leq\ & P[L_1(G(\mathcal{P}_{(1-\varepsilon)n} \cap \Gamma; r_n)) < \beta \log n] \\ & + P[\{\mathcal{P}_{(1-\varepsilon)n} \subseteq \mathcal{X}_n\}^c]. \end{aligned} \tag{11.5}$$

The restriction of $\mathcal{P}_{(1-\varepsilon)n} \cap \Gamma$ to $C(x_0;\delta)$ is a Poisson process on $C(x_0;\delta)$ with intensity function $(1-\varepsilon)nf\mathbf{1}_\Gamma(\cdot)$ which exceeds $n(f\mathbf{1}_\Gamma)_{\max}(1-2\varepsilon)$ on the whole of $C(x_0;\delta)$, and therefore exceeds $(1-3\varepsilon)\lambda r_n^{-d}$ on the whole of $C(x_0;\delta)$ for all n greater than some constant n_1. Therefore by Corollary 9.16, $\mathcal{P}_{(1-\varepsilon)n} \cap \Gamma$ dominates a homogeneous Poisson process, denoted $\mathcal{P}'_n$, on $C(x_0;\delta)$ of intensity $\lambda(1-3\varepsilon)r_n^{-d}$. By scaling (Theorem 9.17), for $n > n_1$,

$$\begin{aligned} P[L_1(G(\mathcal{P}_{(1-\varepsilon)n} \cap \Gamma; r_n)) < \beta \log n] &\leq P[L_1(G(\mathcal{P}'_n; r_n)) < \beta \log n] \\ &= P[L_1(G(\mathcal{H}_{\lambda(1-3\varepsilon), r_n^{-1}\delta}; 1)) < \beta \log n]. \end{aligned} \tag{11.6}$$

By Theorem 10.3,

$$\frac{L_1(G(\mathcal{H}_{\lambda(1-3\varepsilon), r_n^{-1}\delta}; 1))}{d \log(r_n^{-1}\delta)} \xrightarrow{P} \frac{1}{\zeta(\lambda(1-3\varepsilon))} > \beta.$$

Since nr_n^d tends to a positive finite constant, $\log n + d \log r_n$ tends to a limit and $d \log(r_n^{-1}\delta)/\log n$ tends to 1. Hence,

$$P\left[\frac{L_1(G(\mathcal{H}_{\lambda(1-3\varepsilon), r_n^{-1}\delta}; 1))}{\log n} < \beta\right] \to 0.$$

With (11.5) and (11.6), this implies that $P[L_1(G(\mathcal{X}_n; r_n)) < \beta \log n] \to 0$, and combined with (11.4) this gives us the result. $\square$

Later we shall require another lemma, concerning the subcritical limiting regime.

Lemma 11.3 *Suppose $\Gamma \subseteq \mathbb{R}^d$, and suppose $nr_n^d(f\mathbf{1}_\Gamma)_{\max} \to \lambda$ as $n \to \infty$, with $0 < \lambda < \lambda_c$. Let $\varepsilon > 0$, and let F_n be the event that there is a component of $G(\mathcal{X}_n \cap \Gamma; r_n)$ with order greater than εn or with metric diameter greater than ε. Then* $\limsup_{n\to\infty} n^{-1/d} \log P[F_n] < 0$.

Proof Immediate from Proposition 11.2. $\square$

11.2 The supercritical case on the cube

In the supercritical case, we first consider the restriction of the point process $\mathcal{X}_n$ to the cube $B(a)$. The supercriticality condition, in this setting, is that $\liminf_{n\to\infty}(nr_n^d f(x))$ should be bounded away from λ_c on this cube. The notions of a *crossing* geometric graph and a *k-crossing* geometric graph were introduced in Section 10.2.

Proposition 11.4 *Suppose* $d \geq 2$. *Suppose* $a > 0$ *and* $\lim_{n\to\infty}(nr_n^d) = \rho$ *with*

$$\lambda_1 := \inf_{x \in B(a)} \{\rho f(x)\} > \lambda_c.$$

Suppose $(\phi_n, n \geq 1)$ *satisfies* $\phi_n \leq r_n^{-1}a/2$ *and* $\phi_n/\log n \to \infty$ *as* $n \to \infty$. *Let* $E'(n)$ *be the event that* (i) *there is a unique component of* $G(\mathcal{X}_n \cap B(a); r_n)$ *that is crossing for* $B(a)$, *and* (ii) *no other component of* $G(\mathcal{X}_n \cap B(a); r_n)$ *has metric diameter greater than* $\phi_n r_n$. *Then* $\limsup_{n\to\infty} \phi_n^{-1} \log P[E'(n)^c] < 0$.

This asymptotic result is deduced from the following uniform non-asymptotic bound. Given a probability density function f on $\mathbb{R}^d$, and given $n \in \mathbb{N}$, $a > 0$, $b > 0$, and $\mu > 0$, let $E(n, f, a, b, \mu)$ denote the event that for a set $\mathcal{X}_n$ of n independent random d-vectors with common density f, (i) there is a unique component of $G(\mathcal{X}_n \cap B(a); 1)$ that is crossing for $B(a)$; (ii) no other component of $G(\mathcal{X}_n \cap B(a); 1)$ has metric diameter greater than b; and (iii) no component of $G(\mathcal{X}_n \cap B(a); 1)$, other than the crossing component has order greater than $\mu 2^{d+1}\theta b^d$. (Condition (iii) is not relevant to Proposition 11.4 but will be used later on.)

Proposition 11.5 *Suppose* $d \geq 2$, *and* $\mu > \lambda > \lambda_c$. *Then there exist strictly positive finite constants* c, c', *depending only on* λ *and* μ, *such that for all* a, b *with* $2d \leq b \leq a/2$, *for all* $n \in \mathbb{N}$ *and all probability density functions* f *on* $\mathbb{R}^d$ *satisfying*

$$\lambda \leq \inf_{x \in B(a)} \{nf(x)\} \leq \sup_{x \in B(a)} \{nf(x)\} \leq \mu, \tag{11.7}$$

it is the case that

$$1 - P[E(n, f, a, b, \mu)] \leq c' a^d e^{-cb}.$$

Proof of Proposition 11.4 Choose $\lambda_2 \in (\lambda_c, \lambda_1)$ and $\mu > \rho f_{\max}$. Set $r_n^{-1}\mathcal{X}_n := \{r_n^{-1}X : X \in \mathcal{X}_n\}$. The graph $G(\mathcal{X}_n; r_n)$ is isomorphic to $G(r_n^{-1}\mathcal{X}_n; 1)$, and the re-scaled point process $r_n^{-1}\mathcal{X}_n$ is a sample of size n from the probability density function $f_n(x) := r_n^d f(r_n x)$, which lies in the range $[n^{-1}\lambda_2, n^{-1}\mu]$ for all large enough n and all $x \in B(r_n^{-1}a)$. Therefore, by Proposition 11.5, for large n we have

$$P[E'(n)^c] \leq P[E(n, f_n, r_n^{-1}a, \phi_n, \mu)^c] \leq c' r_n^{-d} \exp(-c\phi_n),$$

and by the assumption $\phi_n \gg \log n$ this implies the result. □

Proof of Proposition 11.5 for $d = 2$ Choose λ_3 with $\lambda_c < \lambda_3 < \lambda$. Suppose that the probability density function f and the numbers $n \in \mathbb{N}$ and $a, b \in (0, \infty)$ together satisfy $1 \leq b \leq a/2$ and (11.7). Then, with the Poisson process $\mathcal{P}_{n\lambda_3/\lambda}$ coupled to $\mathcal{X}_n$ in the usual manner described in Section 1.7, the intensity of the restriction of $\mathcal{P}_{n\lambda_3/\lambda}$ to $B(a)$ is at least λ_3. Therefore, by Corollary 9.16, the point process $\mathcal{P}_{n\lambda_3/\lambda}$ dominates the homogeneous Poisson process $\mathcal{H}_{\lambda_3,a}$.

As in the proof of Proposition 10.13, divide $B(a)$ into squares of side a/m, with $m := \lceil 4a/b \rceil$, and define horizontal and vertical dominoes (rectangles with aspect ratio 2) to consist of all pairs of neighbouring squares in this subdivision; let $I_{a,b}$ denote the event that for each such domino D the graph $G(\mathcal{H}_{\lambda_3,a} \cap D; 1)$ includes a component that is crossing the long way for the domino D. By Lemma 10.5, there are constants c, c' (independent of a, b) such that

$$1 - P[I_{a,b}] \leq c'(a/b)^2 \mathrm{e}^{-cb} \leq c' a^2 \mathrm{e}^{-cb}. \tag{11.8}$$

On event $I_{a,b}$ there is a component of $G(\mathcal{H}_{\lambda_3,a}; 1)$ that is crossing for $B(a)$, and no other component has metric diameter greater than b, and moreover, *the second property remains true even if one adds extra points to* $\mathcal{H}_{\lambda_3,a}$. See Fig. 10.4.

By the assumption (11.7), $a^2\lambda \leq \int_{B(a)} nf(x)\mathrm{d}x \leq n$. By Lemma 1.2 and the coupling, there is a constant c such that

$$P[\{\mathcal{P}_{n\lambda_3/\lambda} \subseteq \mathcal{X}_n\}^{\mathrm{c}}] \leq \mathrm{e}^{-cn} \leq \exp(-c\lambda a^2). \tag{11.9}$$

We may assume that $\mathcal{H}_{\lambda_3,a}$ is coupled to $\mathcal{P}_{n\lambda_3/\lambda}$ in such a way that $\mathcal{H}_{\lambda_3,a} \subseteq \mathcal{P}_{n\lambda_3/\lambda}$. Then, if $I_{a,b}$ occurs and also $\mathcal{P}_{n\lambda_3/\lambda} \subseteq \mathcal{X}_n$, there is a component of $G(\mathcal{X}_n \cap B(a); 1)$ that is crossing for $B(a)$, and no other component has metric diameter greater than b.

To check part (iii) of the definition of $E(n, f, a, b)$, choose a minimal set of points $x_1, \ldots, x_\nu$ such that the balls $B(x_1; b), \ldots, B(x_\nu; b)$ cover $B(a)$; observe that $\nu = O(a^d)$. For $1 \leq i \leq \nu$, let F_i be the event that the enlarged ball $B(x_i; 2b)$ contains more than $2^{d+1}\mu\theta b^d$ points of $\mathcal{X}_n \cap B(a)$. By Lemma 1.1, $P[F_i] \leq \exp(-cb^d)$, and hence

$$P\left[\bigcup_{i=1}^{\nu} F_i\right] < c' a^d \exp(-cb^d), \tag{11.10}$$

for some constant c, independent of a, b. But if $G(\mathcal{X}_n \cap B(a); r)$ has a component of metric diameter at most b but containing more than $2^{d+1}\mu\theta b^d$ points, then one of the events $F_1, \ldots, F_\nu$ must occur. Hence, (11.10), (11.8), and (11.9) together give us the result. $\square$

In the case $d \geq 3$, the proof of Proposition 11.5 is divided into steps analogous to those in the proof of Proposition 10.13.

Lemma 11.6 *Suppose $d \geq 3$ and $\lambda > \lambda_c$. Then there exist constants $c > 0, c' > 0$ such that for any $a \geq 1$, $n \in \mathbb{N}$ and $j \in \{1, 2, \ldots, d\}$, and any density function f with $\inf_{x \in B(a)}\{nf(x)\} \geq \lambda$,*

$$P[\{\exists \text{ a } j\text{-crossing component of } G(\mathcal{X}_n \cap B(a); 1) \text{ for } B(a)\}^c] \\ \leq c' \exp(-ca). \tag{11.11}$$

Proof We need only consider the case $j = 1$. Choose $\lambda_3 \in (\lambda_c, \lambda)$, and let $\mathcal{P}_{n\lambda_3/\lambda}$ be coupled to $\mathcal{X}_n$ in the usual manner. By Proposition 10.6, the probability that there is no component of $G(\mathcal{H}_{\lambda_3,a}; 1)$ that is 1-crossing for $B(a)$ decays exponentially in a. Hence, since $\mathcal{P}_{n\lambda_3/\lambda} \cap B(a)$ dominates $\mathcal{H}_{\lambda_3,a}$, the probability that there is no component of $G(\mathcal{P}_{n\lambda_3/\lambda} \cap B(a); 1)$ that is 1-crossing for $B(a)$ decays exponentially in a. But if there is a component for $G(\mathcal{P}_{n\lambda_3/\lambda} \cap B(a); 1)$ that is 1-crossing for $B(a)$, and also $\mathcal{P}_{n\lambda_3/\lambda} \subseteq \mathcal{X}_n$, then there is a component of $G(\mathcal{X}_n \cap B(a); 1)$ that is 1-crossing for $B(a)$. Combined with the argument at (11.9), this gives us the result. □

If $\mathcal{Y} \subset \mathbb{R}^d$ is locally finite, and $x \in \mathbb{R}^d$ (not necessarily in $\mathcal{Y}$), then as in Section 10.2, let $C(x; \mathcal{Y})$ denote the vertex set of the component of $G(\mathcal{Y} \cup \{x\}; 1)$ which contains x, and as in Section 10.3, let $\tilde{C}(x; \mathcal{Y})$ denote the component of $G(\mathcal{Y}; 1)$ which includes at least one vertex in the ball $B(x; \frac{1}{2})$ (or the empty set if no such component exists). For any set $A \subseteq \mathbb{R}^d$, let $C(A; \mathcal{Y}) := \cup_{x \in A} C(x; \mathcal{Y})$ and let $\tilde{C}(A; \mathcal{Y}) := \cup_{x \in A} \tilde{C}(x; \mathcal{Y})$. For $1 \leq k \leq d$, let $\pi_k : \mathbb{R}^d \to \mathbb{R}$ denote projection onto the kth coordinate.

Let $F(n, f, a, k_1, k_2)$ denote the event that $G(\mathcal{X}_n \cap B(a); 1)$ has a component that is k_1-crossing but not k_2-crossing for $B(a)$.

Lemma 11.7 *Suppose $d \geq 3$ and $\mu > \lambda > \lambda_c$. Then there exist constants $c > 0, c' > 0$, such that for any $a \geq 1$, $n \in \mathbb{N}$, any distinct $k_1, k_2 \in \{1, 2, \ldots, d\}$, and any density function f with $\inf_{x \in B(a)}\{nf(x)\} \geq \lambda$ and $\sup_{x \in B(a)}\{nf(x)\} \leq \mu$, we have*

$$P[F(n, f, a, k_1, k_2)] \leq c' a^d \exp(-ca). \tag{11.12}$$

Proof It suffices to consider the case with $k_1 = 1, k_2 = 2$. Set $\beta := 1/(2d)$. Then for every $y \in B(a)$ there exists $x \in \mathbb{Z}^d$ with $\beta x \in B(a)$ and $\|y - \beta x\| \leq \frac{1}{2}$. For $x \in \mathbb{Z}^d$, let $\tilde{C}_x^{(n)}$ denote the component $\tilde{C}(\beta x; \mathcal{X}_n \cap B(a))$. Define the event

$$F_x(n) := \{\tilde{C}_x^{(n)} \text{ is 1-crossing }\} \cap \{\tilde{C}_x^{(n)} \text{ is not 2-crossing }\}.$$

If $F(n, f, a, 1, 2)$ occurs, then $F_x(n)$ must occur for some $x \in \mathbb{Z}^d \cap B(\beta^{-1}a)$ with $\pi_1(x) \leq 0$. Since the number of such x is at most $\lceil \beta^{-1} a \rceil^d$, there exists $c > 0$ such that

$$P[F(n, f, a, 1, 2)] \leq ca^d \sup\{P[F_x(n)] : x \in B(\beta^{-1}a) \cap \mathbb{Z}^d, \pi_1(x) \leq 0\}. \tag{11.13}$$

Fix $x \in B(\beta^{-1}a) \cap \mathbb{Z}^d$ with $\pi_1(x) \leq 0$. Choose λ_4 with $\lambda_c < \lambda_4 < \lambda$. Let $K = K'(\lambda_4)$ as given by Lemma 10.10. Divide $B(a)$ into slabs S_j of thickness K, by setting

$$T_j = \{y \in B(a) : \pi_1(y) \leq \pi_1(\beta x) + 1 + jK\}, \tag{11.14}$$

and setting $S_0 := T_0$ and $S_j = T_j \backslash T_{j-1}$ for $j = 1, 2, 3, \ldots$.

Let $m_a := \lfloor ((a/2) - 1)/K \rfloor$. Let N_{-1} denote the number of points of $\mathcal{X}_n$ in $\mathbb{R}^d \backslash T_{m_a}$. For $0 \leq j \leq m_a$, let N_j denote the number of points of $\mathcal{X}_n$ in the slab S_j. Then $N_{-1}, N_0, N_1, \ldots, N_{m_a}$ are jointly multinomial, and for $0 \leq j \leq m_a$, N_j has a $\mathrm{Bi}(n, p_j)$ distribution, where we set $p_j := \int_{S_j} f(x)\mathrm{d}x$, so that

$$\lambda K a^{d-1} \leq n p_j \leq \mu K a^{d-1}, \qquad 1 \leq j \leq m_a. \tag{11.15}$$

We now use a coupling device. On a suitable probability space, define point processes $\mathcal{X}'_j$, $\mathcal{Q}_{j,0}$ and $\mathcal{Q}_{j,1}$, $0 \leq j \leq m_a$, as follows: for $j = 0, 1, 2, \ldots, m_a$ let X^j be a random d-vector taking values in S_j with density $f(\cdot)/p_j$. Let $X^j_r, r \geq 1, 0 \leq j \leq m_a$, be independent random d-vectors with $X^j, X^j_1, X^j_2, \ldots$ identically distributed for each j. Let $N'_0, N'_1, \ldots, N'_{m_a}$ be random variables with the same multinomial joint distribution as $N_0, N_1, \ldots, N_{m_a}$, independent of (X^j_r).

Let $\zeta_0 = \lambda_4/\lambda < 1$, and choose a constant $\zeta_1 > 1$. Let $M_{j,0}$ and $M_{j,1}$, $0 \leq j \leq m_a$, be Poisson random variables with $EM_{j,i} = n\zeta_i p_j, i = 0, 1$, independent of one another, of $N'_0, N'_1, \ldots, N'_{m_a}$, and of (X^j_m). Then set

$$\mathcal{X}'_j := \{X^j_1, X^j_2, \ldots, X^j_{N'_j}\}, \quad 0 \leq j \leq m_a$$

and for $i = 0, 1$, set

$$\mathcal{Q}_{j,i} := \{X^j_1, X^j_2, \ldots, X^j_{M_{j,i}}\}, \quad 0 \leq j \leq m_a.$$

For $1 \leq j \leq m_a$, if $M_{j,0} \leq N'_j \leq M_{j,1}$, then $\mathcal{Q}_{j,0} \subseteq \mathcal{X}'_j \subseteq \mathcal{Q}_{j,1}$. Define the event

$$W_n := \bigcap_{j=1}^{m_a} \{M_{j,0} \leq N'_j \leq M_{j,1}\}. \tag{11.16}$$

Define $F'_x(n)$ to be the event that $\tilde{C}(x; \cup_{j=0}^{m_a} \mathcal{X}'_j)$ is not 2-crossing for $B(a)$, but does intersect with each of the slabs $S_1, S_2, \ldots, S_{m_a}$. By the construction, the point process $\cup_{j=0}^{m_a} \mathcal{X}'_j$ has the same distribution as $\mathcal{X}_n \cap T_{m_a}$, and so

$$\begin{aligned} P[F_x(n)] &\leq P[F'_x(n)] \\ &\leq P[W_n^{\mathrm{c}}] + P[W_n \cap F'_x(n)]. \end{aligned} \tag{11.17}$$

Let A_j (respectively, $A_{j,1}$) be the event that there is at least one point of $(\mathcal{X}'_j)$ (respectively, $(\mathcal{Q}_{j,1})$) within distance 1 of $\tilde{C}(x; \cup_{q=0}^{j-1} \mathcal{X}'_q)$. Let B_j (respectively $B_{j,0}$) be the event that there is no component of $G(\mathcal{X}'_j; 1)$ (respectively

$G(\mathcal{Q}_{j,0};1))$ which is 2-crossing for $B(a)$ and has non-empty intersection with the 1-neighbourhood of $\tilde{C}(x;\cup_{q=0}^{j-1}\mathcal{X}'_q)$. Then $F'_x(n) \subseteq \bigcap_{j=1}^{m_a}(A_j \cap B_j)$, and

$$(W_n \cap F'_x(n)) \subseteq \bigcap_{j=1}^{m_a}(A_{j,1} \cap B_{j,0}). \tag{11.18}$$

For $0 \leq j \leq m_a$, let $\mathcal{F}_j$ be the σ-field generated by the values of $(N'_q, M_{q,0}, M_{q,1}, X_1^q, X_2^q, X_3^q, \ldots)$, $0 \leq q \leq j$. Then $A_{j,1} \cap B_{j,0} \in \mathcal{F}_j$, and the configuration of $\cup_{q=0}^{j}\mathcal{X}'_q$ is $\mathcal{F}_j$-measurable.

Let $1 \leq j \leq m_a$. The point process $\mathcal{Q}_{j,1}$ is independent of $\mathcal{F}_{j-1}$. It is a Poisson process with intensity $n\zeta_1 f(\cdot)\mathbf{1}_{S_j}(\cdot)$, and therefore is dominated by $\mathcal{H}_{\zeta_1\mu} \cap S_j$. Therefore, defining the $\mathcal{F}_j$-measurable random set $\mathcal{S}(j)$ by

$$\mathcal{S}(j) := \text{vertex set of } \tilde{C}\left(\beta x; \bigcup_{q=0}^{j}\mathcal{X}'_q\right),$$

we have

$$P[A_{j,1}|\mathcal{F}_{j-1}] \leq P[C(\mathcal{S}(j-1); \mathcal{H}_{\zeta_1\mu} \cap S_j) \neq \mathcal{S}(j-1)] \quad \text{a.s.}$$

Similarly, $\mathcal{Q}_{j,0}$ dominates $\mathcal{H}_{\lambda_4} \cap S_j$, and

$$P[B_{j,0}^{\mathrm{c}}|\mathcal{F}_{j-1}] \geq P[C(\mathcal{S}(j-1); \mathcal{H}_{n\zeta_0 f_0} \cap S_j) \text{ is 2-crossing for } S_j] \quad \text{a.s.}$$

Since $K = K'(\lambda_4)$ is given by Lemma 10.10 and S_j is a slab of thickness K, eqn (10.26) from Lemma 10.10 implies that there exists $\gamma > 0$, independent of a or x, such that for $1 \leq j \leq m_a$,

$$P[B_{j,0}^{\mathrm{c}}|\mathcal{F}_{j-1}] \geq \gamma P[A_{j,1}|\mathcal{F}_{j-1}] \quad \text{a.s.}$$

so that

$$\begin{aligned} P[A_{j,1} \cap B_{j,0}|\mathcal{F}_{j-1}] &\leq \min(P[A_{j,1}|\mathcal{F}_{j-1}], 1 - \gamma P[A_{j,1}|\mathcal{F}_{j-1}]) \\ &\leq (1+\gamma)^{-1} \quad \text{a.s.} \end{aligned}$$

and therefore, by (11.18), putting $(1+\gamma)^{-1} = \exp(-\delta)$, we have

$$P[W_n \cap F'_x(n)] \leq \exp(-\delta m_a) \leq \mathrm{e}^{\delta(1+1/K)}\mathrm{e}^{-\delta a/(2K)}. \tag{11.19}$$

To estimate $P[W_n^{\mathrm{c}}]$, choose $\eta \in (\zeta_0, 1)$. By (11.15) and large deviations estimates for the binomial and Poisson distributions (Lemmas 1.1 and 1.2), we can find $c > 0$ such that for all n, j

$$\max\left(P[M_{j,0} \geq np_j\eta], P[N'_j \leq np_j\eta]\right) \leq \exp(-ca^{d-1}).$$

Therefore, $P[M_{j,0} > N'_j] \leq 2\exp(-ca^{d-1})$. There is a similar bound on $P[M_{j,1} < N'_j]$, and so by (11.16),

$$P[W_n^c] \leq 4(a/K)\exp(-ca^{d-1}).$$

Combining this estimate with (11.17) and (11.19), we have for some c, c' that

$$P[F_x(n)] \leq c' \exp(-ca),$$

independently of x. Using (11.13), we obtain the desired bound. □

Next, let $H(n, f, a, b, k)$ denote the event that there exist two distinct components of $G(\mathcal{X}_n \cap B(a); 1)$, denoted C_1 and C_2, say, such that C_1 is k-crossing for $B(a)$ and $\pi_k(C_2)$ has diameter at least b.

Lemma 11.8 *Suppose $d \geq 3$ and $\mu > \lambda > \lambda_c$. There exist strictly positive constants c, c', such that for $a, b \in \mathbb{R}$ with $2 \leq b \leq a/2$, for all $n \in \mathbb{N}$, $k \in \{1, 2, \ldots, d\}$, and for all probability density functions f with* $\inf_{x \in B(a)}\{nf(x)\} \geq \lambda$ *and* $\sup_{x \in B(a)}\{nf(x)\} \leq \mu$, *we have*

$$P[H(n, f, a, b, k)] \leq c' a^d e^{-cb}.$$

Proof It suffices to consider the case $k = 1$. Let $\beta := 1/(2d)$, as in the proof of Lemma 11.7. We have

$$P[H(n, f, a, b, 1)] \leq \sum_{x, z \in \mathbb{Z}^d \cap B(\beta^{-1}a): \pi_1(\beta x) = \pi_1(\beta z) \leq (a/2) - b} P[H_{x,z}(n)],$$

where $H_{x,z}(n)$ denotes the event that $\tilde{C}(\beta x; \mathcal{X}_n \cap B(a))$ and $\tilde{C}(\beta z; \mathcal{X}_n \cap B(a))$ are distinct and are both 1-crossing for $B(a) \cap \pi_1^{-1}([\pi_1(\beta x), \pi_1(\beta x) + b])$.

Fix distinct x and z in $B(\beta^{-1}a) \cap \mathbb{Z}^d$ with $\pi_1(x) = \pi_1(z) \leq (a/2) - b$. Choose λ_5 with $\lambda_c < \lambda_5 < \lambda$. Let $K = K'(\lambda_5)$ as given by Lemma 10.10. As in the proof of Lemma 11.7, define T'_j by (11.14) and define slabs S_j of thickness K by $S_0 := T_0$ and $S_j = T_j \setminus T'_{j-1}$ for $j = 1, 2, 3, \ldots$.

We now use a coupling device similar to the one in the proof of Lemma 11.7. Let the point processes $\mathcal{X}'_j$, $\mathcal{Q}_{j,0}$, $\mathcal{Q}_{j,1}$ and the σ-field $\mathcal{F}_j$ be as defined in that proof. Let A'_j (respectively, $A'_{j,1}$) be the event that $\mathcal{X}'_j$ (respectively, $\mathcal{Q}_{j,1}$) has non-empty intersection with the 1-neighbourhood of each of $\tilde{C}(x; \cup_{q=0}^{j-1} \mathcal{X}'_q)$ and $\tilde{C}(z; \cup_{q=0}^{j-1} X'_q)$. Let B'_j (respectively, $B'_{j,0}$) be the event that there is no component of $G(\mathcal{X}'_j; r_n)$ (respectively, $\mathcal{Q}_{j,0}$) which has non-empty intersection both with $\tilde{C}(x; \cup_{q=0}^{j-1} \mathcal{X}'_q)$ and with $\tilde{C}(z; \cup_{q=0}^{j-1} \mathcal{X}'_q)$. By (10.27) from Lemma 10.10, there exists a constant $\gamma' > 0$, independent of x, z, a or b, such that for all $j = 1, 2, \ldots, \lfloor (b-1)/K \rfloor$ we have

$$P[(B'_{j,0})^c | \mathcal{F}'_{j-1}] \geq \gamma' P[A'_{j,1} | \mathcal{F}'_{j-1}] \qquad \text{a.s.}$$

and the remainder of the proof is much the same as for Lemma 11.7, since

$$P[H_{x,z}(n)] \leq P[W_n^{\mathrm{c}}] + P\left[\bigcap_{j=1}^{m_n-1}(A'_j \cap B'_j) \cap W_n\right]$$

$$\leq P[W_n^{\mathrm{c}}] + P\left[\bigcap_{j=1}^{m_n-1}(A'_{j,1} \cap B'_{j,0})\right]. \qquad \square$$

Proof of Proposition 11.5 The proof for $d = 2$ was given earlier, so we now assume $d \geq 3$. Consider $G(\mathcal{X}_n \cap B(a); 1)$. By Lemma 11.6, with high probability there is a 1-crossing component and, by Lemma 11.7, this component is actually crossing. By Lemma 11.8, it is unique and there is no other component of metric diameter greater than br. Finally, by the argument used in the case $d = 2$ (see eqn (11.10)), there is no component of metric diameter at most b and with order greater than $2^{d+1}\mu\theta b^d$. All these statements happen with high probability, in the sense that their complements hold with probability bounded by $c'a^d \exp(-cb)$. $\square$

11.3 Fractional consistency of single-linkage clustering

For $h > 0$, set $f^{-1}([h,\infty)) := \{x \in \mathbb{R}^d : f(x) \geq h\}$. Following Hartigan (1975, 1981) we define the high-density *population clusters* (also known as density-contour clusters, or as high-density clusters) at level h to be the connected components of $f^{-1}([h,\infty))$, that is, the regions inside 'contours' of the probability density function f at level h. When $nr_n^d \to \rho$ with $\rho f_{\max} > \lambda_{\mathrm{c}}$, there will be big components of $G(\mathcal{X}_n; r_n)$ with a positive fraction of points of $\mathcal{X}_n$; the number of big components depends on the number of high-density population clusters at level $\lambda_{\mathrm{c}}/\rho$. Asymptotically, there will be one big component of $G(\mathcal{X}_n; r_n)$ for each such population cluster. This means that one can hope to use the big components of $G(\mathcal{X}_n; r_n)$ as consistent estimators for population clusters. However, for each population cluster at level $\lambda_{\mathrm{c}}/\rho$, the associated big component contains not *all* the sample points in D, but *a positive proportion* of the sample points in D; this property is called *fractional consistency* of the big components (i.e. the big single-linkage clusters) as estimators of the population clusters.

This section is concerned with establishing, and making precise, the preceding assertions. Given $D \subseteq \mathbb{R}^d$, and $\rho > 0$, define the integral $I(D;\rho)$ by

$$I(D;\rho) := \int_D p_\infty(\rho f(x)) f(x) \mathrm{d}x. \qquad (11.20)$$

If D is a population cluster for f at level $\lambda_{\mathrm{c}}/\rho$, the asymptotic proportionate order of the big component of $G(\mathcal{X}_n; r_n)$ associated with D is expressed in terms of the integral $I(D;\rho)$. Recall that $L_j(G)$ denotes the order of the jth-largest component of a graph G.

Theorem 11.9 *Suppose that the density function f is continuous. Suppose that* $\lim_{n\to\infty}(nr_n^d) = \rho \in (0,\infty)$, *and that there exists* $h \in (0, \lambda_{\mathrm{c}}/\rho)$ *such that*

$f^{-1}([h,\infty))$ is bounded in $\mathbb{R}^d$. Suppose there are finitely many population clusters at level λ_c/ρ, denoted $R_1,\ldots,R_k$, with $I(R_1;\rho) \geq I(R_2;\rho) \geq \cdots \geq I(R_k;\rho)$. Suppose also for $i = 1,2,\ldots,k$ that R_i is the closure of its interior, that it has connected interior, that its boundary has zero Lebesgue measure, and that $f(x) > \lambda_c/\rho$ for x in the interior of R_i. Then, for all $\varepsilon > 0$,

$$\limsup_{n\to\infty} n^{-1/d} \log P\left[\left|\frac{L_j(G(\mathcal{X}_n;r_n))}{nI(R_j;\rho)} - 1\right| \geq \varepsilon\right] < 0, \qquad 1 \leq j \leq k,$$

and

$$\limsup_{n\to\infty} n^{-1/d} \log P[L_{k+1}(G(\mathcal{X}_n;r_n)) > \varepsilon n] < 0.$$

Hence, for $1 \leq j \leq k$, $n^{-1}L_j(G(\mathcal{X}_n;r_n)) \to I(R_j)$ and $n^{-1}L_{k+1}(G(\mathcal{X}_n;r_n)) \to 0$ as $n \to \infty$ with complete convergence.

Theorem 11.9 is a corollary of Theorem 11.13 below, which establishes that with high probability (i.e. except on an event whose probability decays exponentially in $n^{1/d}$), there is a big cluster associated with each population cluster at level λ_c/ρ. The proof requires various extensions of Proposition 11.4. The first of these gives upper and lower bounds for the order of the biggest component of a random geometric graph on points in a cube.

Proposition 11.10 *Suppose $a > 0$, and set $f_0 := \inf_{x\in B(a)} f(x)$ and $f_1 := \sup_{x\in B(a)} f(x)$. Suppose $\lim_{n\to\infty}(nr_n^d) = \rho$ with $\rho f_0 > \lambda_c$. Set*

$$I_0 := a^d f_0 p_\infty(\rho f_0), \quad I_1 := a^d f_1 p_\infty(\rho f_1).$$

Let $0 < \varepsilon < \min(a/2, 1)$. Let H_n denote the event that (i) *the graph $G(\mathcal{X}_n \cap B(a); r_n)$ has a unique component, denoted $C_n(B(a))$, of metric diameter exceeding ε, and* (ii) *the proportion $Z_n := n^{-1}\mathrm{order}(C_n(B(a)))$ of sample points in $C_n(B(a))$ satisfies*

$$(1-\varepsilon)I_0 \leq Z_n \leq (1+\varepsilon)I_1. \tag{11.21}$$

Then $\limsup_{n\to\infty} n^{-1/d} \log P[H_n^c] < 0$.

Proof By the continuity of $p_\infty(\cdot)$ (Theorem 9.20) we can (and do) choose $\zeta_0 < \zeta_0' < 1 < \zeta_1' < \zeta_1$, such that $\zeta_0\rho f_0 > \lambda_c$, and such that

$$\zeta_0 p_\infty(\rho f_0 \zeta_0) > (1-\varepsilon/2)p_\infty(\rho f_0), \qquad \zeta_1 p_\infty(\rho f_1 \zeta_1) < (1+\varepsilon/2)p_\infty(\rho f_1).$$

For $i = 0,1$, set $\mathcal{Y}_n^i := \mathcal{P}_{n\zeta_i'} \cap B(a)$, a Poisson process with intensity function $n\zeta_i' f(\cdot)\mathbf{1}_{B(a)}(\cdot)$. Then since $nr_n^d \to \rho$, for n large enough $\mathcal{Y}_n^0$ dominates the homogeneous Poisson process $\mathcal{H}_{r_n^{-d}\zeta_0\rho f_0, a}$, and by scaling (Theorem 9.17), $r_n^{-1}\mathcal{Y}_n^0$ dominates $\mathcal{H}_{\zeta_0\rho f_0, r_n^{-1}a}$. Similarly, $r_n^{-1}\mathcal{Y}_n^1$ is dominated by $\mathcal{H}_{\zeta_1\rho f_1, r_n^{-1}a}$.

Let E_n denote the event that $G(\mathcal{X}_n \cap B(a); r_n)$ has a component of metric diameter at least εa and of order at least $n(1-\varepsilon)I_0$. Let E_n' denote the corresponding event when $\mathcal{X}_n$ is replaced by $\mathcal{Y}_n^0$. By considering $r_n^{-1}\mathcal{Y}_n^0$, one sees that $P[E'_n]$ is at least the probability that $G(\mathcal{H}_{\zeta_0 \rho f_0, r_n^{-1} a}; 1)$ has a component of metric diameter at least $\varepsilon a r_n^{-1}$ and of order at least $n(1-\varepsilon)I_0$. The probability that this last event fails to happen decays exponentially in $n^{1/d}$, by Theorem 10.20 and the definitions of I_0 and ζ_0. Also,

$$P[E_n] \geq P[E'_n] - P[\{\mathcal{P}_{n\zeta'_0} \subseteq \mathcal{X}_n\}^{\mathrm{c}}],$$

and hence, by Lemma 1.1, E_n holds with high probability. Combining this fact with the uniqueness result of Proposition 11.4, we have (i) and the lower bound in (ii).

The proof of the upper bound in (ii) is similar. If $Z_n \geq (1+\varepsilon)I_1$, then either $\mathcal{X}_n$ is not contained in $\mathcal{P}_{n\zeta'_1}$, or there is a component of $G(\mathcal{Y}^1_n; r_n)$ with more than $n(1+\varepsilon)I_1$ vertices. The latter event has probability decaying exponentially in $n^{1/d}$ by Theorem 10.20 and the definitions of I_1 and ζ_1; also $P[\{\mathcal{X}_n \subseteq \mathcal{P}_{n\zeta_1}\}^{\mathrm{c}}]$ decays exponentially in n by Lemma 1.1. Thus (ii) is proved. □

For $a > 0$, let $\mathcal{A}_a$ denote the class of sets A of the form $A = \cup_{i=1}^k (B(a) \oplus \{az_i\})$, with $\{z_1, \ldots, z_k\} \subset \mathbb{Z}^d$, such that A has connected interior, that is, such that $\{z_1, \ldots, z_m\}$ is a connected subset of $\mathbb{Z}^d$ (an 'animal'). See Fig. 11.1 for an example with $k = 6$. We extend Proposition 11.10 to sets in $\mathcal{A}_a$ as follows.

Proposition 11.11 *Suppose f is continuous, suppose $a > 0$ and $A \in \mathcal{A}_a$, with A non-empty. Suppose $\lim_{n\to\infty}(nr_n^d) = \rho$ with $\inf_{x\in A}\{\rho f(x)\} > \lambda_{\mathrm{c}}$. Let $0 < \varepsilon < \min(a/2, 1)$. Let $F_n(\varepsilon)$ denote the event that the graph $G(\mathcal{X}_n \cap A; r_n)$ has a unique component, denoted $C_n(A)$, having metric diameter exceeding ε.*

Let $F'_n(\varepsilon)$ be the event that in addition to event $F_n(\varepsilon)$ occurring, (i) *no component of $G(\mathcal{X}_n \cap A; r_n)$, other than $C_n(A)$, has order greater than $n\varepsilon$ and* (ii)

$$(1-\varepsilon)nI(A;\rho) < \mathrm{order}(C_n(A)) < (1+\varepsilon)nI(A;\rho). \qquad (11.22)$$

Then $\limsup_{n\to\infty} n^{-1/d} \log P[F'_n(\varepsilon)^{\mathrm{c}}] < 0$.

Proof Choose $\eta > 0$ such that

$$(1-\eta)^2 > 1-\varepsilon, \quad (1+\eta)^2 + \eta < 1+\varepsilon. \qquad (11.23)$$

Given $m \in \mathbb{N}$, divide A into cubes of side $m^{-1}a$. For x in one of these cubes, let $f_{0,m}(x)$ (respectively, $f_{1,m}(x)$) denote the infimum (respectively, the supremum) of $f(\cdot)$ over that cube, so that $f_{0,m}$ and $f_{1,m}$ are step functions on A. By the continuity of f and of $p_\infty(\cdot)$ (Theorem 9.20), the function $f(\cdot)p_\infty(\rho f(\cdot))$ is Riemann integrable over A (see, e.g., Hoffman (1975)), so we can (and do) take $m_0 \geq 3$ to be so large that

$$\left| \int_A f_{i,m_0}(x) p_\infty(\rho f_{i,m_0}(x))\mathrm{d}x - I(A;\rho) \right| \leq \eta I(A;\rho), \qquad i = 0, 1. \qquad (11.24)$$

Let the constituent cubes (of side $m_0^{-1}a$) of A be denoted $B_1, B_2, \ldots, B_\nu$, taken in an order such that for some $\mu < \nu$ the last $\nu - \mu$ of these cubes in

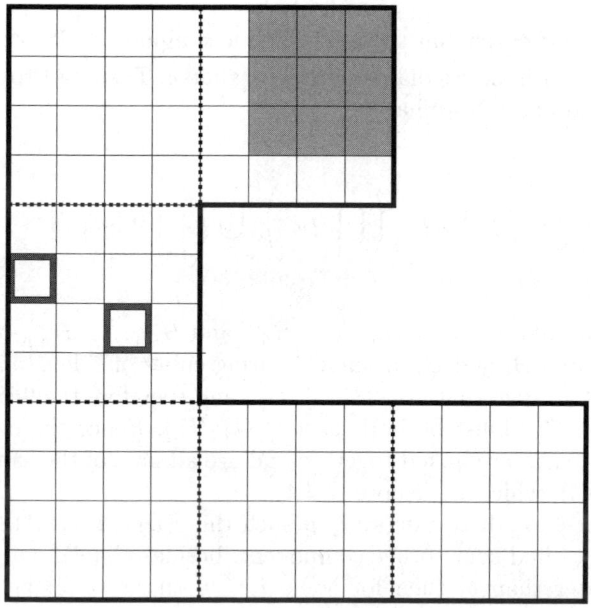

FIG. 11.1. The bold line is the boundary of a set $A \in \mathcal{A}_a$, the dotted lines are of length a, and the grid represents the squares B_i subdividing A; also one of the squares B_i^+ and two of the annular regions $B_i \setminus B_i^-$ are shaded.

the ordering are the ones lying adjacent to the complement of A. For each i, let B_i^+ be the rectilinear cube of side $3m_0^{-1}a$; with the same centre as B_i; then $\cup_{i=1}^{\mu} B_i^+ = A$.

Given $\delta \in (0, a/(4m_0))$, let B_i^- be the rectilinear cube of side $(a/m_0) - 4\delta$ with the same centre as B_i. See Fig. 11.1 for an illustration of both B_i^+ and B_i^-. Since the density function f is assumed bounded, we can (and do) choose $\delta \in (0, \min(a/(4m_0), \varepsilon))$ in such a way that we have

$$F(B_i \setminus B_i^-) < \eta I(A;\rho)/(2\nu), \qquad 1 \le i \le m_0, \tag{11.25}$$

$$F(B(x;2\delta)) < \varepsilon/2, \qquad x \in \mathbb{R}^d. \tag{11.26}$$

For $1 \le i \le \nu$, let $f_i := \inf_{B_i} f(\cdot)$ and $f^i := \sup_{B_i} f(\cdot)$. Let $H_{n,i}$ denote the event that (i) the graph $G(\mathcal{X}_n \cap B_i; r_n)$ has a unique component, denoted $C_n(B_i)$, of metric diameter exceeding δ; (ii) the order of $C_n(B_i)$ satisfies

$$\begin{aligned}(1-\eta)(a/m_0)^d f_i p_\infty(\rho f_i) &\le n^{-1}\text{order}(C_n(B_i)) \\ &\le (1+\eta)(a/m_0)^d f^i p_\infty(\rho f^i);\end{aligned} \tag{11.27}$$

and (iii) $\mathcal{X}_n(B_i \setminus B_i^-) \le n\eta I(A;\rho)/\nu$.

Let $H^+_{n,i}$ be the event that there is a unique component (denoted $C_n(B^+_i)$) in $G(\mathcal{X}_n \cap B^+_i; r_n)$ with metric diameter greater than δ. Then by Proposition 11.10, Lemma 1.1, and (11.25), we have

$$\limsup_{n\to\infty} n^{-1/d} \log P\left[\left(\bigcap_{i=1}^{\nu} H_{n,i}\right)^{\mathrm{c}} \cup \left(\bigcap_{j=1}^{\mu} H^+_{n,j}\right)^{\mathrm{c}}\right] < 0. \tag{11.28}$$

Suppose that all of the events $H_{n,1}, \ldots, H_{n,\nu}$ and $H^+_{n,1}, \ldots, H^+_{n,\mu}$ occur. Then, if $1 \leq i < j \leq \nu$ and B_i and B_j are neighbouring cubes of side a/m_0, there exists k with $1 \leq k \leq \mu$, such that $(B_i \cup B_j) \subseteq B^+_k$, and therefore the 'big' components $C_n(B_i)$ and $C_n(B_j)$ must be both part of $C_n(B^+_k)$. Hence the components for $C_n(B_1), \ldots, C_n(B_\nu)$ are linked together and are all part of the same component of $G(\mathcal{X}_n \cap A; r_n)$, which we denote $C_n(A)$.

For every $x \in A$, there exists $i \leq \mu$ such that $(B(x; 2\delta) \cap A) \subseteq B^+_i$. Hence, if $G(\mathcal{X}_n \cap A; r_n)$ had some other component, besides $C_n(A)$, which had metric diameter greater than δ, then for some $i \leq \mu$ there would be a component of $(\mathcal{X}_n \cap B^+_i; r_n)$, besides $C_n(B^+_i)$, that had metric diameter greater than δ, a contradiction. Hence, if all events $H_{n,i}$ and $H^+_{n,j}$ $(i \leq \nu, j \leq \mu)$ occur, then no component of $G(\mathcal{X}_n \cap A; r_n)$, other than $C_n(A)$ has metric diameter greater than δ, and in particular $C_n(A)$ is the unique component of metric diameter greater than ε, so $F_n(\varepsilon)$ occurs.

To prove (i) in the definition of $F'_n(\varepsilon)$, take points $x_1, x_2, \ldots, x_r$, such that $A \subseteq \cup_{i=1}^r B(x_i; 2\delta)$. Since every component of $G(\mathcal{X}_n \cap A; r_n)$, other than $C_n(A)$, has metric diameter at most δ, every such component has all vertices in the ball $B(x_i; 2\delta)$ for some i. By (11.26) and large deviations for the binomial distribution (Lemma 1.1), every such ball contains at most $n\varepsilon$ points of $\mathcal{X}_n$ with high probability, so condition (i) in the definition of event $F'_n(\varepsilon)$ holds with high probability.

Finally, we establish (11.22). For $1 \leq i \leq \nu$, while $C_n(A) \cap B_i$ may have several components, by the uniqueness condition (i) in the definition of event $H_{n,i}$ all of these components except for $C_n(B_i)$ are contained in the annular region $B_i \setminus B^-_i$ (shown in Fig. 11.1). Hence by condition (iii) in the definition of event $H_{n,i}$,

$$\begin{aligned} 0 &\leq \mathrm{order}(C_n(A)) - \sum_{i=1}^{\nu} \mathrm{order}(C_n(B_i)) \\ &\leq \sum_{i=1}^{\nu} \mathcal{X}_n(B_i \setminus B^-_i) \leq \eta n I(A; \rho). \end{aligned} \tag{11.29}$$

By (11.27) and (11.24),

$$
\begin{aligned}
n^{-1}\sum_{i=1}^{\nu}\mathrm{order}(C_n(B_i)) &\geq (1-\eta)\int_A f_{0,m_0}(x)p_\infty(\rho f_{0,m_0}(x))\mathrm{d}x \\
&\geq (1-\eta)^2 I(A;\rho)
\end{aligned}
$$

and

$$
\begin{aligned}
n^{-1}\sum_{i=1}^{\nu}\mathrm{order}(C_n(B_i)) &\leq (1+\eta)\int_A f_{1,m_0}(x)p_\infty(\rho f_{1,m_0}(x))\mathrm{d}x \\
&\leq (1+\eta)^2 I(A;\rho),
\end{aligned}
$$

and combining these with (11.29), and using (11.23), we obtain (11.22). Therefore the result follows from (11.28). □

Lemma 11.12 *Suppose D is a bounded, connected, open set in $\mathbb{R}^d$, with $\mathbf{0} \in D$. For integer m, let A_m be the maximal element A of $\mathcal{A}_{2^{-m}}$ (possibly the empty set) such that $\mathbf{0} \in A$ and $A \subseteq D$; let A_m^o denote the interior of A_m. Then $A_1 \subseteq A_2 \subseteq A_3 \subseteq \cdots$ and $\cup_{m=1}^{\infty} A_m^o = D$.*

Proof The inclusion $A_m \subseteq A_{m+1}$ is obvious. Since D is open and connected, it is path connected; see Dugundji (1966, Chapter V, Corollary 5.6). Given $x \in D$, take a continuum path γ in D from $\mathbf{0}$ to x. By a compactness argument, this path is bounded away from the boundary of Ω, so lies in the union of the sets A_m^o, $m \geq 1$. Hence $x \in \cup_{m\geq 1} A_m^o$. □

For any set $\Delta \subseteq \mathbb{R}^d$, and $r > 0$, let $U_r(\Delta)$ be the set $\cup_{x\in\Delta} B(x;r)$. Also let $U_{-r}(\Delta)$ be the set of $x \in \mathbb{R}^d$ such that $B(x;r) \subseteq \Delta$ (the r-interior of Δ).

Theorem 11.13 *Suppose that $\lim_{n\to\infty}(nr_n^d) = \rho \in (0,\infty)$, that f is continuous, and that $\Delta \subset \mathbb{R}^d$, with interior D, is a bounded population cluster at level λ_c/ρ. Suppose that Δ is the closure of D, $\Delta \setminus D$ has zero Lebesgue measure, D is connected, and $f(x) > \lambda_c/\rho$ for $x \in D$. Suppose also that there exists $\delta > 0$ such that $f(x) < \lambda_c/\rho, x \in U_\delta(\Delta)\setminus\Delta$.*

For $\varepsilon > 0, \eta > 0$, let $E_n(\varepsilon,\eta)$ be the event that (i) *there is a unique big component of $G(\mathcal{X}_n;r_n)$, denoted $C_n^*(D)$, of order greater than $n\varepsilon$, including at least one vertex in D;* (ii) *no other component having at least one vertex in $U_\eta(D)$ has order greater than $n\varepsilon$, and* (iii)

$$
(1-\varepsilon)nI(D;\rho) < \mathrm{order}(C_n^*(D)) < (1+\varepsilon)nI(D;\rho). \tag{11.30}
$$

Let $0 < \varepsilon < \min(I(D;\rho),1)$. Then there exists $\eta_0 > 0$ such that for $0 < \eta < \eta_0$,

$$
\limsup_{n\to\infty} n^{-1/d}\log P[E_n(\varepsilon,\eta)^{\mathrm{c}}] < 0.
$$

Proof Since Δ is the closure of D, D is non-empty. Assume without loss of generality that $\mathbf{0} \in D$. Let $\varepsilon \in (0,\min(I(D;\rho),1))$. Choose $\eta \in (0,\varepsilon/3)$ such

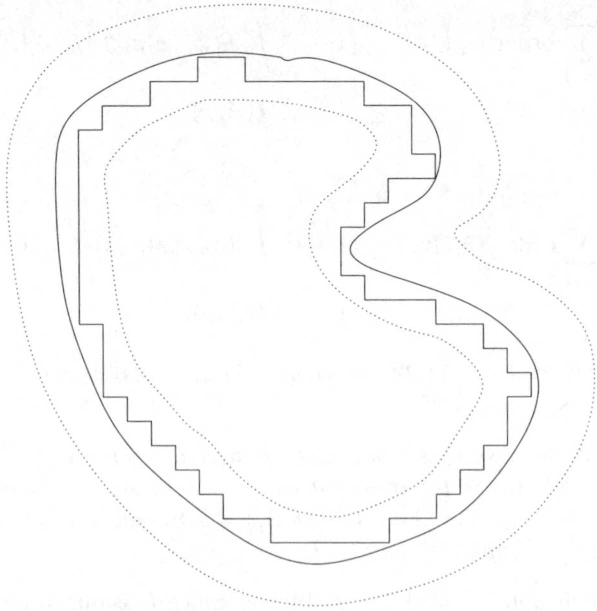

FIG. 11.2. The solid lines represent the boundaries of D and of A_{m_1}. The dotted lines represent the boundary of the set $U_\eta(\Delta \setminus D)$.

that $f(x) < \lambda_c/\rho$ for $x \in U_{3\eta}(D)\backslash\Delta$, such that $B(\mathbf{0}; 3\eta) \subseteq D$, and such that $F(U_{2\eta}(\Delta\backslash D)) < \varepsilon I(D;\rho)/3$, which implies, by Lemma 1.1, that

$$\lim_{n\to\infty} n^{-1} \log P[\mathcal{X}_n(U_{2\eta}(\Delta \setminus D)) > n\varepsilon I(D;\rho)/2] < 0. \tag{11.31}$$

Since $\sup\{f(x) : x \in U_{2\eta}(D) \setminus U_\eta(D)\} < \lambda_c/\rho$, by Lemma 11.3, with high probability no component of $G(\mathcal{X}_n; r_n)$ has vertices both in $\mathbb{R}^d \setminus U_{2\eta}(D)$ and in $U_\eta(D)$. Then all components that include at least one vertex $U_\eta(D)$ are either 'boundary components' having all vertices in $U_{2\eta}(\Delta\backslash D)$, or are 'interior components' having at least one vertex in $U_{-2\eta}(D)$. By (11.31), all boundary components have order at most εn with high probability.

For integer m, let A_m, with interior A_m^o, be the maximal element A of $\mathcal{A}_{2^{-m}}$ such that $\mathbf{0} \in A \subseteq D$ (or $A_m = \emptyset$ if there is no such A). Then, by Lemma 11.12, $A_1 \subseteq A_2 \subseteq A_3 \subseteq \cdots$ and $\cup_{m=1}^\infty A_m^o = D$. By a compactness argument, there exists m_1 such that $U_{-\eta}(A) \subset A_{m_1}$, and such that in addition

$$(1 - \varepsilon/4)I(D;\rho) < I(A_{m_1};\rho) \leq I(D;\rho). \tag{11.32}$$

These sets are shown in Fig. 11.2.

Since $\inf_{A_{m_1}} f(\cdot) > \lambda_c/\rho$, Proposition 11.11 shows that with high probability there is a component $C_n(A_{m_1})$ of $G(\mathcal{X}_n \cap A_{m_1}; r_n)$ with metric diameter greater than $\eta/2$, and no other component with order greater than εn, and also

$$(1-\varepsilon/4)I(A_{m_1};\rho) \leq n^{-1}\mathrm{order}(C_n(A_{m_1})) \leq (1+\varepsilon/4)I(A_{m_1};\rho). \quad (11.33)$$

Let $C_n^*(D)$ denote the component of $G(\mathcal{X}_n;r_n)$ which contains $C_n(A_{m_1})$.

Since $U_{-2\eta}(D) \subseteq U_{-\eta}(A_{m_1})$, every interior component of $G(\mathcal{X}_n;r_n)$, other than $C_n^*(D)$, actually has all of its vertices in A_{m_1}, and so has order at most εn. Therefore, no boundary or interior component of $G(\mathcal{X}_n;r_n)$, other than $C_n^*(D)$, has order more than $n\varepsilon$, with high probability; thus, $C_n^*(D)$ is the unique component with at least one vertex in D and with order more than $n\varepsilon$, as asserted.

All vertices of $C_n^*(D) \setminus C_n(A_{m_1})$ lie in $U_{2\eta}(\Delta \setminus D)$, and so by the boundary estimate (11.31) the total number of such vertices is at most $\varepsilon I(D;\rho)n/2$, with high probability. Combined with (11.33) and (11.32), this gives us (11.30). □

Proof of Theorem 11.9 By continuity and the assumption that there exits $h < \lambda_c/\rho$ such that $f^{-1}([h,\infty))$ is bounded, the population clusters $R_1,\ldots,R_k$ are disjoint compact sets in $\mathbb{R}^d$; hence there exists $\delta > 0$ such that for $1 \leq j \leq k$, $f(x) < \lambda_c/\rho$ for $x \in U_\delta(R_j) \setminus R_j$. Also, for any $\varepsilon > 0$ the supremum of f over the region $\mathbb{R}^d \setminus \cup_{j=1}^k U_\varepsilon(R_j)$ is strictly less than ρ. Then the result is immediate from Theorem 11.13 and Lemma 11.3. □

11.4 Consistency of the RUNT test for unimodality

Given a finite set $\mathcal{X} \subset \mathbb{R}^d$, consider $L_2(G(\mathcal{X};r))$, the order of the second-largest component of $G(\mathcal{X};r)$, as a function of r. As r grows from 0, this function will tend to grow when r is small, but after a while smaller components will tend to get sucked into the biggest component and the order of the second-largest component will tend to shrink, finally becoming zero when r is big enough for the graph to be connected.

In this section, we consider the *maximum* order of the second-largest component, as r varies, We denote this statistic $S(\mathcal{X})$; formally,

$$S(\mathcal{X}) := \max_{r\geq 0} L_2(G(\mathcal{X};r)).$$

We consider $S(\mathcal{X}_n)$, where as usual $\mathcal{X}_n$ is an n-sample from a d-dimensional density function f. We say f is *unimodal* if for every $h > 0$ there is a single population cluster at level h, and *multimodal* otherwise. Hartigan (1981) suggested, and Hartigan and Mohanty (1992) explored further, the idea that $S(\mathcal{X}_n)$ could be used as a test statistic for unimodality, with large values indicating multimodality of f. They call $S(\mathcal{X})$ the 'RUNT'.

This section contains consistency results for the RUNT test (Theorems 11.14 and 11.15). These show that for any density function f that is 'well-behaved' in a sense to be made precise below, the limit of $n^{-1}S(\mathcal{X}_n)$ exists almost surely, and is zero if f is unimodal but is strictly positive if f is multimodal. .

We shall say that height $h > 0$ is *regular for the density* f if it has finitely the population clusters at level h, all satisfying the conditions of Theorem 11.13. That is, h is regular for f if there are finitely many population clusters at level

h, each of which is bounded, is the closure of its interior, has connected interior, has boundary of zero Lebesgue measure, and has $f > h$ on its interior.

We shall say that f is *nowhere constant* if for any $h > 0$, the level set $f^{-1}(\{h\})$ has zero Lebesgue measure.

Theorem 11.14 *Suppose f is continuous, unimodal and nowhere constant, and the set of h that are regular for f is dense in $[0, f_{\max}]$. Then $n^{-1}S(\mathcal{X}_n) \to 0$ as $n \to \infty$, almost surely.*

Proof Define $I(\rho) := I(\mathbb{R}^d; \rho)$ from (11.20), that is,

$$I(\rho) := \int_{\mathbb{R}^d} p_\infty(\rho f(x)) f(x) \mathrm{d}x.$$

By Theorem 9.20 and the dominated convergence theorem, $I(\rho)$ is monotone nondecreasing and continuous in ρ; also $I(\rho) = 0$ for $\rho < \lambda_c/f_{\max}$, and $I(\rho) \to 1$ as $\rho \to \infty$ (by Proposition 9.21).

Let $\varepsilon > 0$. Choose $\rho_1 < \rho_2 < \cdots < \rho_k$ such that λ_c/ρ_i is regular for f for each $i \in \{1, 2, \ldots, k\}$, and such that

$$\varepsilon/2 < I(\rho_1) < 3\varepsilon/4, \qquad I(\rho_k) > 1 - \varepsilon/4, \tag{11.34}$$

and

$$I(\rho_{i+1}) - I(\rho_i) < \varepsilon/4 \quad \text{for } i = 1, 2, \ldots, k-1. \tag{11.35}$$

Set $r_{j,n} = (\rho_j/n)^{1/d}$, for $j = 1, 2, \ldots, k$.

By the assumption of unimodality, for each $j = 1, \ldots, k$ there is a single population cluster at level λ_c/ρ_j. By Theorem 11.9, there exists an almost surely finite random variable N such that for $n \geq N$ and $j = 1, 2, \ldots, k$ we have

$$|n^{-1}L_1(G(\mathcal{X}_n; r_{j,n})) - I(\rho_j)| < \varepsilon/4 \tag{11.36}$$

and

$$n^{-1}L_2(G(\mathcal{X}_n; r_{j,n})) < \varepsilon/4. \tag{11.37}$$

For any geometric graph G and $i = 1, 2, \ldots$, let $C_i(G)$ be the ith-largest component of G (using an arbitrary deterministic ordering in the case of ties). Then, for $i = 1, 2, \ldots, k-1$,

$$C_1(G(\mathcal{X}_n; r_{i,n})) \subseteq C_1(G(\mathcal{X}_n; r_{i+1,n})), \tag{11.38}$$

since if not, and if (11.36) holds for $j = i$ then (11.37) fails for $j = i + 1$, by the hierarchical property of single linkage clustering (see Section 1.2).

Suppose $n \geq N$ and for some $j \in \{1, 2, \ldots, k-1\}$ we have $r \in (r_{j,n}, r_{j+1,n}]$. If $L_2(G(\mathcal{X}_n; r))$ were greater than $n\varepsilon$, then $C_1(G(\mathcal{X}_n; r))$, $C_2(G(\mathcal{X}_n; r))$ would both be contained in $C_1(G(\mathcal{X}_n; r_{j+1,n}))$ (otherwise the second-largest component of

$G(\mathcal{X}_n; r_{j+1,n})$ would be too big) but at least one of $C_1(G(\mathcal{X}_n; r))$, $C_2(G(\mathcal{X}_n; r))$ would be disjoint from $C_1(G(\mathcal{X}_n; r_{j,n}))$ (else they would be connected in $G(\mathcal{X}_n; r)$), and therefore by (11.38) we would have

$$\begin{aligned} n^{-1}L_1(G(\mathcal{X}_n; r_{j+1,n})) &> n^{-1}L_1(G(\mathcal{X}_n; r_{j,n})) + \varepsilon \\ &> I(\rho_j) + 3\varepsilon/4 \qquad \text{by (11.36)} \\ &> I(\rho_{j+1}) + \varepsilon/2 \qquad \text{by (11.35)}, \end{aligned}$$

which contradicts (11.36).

Suppose $n \geq N$ and $r \in (0, r_{1,n}]$. Then $L_2(G(\mathcal{X}_n; r)) \leq L_1(G(\mathcal{X}_n; r_{1,n}))$, which is at most $n\varepsilon$ by (11.34) and (11.36).

Suppose $n \geq N$ and $r > r_{k,n}$. Then $L_1(G(\mathcal{X}_n; r)) \geq L_1(G(\mathcal{X}_n; r_{k,n})) \geq n(1 - \varepsilon)$, the last condition coming from (11.36) and (11.34). Hence $L_2(G(\mathcal{X}_n; r)) \leq n\varepsilon$.

Thus, for $n \geq N$ we have $L_2(G(\mathcal{X}_n; r)) \leq n\varepsilon$ for all r simultaneously, that is, $n^{-1}S(\mathcal{X}_n) < \varepsilon$. Since $\varepsilon > 0$ is arbitrary, we have the result. □

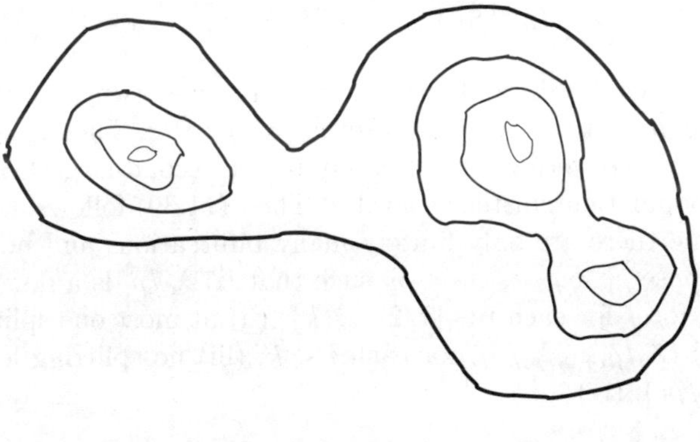

FIG. 11.3. Contour map of a density function with two bifurcations and no trifurcations.

In the multimodal case, a little further examination of population clusters is useful. These clusters have a hierarchical tree structure: if Δ_i is a population cluster at level h_i for $i = 1, 2$, and $h_1 \leq h_2$, then either Δ_1 and Δ_2 are disjoint, or $\Delta_2 \subseteq \Delta_1$. In the latter case let us say Δ_1 is an *ancestor* of Δ_2 and Δ_2 is a *descendant* of Δ_1. Given $h_1 < h_2$, every population cluster at level h_2 has a unique ancestor at level h_1.

If Δ is a population cluster at level h, then as h decreases Δ grows, and may coalesce with one or more other clusters at a splitting level h^*. We shall refer to the merging of two clusters at level h^* as a *bifurcation* and the merging of three or more clusters at level h^* as a *trifurcation* (see Fig. 11.3.) Formally, these are defined as follows.

A *splitting* (respectively, a *bifurcation*) at level $h^* > 0$ is a family of sets $(\Delta_h, h < h^*)$ such that (i) for each h, Δ_h is a population cluster at level h;

(ii) Δ_g is an ancestor of Δ_h for each $g < h < h^*$, and (iii) there exists $\varepsilon > 0$ such that for each h_1, h_2 with $h^* - \varepsilon < h_1 < h^* < h_2 < h^* + \varepsilon$ the population cluster Δ_{h_1} has at least two (respectively, exactly two) descendants at level h_2. A splitting (respectively, a bifurcation) at level 0 occurs if there exists $\varepsilon > 0$ such that for each h with $0 < h < \varepsilon$ there are at least two (respectively, exactly two) population clusters at level h. A *trifurcation* is a splitting that is not a bifurcation.

Theorem 11.15 *Suppose f is continuous, bounded, multimodal and nowhere constant, and the set of regular h for f is dense in $[0, f_{\max}]$. Then*

$$\liminf_{n\to\infty}(n^{-1}S(\mathcal{X}_n)) \geq I_2^{\max} \qquad \text{a.s.,} \tag{11.39}$$

where $I_2^{\max}$ is the supremum, over all regular h such that there are at least two population clusters at level h, of the second-largest of the integrals $I(\Delta; \lambda_c/h)$, Δ a population cluster at level h. Also, if f has finitely many bifurcations and no trifurcations, then

$$\lim_{n\to\infty}(n^{-1}S(\mathcal{X}_n)) = I_2^{\max} \qquad \text{a.s.} \tag{11.40}$$

Proof Choose regular $h > 0$ such that there exist two or more population clusters at level h. Put $\rho = \lambda_c/h$, and set $r_n = (\rho/n)^{1/d}$. Then by Theorem 11.9, $n^{-1}L_2(G(\mathcal{X}_n; r_n))$ converges almost surely to the second-largest of the integrals $I(\Delta; \rho)$, Δ a population cluster at level h. Then (11.39) follows.

Now assume there are only finitely many bifurcations and no trifurcations. Let $\varepsilon > 0$. Choose $\rho_1 < \rho_2 < \cdots < \rho_k$ such that (i) λ_c/ρ_i is a non-splitting level and is regular for f for each $i \in \{1, 2, \ldots, k\}$; (ii) at most one splitting level lies in the interval $(\lambda_c/\rho_{i+1}, \lambda_c/\rho_i)$ for each $i < k$; (iii) no splitting level lies in the interval $(0, \lambda_c/\rho_k)$; (iv)

$$\varepsilon/2 < I(\rho_1) < 3\varepsilon/4, \qquad I(\rho_k) > 1 - \varepsilon/4, \tag{11.41}$$

and (v)

$$I(\rho_{i+1}) - I(\rho_i) < \varepsilon/4 \quad \text{for } i = 1, 2, \ldots, k-1. \tag{11.42}$$

Set $r_{j,n} := (\rho_j/n)^{1/d}$, for $j = 1, 2, \ldots, k$.

For $i = 1, 2, \ldots, k$, let $\Delta_{i,1}, \ldots, \Delta_{i,m(i)}$ be the population clusters at level λ_c/ρ_i. By Theorem 11.13, Lemma 11.3 and the Borel–Cantelli lemma, there exists an almost surely finite random variable N such that for $n \geq N$ and $i = 1, 2, \ldots, k$ there exists a collection of distinct components $C_{i,1,n}, \ldots, C_{i,m(i),n}$ of $G(\mathcal{X}_n; r_{i,n})$ such that for each $j = 1, 2, \ldots, m(i)$ the vertex set of the component $C_{i,j,n}$ has non-empty intersection with $\Delta_{i,j}$ and its order satisfies

$$|(n^{-1}\text{order}(C_{i,j,n})) - I(\Delta_{i,j}; \rho_i)| < \varepsilon/4 \tag{11.43}$$

and moreover no other component of $G(\mathcal{X}_n; r_{i,n})$ has order exceeding $n\varepsilon/4$, so that

$$n^{-1}L_2(G(\mathcal{X}_n;r_{i,n})) \le I_2^{\max} + \varepsilon/4. \tag{11.44}$$

Suppose $r_{i-1,n} < r \le r_{i,n}$ and suppose there are two components of $G(\mathcal{X}_n;r)$, denoted C', C'' say, of order greater than $n(I_2^{\max} + \varepsilon)$. By (11.44), they are contained in the same component, denoted C say, of $G(\mathcal{X}_n;r_{i,n})$; let us say that C has non-empty intersection with the population cluster Δ at level λ_c/ρ_i.

The population cluster Δ contains at most two descendants at level λ_c/ρ_{i-1}. If it has two descendants at level λ_c/ρ_{i-1}, let them be denoted Δ', Δ'', with $I(\Delta';\rho_{i-1}) \ge I(\Delta'';\rho_{i-1})$. If it has one descendant at level λ_c/ρ_{i-1}, let this descendant population cluster be denoted Δ', and let Δ'' be the empty set. If it has no descendants at level λ_c/ρ_{i-1}, let both Δ' and Δ'' be the empty set. In all of these cases, by (11.42),

$$\begin{aligned} I(\Delta;\rho_i) &< I(\Delta;\rho_{i-1}) + \varepsilon/4 \\ &= I(\Delta';\rho_{i-1}) + I(\Delta'';\rho_{i-1}) + \varepsilon/4 \\ &\le I(\Delta';\rho_{i-1}) + I_2^{\max} + \varepsilon/4. \end{aligned} \tag{11.45}$$

At least one of C', C'' (say, C'') is disjoint from the component of $G(\mathcal{X}_n;r_{i-1,n})$ associated with Δ'; the latter component is contained in C and has order at least $n(I(\Delta';\rho_{i-1}) - \varepsilon/4)$ by (11.43). Therefore,

$$\begin{aligned} I(\Delta;\rho_i) &\ge n^{-1}\text{order}(C) - \varepsilon/4 \qquad \text{by (11.43)} \\ &\ge I(\Delta', \rho_{i-1}) - \varepsilon/2 + n^{-1}\text{order}(C'') \\ &\ge I(\Delta';\rho_{i-1}) + I_2^{\max} + \varepsilon/2, \end{aligned}$$

which contradicts (11.45). Hence, $L_2(G(\mathcal{X}_n;r)) \le n(I_2^{\max}+\varepsilon)$ for $r_{1,n} < r \le r_{k,n}$, $n \ge N$.

If $r \le r_{1,n}$ then $L_2(G(\mathcal{X}_n;r)) \le L_1(G(\mathcal{X}_n;r_{1,n})) \le \varepsilon$, by (11.41) and (11.43). Also, if there is no splitting at level 0 then there is just one population cluster at level λ_c/ρ_k, and by (11.41) and (11.43), $L_1(G(\mathcal{X}_n;r_{k,n})) \ge n(1-\varepsilon)$; hence, if $n \ge N$ and $r > r_{k,n}$, $L_1(G(\mathcal{X}_n;r)) \ge n(1-\varepsilon)$ so that $L_2(G(\mathcal{X}_n;r)) \le n\varepsilon$.

Now suppose there is a splitting (assumed to be a bifurcation) at level 0. Then there are two population clusters at level λ_c/ρ_k; let them be denoted Δ and Δ' with $I(\Delta;\rho_k) \ge I(\Delta';\rho_k)$. Then, by (11.41),

$$1 - \varepsilon/4 < I(\rho_k) \le I(\Delta;\rho_k) + I_2^{\max}. \tag{11.46}$$

If there exists $r > r_{k,n}$, and distinct components C', C'' of $G(\mathcal{X}_n;r)$ both with order greater than $n(I_2^{\max} + \varepsilon)$, then at least one of them (say, C'') is disjoint from the component of $G(\mathcal{X}_n;r_{k,n})$ associated with Δ', which has order at least $n(I(\Delta';\rho_k) - \varepsilon/4)$ by (11.43), so that

$$\begin{aligned} n &\ge n(I(\Delta';\rho_k) - \varepsilon/4) + \text{order}(C'') \\ &\ge n(I(\Delta';\rho_k) - \varepsilon/4) + n(I_2^{\max} + \varepsilon), \end{aligned}$$

which contradicts (11.46).

Thus for all r we have $L_2(G(\mathcal{X}_n;r)) \le n(I_2^{\max}+\varepsilon)$, and so $S(\mathcal{X}_n) \le n(I_2^{\max} + \varepsilon)$, $n \ge N$. Since $\varepsilon > 0$ is arbitrary, (11.40) follows. □

11.5 Fluctuations of the giant component

This section contains a central limit theorem for the order of the largest component of $G(\mathcal{X}_n; r_n)$ in the supercritical thermodynamic limit, obtained by de-Poissonizing Theorem 10.22. We consider only the uniform case, assuming throughout this section that $f = f_{\mathrm{U}}$, $d \geq 2$. Let λ be a constant with $\lambda > \lambda_{\mathrm{c}}$, and set $b_n := (n/\lambda)^{1/d}$. Set $H_n(\mathcal{X}) := L_1(G(\mathcal{X}; b_n^{-1}))$. The central limit theorem in this section is for $H_n(\mathcal{X}_n)$.

Theorem 11.16 *Let $\sigma^2 \geq 0$ be the constant appearing as a limiting variance in Theorem 10.22. There is a constant τ^2, with $0 < \tau^2 \leq \sigma^2$, such that*

$$(\lambda/n)\mathrm{Var}(H_n(\mathcal{X}_n)) \to \tau^2 \tag{11.47}$$

and

$$(\lambda/n)^{1/2}(H_n(\mathcal{X}_n) - EH_n(\mathcal{X}_n)) \xrightarrow{\mathcal{D}} \mathcal{N}(0, \tau^2). \tag{11.48}$$

Since $\tau^2 > 0$ in the above result, this shows also that $\sigma^2 > 0$ in Theorem 10.22.

To prove Theorem 11.16, the goal is to use the general de-Poissonization results in Section 2.5. As at (10.79), let $H(\mathcal{X}) := L_1(G(\mathcal{X}; 1))$. We cannot use Theorem 2.16 directly because the functional $H(\cdot)$ is not strongly stabilizing in the sense of Definition 2.15. However, as we shall see, with some effort we can use Lemma 2.13. The main ingredient enabling us to apply this is Lemma 11.17.

Let B_n denote the box $B(b_n)$, and let $U_{i,n} := b_n X_i$. Then let $\mathcal{U}_{m,n} := \{U_{1,n}, \ldots, U_{m,n}\}$, a set of m independent identically distributed uniform random d-vectors on B_n, and define

$$R_{m,n} := H_n(\mathcal{X}_{m+1}) - H_n(\mathcal{X}_m) = H(\mathcal{U}_{m+1,n}) - H(\mathcal{U}_{m,n}).$$

Let Δ be the increment in the order of the infinite component caused by an insertion at the origin. That is, let Δ be the number of points of $\mathcal{H}_{\lambda,0} := \mathcal{H}_\lambda \cup \{\mathbf{0}\}$ (including $\mathbf{0}$ itself) which lie in the infinite component of $G(\mathcal{H}_{\lambda,0}; 1)$ but not in the infinite component of $G(\mathcal{H}_\lambda; 1)$. This is almost surely finite, because there are at most finitely many finite components of $G(\mathcal{H}_\lambda; 1)$ that have at least one vertex in $B(\mathbf{0}; 1)$, and only vertices of such components get (possibly) added to the infinite component as a result of an insertion at the origin.

Lemma 11.17 *Let $\varepsilon > 0$. Then there exists $\delta > 0$ and $n_0 \geq 1$ such that for all $n \geq n_0$ and all $m, m' \in [(1-\delta)n, (1+\delta)n]$ with $m < m'$, there exists a coupled family of random variables D, D', R, R' with following properties:*

- *D and D' are independent, and each have the same distribution as Δ;*
- *(R, R') have the same joint distribution as $(R_{m,n}, R_{m',n})$;*
- *$P[\{D \neq R\} \cup \{D' \neq R'\}] < \varepsilon$.*

Proof By continuity of the percolation probability $p_\infty(\cdot)$ (see Theorem 9.20), we can choose $\delta_1 > 0$ such that for $0 < \delta \leq \delta_1$ we have $\lambda(1-2\delta) > \lambda_c$ and

$$p_\infty(\lambda(1-2\delta)) > p_\infty(\lambda(1+2\delta)) - \varepsilon/18. \tag{11.49}$$

For any locally finite point set $\mathcal{X} \subset \mathbb{R}^d$, let $C_\infty(\mathcal{X})$ be the set of points of $\mathcal{X}$ which are vertices in an infinite component of $G(\mathcal{X};1)$. Let S_0 be the maximum distance from the origin at which a point joins $C_\infty(\mathcal{H}_\lambda)$ as a result of an insertion of a point at the origin, that is, set

$$S_0 := \max\{\|X\| : X \in C_\infty(\mathcal{H}_{\lambda,0}) \setminus C_\infty(H_\lambda)\}.$$

Then S_0 is almost surely finite. Choose K such that

$$\max(P[S_0 > K-2], P[C_\infty(\mathcal{H}_{\lambda(1-2\delta_1)}) \cap B(\mathbf{0};K) = \emptyset]) < \varepsilon/18. \tag{11.50}$$

Set $\delta_2 := \varepsilon K^{-d}/(72\theta\lambda)$, and assume from now on that $\delta < \min(\delta_1, \delta_2)$.

On a suitable probability space, let $\mathcal{H}_{\lambda(1-2\delta)}$, $\mathcal{H}'_{\lambda(1-2\delta)}$, $\mathcal{H}_{4\lambda\delta}$, and $\mathcal{H}'_{4\lambda\delta}$ be four independent homogeneous Poisson processes on $\mathbb{R}^d$ of intensities indicated by the subscripts. Also, given n, let $U, U', V_1, V_2, \ldots$ be independent random d-vectors uniformly distributed over B_n, independent of all these Poisson processes. The random d-vectors U and U' will play the roles of $U_{m+1,n}$ and $U_{m'+1,n}$, respectively.

Let $\mathcal{H}_{\lambda(1+2\delta)}$ be the union of Poisson processes $\mathcal{H}_{\lambda(1-2\delta)} \cup \mathcal{H}_{4\lambda\delta}$, and let $\mathcal{H}'_{\lambda(1+2\delta)}$ be the union $\mathcal{H}'_{\lambda(1-2\delta)} \cup \mathcal{H}'_{4\lambda\delta}$. Then $\mathcal{H}_{\lambda(1+2\delta)}$ and $\mathcal{H}'_{\lambda(1+2\delta)}$ are independent homogeneous Poisson processes, both of intensity $\lambda(1+2\delta)$, by the superposition theorem (Theorem 9.14).

Let $\mathcal{H}_\lambda$ be the union of $\mathcal{H}_{\lambda(1-2\delta)}$ with a thinned modification of $\mathcal{H}_{4\lambda\delta}$ in which each point of $\mathcal{H}_{4\lambda\delta}$ is included with probability $\frac{1}{2}$, independently of the other points. By the thinning and superposition theorems, $\mathcal{H}_\lambda$ is a homogeneous Poisson process of intensity λ. Similarly let $\mathcal{H}'_\lambda$ be the union of $\mathcal{H}'_{\lambda(1-2\delta)}$ with a thinned modification of $\mathcal{H}'_{4\lambda\delta}$ in which each point of $\mathcal{H}_{4\lambda\delta}$ is included with probability $\frac{1}{2}$, independently of the other points (a homogeneous Poisson process of intensity λ). With probability 1, $\mathcal{H}_{\lambda(1-2\delta)} \subseteq \mathcal{H}_\lambda \subseteq \mathcal{H}_{\lambda(1+2\delta)}$ and $\mathcal{H}'_{\lambda(1-2\delta)} \subseteq \mathcal{H}'_\lambda \subseteq \mathcal{H}'_{\lambda(1+2\delta)}$.

Let $\mathcal{H}''_{\lambda(1-2\delta)}$ be the point process consisting of those points of $\mathcal{H}_{\lambda(1-2\delta)}$ which lie closer to U than to U' (in the Euclidean norm), together with those points of $\mathcal{H}'_{\lambda(1-2\delta)}$ which lie closer to U' than to U. Clearly, $\mathcal{H}''_{\lambda(1-2\delta)}$ is a Poisson process of intensity $\lambda(1-2\delta)$ on $\mathbb{R}^d$, and moreover, it is independent of U and of U', because the conditional distribution of the point process $\mathcal{H}''_{\lambda(1-2\delta)}$, given (U, U'), does not depend on the values taken by U, U'. Define $\mathcal{H}''_{4\lambda\delta}$ similarly, and set $\mathcal{H}''_{\lambda(1+2\delta)} := \mathcal{H}''_{\lambda(1-2\delta)} \cup \mathcal{H}''_{4\lambda\delta}$.

Let N^- (respectively, N^*) denote the number of points of $\mathcal{H}''_{\lambda(1-2\delta)}$ (respectively, $\mathcal{H}''_{4\lambda\delta}$) lying in B_n, a Poisson variable with mean $n(1-2\delta)$ (respectively,

$4n\delta$). Let $N^+ := N^- + N^*$, a Poisson variable with mean $n(1+2\delta)$. Choose an ordering on the points of $\mathcal{H}''_{\lambda(1-2\delta)}$ lying in B_n, uniformly at random from all $N^-!$ possible such orderings, and similarly choose an ordering on the points of $\mathcal{H}''_{4\lambda\delta}$ lying in B_n, uniformly at random from all $N^*!$ possible such orderings. Use these ordering to list the points of $\mathcal{H}''_{\lambda(1-2\delta)}$ in B_n as $W_1, W_2, \ldots, W_{N^-}$, and the points of $\mathcal{H}''_{4\lambda\delta}$ in B_n as $W_{N^-+1}, W_{N^-+2}, \ldots, W_{N^+}$. Also, set $W_{N^++1} := V_1$, $W_{N^++2} := V_2$, $W_{N^++3} := V_3$, and so on.

Given $m < m'$, let $\mathcal{U}'_{m,n} := \{W_1, \ldots, W_m\}$ and $\mathcal{U}'_{m+1,n} := \mathcal{U}'_{m,n} \cup \{U\}$; let $\mathcal{U}'_{m',n} := \{W_1, \ldots, W_{m'-1}, U\}$ and let $\mathcal{U}'_{m'+1,n} := \mathcal{U}'_{m',n} \cup \{U'\}$. Let $R := H(\mathcal{U}'_{m+1,n}) - H(\mathcal{U}'_{m,n})$, and let $R' := H(\mathcal{U}'_{m'+1,n}) - H(\mathcal{U}'_{m',n})$. The random d-vectors $U, U', W_1, W_2, W_3, \ldots$, are independent and uniformly distributed on B_n, and therefore the pairs (R, R') and $(R_{m,n}, R_{m',n})$ have the same joint distribution as asserted.

Let D be the number of points of $\mathcal{H}_\lambda \cup \{U\}$ which lie in $C_\infty(\mathcal{H}_\lambda \cup \{U\}) \setminus C_\infty(\mathcal{H}_\lambda)$, and let D' be the number of points of $\mathcal{H}'_\lambda \cup \{U'\}$ which lie in $C_\infty(\mathcal{H}'_\lambda \cup \{U'\}) \setminus C_\infty(\mathcal{H}'_\lambda)$. Then D, D' are independent, and each have the same distribution as Δ, as asserted.

It remains to show that $(D, D') = (R, R')$ with high probability. Let S be the largest distance from U at which a point of $\mathcal{H}_\lambda$ joins the infinite component as a result of the addition of U, an almost surely finite random variable, and let S' be defined similarly in terms of $\mathcal{H}'_\lambda$, U'. That is, set

$$S := \max\{\|X - U\| : X \in C_\infty(\mathcal{H}_\lambda \cup \{U\}) \setminus C_\infty(\mathcal{H}_\lambda)\},$$
$$S' := \max\{\|X - U\| : X \in C_\infty(\mathcal{H}'_\lambda \cup \{U'\}) \setminus C_\infty(\mathcal{H}'_\lambda)\}.$$

Let T_n be the trapezoidal (if $d = 2$) set given by the intersection of B_n with the half-space of points in $\mathbb{R}^d$ lying closer to U than to U', and let T'_n be the set $B_n \setminus T_n$ (see Fig. 11.4).

Given K and δ, define the exceptional events $E_i = E_i(n), 1 \le i \le 6$ as follows:

$$
\begin{aligned}
E_1 &:= \{N^- > n(1-\delta)\} \cup \{N^+ < n(1+\delta)\},\\
E_2 &:= \{B(U; b_n^{1/2}+1) \subseteq T_n\}^{\mathrm{c}} \cup \{B(U'; b_n^{1/2}+1) \subseteq T'_n\}^{\mathrm{c}},\\
E_3 &:= \{C_\infty(\mathcal{H}_{\lambda(1-2\delta)} \cap B(U;K)) = \emptyset\} \cup \{C_\infty(\mathcal{H}'_{\lambda(1-2\delta)} \cap B(U';K)) = \emptyset\},\\
E_4 &:= \{S > K-2\} \cup \{S' > K-2\},\\
E_5 &:= \{\mathcal{H}_{4\lambda\delta}(B(U;K)) > 0\} \cup \{\mathcal{H}'_{4\lambda\delta}(B(U';K)) > 0\},\\
E_6 &:= \{U \in C_\infty(\mathcal{H}_{\lambda(1+2\delta)}) \setminus C_\infty(\mathcal{H}_{\lambda(1-2\delta)})\}\\
&\qquad \cup \{U' \in C_\infty(\mathcal{H}'_{\lambda(1+2\delta)}) \setminus C_\infty(\mathcal{H}'_{\lambda(1-2\delta)})\}.
\end{aligned}
$$

Given also some choice of $m, m' \in [(1-\delta)n, (1+\delta)n]$ with $m < m'$, let $E_7 = E_7(n, m)$ be the complement of the event that $G(\mathcal{U}'_{m,n}; 1)$ has a unique crossing component and no other component of metric diameter greater than $b_n^{1/(2d)}$ or of

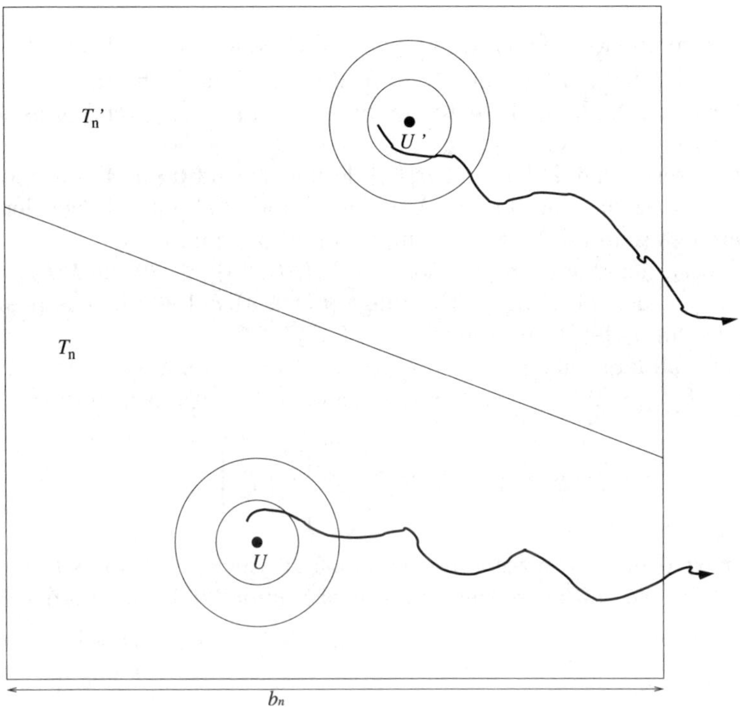

FIG. 11.4. Illustration of the event $(E_2 \cup E_3)^c$. The smaller circles have radius K, while the larger ones have radius $b_n^{1/2} + 1$. The arrows represent paths to infinity in the geometric graph.

order greater than $b_n^{1/3}$. Similarly, let $E_8 = E_8(n, m')$ be the complement of the event that $G(\mathcal{U}'_{m',n}; 1)$ has a unique crossing component and no other component of metric diameter greater than $b_n^{1/(2d)}$ or of order greater than $b_n^{1/3}$.

Suppose none of the events E_i $(1 \leq i \leq 8)$ occurs. Then, by the definitions of E_2 and E_3, there is at least one path in $G(\mathcal{H}_{\lambda(1-2\delta)} \cap T_n; 1)$ from $B(U; K)$ to the region $\mathbb{R}^d \setminus B(U; b_n^{1/2})$, and by the definition of E_1 and E_7, all points lying in any such path must be part of the biggest component of $G(\mathcal{U}'_{m,n}; 1)$. Also, by definition of E_5 and E_1,

$$\mathcal{U}'_{m,n} \cap B(U; K) = \mathcal{H}_{\lambda(1-2\delta)} \cap B(U; K).$$

By definition of E_4, E_5, and E_6, adding the point at U causes precisely D points, all of them in $B(U; K-2)$ to join the infinite component of $G(\mathcal{H}_{\lambda(1-2\delta)}; 1)$, and these are also added to the biggest component of $G(\mathcal{U}'_{m,n}; 1)$; hence $D = R$ if none of the events E_i occurs.

By an analogous argument at U', if none of the events E_i occur, then adding the point at U' causes precisely D' points, all of them in $B(U'; K-2)$ to join the infinite component of $G(\mathcal{H}'_{\lambda(1-2\delta)};1)$, and these are also added to the biggest component of $G(\mathcal{U}'_{m',n};1)$; hence $D' = R'$ if none of the events E_i occurs.

By Lemma 1.2, $P[E_1]$ tends to 0 as $n \to \infty$. Also $P[E_2]$ tends 0 as $n \to \infty$, since $b_n \to \infty$.

Next, observe that $P[E_3]$ and $P[E_4]$ do not depend on n, and are less than $\varepsilon/9$ by the choice of K at (11.50). Also, $P[E_5]$ and $P[E_6]$ also do not depend on n, and are less than $\varepsilon/9$ by the assumption that $\delta < \min(\delta_1, \delta_2)$.

Choose n_2 such that for $n \geq n_2$ we have $P[E_7(n,m)] < \varepsilon/9$ and $P[E_8(n,m')] < \varepsilon/9$, for any choice of m, m' in the range $[n(1-\delta), n(1+\delta)]$. It is possible to choose such an n_2 by Proposition 11.5.

Thus for all large enough n and all $m < m'$ in the range $[(1-\delta)n, (1+\delta)n]$, we have $P[E_j] \leq \varepsilon/9$ for $1 \leq j \leq 8$, and hence by Boole's inequality,

$$P[(D, D') \neq (R, R')] \leq P\left[\bigcup_{j=1}^{8} E_j\right] < \varepsilon. \qquad \square$$

The next lemma provides a moments bound which will help us to check the conditions for the de-Poissonization result of Section 2.5 in the present context.

Lemma 11.18 *There exists $\delta > 0$ such that the functional H satisfies the moments condition*

$$\limsup_{n\to\infty} \sup_{m\in[(1-\delta)n,(1+\delta)n]} \sup_{x\in B_n} \{E[(H(\mathcal{U}_{m,n} \cup \{x\}) - H(\mathcal{U}_{m,n}))^4]\} < \infty. \quad (11.51)$$

Proof Choose $\delta > 0$ so that $\lambda(1-\delta) > \lambda_c$. For $r > 0$, let $E'(n,m;x,r)$ denote the event that there exist two disjoint components in $G(\mathcal{U}_{m,n};1)$ both of which have at least one vertex in $B(x;1)$ and have metric diameter greater than r.

If $(n/\lambda)^{1/d} \leq r/d$, then $E'(n,m;x,r)$ cannot happen. If $(n/\lambda)^{1/d} > r/d$, take a box of side r/d centred at x, and if it extends beyond the edges of B_n, then translate it just enough so it does not, to obtain a box B' (see the proof of Lemma 10.23). For the event $E'(n,m;x,r)$ to occur there must be two disjoint components of $G(\mathcal{U}_{m,n} \cap B';1)$ of metric diameter at least $(r/d) - 2$, and by Proposition 11.5 the probability of this is bounded by $c'\mathrm{e}^{-cr}$, uniformly in n; Proposition 11.5 applies because we assume $m \geq n(1-\delta)$, and the probability density function f_n of a single point uniformly distributed over B_n is λ/n times the indicator function of B_n, so that $mf_n \geq (1-\delta)\lambda$ on the box B'.

The number of points of $\mathcal{U}_{m,n}$ in $B(x;2r)$, is binomial with mean satisfying

$$E[\mathcal{U}_{m,n}(B(x;2r))] \leq \theta m(\lambda/n)(2r)^d \leq 2^d\theta\lambda(1+\delta)r^d,$$

and hence, by Lemma 1.1, there is a constant c such that

$$P[\mathcal{U}_{m,n}(B(x;2r)) \geq 2^{d+1}\theta\lambda(1+\delta)r^d] \leq \exp(-cr^d).$$

If $H(\mathcal{U}_{m,n} \cup \{x\}) - H(\mathcal{U}_{m,n})$ is to exceed $2^{d+1}\theta\lambda(1+\delta)r^d + 1$, then either we must have $\mathcal{U}_{m,n}(B(x;2r)) \geq 2^{d+1}\theta\lambda(1+\delta)r^d$, or event $E'(n,m;x,r)$ must occur.

Therefore, by the preceding estimates there are constants c, c' such that for all t,

$$P[H(\mathcal{U}_{m,n} \cup \{x\}) - H(\mathcal{U}_{m,n}) > t] < c' \exp(-ct^{1/d}),$$

uniformly over $n \geq 1$ and $m \in [n(1-\delta), n(1+\delta)]$. By the integration by parts formula for expectation, this uniformly sub-exponentially decaying tail behaviour is enough to yield the uniformly bounded fourth moments (11.51). □

Proof of Theorem 11.16 We apply Lemma 2.13. Let the binomial and Poisson point processes $\mathcal{X}_n$ and $\mathcal{P}_n$ be defined as in Sections 1.5 and 1.7, using the uniform density $f = f_U$. By scaling (Theorem 9.17), the point process $(n/\lambda)^{1/d}\mathcal{P}_n$ has the same distribution as $\mathcal{H}_{\lambda,(n/\lambda)^{1/d}}$, so $H_n(\mathcal{P}_n)$ has the same distribution as $H(\mathcal{H}_{\lambda,(n/\lambda)^{1/d}})$. Therefore, with σ^2 defined in Theorem 10.22, that result gives us

$$(\lambda/n)\mathrm{Var}(H_n(\mathcal{P}_n)) \to \sigma^2, \quad (\lambda/n)^{1/2}(H_n(\mathcal{P}_n) - EH_n(\mathcal{P}_n)) \xrightarrow{\mathcal{D}} \mathcal{N}(0, \sigma^2),$$

so that

$$n^{-1}\mathrm{Var}(H_n(\mathcal{P}_n)) \to \sigma^2/\lambda, \quad n^{-1/2}(H_n(\mathcal{P}_n) - EH_n(\mathcal{P}_n)) \xrightarrow{\mathcal{D}} \mathcal{N}(0, \sigma^2/\lambda).$$

If $(\nu(n), \nu'(n))$ is an $(\mathbb{N} \times \mathbb{N})$-valued sequence satisfying $\nu(n) < \nu'(n)$ for all n and $n^{-1}\nu(n) \to 1$ and $n^{-1}\nu'(n) \to 1$ as $n \to \infty$, then it follows from Lemma 11.17 that $(R_{\nu(n),n}, R_{\nu'(n),n}) \xrightarrow{\mathcal{D}} (\Delta, \Delta')$, where Δ' is an independent copy of Δ. In other words, the first condition (2.46) in Lemma 2.13 is satisfied. The second condition (2.47) is also satisfied by Lemma 11.18.

Thus, Lemma 2.13 is applicable, and shows that conditions (2.38)–(2.40) in Theorem 2.12 hold here, with $\alpha := E\Delta$ and $\gamma = \frac{3}{4}$. Also, condition (2.41) holds because $H_n(\mathcal{X}_m) \leq m$ trivially. Thus, H_n satisfies all the conditions for the de-Poissonization result (Theorem 2.12), which gives us (11.47) and (11.48) as asserted, with $\tau^2 = \sigma^2 - \lambda(E\Delta)^2$.

An argument similar to the proof of Lemma 11.17 (we omit the details) shows that the functional H_n of this section satisfies the conditions (2.52) and (2.53) in Lemma 2.14. Clearly the variable Δ defined just before Lemma 11.17 has a non-degenerate distribution, and therefore it follows from Lemma 2.14 that $\liminf_{n\to\infty}(n^{-1}\mathrm{Var}H_n(\mathcal{X}_n)) > 0$; hence $\tau > 0$. □

11.6 Notes and open problems

Notes Theorem 11.13 appears in Penrose (1995). The other results in this chapter are new. The argument in Section 11.5 is related to the one used in Penrose and Yukich (2001) to de-Poissonize various other central limit theorems arising in geometrical probability.

Open problems In the setting of Theorem 11.9 one may be able to show that if $j > k$, then $L_j(G(\mathcal{X}_n; r_n))$ grows logarithmically in n, almost surely.

In Theorem 11.9, the order of the large deviations estimate is $n^{1/d}$ (i.e. the probability of the exceptional event decays exponentially in $n^{1/d}$). This is the correct order because of the possibility of a 'bridge' between distinct population clusters. However, in the case where there is only a single population cluster at level λ_c/ρ, it may be possible to improve the order of the large deviations estimate to $n^{(d-1)/d}$ instead of $n^{1/d}$. In the Poisson case, at least for $f = f_U$, this is the order of magnitude of the large deviations given by Theorem 10.19; it is an open problem to de-Poissonize that result.

When the 'nowhere constant' condition of Theorem 11.14 fails, for example when f is the uniform density f_U, the behaviour of the RUNT statistic $S(\mathcal{X}_n)$ is much more delicate, and its analysis is left as an open problem. As mentioned in a related context in the preceding chapter, some kind of continuum version of results for lattice percolation in Borgs *et al.* (2001) could be helpful here.

Theorem 11.16 yields a central limit theorem for $L_1(G(\mathcal{X}_n; (\lambda/n)^{1/d}))$ in the uniform case $f = f_U$; proving an analogous result in the non-uniform case remains open.

Recall from Theorem 10.18 that for $\lambda > \lambda_c$, $L_2(G(\mathcal{H}_\lambda; 1))$ is $\Theta(\log s)^{d/(d-1)}$ in probability. A de-Poissonized version of this result would say that for $f = f_U$, and for $\lim_{n\to\infty}(nr_n^d) = \lambda \in (\lambda_c, \infty)$, there are positive finite constants c_1, c_2 such that

$$\lim_{n\to\infty} P\left[c_1(\log n)^{d/(d-1)} < L_2(G(\mathcal{X}_n; r_n)) < c_2(\log n)^{d/(d-1)}\right] = 1.$$

The author believes that this can probably be proved by first showing that $L_2(G(\mathcal{P}_{n+n^{3/4}}); 1)$ has the desired behaviour, and then removing randomly selected points from $\mathcal{P}_{n+n^{3/4}}$ to get the point process $\mathcal{X}_n$; however, the first step in such an argument would require many of the arguments in Chapter 10, concerning $G(\mathcal{H}_{\lambda,s}; 1)$ with λ fixed with $\lambda > \lambda_c$, to be generalized to $G(\mathcal{H}_{\lambda(s),s}; 1)$ when $\lambda(s)$ is a function of s tending to a limit $\lambda > \lambda_c$. One needs to check that all the relevant arguments in Chapter 10 can be modified to this more general case, and the author has not done so.

A tighter version of the preceding conjecture would say that under the above hypotheses concerning f and r_n, $(\log n)^{-d/(d-1)} L_2(G(\mathcal{X}_n; r_n))$ converges to a positive finite limit in probability as $n \to \infty$.

12

ORDERING AND PARTITIONING PROBLEMS

This chapter contains an investigation of asymptotic growth rates for the optimal costs of various layout problems, of the type described in Section 1.3, on random geometric graphs $G(\mathcal{X}_n; r_n)$, in the thermodynamic limiting regime $nr_n^d \to \lambda$ or the dense limiting regime $nr_n^d \to \infty$. Throughout the chapter, we assume that *the underlying density is* $f = f_{\mathrm{U}}$ (i.e. the uniform density on the unit cube $[0, 1]^d$), and that *the norm of choice* $\|\cdot\|$ *is one of the* l_p *norms,* $1 \leq p \leq \infty$.

It turns out that these layout problems exhibit a phase transition at $\lambda = \lambda_{\mathrm{c}}$, where we recall, from Section 9.6, the definition of the continuum percolation probability $p_\infty(\lambda)$ in terms of the infinite homogeneous Poisson process $\mathcal{H}_\lambda$, and the critical value $\lambda_{\mathrm{c}} := \inf\{\lambda : p_\infty(\lambda) > 0\}$. The subcritical case with $\lambda < \lambda_{\mathrm{c}}$ is considered in Section 12.2, the subcritical case with $\lambda < \lambda_{\mathrm{c}}$ is considered in Section 12.3, and in Section 12.4 sharper asymptotic bounds are obtained in the superconnective regime with $nr_n^d / \log n \to \infty$.

12.1 Background on layout problems

The layout problems considered here are formally defined as follows. Given a finite graph $G = (V, E)$, a *layout* or *ordering* φ on G is a one-to-one function $\varphi : \{1, 2, \ldots, n\} \to V$, with $n = |V|$ and $|\cdot|$ denoting cardinality. Given such a layout φ, for each edge $e = \{u, v\} \in E$ the associated weight is $\sigma(e, \varphi) := |\varphi^{-1}(u) - \varphi^{-1}(v)|$. For $v \in V$, define $R(v, \varphi) := \{u \in V : \varphi^{-1}(u) > \varphi^{-1}(v)\}$ (the vertices to the 'right' of v, in the sense that they succeed v in the ordering) and $L(v, \varphi) := V \backslash R(v, \varphi)$ (the vertices to the 'left' of v, including v itself). Define the edge-boundary $\chi(v, \varphi) = \chi(v, \varphi, G)$ and interior vertex-boundary $\Delta(v, \varphi) = \Delta(v, \varphi, G)$ of $L(v, \varphi)$ by

$$\chi(v, \varphi) := |\{\{u, w\} \in E \ : \ u \in L(v, \varphi) \text{ and } w \in R(v, \varphi)\}|, \tag{12.1}$$

$$\Delta(v, \varphi) := |\{u \in L(v, \varphi) \ : \ \exists w \in R(v, \varphi) \text{ with } \{u, w\} \in E\}|. \tag{12.2}$$

For the minimum linear arrangement (MLA) problem, the cost $\mathrm{LA}(\varphi)$ of a layout φ is given by $\mathrm{LA}(\varphi) := \sum_{e \in E} \sigma(e, \varphi)$. An alternative formulation is $\mathrm{LA}(\varphi) := \sum_{v \in V} \chi(v, \varphi)$, which is equivalent because

$$\begin{aligned} \sum_{e \in E} \sigma(e, \varphi) &= \sum_{i=1}^{n-1} \sum_{j=i+1}^{n} \sum_{k=i}^{j-1} \mathbf{1}_{\{\{\varphi(i), \varphi(j)\} \in E\}} \\ &= \sum_{k=1}^{n-1} \sum_{i=1}^{k} \sum_{j=k+1}^{n} \mathbf{1}_{\{\{\varphi(i), \varphi(j)\} \in E\}} = \sum_{v \in V} \chi(v, \varphi). \end{aligned}$$

As well as MLA, the minimum bandwidth (MBW) and minimum bisection (MBIS) problems were mentioned in Section 1.3. In addition to these problems, we study the problems of *minimum cut* (MCUT), *minimum sum cut* (MSC), and *minimum vertex separation* (MVS). In each of the six problems, given a graph G the object is to minimize some cost functional over the collection $\Phi(G)$ of all layouts on G. The respective cost functionals for a given layout φ are denoted $\text{LA}(\varphi)$, $\text{BW}(\varphi)$, $\text{BIS}(\varphi)$, $\text{CUT}(\varphi)$, $\text{SC}(\varphi)$, $\text{VS}(\varphi)$, respectively, defined as follows:

$$\begin{aligned}
\text{MLA}(G) &= \min_{\varphi\in\Phi(G)} \text{LA}(\varphi) \quad \text{with} \quad \text{LA}(\varphi) = \sum_{e\in E}\sigma(e,\varphi) = \sum_{v\in V}\chi(v,\varphi),\\
\text{MBW}(G) &= \min_{\varphi\in\Phi(G)} \text{BW}(\varphi) \quad \text{with} \quad \text{BW}(\varphi) = \max_{e\in E}\sigma(e,\varphi),\\
\text{MBIS}(G) &= \min_{\varphi\in\Phi(G)} \text{BIS}(\varphi) \quad \text{with} \quad \text{BIS}(\varphi) = \chi(\varphi(\lfloor n/2\rfloor),\varphi),\\
\text{MCUT}(G) &= \min_{\varphi\in\Phi(G)} \text{CUT}(\varphi) \quad \text{with} \quad \text{CUT}(\varphi) = \max_{v\in V}\chi(v,\varphi),\\
\text{MSC}(G) &= \min_{\varphi\in\Phi(G)} \text{SC}(\varphi) \quad \text{with} \quad \text{SC}(\varphi) = \sum_{v\in V}\Delta(v,\varphi),\\
\text{MVS}(G) &= \min_{\varphi\in\Phi(G)} \text{VS}(\varphi) \quad \text{with} \quad \text{VS}(\varphi) = \max_{v\in V}\Delta(v,\varphi).
\end{aligned}$$

Motivation for studying layout problems was briefly described in Section 1.3; we continue this discussion here. A more extensive discussion can be found in Petit (2001); also in Díaz *et al.* (2001*a*).

In the study of very large scale integration (VLSI) problems, one may represent an integrated circuit by means of a graph. One possible aim is to lay out the nodes and edges of a specified input graph onto a board in an efficient manner. The possible node positions may lie in a one- or two-dimensional array. If the array is one-dimensional and the aim is to minimize the total length of wire connecting nodes, this is precisely the MLA problem. Finding the minimax wire length for one-dimensional arrays is the BW problem. For further discussion, see the surveys of Bhatt and Leighton (1984) and Sangionvanni-Vincentelli (1987).

Layout problems also arise in parallel computing. Given two parallel processors with which to attack some problem with a graph representation, it may be beneficial to minimize the interaction between the two processors, and MBIS and related problems are relevant (see Leighton (1992) and Diekmann *et al.* (1995)). Given a larger collection of processors, *embeddings* (i.e. injections of the vertices) of a specified graph G into a host graph H are an important object of study (see Monien and Sudborough (1990)) and when the host graph is a one- or two-dimensional array the study of efficient embeddings resembles the layout problems arising in VLSI.

In numerical analysis, there are computations and information storage procedures on sparse symmetric matrices which are most efficiently carried out when all non-zero entries lie near the diagonal. The *bandwidth* of a matrix is the maximal distance from the diagonal of non-zero entries. A symmetric matrix may be represented by a labelled graph with edges representing non-zero entries, and the MBW problem amounts to the relabelling of the matrix to minimize this bandwidth. Also of interest is minimizing the *profile* of the matrix, which is the maximum distance from the diagonal of non-zero sub-diagonal entries in a given row, summed over all rows. The profile of a matrix is equal to the sum-cut of the reverse ordering to that of the corresponding ordered graph (we leave the proof of this as an exercise). Therefore, relabelling a matrix to minimize the profile is equivalent to the MSC problem. See, for example, Gibbs *et al.* (1976) and Saad (1996) for more information on these applications.

Ordering problems also arise in the reconstruction of DNA sequences from fragments, given information on overlaps of genes between fragments that may sometimes be usefully expressed graphically; see Karp (1993). MLA has been used in brain cortex modelling Mitchison and Durbin (1986). For numerous other applications of these problems, see Petit (2001) and Díaz *et al.* (2001*a*).

As mentioned in Section 1.3, many of the graphs arising in these applications are geometrical in nature, and random geometric graphs provide a natural testing ground for comparing heuristics for these problems. Simulation studies based on random geometric graphs for these kinds of problems include Berry and Goldberg (1999), Johnson *et al.* (1989), and Lang and Rao (1993).

Except for MBIS, the minimal costs have a *monotone* property given by the following lemma. The proof is trivial and is omitted.

Lemma 12.1 *If G is a subgraph of G' then* $\mathrm{MSC}(G) \leq \mathrm{MSC}(G')$, $\mathrm{MLA}(G) \leq \mathrm{MLA}(G')$, $\mathrm{MCUT}(G) \leq \mathrm{MCUT}(G')$, $\mathrm{MBW}(G) \leq \mathrm{MBW}(G')$, *and* $\mathrm{MVS}(G) \leq \mathrm{MVS}(G')$.

The cost for MBIS is not monotone, but satisfies

$$\mathrm{MBIS}(G) \leq \mathrm{MCUT}(G). \tag{12.3}$$

The next lemma provides inequalities relating the different layout problems to one another.

Lemma 12.2 *For any graph G with n vertices and maximum degree D,*

$$\mathrm{MSC}(G) \leq \mathrm{MLA}(G) \leq n\mathrm{MCUT}(G) \leq D \times n\mathrm{MBW}(G), \tag{12.4}$$

$$\mathrm{MSC}(G) \leq n\mathrm{MVS}(G) \leq n\mathrm{MBW}(G). \tag{12.5}$$

Proof It suffices to prove that for any layout φ on G we have

$$\mathrm{SC}(\varphi) \leq \mathrm{LA}(\varphi) \leq n \times \mathrm{CUT}(\varphi) \leq D \times n\mathrm{BW}(\varphi); \tag{12.6}$$

$$\mathrm{SC}(\varphi) \leq n \times \mathrm{VS}(\varphi) \leq n \times \mathrm{BW}(\varphi). \tag{12.7}$$

To prove the second inequality of (12.7), choose a layout φ, and $v \in V$ such that $\Delta(v, \varphi) = \mathrm{VS}(\varphi)$. Then there are $\mathrm{VS}(\varphi)$ vertices in $L(v, \varphi)$ that are connected to vertices in $R(\varphi)$. The first of these in the ordering must have an edge that jumps at least $\mathrm{VS}(\varphi)$ nodes. In other words, there is an edge $e \in E$ with weight $\sigma(e, \varphi) \geq \mathrm{VS}(\varphi)$ so that $\mathrm{BW}(\varphi) \geq \mathrm{VS}(\varphi)$. The other inequalities in (12.6) and (12.7) are proved by similar elementary arguments that we leave as exercises. □

12.2 The subcritical case

This section is concerned with the asymptotic behaviour of layout problems on $G(\mathcal{X}_n; r_n)$ in the subcritical thermodynamic limit $nr_n^d \to \lambda$ with $0 < \lambda < \lambda_c$. Recall from Section 9.6 that $p_k(\lambda)$ is the probability that the component of $G(\mathcal{H}_\lambda \cup \{\mathbf{0}\}; 1)$ containing the origin is of order k, and $p_\infty(\lambda) = 1 - \sum_{k=1}^{\infty} p_k(\lambda)$. For each finite graph Γ let $|\Gamma|$ denote its order and let $p_\Gamma(\lambda)$ be the probability that the component containing the origin of $G(\mathcal{H}_\lambda \cup \{\mathbf{0}\}; 1)$ is isomorphic to Γ.

Theorem 12.3 *Suppose* $\lim_{n\to\infty}(nr_n^d) = \lambda$ *with* $0 < \lambda < \lambda_c$. *Then, as* $n \to \infty$,

$$n^{-1}\mathrm{MLA}(G(\mathcal{X}_n; r_n)) \xrightarrow{P} \beta_{\mathrm{LA}}(\lambda) := \sum_{\Gamma} |\Gamma|^{-1} p_\Gamma(\lambda)\mathrm{MLA}(\Gamma) \qquad (12.8)$$

and

$$n^{-1}\mathrm{MSC}(G(\mathcal{X}_n; r_n)) \xrightarrow{P} \beta_{\mathrm{SC}}(\lambda) := \sum_{\Gamma} |\Gamma|^{-1} p_\Gamma(\lambda)\mathrm{MSC}(\Gamma), \qquad (12.9)$$

where in each case the sum is over all finite graphs Γ *(more accurately, over all isomorphism-equivalence classes of such graphs). Also,* $\beta_{\mathrm{LA}}(\lambda)$ *and* $\beta_{\mathrm{SC}}(\lambda)$ *are finite.*

Proof For any finite graph Γ,

$$\mathrm{MLA}(\Gamma) \le |\Gamma|^3, \quad \mathrm{MSC}(\Gamma) \le |\Gamma|^2. \qquad (12.10)$$

Therefore $\beta_{\mathrm{LA}}(\lambda) \le \sum_{k=1}^{\infty} k^2 p_k(\lambda)$, which is finite by exponential decay in the subcritical regime (Lemma 10.2), and similarly $\beta_{\mathrm{SC}}(\lambda) < \infty$.

Given a finite graph $G = (V, E)$ with components $G_1, \ldots, G_m$, we have $\mathrm{MLA}(G) = \sum_{i=1}^{m} \mathrm{MLA}(G_i)$. For $k \in \mathbb{N}$, let $\mathrm{MLA}_k(G)$ be the contribution to this sum from components of order at most k. Also, let $\mathrm{MLA}^k(G)$ denote the remainder $\mathrm{MLA}(G) - \mathrm{MLA}_k(G)$. For each vertex $v \in V$, let $q(v, G)$ be the order of the component of G containing v. Let

$$U_k(G) := \sum_{i=1}^{m} |G_i|^3 \mathbf{1}_{\{|G_i|>k\}} = \sum_{v\in V} q(v; G)^2 \mathbf{1}_{\{q(v;G)>k\}}.$$

Then by (12.10), $\mathrm{MLA}^k(G) \le U_k(G)$. Also, $U_k(\cdot)$ is monotone in the sense that if G is a subgraph of G' then $U_k(G) \le U_k(G')$. By Theorem 3.15, for any k we have

$$n^{-1}\mathrm{MLA}_k(G(\mathcal{X}_n; r_n)) \xrightarrow{P} \beta_{\mathrm{LA}}(\lambda; k) := \sum_{\Gamma:|\Gamma|\le k} |\Gamma|^{-1} p_\Gamma(\lambda)\mathrm{MLA}(\Gamma). \quad (12.11)$$

Choose μ, ν with $\lambda < \mu < \nu < \lambda_c$, and let D denote the order of the component containing the origin of $G(\mathcal{H}_\nu \cup \{\mathbf{0}\}; 1)$. Let $\varepsilon > 0$ and, using exponential decay (Lemma 10.2) once more, choose k to be so large that $\beta_{\text{LA}}(\lambda) - \beta_{\text{LA}}(\lambda; k) < \varepsilon$ and $\nu E[D^2 \mathbf{1}_{\{D>k\}}] < \varepsilon^2 \lambda$.

With $N_{n\mu/\lambda}$ denoting the cardinality of the coupled Poisson process $\mathcal{P}_{n\mu/\lambda}$ as described in Section 1.7, and $\mathcal{H}_{\lambda,s}$ denoting a homogeneous Poisson process on the box $[-s/2, s/2]^d$ as at (9.11), for n large we have

$$\begin{aligned} P[n^{-1}U_k(G(\mathcal{X}_n; r_n)) > \varepsilon] &\leq P[n^{-1}U_k(G(\mathcal{P}_{n\mu/\lambda}; r_n)) > \varepsilon] + P[N_{n\mu/\lambda} < n] \\ &= P[U_k(G(\mathcal{H}_{nr_n^d\mu/\lambda, r_n^{-1}}; 1)) > n\varepsilon] + P[N_{n\mu/\lambda} < n] \\ &\leq P[U_k(G(\mathcal{H}_{\nu, r_n^{-1}}; 1)) > n\varepsilon] + P[N_{n\mu/\lambda} < n]. \quad (12.12) \end{aligned}$$

The second term in (12.12) tends to zero, while by Markov's inequality and Palm theory for the Poisson process (Theorem 1.6), the first is bounded by

$$\begin{aligned} &(n\varepsilon)^{-1} \int_{B(r_n^{-1})} E[q(x; G(\mathcal{H}_{\nu, r_n^{-1}} \cup \{x\}; 1))^2 \mathbf{1}_{\{q(x; G(\mathcal{H}_{\nu, r_n^{-1}} \cup \{x\}; 1)) > k\}}] \nu dx \\ &\qquad \leq (n\varepsilon r_n^d)^{-1} \nu E[D^2 \mathbf{1}_{\{D>k\}}] < \varepsilon, \qquad n \text{ large}. \end{aligned}$$

Combined with (12.11), this gives us

$$P[|n^{-1}(\text{MLA}_k(G(\mathcal{X}_n; r_n)) + \text{MLA}^k(G(\mathcal{X}_n; r_n))) - \beta_{\text{LA}}| > 3\varepsilon] < 2\varepsilon$$

for large enough n, and this in turn gives us (12.8). The proof of (12.9) is similar. □

Theorem 12.4 *Suppose* $\lim_{n\to\infty}(nr_n^d) = \lambda$ *with* $p_\infty(\lambda) < \frac{1}{2}$. *Then*

$$\text{MBIS}(G(\mathcal{X}_n; r_n)) \xrightarrow{P} 0. \qquad (12.13)$$

Proof By assumption, $p_\infty(\lambda) < \frac{1}{2}$ so $\sum_{k=1}^{\infty} p_k(\lambda) > \frac{1}{2}$. Moreover, $p_k(\lambda) > 0$ for all k (see Lemma 9.23). Pick k_1 such that $\sum_{k=1}^{k_1} p_k(\lambda) > \frac{1}{2}$.

Let $N_k(n)$ be the number of components of $G(\mathcal{X}_n; r_n)$ of order k. Let E_n be the event that, firstly, $N_k(n) > 0$ for all $k \leq k_1$, and secondly, $\sum_{k=1}^{k_1} N_k(n) > n/2$. By Theorem 3.15 for any finite k, $n^{-1}N_k(n)$ converges almost surely to $p_k(\lambda)$, so E_n occurs for all but finitely many n, almost surely.

It suffices to prove that on event E_n we have $\text{MBIS}(G(\mathcal{X}_n; r_n)) = 0$. If E_n occurs, generate a subset W of $\mathcal{X}_n$ as follows. First take the union of all components of order greater than k_1. Then add components of order k_1 until there are none left. Then add components of order $k_1 - 1$ until there are none left. Continue in this way. At some point, having just added a set of i points, we will have a set of $\lfloor n/2 \rfloor - m$ points, with $0 \leq m < i$. If $m = 0$, then stop. If $m > 0$, add a component of order m and stop. Let W be the union of added components. Then $|W| = \lfloor n/2 \rfloor$, and there are no edges of $G(\mathcal{X}_n; r_n)$ connecting W to $\mathcal{X}_n \setminus W$, which shows that $\text{MBIS}(G(\mathcal{X}_n; r_n)) = 0$. □

The known results for the vertex separation and minimum cut costs in the subcritical regime are less precise, and just give order-of-magnitude growth rates, in the case $d = 2$.

Theorem 12.5 *Suppose $d = 2$, and suppose $nr_n^2 \to \lambda$ with $0 < \lambda < \lambda_c$. Then with probability 1,*

$$\mathrm{MVS}(G(\mathcal{X}_n; r_n)) = \Theta(\log n / (\log \log n)) \tag{12.14}$$

and

$$\mathrm{MCUT}(G(\mathcal{X}_n; r_n)) = \Theta((\log n / (\log \log n))^2), \tag{12.15}$$

as $n \to \infty$.

In the proof of this we shall use the notation $\log_2 n$ for $\log \log n$. The first step in the proof is the following deterministic upper bound on the worst-case cut value in the lattice.

Lemma 12.6 *Suppose $d = 2$ and $r > 0$. Then, for any geometric graph $G(A; r)$ with A a finite subset of $\mathbb{Z}^2$, and for any $k \in \{1, 2, \ldots, |A|\}$, there exists an ordering φ on $G(A; r)$ with $\chi(\varphi(k), \varphi) \leq 12(r+1)^4 |A|^{1/2}$.*

Proof Let $n \geq 1$ and let $A \subseteq \mathbb{Z}^d$ with $|A| = n$. We need to find $S \subseteq A$ with $|S| = k$, connected to $A \setminus S$ by at most $12(r+1)^4 n^{1/2}$ edges. Note first that since we are using an l_p norm, every vertex in $G(A; r)$ has degree at most $(2r+1)^2 - 1 = 4r(r+1)$.

For $x \in \mathbb{Z}$, let $S_x = \{y \in \mathbb{Z} : (x, y) \in A\}$, and define the sets

$$W := \{u \in \mathbb{Z} : |S_u| \geq n^{1/2}\}, \quad V := \{x \in \mathbb{Z} : [x - r + 1, x] \cap W \neq \emptyset\}.$$

For $i \in \mathbb{Z}$, let H_i denote the half-space $(-\infty, i] \times \mathbb{R}$. Set

$$i_1 := \max\{i \in \mathbb{Z} \setminus V : |A \cap H_i| < k\}, \quad i_2 := \min\{i \in \mathbb{Z} \setminus V : |A \cap H_i| \geq k\}.$$

Then $i_1 \notin V$ and $i_2 \notin V$. Also, $i_2 - i_1 - 1 \leq rn^{1/2}$ since $|W| \leq n^{1/2}$ and hence $|V| \leq rn^{1/2}$. Also, $|A \cap H_{i_1}| < k \leq |A \cap H_{i_2}|$. For $j \in \mathbb{Z}$, let $T_j = [i_1 + 1, i_2] \times (-\infty, j]$. Choose j_0 so that

$$|A \cap (H_{i_1} \cup T_{j_0 - 1})| < k \leq |A \cap (H_{i_1 - 1} \cup T_{j_0})|,$$

and let

$$S := A \cap (H_{i_1} \cup T_{j_0 - 1} \cup ([i_1, i_3] \times \{j_0\})),$$

with $i_3 \in [i_1, i_2]$ chosen so that S has precisely k elements (see Fig. 12.1).

Of the vertices in $A \cap H_{i_1}$, only those in the strip $[i_1 - r + 1, i_1] \times \mathbb{R}$ can possibly be connected to vertices in $A \setminus S$, and since $i_1 \notin V$ the number of such vertices is at most $rn^{1/2}$; hence by the uniform bound on degrees, the number of edges of $G(A; r)$ between points in $A \cap H_{i_1}$ and in $A \setminus S$ is at most $4r^2(r+1)n^{1/2}$. Similarly, since $i_2 \notin V$, the number of edges between points in S and points in $A \setminus H_{i_2}$ is at most $4r^2(r+1)n^{1/2}$. Finally, since $i_2 - i_1 - 1 \leq rn^{1/2}$, there are at most $(r+1)^2 n^{1/2}$ points in $S \cap (H_{i_2} \setminus H_{i_1})$ that could possibly be connected to points in $(A \setminus S) \cap (H_{i_2} \setminus H_{i_1})$. Hence, by the uniform degree bound, the number of edges between points in $S \cap (H_{i_2} \setminus H_{i_1})$ and points in $(A \setminus S) \cap (H_{i_2} \setminus H_{i_1})$ is at most $4r(r+1)^3 n^{1/2}$. Combining these three estimates gives us the result. □

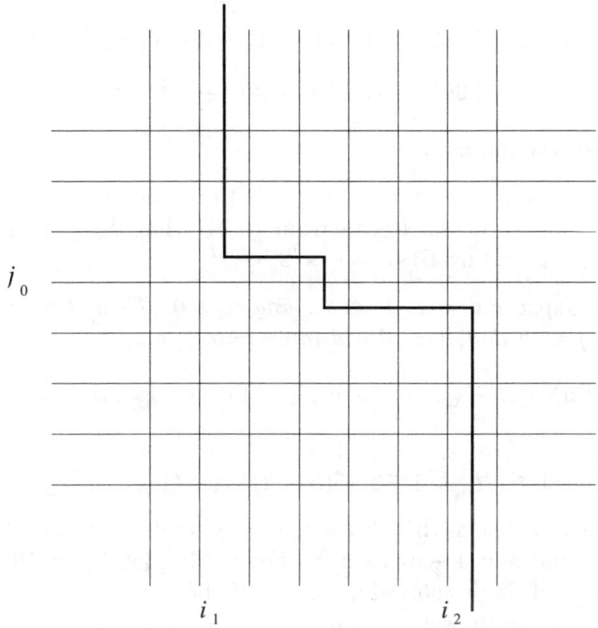

FIG. 12.1. The set S, arising in the proof of Lemma 12.6, lies below and to the left of the bold line.

Lemma 12.6 yields the following deterministic upper bound on the MCUT cost in the lattice.

Lemma 12.7 *Suppose $d = 2$ and $r > 0$. Then, for any geometric graph $G(A; r)$ with A a subset of $\mathbb{Z}^2$, we have*

$$\text{MCUT}(G(A; r)) \leq 48(r+1)^4 |A|^{1/2}. \tag{12.16}$$

Proof Clearly the result holds for $|A| \leq 7$. We extend it to $|A| = n, n \geq 8$, by induction on n. Let $n \geq 8$ and assume (12.16) holds for $|A| < n$. Then take $A \subset \mathbb{Z}^2$ with $|A| = n$.

By Lemma 12.6, we can partition A into two sets A_1, A_2, of cardinality $\lceil \frac{n}{2} \rceil$ and $\lfloor \frac{n}{2} \rfloor$, respectively, which are connected by at most $12(r+1)^4 n^{1/2}$ edges. Since $n \geq 8$ we have $\lceil \frac{n}{2} \rceil \leq \frac{9n}{16}$. By the inductive hypothesis, we can take optimal orderings φ_1 and φ_2 on A_1, A_2, respectively, such that for $i = 1, 2$,

$$\text{CUT}(\varphi_i) \leq 48(r+1)^4 \left\lceil \frac{n}{2} \right\rceil^{1/2} \leq 36(r+1)^4 n^{1/2}.$$

Combine these to make an ordering φ on A given by $\varphi(i) = \varphi_1(i)$ for $i \leq \lceil \frac{n}{2} \rceil$, and $\varphi(i) = \varphi_2(i - \lceil \frac{n}{2} \rceil)$ for $\lceil \frac{n}{2} \rceil < i \leq n$. Then

$$\begin{aligned}\text{CUT}(\varphi) &\le \max(\text{CUT}(\varphi_1), \text{CUT}(\varphi_2)) + \chi(\varphi(\lceil \frac{n}{2} \rceil), \varphi) \\ &\le 36(r+1)^4 n^{1/2} + 12(r+1)^4 n^{1/2},\end{aligned}$$

which completes the induction. □

The next lemma gives an upper bound of the form needed in Theorem 12.5, for a *Poisson* point process. Recall from (9.11) that $\mathcal{H}_{\lambda,s}$ is a homogeneous Poisson process on the box $B(s) := [-s/2, s/2]^d$.

Lemma 12.8 *Suppose $d = 2$, $\lambda < \lambda_c$, and $\alpha > 0$. Then there exist constants c, m_0, in $(0, \infty)$ such that, for all odd integers $m \ge m_0$,*

$$P[\text{MVS}(G(\mathcal{H}_{\lambda,m}; 1)) > 5(\alpha+2)\log m / \log_2 m] \le cm^{-\alpha}$$

and

$$P[\text{MCUT}(G(\mathcal{H}_{\lambda,m}; 1)) > (5(\alpha+2)\log m / \log_2 m)^2] \le cm^{-\alpha}.$$

Proof Choose $\varepsilon > 0$ such that $\lambda(1+8\varepsilon)^2 < \lambda_c$ and ε^{-1} is an odd integer. Set $l := (1+4\varepsilon)/\varepsilon$, and $p := 1 - \exp(-\lambda\varepsilon^2)$. For $z \in \mathbb{Z}^2$, let $B_z := B(\varepsilon) \oplus \{\varepsilon z\}$, the rectilinear square of side ε centred at εz. Let $\mathcal{B}_p$ be the Bernoulli site percolation process on $\mathbb{Z}^2$, that is, the set of open sites, obtained by setting each site $z \in \mathbb{Z}^2$ to be open if $\mathcal{H}_\lambda(B_z) > 0$ and closed otherwise. As explained at the start of the proof of Lemma 10.2, $p < p_c(l)$ (the critical parameter for site percolation on $(\mathbb{Z}^2, \sim_l)$). Let C_0 denote the l-cluster at the origin for $\mathcal{B}_p$; by exponential decay (Theorem 9.7), there are constants $\mu > 0, n_0 > 0$ such that, for all $n \ge n_0$,

$$P[|C_0| \ge n] \le \mathrm{e}^{-\mu n}. \tag{12.17}$$

With $B'_{\mathbb{Z}}(m)$ (m an odd integer) denoting the lattice m-box centred at the origin as at (10.51), set $\mathcal{B}'_{m,p} := \mathcal{B}_p \cap B'_{\mathbb{Z}}(m/\varepsilon)$. Let $\mathcal{G}_m$ denote the graph $G(\mathcal{B}'_{m,p}; l)$. By Boole's inequality and (12.17), the probability that $\mathcal{G}_m$ has a connected component of order greater than $((\alpha+2)/\mu)\log m$ is bounded by $(m/\varepsilon)^2 m^{-(\alpha+2)}$, so by Lemma 12.7,

$$P[\text{MCUT}(\mathcal{G}_m) > 48(l+1)^4(\alpha+2)^{1/2}\mu^{-1/2}(\log m)^{1/2}] \le \varepsilon^{-2} m^{-\alpha}. \tag{12.18}$$

Given the configuration of $\mathcal{B}'_{m,p}$, let φ be an ordering on $\mathcal{G}_m$ such that $\text{CUT}(\varphi) = \text{MCUT}(\mathcal{G}_m)$, and let $\tilde{\varphi}$ be the reverse ordering $\tilde{\varphi}(i) = \varphi(|\mathcal{B}'_{m,p}|+1-i)$. Recall the definition of $\Delta(v, \varphi)$ at (12.2). For each $z \in \mathcal{B}'_{m,p}$, let

$$U_z := \sum_{x \in \Delta(z,\varphi)} \mathcal{H}_\lambda(B_x), \quad V_z := \sum_{x \in \Delta(z,\tilde{\varphi})} \mathcal{H}_\lambda(B_x), \quad W_z := \sum_{x \in \Delta(z,\varphi) \cup \Delta(z,\tilde{\varphi})} \mathcal{H}_\lambda(B_z).$$

Then

$$\text{MVS}(G(\mathcal{H}_{\lambda,m}; 1)) \le \max\{U_z : z \in \mathcal{B}'_{m,p}\} \le \max\{W_z : z \in \mathcal{B}'_{m,p}\}, \tag{12.19}$$

$$\text{MCUT}(G(\mathcal{H}_{\lambda,m}; 1)) \le \max\{U_z V_z : z \in \mathcal{B}'_{m,p}\} \le \max\{W_z^2 : z \in \mathcal{B}'_{m,p}\}. \tag{12.20}$$

Conditional on $\mathcal{B}'_{m,p}$, the variables $\mathcal{H}_\lambda(B_z), z \in \mathcal{B}'_{m,p}$, are independent and each have the distribution of a Poisson variable with parameter $\nu := \lambda\varepsilon^2$, conditioned to be at least 1. Let

$$j(m) := \lfloor 48(l+1)^4(\alpha+2)^{1/2}\mu^{-1/2}(\log m)^{1/2}\rfloor,$$

and suppose the configuration of $\mathcal{B}'_{m,p}$ is such that $\mathrm{MCUT}(\mathcal{G}_m) \le j(m)$. Given $\mathcal{B}'_{m,p}$ and given φ, for each $z \in \mathcal{B}'_{m,p}$ the conditional probability that W_z exceeds $5(\alpha+2)\log m/\log_2 m$ is bounded by

$$P\left[\sum_{i=1}^{2j(m)} Y_i \ge 5(\alpha+2)\log m/\log_2 m \,\middle|\, \bigcap_{i=1}^{2j(m)} \{Y_i > 0\}\right],$$

where Y_i are independent $\mathrm{Po}(\nu)$ variables. By (1.12), this probability is bounded by

$$(1-\mathrm{e}^{-\nu})^{-2j(m)} \exp\left(-\left(\frac{5(\alpha+2)\log m}{2\log_2 m}\right)\log\left(\frac{5(\alpha+2)\log m}{2\nu j(m)\log_2 m}\right)\right)$$
$$\le \exp\left(-((\alpha+2)\log m)\right), \qquad m \text{ large}.$$

Therefore, by Boole's inequality, if the configuration of $\mathcal{B}'_{m,p}$ is such that $\mathrm{MCUT}(\mathcal{G}_m) \le j(m)$, the conditional probability of the event

$$F_m := \{\max\{W_z : z \in \mathcal{B}'_{m,p}\} \ge 5(\alpha+2)\log m/\log_2 m\}$$

is bounded by $\varepsilon^{-2}m^{-\alpha}$. Hence, by (12.18), the (unconditional) probability of the event F_m is at most $2\varepsilon^{-2}m^{-\alpha}$. By (12.19) and (12.20), unless F_m occurs we have $\mathrm{MVS}(G(\mathcal{H}_{\lambda,m};1)) \le 5(\alpha+2)\log m/\log_2 m$ and $\mathrm{MCUT}(G(\mathcal{H}_{\lambda,m};1)) \le (5(\alpha+2)\log m/\log_2 m)^2$, giving us the result. □

Proof of Theorem 12.5 We need to de-Poissonize the preceding lemma to get the upper bound. Take λ_0, μ, ν with $0 < \lambda_0 < \lambda < \mu < \nu < \lambda_c$. Couple $\mathcal{X}_n$ to the Poissonized process $\mathcal{P}_{n\mu/\lambda}$ in the usual way described in Section 1.7. Then by Lemmas 1.2 and 12.1 and the Borel–Cantelli lemma, with probability 1 we have for all but finitely many n that

$$\mathrm{MVS}(G(\mathcal{X}_n; r_n)) \le \mathrm{MVS}(G(\mathcal{P}_{n\mu/\lambda}; r_n)),$$
$$\mathrm{MCUT}(G(\mathcal{X}_n; r_n)) \le \mathrm{MCUT}(G(\mathcal{P}_{n\mu/\lambda}; r_n)).$$

For large enough n we have $nr_n^2\mu/\lambda < \nu$ and $r_n^{-1} < (n/\lambda_0)^{1/2}$, so that by scaling (Theorem 9.17), for $\beta > 0$ we have

$$P\left[\mathrm{MVS}(G(\mathcal{P}_{n\mu/\lambda}; r_n)) > \frac{\beta\log n}{\log_2 n}\right] = P\left[\mathrm{MVS}(G(\mathcal{H}_{nr_n^2\mu/\lambda; r_n^{-1}}; 1)) > \frac{\beta\log n}{\log_2 n}\right]$$
$$\le P\left[\mathrm{MVS}(G(\mathcal{H}_{\nu;(n/\lambda_0)^{1/2}}; 1)) > \frac{\beta\log n}{\log_2 n}\right],$$

and by Lemma 12.8, with a suitable choice of β this is less than cn^{-2} (the restriction in the statement of Lemma 12.8 to $\mathcal{H}_{\lambda,s}$ with s an odd integer is easily overcome using Lemma 12.1). Similarly, for suitable β and large n we have

$$P\left[\mathrm{MCUT}(G(\mathcal{P}_{n\mu/\lambda};r_n)) > \left(\frac{\beta \log n}{\log_2 n}\right)^2\right] < cn^{-2}.$$

Hence, by the Borel–Cantelli lemma, we can choose β such that with probability 1, for all but finitely many n,

$$\mathrm{MVS}(G(\mathcal{X}_n;r_n)) \leq \frac{\beta \log n}{\log_2 n}, \quad \mathrm{MCUT}(G(\mathcal{X}_n;r_n)) \leq \left(\frac{\beta \log n}{\log_2 n}\right)^2. \qquad (12.21)$$

To prove lower bounds of the same form, we apply Theorem 6.10 and the subsequent remark at (6.31), which show that in the limiting regime under consideration here, with probability 1, the clique number $\mathcal{C}(G(\mathcal{X}_n;r_n))$ satisfies

$$\liminf_{n\to\infty}\left(\frac{\mathcal{C}(G(\mathcal{X}_n;r_n))\log_2 n}{\log n}\right) \geq 1.$$

Since the complete graph Γ_k on k vertices satisfies $\mathrm{MVS}(\Gamma_k) \geq k-1$ and $\mathrm{MCUT}(\Gamma_k) \geq \lfloor k/2 \rfloor^2$, this implies that with probability 1 there exists n_0 such that, for $n \geq n_0$,

$$\mathrm{MVS}(G(\mathcal{X}_n;r_n)) \geq \left(\frac{\log n}{2\log_2 n}\right), \qquad \mathrm{MCUT}(G(\mathcal{X}_n;r_n)) \geq \frac{1}{5}\left(\frac{\log n}{\log_2 n}\right)^2.$$

Combined with the preceding upper bounds at (12.21), this completes the proof. □

12.3 The supercritical case

In the supercritical limiting regime, where $\lambda := \lim_{n\to\infty} nr_n^d > \lambda_c$, we have the following order of magnitude bounds for the optimal costs of layout problems. These orders of magnitude are different from those seen in Section 12.2 for the the subcritical case $\lambda < \lambda_c$, and also different from the case $p_\infty(\lambda) < \frac{1}{2}$ for MBIS.

Theorem 12.9 *Suppose* $\lim_{n\to\infty} nr_n^d = \lambda \in (\lambda_c, \infty]$. *Then with probability* 1,

$$\mathrm{MBW}(G(\mathcal{X}_n;r_n)) = \Theta(nr_n), \qquad (12.22)$$
$$\mathrm{MVS}(G(\mathcal{X}_n;r_n)) = \Theta(nr_n), \qquad (12.23)$$
$$\mathrm{MSC}(G(\mathcal{X}_n;r_n)) = \Theta(n^2r_n), \qquad (12.24)$$
$$\mathrm{MLA}(G(\mathcal{X}_n;r_n)) = \Theta(n^3r_n^{d+1}), \qquad (12.25)$$
$$\mathrm{MCUT}(G(\mathcal{X}_n;r_n)) = \Theta(n^2r_n^{d+1}), \qquad (12.26)$$

and if also $p_\infty(\lambda) > \frac{1}{2}$ *or* $\lambda = \infty$, *then*

$$\mathrm{MBIS}(G(\mathcal{X}_n;r_n)) = \Theta(n^2r_n^{d+1}). \qquad (12.27)$$

This is one of the principal results of this chapter and the proof is fairly lengthy. The proof given here is restricted to the case with $\lambda < \infty$. For the case $\lambda = \infty$, see Penrose (2000*b*).

The upper bounds implicit in Theorem 12.9 are rather crude in the sense that they are established by simply looking at the lexicographic ordering (henceforth called the *projection layout*) with points of $\mathcal{X}_n$ ordered by their first coordinate. We shall demonstrate this in detail below, but informally, the reason it gives these orders of magnitude is as follows.

The bandwidth cost BW and also the vertex separation cost VS for the projection layout, would be expected to behave like the number of points in a slab of width r_n, which in turn behaves like nr_n. The sum-cut cost SC is the sum of n expressions of this form, and so should behave like n^2r_n. Both CUT and BIS behave like the number of edges connecting points to the 'left' of a given point to points to its 'right'; this behaves like the number in a vertical slab (which should behave like nr_n as before), multiplied by the typical number of connections from a point in the slab to points in the neighbouring slab to its right (which should behave like nr_n^d), giving overall behaviour like $n^2r_n^{d+1}$; the linear arrangement cost, using the alternative expression for LA, is given by the sum of n expressions of this form, giving the correct order of magnitude of $n^3r_n^{d+1}$ for LA.

Since the orders of magnitude for the costs of layout problems, as given by Theorem 12.9 are achieved by the projection layout, this shows that for each of these problems, in the supercritical regime the cost of the projection layout stays within a constant factor of being optimal; that is, it is a *constant approximation algorithm* for these problems.

The first step towards a proof of Theorem 12.9 is a deterministic lower bound for the sum cut cost of an arbitrary graph in terms of a measure of its level of connectivity.

Lemma 12.10 *Suppose $G = (V, E)$ is a connected graph with n vertices. Suppose k and ν are positive integers with $k \leq n/2$, such that for any two disjoint subsets A, B of V, with $|A| \geq k$ and $|B| \geq k$, there exists a collection of ν vertex-disjoint paths in G, with each path starting in A and ending in B. Then*

$$\mathrm{MSC}(G) \geq (n - 2k)\nu. \qquad (12.28)$$

Furthermore, if $G' = (V', E')$ is a graph with G as a subgraph, and $n' := |V'|$ satisfies $k + (n'/2) + 1 \leq n$, then $\mathrm{MBIS}(G') \geq \nu$.

Proof Let φ be an arbitrary ordering on the vertices of G. Let A consist of the first k vertices in the ordering, and let B consist of the last k vertices. Take a collection of ν vertex-disjoint paths in G, with each path starting in A and ending in B.

Pick a vertex $v \in V \backslash (A \cup B)$. Each of the paths has a first crossing of v, that is, a first edge from a vertex preceding or equalling v in the ordering, to one following v in the ordering. This implies that $\Delta(v, \varphi) \geq \nu$; summing over all vertices in $V \backslash (A \cup B)$, we obtain (12.28).

Suppose $G' = (V', E')$ is a graph with G as a subgraph, and $n' := |V'|$ satisfies $k + (n'/2) + 1 \le n$. Each ordering on G' determines a bisection, that is, a partition (A_0, A_1) of V' with $|(|A_0| - |A_1|)| \le 1$. For $i = 0, 1$ we have $|A_i| \le (n'/2) + 1$, so that

$$|V \cap A_{1-i}| = |V \backslash A_i| \ge n - \frac{n'}{2} - 1 \ge k.$$

Hence, there are at least ν disjoint edges connecting $V \cap A_0$ to $V \cap A_1$, and $\mathrm{MBIS}(G') \ge \nu$. □

The next step towards Theorem 12.9 is a Poisson analogue, comprising lower and upper bounds on the costs for layout problems on the graph $G(\mathcal{H}_{\lambda,s}; 1)$, with $\lambda > \lambda_c$ (recall from (9.11) that $\mathcal{H}_{\lambda,s}$ is a homogeneous Poisson process on the box $[-s/2, s/2]^d$).

Theorem 12.11 *Suppose $0 < \lambda < \infty$. Then there exists a finite constant K such that, except on an event of probability decaying exponentially in s^{d-1} as $s \to \infty$,*

$$\mathrm{MBW}(G(\mathcal{H}_{\lambda,s}; 1)) \le K s^{d-1}, \tag{12.29}$$

$$\mathrm{MVS}(G(\mathcal{H}_{\lambda,s}; 1)) \le K s^{d-1}, \tag{12.30}$$

$$\mathrm{MSC}(G(\mathcal{H}_{\lambda,s}; 1)) \le K s^{2d-1}, \tag{12.31}$$

and, except on an event of probability decaying exponentially in $s^{(d-1)/2}$,

$$\mathrm{MLA}(G(\mathcal{H}_{\lambda,s}; 1)) \le K s^{2d-1}, \tag{12.32}$$

$$\mathrm{MCUT}(G(\mathcal{H}_{\lambda,s}; 1)) \le K s^{d-1}, \tag{12.33}$$

$$\mathrm{MBIS}(G(\mathcal{H}_{\lambda,s}; 1)) \le K s^{d-1}. \tag{12.34}$$

Proof Note first that MBIS satisfies (12.3), so that (12.34) will follow from (12.33). By Lemma 12.1 it suffices to prove the results (12.29)–(12.33) as s runs through the integers. Hence, from now on we assume s runs only through the integers, and write m instead of s. Also, we consider $\mathcal{H}'_{\lambda,m} := \mathcal{H}_\lambda \cap (0, m]^d$, instead of $\mathcal{H}_{\lambda,m}$: clearly, this does not affect the probabilities.

Let φ_{LEX} be the projection layout, that is, let φ_{LEX} be the lexicographic ordering on the vertices of $G(\mathcal{H}'_{\lambda,m}; 1)$ with points simply ordered by their first coordinate. The result is established by showing that suitable upper bounds hold with high probability for the cost of φ_{LEX}, for each of the six problems in question.

Divide $(0, m]^d$ into slabs $S_{1,m}, S_{2,m}, \ldots, S_{m,m}$ defined by $S_{j,m} = (j-1, j] \times (0, m]^{d-1}$. Then for $i < j$ the points in $S_{i,m}$ precede those in $S_{j,m}$ in the ordering φ_{LEX}. Also, points in $S_{i,m}$ and $S_{j,m}$ are not connected by edges of $G(\mathcal{H}'_{\lambda,m}; 1)$ for $|i - j| \ge 2$.

Let E_m be the event $\cap_{j=1}^m \{\mathcal{H}_\lambda(S_{j,m}) \le 2\lambda m^{d-1}\}$, that each slab $S_{j,m}$ contains at most $2\lambda m^{d-1}$ points of $\mathcal{H}_{\lambda_m}$. Then $P[E_m^c]$ decays exponentially in m^{d-1} by

Lemma 1.2. Also, when event E_m occurs the lexicographic ordering satisfies $\mathrm{BW}(\varphi_{\mathrm{LEX}}) \leq 4\lambda m^{d-1}$, giving us (12.29); then by (12.5) we also have (12.30) and (12.31).

The proof for (12.32)–(12.34) is more involved but is still based on the projection layout. For $\boldsymbol{i} \in B_{\mathbb{Z}}(m)$, set $Q_{\boldsymbol{i}} := B(2) \oplus \{\boldsymbol{i}\}$, the cube of side 2 centred at $\boldsymbol{i}$. Then for each edge $\{X, Y\}$ of $G(\mathcal{H}'_{\lambda,m}; 1)$, there exists $\boldsymbol{i} \in B_{\mathbb{Z}}(m)$ such that $X \in Q_{\boldsymbol{i}}$ and $Y \in Q_{\boldsymbol{i}}$. Let $i \in \{1, 2, \ldots, m\}$, and define the event

$$F_i := \left\{ \sum_{\boldsymbol{j} \in (\mathbb{Z} \cap [1,m])^{d-1}} \mathcal{H}_\lambda(Q_{i,\boldsymbol{j}})^2 \leq m^{d-1} 4^d (2\lambda^2 + \lambda) \right\}.$$

For $\boldsymbol{j} \in (\mathbb{Z} \cap [1,m])^{d-1}$, set $W_{i,\boldsymbol{j}} := \mathcal{H}_\lambda(Q_{i,\boldsymbol{j}})$. Observe that $W_{\boldsymbol{j}}$ is independent of $W_{\boldsymbol{k}}$ for $\|\boldsymbol{j} - \boldsymbol{k}\|_\infty \geq 2$. Since the chromatic number of $G(\mathbb{Z}^{d-1}; 1)$, using the l_∞ norm, is 2^{d-1} (choose one colour for each integer translate of $2\mathbb{Z}^{d-1}$), we can (and do) partition $(\mathbb{Z} \cap [1,m])^{d-1}$ into 2^{d-1} pieces $U_1^m, \ldots, U_{2^{d-1}}^m$ with $\{W_{\boldsymbol{j}}, \boldsymbol{j} \in U_r^m\}$ mutually independent for each r, and with $\lfloor m/2 \rfloor^{d-1} \leq |U_r^m| \leq m^{d-1}$ for each r. Since $EW_{\boldsymbol{j}}^2 \leq 4^d(\lambda^2 + \lambda)$, we have

$$\begin{aligned} P[F_i^c] &\leq P\left[\sum_{\boldsymbol{j} \in (\mathbb{Z} \cap [1,m])^{d-1}} (W_{\boldsymbol{j}}^2 - EW_{\boldsymbol{j}}^2) > m^{d-1} 4^d \lambda^2 \right] \\ &\leq \sum_{r=1}^{2^{d-1}} P\left[\sum_{\boldsymbol{j} \in U_r^m} (W_{\boldsymbol{j}}^2 - EW_{\boldsymbol{j}}^2) > m^{d-1} \lambda^2 \right], \end{aligned}$$

so that, by Lemma 2.11, $P[F_i^c]$ decays exponentially in $m^{(d-1)/2}$, and hence so does $P[\cup_{i=0}^{m-1} F_i^c]$.

Next, we show that

$$\bigcap_{i=0}^{m-1} F_i \subseteq \{\mathrm{MCUT}(G(\mathcal{H}'_{\lambda,m}; 1)) \leq 2^{2d+1}(2\lambda^2 + \lambda) m^{d-1}\}. \tag{12.35}$$

Suppose X, Y, and Z are points of $\mathcal{H}'_{\lambda,m}$ such that $\{Y, Z\}$ contributes to to $\chi(X, \varphi_{\mathrm{LEX}})$, so that $\pi_1(Y) \leq \pi_1(X) < \pi_1(Z)$ with π_1 denoting projection onto the first coordinate, and also $\|Y - Z\| \leq 1$. Then for some $\boldsymbol{i} = (i_1, \boldsymbol{j}) \in B_{\mathbb{Z}}(m)$, we have $Y \in Q_{\boldsymbol{i}}$ and $Z \in Q_{\boldsymbol{i}}$ (see Fig. 12.2).

Furthermore, if i is taken so that X lies in the slab S_i, we must have $i = i_1$ or $i = i_1 + 1$, so that

$$\chi(X, \varphi_{\mathrm{LEX}}) \leq \sum_{i_1 = i-1}^{i} \sum_{\boldsymbol{j} \in (\mathbb{Z} \cap [1,m])^{d-1}} \mathcal{H}_\lambda(Q_{i_1,\boldsymbol{j}})^2,$$

and (12.35) follows. This completes the proof of (12.33), and (12.32) follows by (12.4), while (12.34) follows by (12.3). □

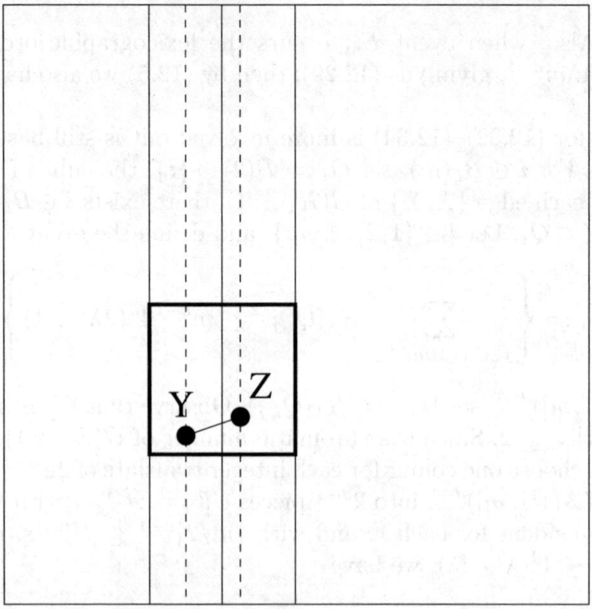

FIG. 12.2. The point X must lie between the dashed lines, so lies in one of the two strips shown.

Next, we give lower bounds of the same form as the upper bounds appearing in Theorem 12.11.

Theorem 12.12 *Let $\lambda \in (\lambda_c, \infty)$. Then:*

(a) *there exists a constant $\eta > 0$ such that except on an event of probability decaying exponentially in s^{d-1} as $s \to \infty$,*

$$\mathrm{MBW}(G(\mathcal{H}_{\lambda,s};1)) \geq \eta s^{d-1}, \tag{12.36}$$

$$\mathrm{MVS}(G(\mathcal{H}_{\lambda,s};1)) \geq \eta s^{d-1}, \tag{12.37}$$

$$\mathrm{MSC}(G(\mathcal{H}_{\lambda,s};1)) \geq \eta s^{2d-1}, \tag{12.38}$$

$$\mathrm{MLA}(G(\mathcal{H}_{\lambda,s};1)) \geq \eta s^{2d-1}, \tag{12.39}$$

$$\mathrm{MCUT}(G(\mathcal{H}_{\lambda,s};1)) \geq \eta s^{d-1}; \tag{12.40}$$

(b) *if also $p_\infty(\lambda) > \frac{1}{2}$, then there exists a constant $\eta > 0$ such that, except on an event of probability decaying exponentially in s^{d-1},*

$$\mathrm{MBIS}(G(\mathcal{H}_{\lambda,s};1)) \geq \eta s^{d-1}. \tag{12.41}$$

The proof is based on combining the following lemma with Lemma 12.10. In this result, $|\cdot|$ denotes cardinality.

Lemma 12.13 *Let* $\lambda \in (\lambda_c, \infty)$ *and* $\varepsilon \in (0, \lambda p_\infty(\lambda)/5)$. *For* $\delta > 0$, *let* $E_{\varepsilon,s,\delta}$ *denote the event that* (i) *there is a unique component* C *of* $G(\mathcal{H}_{\lambda,s}; 1)$ *of order exceeding* $(\lambda p_\infty(\lambda) - \varepsilon)s^d$, *and* (ii) *for any pair of disjoint subsets* A, B *of the vertex set of* C *with* $|A| \geq 2\varepsilon s^d$ *and* $|B| \geq 2\varepsilon s^d$, *there are at least* δs^{d-1} *vertex-disjoint paths in* C *from* A *to* B.

Then there exists $\delta = \delta(\lambda, \varepsilon) > 0$, *such that* $P[E^{\mathrm{c}}_{\varepsilon,s,\delta}]$ *decays exponentially in* s^{d-1}.

Proof Take $\mu \in (0, \lambda)$ such that $\mu p_\infty(\mu) > \lambda p_\infty(\lambda) - \varepsilon$. Such a μ exists by continuity of the continuum percolation probability above the critical point (Theorem 9.20). By Theorem 10.19, there exists $\gamma > 0$ such that, for large enough s,

$$P[L_1(G(\mathcal{H}_{\mu,s}; 1)) \leq (\lambda p_\infty(\lambda) - \varepsilon)s^d] < \exp(-\gamma s^{d-1}). \tag{12.42}$$

Take $\delta > 0$ such that $\delta \log(\lambda/(\lambda - \mu)) < \gamma$. Let F_s denote the event that (i) there is a unique component C of $G(\mathcal{H}_{\lambda,s}; 1)$ of order exceeding $(\lambda p_\infty(\lambda) - \varepsilon)s^d$; (ii) the order of this component is less than $(\lambda p_\infty(\lambda) + \varepsilon)s^d$; and (iii) there exist disjoint subsets A, B of the vertex set of C, with $|A| \geq 2\varepsilon s^d$ and $|B| \geq 2\varepsilon s^d$, such that there exist at most δs^{d-1} vertex-disjoint paths in C from A to B.

If F_s occurs, then by Menger's theorem (see Section 1.5), it is possible by removing at most δs^{d-1} vertices to disconnect A from B; to use Menger's theorem directly, add a vertex connected to each vertex of A, and likewise for B, and consider independent paths between the two added vertices. By the uniqueness of C, and the fact that after removing these vertices no sub-component of C has order greater than $(\lambda p_\infty(\lambda) + \varepsilon - 2\varepsilon)s^d$, after this removal of vertices there is no component of $G(\mathcal{H}_{\lambda,s}; 1)$ of order greater than $(\lambda p_\infty(\lambda) - \varepsilon)s^d$.

By Theorem 9.24 and (12.42), for large enough s we have

$$\begin{aligned} P[F_s] &\leq \left(\frac{\lambda}{\lambda - \mu}\right)^{\delta s^{d-1}} P[L_1(G(\mathcal{H}_{\mu,s}; 1)) \leq (\lambda p_\infty(\lambda) - \varepsilon)s^d] \\ &\leq \exp\left[s^{d-1}\left(\delta \log\left(\frac{\lambda}{\lambda - \mu}\right) - \gamma\right)\right], \end{aligned}$$

which decays exponentially in s^{d-1} by the choice of δ.

If conditions (i) and (ii) but not (iii) in the definition of event F_s occur, then event $E_{\varepsilon,s,\delta}$ occurs. Hence, if $E^{\mathrm{c}}_{\varepsilon,s,\delta} \cap F^{\mathrm{c}}_s$ occurs, then either condition (i) or (ii) in the definition of event F_s fails. Hence, by Theorem 10.19, $P[E^{\mathrm{c}}_{\varepsilon,s,\delta} \backslash F_s]$ also decays exponentially in s^{d-1}, completing the proof. □

Proof of Theorem 12.12 Assume $\lambda > \lambda_c$. Using Lemma 12.13, choose $\varepsilon_1 \in (0, \lambda p_\infty(\lambda)/6)$, and $\delta = \delta(\lambda, \varepsilon_1) > 0$, so that $P[E^{\mathrm{c}}_{\varepsilon_1,s,\delta}]$ decays exponentially in s^{d-1}. Suppose $E_{\varepsilon_1,s,\delta}$ occurs, and let C be the unique component of order exceeding $(\lambda p_\infty(\lambda) - \varepsilon_1)s^d$. Then, by Lemma 12.10,

$$\mathrm{MSC}(G(\mathcal{H}_{\lambda,s}; 1)) \geq \mathrm{MSC}(C) \geq ((\lambda p_\infty(\lambda) - \varepsilon_1) - 4\varepsilon_1)s^d \delta s^{d-1},$$

giving us (12.38). Then (12.39) and (12.40) follow by (12.4), and (12.37) and (12.36) follow by (12.5).

For (b), assume additionally that $p_\infty(\lambda) > \frac{1}{2}$. Take $\varepsilon_2 \in (0, \lambda p_\infty(\lambda)/5)$ with

$$4\varepsilon_2 + \lambda/2 < \lambda p_\infty(\lambda). \tag{12.43}$$

Using Lemma 12.13, take $\delta > 0$ such that $P[E^{\mathrm{c}}_{\varepsilon_2,s,\delta}]$ decays exponentially in s^{d-1}. By Lemma 1.2, $P[|\mathcal{H}_{\lambda,s}| > (\lambda+\varepsilon_2)s^d]$ decays exponentially in s^d. Suppose $E_{\varepsilon_2,s,\delta}$ occurs, and also $|\mathcal{H}_{\lambda,s}| \leq (\lambda + \varepsilon_2)s^d$. Let C be the vertex set of the unique component of $G(\mathcal{H}_{\lambda,s}; 1)$ of order exceeding $(\lambda p_\infty(\lambda) - \varepsilon_2)s^d$. Then, by (12.43) and elementary algebra, $\lceil 2\varepsilon_2 s^d \rceil + \frac{1}{2}|\mathcal{H}_{\lambda,s}| + 1 \leq |C|$, so by Lemma 12.10, $\mathrm{MBIS}(G(\mathcal{H}_{\lambda,s}; 1)) \geq \delta s^{d-1}$. □

We now complete the proof of Theorem 12.9, by de-Poissonizing Theorems 12.11 and 12.12.

Theorem 12.14 *Suppose* $\lim_{n\to\infty} nr_n^d = \lambda \in (\lambda_{\mathrm{c}}, \infty)$. *Then there exist constants* $0 < \eta < K$ *such that, except on an event of probability decaying exponentially in* $n^{(d-1)/d}$,

$$\eta r_n^{1-d} \leq \mathrm{MBW}(G(\mathcal{X}_n; r_n)) \leq K r_n^{1-d}, \tag{12.44}$$

$$\eta r_n^{1-d} \leq \mathrm{MVS}(G(\mathcal{X}_n; r_n)) \leq K r_n^{1-d}, \tag{12.45}$$

$$\eta r_n^{1-2d} \leq \mathrm{MSC}(G(\mathcal{X}_n; r_n)) \leq K r_n^{1-2d}, \tag{12.46}$$

and, except on an event of probability decaying exponentially in $n^{(d-1)/(2d)}$,

$$\eta r_n^{1-2d} \leq \mathrm{MLA}(G(\mathcal{X}_n; r_n)) \leq K r_n^{1-2d}, \tag{12.47}$$

$$\eta r_n^{1-d} \leq \mathrm{MCUT}(G(\mathcal{X}_n; r_n)) \leq K r_n^{1-d}, \tag{12.48}$$

and if $p_\infty(\lambda) > \frac{1}{2}$, *then*

$$\eta r_n^{1-d} \leq \mathrm{MBIS}(G(\mathcal{X}_n; r_n)) \leq K r_n^{1-d}. \tag{12.49}$$

Proof Let $\lambda_{\mathrm{c}} < \lambda_0 < \lambda_1 < \lambda < \mu_1 < \mu_2$. Let MGEN($G$) stand for any of MLA($G$), MBW($G$), MCUT($G$), MSC($G$), or MVS($G$). Then for any sequence of constants b_n, by monotonicity (Lemma 12.1), and the usual coupling from Section 1.7 of $\mathcal{X}_n$ to a Poisson process $\mathcal{P}_{n\mu_1/\lambda}$ on the unit cube with $N_{n\mu_1/\lambda}$ points, and scaling (Theorem 9.17),

$$\begin{aligned} P[\mathrm{MGEN}(G(\mathcal{X}_n; r_n)) > b_n] &\leq P[\mathrm{MGEN}(G(\mathcal{P}_{n\mu_1/\lambda}; r_n)) > b_n] + P[N_{n\mu_1/\lambda} < n] \\ &= P[\mathrm{MGEN}(G(\mathcal{H}_{nr_n^d\mu_1/\lambda, r_n^{-1}}; 1)) > b_n] + P[N_{n\mu_1/\lambda} < n] \\ &\leq P[\mathrm{MGEN}(\mathcal{H}_{\mu_2, r_n^{-1}}; 1) > b_n] + P[N_{n\mu_1/\lambda} < n]. \end{aligned}$$

Similarly, for any sequence a_n,

$$P[\mathrm{MGEN}(G(\mathcal{X}_n; r_n)) < a_n] \leq P[\mathrm{MGEN}(G(\mathcal{H}_{\lambda_0, r_n^{-1}}; 1)) < a_n] + P[N_{n\lambda_1/\lambda} > n].$$

We can then use Theorems 12.11 and 12.12 to obtain (12.44)–(12.48). For example, in the case of MBW, we set $a_n = \eta r_n^{1-d}$ and $b_n = K r_n^{1-d}$; for suitable choices of η, K, Theorems 12.11 and 12.12 show that (12.44) holds except on an event of probability decaying exponentially in $n^{(d-1)/d}$. The arguments for the other upper and lower bounds are similar.

In the case of the MBIS cost, we need to do extra work because it is not monotone. By (12.3) and (12.48), we have the required upper bound for the MBIS cost, so it remains only to prove the lower bound in (12.49).

Assume that $p_\infty(\lambda) > \frac{1}{2}$. Using the continuity of the continuum percolation probability above the critical point, take $\lambda_1 < \lambda$, and $\varepsilon_3 \in (0, \lambda_1 p_\infty(\lambda_1)/5)$, such that

$$\lambda_1 p_\infty(\lambda_1) - 4\varepsilon_3 > \lambda/2. \tag{12.50}$$

By Lemma 12.13 and scaling, there exists $\delta > 0$ such that except on an event of probability decaying exponentially in r_n^{1-d}, the graph $G(\mathcal{H}_{\lambda_1, r_n^{-1}}; 1)$, and hence also the graph $G(\mathcal{P}_{\lambda_1 r_n^{-d}}; r_n)$, includes a component C of order at least $(\lambda_1 p_\infty(\lambda_1) - \varepsilon_3) r_n^{-d}$, such that for any two subsets of C of order at least $2\varepsilon_3 r_n^{-d}$, there are at least δr_n^{1-d} edge-disjoint paths connecting them.

Since $n \sim \lambda r_n^{-d}$, by Lemma 1.2 we have with high probability that $\mathcal{P}_{\lambda_1 r_n^{-d}} \subseteq \mathcal{X}_n$, and also by (12.50), for large n we have $\lceil 2\varepsilon_3 r_n^{-d} \rceil + \frac{n}{2} + 1 < (\lambda_1 p_\infty(\lambda_1) - \varepsilon_3) r_n^{-d}$, so by the last part of Lemma 12.10, $\mathrm{MBIS}(G(\mathcal{X}_n; r_n)) \geq \delta r_n^{1-d}$, giving us the lower bound in (12.49). □

Proof of Theorem 12.9 for $\lambda < \infty$ Immediate from Theorem 12.14 and the Borel–Cantelli lemma, together with the assumption that $n = \Theta(r_n^{-d})$ when $\lambda < \infty$. □

12.4 The superconnectivity regime

The results in this section are concerned with the case where $d = 2$ and $nr_n^2/\log n \to \infty$. They improve on Theorem 12.9, for this case, by giving explicit constants in the asymptotic upper and lower bounds for the costs of ordering problems on random geometric graphs. We assume for this section that *the norm of choice is the l_∞ norm,* that is, that $\|\cdot\| = \|\cdot\|_\infty$.

Theorem 12.15 *Suppose $d = 2$, suppose $r_n \to 0$ and $nr_n^2/\log n \to \infty$ as $n \to \infty$. Then, with probability* 1,

$$\lim_{n\to\infty}\left(\frac{\mathrm{MVS}(G(\mathcal{X}_n; r_n))}{nr_n}\right) = \lim_{n\to\infty}\left(\frac{\mathrm{MBW}(G(\mathcal{X}_n; r_n))}{nr_n}\right) = 1,$$

$$\frac{5}{6} \leq \liminf_{n\to\infty}\left(\frac{\mathrm{MSC}(G(\mathcal{X}_n; r_n))}{n^2 r_n}\right) \leq \limsup_{n\to\infty}\left(\frac{\mathrm{MSC}(G(\mathcal{X}_n; r_n))}{n^2 r_n}\right) \leq 1.$$

For the other three problems, a similar result holds, giving upper and lower bounds that are reasonably close. However, they are not as close as in the previous case.

Theorem 12.16 *Suppose $d = 2$, suppose $r_n \to 0$ and $nr_n^2/\log n \to \infty$ as $n \to \infty$. Then, with probability 1,*

$$1 \geq \limsup_{n\to\infty} \frac{\text{MCUT}(G(\mathcal{X}_n; r_n))}{n^2 r_n^3} \geq \liminf_{n\to\infty} \frac{\text{MCUT}(G(\mathcal{X}_n; r_n))}{n^2 r_n^3} \geq \frac{3}{8\sqrt{2}} \approx 0.265,$$

$$1 \geq \limsup_{n\to\infty} \frac{\text{MBIS}(G(\mathcal{X}_n; r_n))}{n^2 r_n^3} \geq \liminf_{n\to\infty} \frac{\text{MBIS}(G(\mathcal{X}_n; r_n))}{n^2 r_n^3} \geq \frac{3}{8\sqrt{2}} \approx 0.265,$$

$$1 \geq \limsup_{n\to\infty} \frac{\text{MLA}(G(\mathcal{X}_n; r_n))}{n^2 r_n^3} \geq \liminf_{n\to\infty} \frac{\text{MLA}(G(\mathcal{X}_n; r_n))}{n^2 r_n^3} \geq \frac{1}{4\sqrt{2}} \approx 0.176.$$

We give a proof only of Theorem 12.15; the proof of Theorem 12.16 uses similar ideas, and may be found in Díaz *et al.* (2001*a*). For the proof, we introduce a concept of a point set being evenly spread over the unit square, as follows.

Definition 12.17 *Suppose $d = 2$ and suppose $(r_n)_{n\geq 1}$ is given. Given $\gamma \in (0, 1)$, set $m_n = m_n(\gamma) := \lceil 1/(\gamma r_n) \rceil$, and divide the unit square $B(1)$ into m_n^2 boxes (i.e. squares), each of side $1/m_n$. We shall say that a configuration $\mathcal{X}$ of n points in $B(1)$ is γ-good if every box contains at least $(1-\gamma)n(\gamma r_n)^2$ points and at most $(1+\gamma)n(\gamma r_n)^2$ points.*

Lemma 12.18 *Suppose $d = 2$, suppose $r_n \to 0$ and $nr_n^2/\log n \to \infty$ as $n \to \infty$. Given any $\gamma \in (0,1)$, with probability 1 the point set $\mathcal{X}_n$ is γ-good for all but finitely many n.*

Proof Let X be the number of points in a box. Then X has the binomial $\text{Bi}(n, m_n^{-2})$ distribution, and by definition $m_n \sim (\gamma r_n)^{-1}$ so $\mu_n := E[X] \sim n\gamma^2 r_n^2$. By Lemma 1.1, with $H(a) = 1 - a + a \log a$, for large enough n we have

$$P[X < (1-\gamma)n(\gamma r_n)^2] \leq \exp\left(-\mu_n H(1-\gamma/2)\right),$$
$$P[X > (1+\gamma)n(\gamma r_n)^2] \leq \exp\left(-\mu_n H(1+\gamma/2)\right).$$

Since we assume $nr_n^2 \gg \log n$, each of these upper bounds is bounded by n^{-3} for large n. The number of boxes is smaller than n, so by Boole's inequality, the probability that for some box the number of points in the box is less than $(1-\gamma)n(\gamma r_n)^2$ or more than $(1+\gamma)n(\gamma r_n)^2$, is bounded by $2n^{-2}$, which is summable in n, so the result follows from the Borel–Cantelli lemma. □

Lemma 12.19 *Suppose $d = 2$, and $\gamma \in (0,1)$, and suppose $(r_n)_{n\geq 1}$ is a sequence of positive numbers with $\lim_{n\to\infty}(r_n) = 0$. Then there exists n_0 such that for any integer $n \geq n_0$, any $i \in \{1, 2, \dots, n\}$, any γ-good configuration $\mathcal{X}$ of n points, and for any ordering φ on the the vertices of the graph $G_n := G(\mathcal{X}_n; r_n)$,*

$$\Delta(\varphi(i), \varphi, G_n) \geq (1-3\gamma)nr_n(h(i/n) - 2\gamma^{1/2}) - 6nr_n^2, \qquad (12.51)$$

where for $x \in [0,1]$ we set

$$h(x) = \min\{2\sqrt{x}, 2\sqrt{1-x}, 1\}.$$

Proof With $B(1)$ divided into boxes of side $1/m_n(\gamma)$, let two boxes be deemed *adjacent* if the l_∞ distance between their centres is at most $(1-\gamma)r_n$. Then any two points in adjacent boxes are an l_∞ distance at most r_n from each other.

Given an ordering φ on $\mathcal{X}_n$, let the first i points be denoted *red* and the others *blue.* Then $\Delta(\varphi(i), \varphi, G_n)$ is the number of red points of $\mathcal{X}_n$ having one or more blue point within a distance r_n. Let $\Delta'(\varphi(i), \varphi, G_n)$ be number of red points X such that there is at least one blue point lying either in the box containing X or in a box adjacent to the box containing X. Then $\Delta'(\varphi(i), \varphi, G_n) \leq \Delta(\varphi(i), \varphi, G_n)$. We shall show that the right-hand side of (12.51) is a lower bound for $\Delta'(\varphi(i), \varphi, G_n)$.

Given φ, let boxes containing only red points be denoted *red,* let boxes containing only blue points be denoted *blue,* and let other boxes be denoted *yellow.* Then $\Delta'(\varphi(i), \varphi, G_n)$ is the number of red points X for which the box containing X is either itself not red, or has some non-red box adjacent to it.

We assert that given $\mathcal{X}_n$, there is an ordering φ on $\mathcal{X}_n$ minimizing $\Delta'(\varphi(i), \cdot, G_n)$ such that φ induces at most one yellow box. Indeed, given an ordering φ inducing more than one yellow box, choose an ordering on yellow boxes. It is then possible to modify φ to an ordering φ' on points which respects the chosen ordering on yellow boxes of φ, and which satisfies $\Delta'(\varphi'(i), \varphi', G_n) \leq \Delta'(\varphi(i), \varphi, G_n)$. This can be done by successively swapping red and blue points, with each swap not increasing Δ'.

Thus, without loss of generality, we can (and do) assume that φ induces at most one yellow box. Set $\alpha := i/n$. Let N_{R} be the number of red boxes. Then by γ-goodness and the fact that there are a total of αn red points,

$$\frac{\alpha}{(1+\gamma)(\gamma r_n)^2} - 1 \leq N_{\mathrm{R}} \leq \frac{\alpha}{(1-\gamma)(\gamma r_n)^2}. \tag{12.52}$$

Let A_{R} be union of the red boxes and let $A_{\mathrm{B}} = [0,1]^2 \setminus A_{\mathrm{R}}$, the union of blue and yellow boxes. Since each box has area $(m_n(\gamma))^{-2} \leq (\gamma r_n)^2$, and since $m_n(\gamma) \sim (\gamma r_n)^{-1}$ as $n \to \infty$, by (12.52) we have for large n that, with $|\cdot|$ denoting area,

$$|A_{\mathrm{R}}| \geq \frac{\alpha}{1+2\gamma} - (\gamma r_n)^2, \qquad |A_{\mathrm{B}}| \geq 1 - \frac{\alpha}{1-\gamma}. \tag{12.53}$$

Let DB be the union of red boxes that are adjacent to blue or yellow boxes. Then using the notation of Proposition 5.13, $DB = (A_{\mathrm{B}})_{(1-\gamma)r_n} \setminus A_{\mathrm{B}}$, and by that result,

$$|DB| \geq \min\{2(1-\gamma)r_n|A_{\mathrm{B}}|^{1/2}, 2(1-\gamma)r_n|A_{\mathrm{R}}|^{1/2} - 4((1-\gamma)r_n)^2, (1-\gamma)r_n\}.$$

Using (12.53) and the fact that $a > b$ implies $\sqrt{a-b} \geq \sqrt{a} - \sqrt{b}$, we have

$$|A_{\mathrm{B}}|^{1/2} \geq (1-\gamma)^{-1/2}(1-\alpha-\gamma)^{1/2} \geq (1-\alpha)^{1/2} - \gamma^{1/2}.$$

Using also the fact that $(1+2\gamma)^{-1/2} \geq 1-\gamma$, we have

$$|A_{\mathrm{R}}|^{1/2} \geq \left(\frac{\alpha}{1+2\gamma}\right)^{1/2} - r_n \geq (1-\gamma)\alpha^{1/2} - r_n.$$

Combining these and using the inequality $(1-\gamma)^2 \geq 1-2\gamma$, we obtain

$$|DB| \geq 2(1-2\gamma)r_n(\min\{(1-\alpha)^{1/2}, \alpha^{1/2}, \tfrac{1}{2}\} - \gamma^{1/2}) - 6r_n^2. \quad (12.54)$$

For each box, the area is at most $(\gamma r_n)^2$, and the number of points is at least $(1-\gamma)n(\gamma r_n)^2$ by γ-goodness, so that the number of points per unit area in each box is at least $(1-\gamma)n$. Hence, since DB is a union of boxes, the number of points per unit area in DB is at least $(1-\gamma)n$, so that by (12.54) we obtain

$$\Delta'(\varphi(i), \varphi, G_n) \geq (1-3\gamma)nr_n(h(\alpha) - 2\gamma^{1/2}) - 6nr_n^2,$$

which gives us (12.51). □

Lemma 12.20 (Lower bounds). *Let $\varepsilon \in (0,1)$, and suppose $(r_n)_{n\geq 1}$ is a sequence of positive numbers that tends to zero as $n \to \infty$. Then there exists $\gamma > 0$, such that for all large enough n, if the configuration of $\mathcal{X}_n$ is γ-good, then*

$$\mathrm{MVS}(G(\mathcal{X}_n; r_n)) \geq (1-\varepsilon)nr_n, \quad (12.55)$$

$$\mathrm{MBW}(G(\mathcal{X}_n; r_n)) \geq (1-\varepsilon)nr_n, \quad (12.56)$$

$$\mathrm{MSC}(G(\mathcal{X}_n; r_n)) \geq (\tfrac{5}{6}-\varepsilon)n^2 r_n. \quad (12.57)$$

Proof The proof of (12.55) is obtained directly from Lemma 12.19 by taking i with $\frac{1}{4} \leq (i/n) \leq \frac{3}{4}$, so that $h(i/n) = 1$. By (12.5), the bound (12.56) follows from (12.55).

To prove (12.57), consider any layout φ on $G_n := G(\mathcal{X}_n; r_n)$. Then, using Lemma 12.19, we have for large enough n that

$$\begin{aligned}
\mathrm{SC}(\varphi) &= \sum_{1\leq i\leq n} \Delta(\varphi(i), \varphi, G_n) \\
&\geq (1-3\gamma)n^2 r_n \left(\left(\sum_{i=1}^{n} n^{-1}h(i/n)\right) - 2\gamma^{1/2}\right) - 6n^2r_n^2 \\
&\geq (1-3\gamma)n^2 r_n \left(\left(\int_0^1 h(x)\mathrm{d}x\right) - 3\gamma^{1/2}\right) - 6n^2r_n^2 \\
&= (1-3\gamma)n^2 r_n(\tfrac{5}{6} - 3\gamma^{1/2}) - 6n^2r_n^2. \quad (12.58)
\end{aligned}$$

Choose γ so that $(1-3\gamma)(\frac{5}{6} - 3\gamma^{1/2}) > \frac{5}{6} - \varepsilon$; then (12.58) gives us (12.57). □

Lemma 12.21 (Upper bounds) *Let $\varepsilon \in (0,1)$. Then there exists $\gamma > 0$, such that for all large enough n, if the configuration of $\mathcal{X}_n$ is γ-good, then*

$$\mathrm{MVS}(G(\mathcal{X}_n; r_n)) \leq (1+\varepsilon)nr_n, \quad (12.59)$$

$$\mathrm{MBW}(G(\mathcal{X}_n; r_n)) \leq (1+\varepsilon)nr_n, \quad (12.60)$$

$$\mathrm{MSC}(G(\mathcal{X}_n; r_n)) \leq (1+\varepsilon)n^2 r_n. \quad (12.61)$$

Proof Choose $\gamma > 0$ so that $(1+\gamma)^3(1+2\gamma) < 1+\varepsilon$. Let φ_{LEX} be the projection layout, that is, the lexicographic ordering, on $\mathcal{X}_n$. Then $\mathrm{BW}(\varphi_{\mathrm{LEX}})$ is bounded above by the maximum number of points of $\mathcal{X}_n$ contained in any set of the form $[a, a+r_n] \times [0,1]$ with $0 \leq a \leq 1 - r_n$. Each set of this form is contained in a union of boxes described in Definition 12.17, of total area at most $r_n + 2/m_n$, that is, a total of at most $r_n m_n^2 + 2m_n$ such boxes. Assuming $\mathcal{X}_n$ is γ-good, the number of points in any such collection of boxes is bounded by

$$(r_n m_n^2 + 2m_n)(1+\gamma)n(\gamma r_n)^2 = (1+\gamma)\gamma^2 n r_n (r_n^2 m_n^2 + 2r_n m_n). \quad (12.62)$$

Assuming n is large enough to yield $m_n \leq (1+\gamma)(\gamma r_n)^{-1}$, the expression (12.62) is in turn bounded by $(1+\gamma)^3 n r_n (1+2\gamma)$, and by the choice of γ, this is less than $(1+\varepsilon)nr_n$. Since the above expression is an upper bound for $\mathrm{MBW}(G(\mathcal{X}_n; r_n))$ when $\mathcal{X}_n$ is γ-good, this gives us (12.60). Then (12.59) and (12.61) both follow by (12.5). □

Proof of Theorem 12.15 Immediate from Lemmas 12.20 and 12.21. □

12.5 Notes and open problems

Notes *Section 12.2.* The results in this section come from Díaz *et al.* (2000), although the proofs are not all the same. An alternative proof of Theorem 12.3 is by the general result of Penrose and Yukich (2003), which can also be used to generalize Theorem 12.3 to the case of an arbitrary underlying density function f satisfying $\lambda f_{\max} < \lambda_c$.

Section 12.3. Theorem 12.9 is from Penrose (2000*b*).

Section 12.4. The results in this section are from Díaz *et al.* (2001*a*).

Open problems *Section 12.2.* In view of Theorem 12.5, one would expect that under the (subcritical) conditions of that result, there should be constants $\beta_{\mathrm{VS}}(\lambda)$ and $\beta_{\mathrm{CUT}}(\lambda)$ such that

$$\mathrm{MVS}(G(\mathcal{X}_n; r_n))(\log\log n)/\log n \xrightarrow{P} \beta_{\mathrm{SC}}(\lambda),$$
$$\mathrm{MCUT}(G(\mathcal{X}_n; r_n))(\log\log n)^2/(\log n)^2 \xrightarrow{P} \beta_{\mathrm{CUT}}(\lambda).$$

Proving this is an open problem, as is extending Theorem 12.5 to higher dimensions $d > 2$, and also obtaining the order of magnitude of the bandwidth cost $\mathrm{MBW}(G(\mathcal{X}_n; r_n))$ in the subcritical case.

Sections 12.3 and 12.4. We conjecture that throughout the supercritical phase, for each of the six problems the random optimal ordering cost, divided by the order of magnitude given by Theorem 12.9, converges in probability to a limit. If true, this would be analogous to the well-known result of Beardwood *et al.* (1959) for the travelling salesman problem and various analogous results for other problems described in Steele (1997) and Yukich (1998).

In cases where $d = 2$ and $nr_n^d/\log n \to \infty$, Theorem 12.15 shows that this conjecture is true for MBW and MBIS, and Theorem 12.16 takes some steps towards proving the conjecture for MCUT, MBIS, and MLA, by providing explicit asymptotic upper and lower bounds. Methods based on subadditivity, heavily used in Steele (1997) and Yukich (1998), do not seem to be useful for the ordering problems of this chapter, at least in the supercritical phase.

13

CONNECTIVITY AND THE NUMBER OF COMPONENTS

A fundamental question about any graph is whether or not it is connected. Since connectedness is a monotone property, a natural object of study is the *connectivity threshold* for a finite set $\mathcal{X} \subset \mathbb{R}^d$, defined to be the minimum value of r such that $G(\mathcal{X}; r)$ is connected. The connectivity threshold for $\mathcal{X}$ is also the *longest edge length of the minimal spanning tree on* $\mathcal{X}$; see, for example, Penrose (1997). Applications include (i) Rohlf's (1975) test for outliers, which is discussed further in the notes at the end of this chapter; (ii) wireless networks (Gupta and Kumar 1998) and (iii) estimation of a set from a random sample of points in that set (Baillo and Cuevas 2001).

It turns out for a large class of connected domains in two or more dimensions, the asymptotics for the connectivity threshold (denoted T_1) on $\mathcal{X}_n$ are the same as for the largest nearest-neighbour link (M_1), which has already been considered. This asymptotic equivalence can take the form of the ratio T_1/M_1 tending to 1 in probability, or (at least for certain particular density functions f) the stronger form that $P[T_1 \neq M_1] \to 0$ as $n \to \infty$. Therefore, we can aim to obtain laws of large numbers and weak convergence results for T_1, similar to those already derived for M_1. These are the main subject of this chapter.

A related topic is the *total number of components* of the geometric graph. Let K_n (respectively, K'_n) denote the total number of components of $G(\mathcal{X}_n; r_n)$ (respectively, the total number of components of $G(\mathcal{P}_n; r_n)$). We shall give some results on their limit distributions, in some particular limiting regimes of interest, without considering exhaustively all possible limiting regimes. In particular, we give a Poisson limit for the number of components in the connectivity regime for uniformly distributed points (Theorem 13.11), and for normally distributed points (Theorem 13.23), and a normal limit for the number of components in the thermodynamic limiting regime (Theorems 13.27 and 13.26).

In this chapter, Ω denotes the support of the underlying probability density function f on $\mathbb{R}^d$. Also, f_0 denotes the essential infimum of the restriction $f|_\Omega$ of f to Ω, and θ is the volume of the unit ball in the chosen norm, as usual. For bounded $U \subseteq \mathbb{R}^d$ set

$$\mathrm{diam}_\infty(U) := \sup\{\|x-y\|_\infty : x, y \in U\} = \inf\{r : \exists x \text{ with } U \subseteq x \oplus B(r)\},$$

where $B(r)$ is the cube of side r centred at the origin as at (9.11). Thus, diam_∞ is diameter defined in terms of the l_∞ norm even though the geometric graphs under consideration might be defined using some other norm.

13.1 Multiple connectivity

If k is a positive integer, a graph G of order greater than $k+1$ is said to be k-connected if it cannot be disconnected by the removal of $k-1$ or fewer vertices. Equivalently, G is k-connected if for each pair of distinct vertices there exist at least k independent paths in the graph connecting them. This equivalence follows from Menger's theorem.

The *connectivity* of G, here denoted κ, is the maximum k such that G is k-connected; if the graph is not connected we put $\kappa = 0$. For a finite set $\mathcal{X}$ in $\mathbb{R}^d$, and a positive integer k, define the *k-connectivity threshold* $T_k(\mathcal{X})$, using notation $\rho(\mathcal{X}; Q)$ from Section 1.4, by $T_k(\mathcal{X}) := \rho(\mathcal{X}; \kappa \geq k)$, the threshold value of r above which $G(\mathcal{X}; r)$ is k-connected.

A second notion of multiple connectivity is *edge-connectivity.* A graph G is said to be k-edge-connected. if it cannot be disconnected by the removal of $k-1$ or fewer edges. Equivalently, it is k-edge-connected if for each pair of vertices there exist at least k edge-disjoint paths connecting them (paths are edge-disjoint if they have no edges in common). This equivalence follows from the edge version of Menger's theorem which can be found in Bollobás (1985).

The *edge-connectivity* of G, here denoted κ^{e}, is the maximum k such that G is k-edge-connected. For a finite set $\mathcal{X}$ in $\mathbb{R}^d$, and a positive integer k, define the k-edge-connectivity threshold $T_k^{\mathrm{e}}(\mathcal{X})$ by $T_k^{\mathrm{e}}(\mathcal{X}) := \rho(\mathcal{X}; \kappa^{\mathrm{e}} \geq k)$, the threshold value of r above which $G(\mathcal{X}; r)$ is k-edge-connected.

Recall from Chapter 7 that the largest k-nearest-neighbour link $M_k(\mathcal{X}) := \rho(\mathcal{X}; \delta \geq k)$ is the threshold value of r above which $G(\mathcal{X}; r)$ has minimal degree at least k. It is easy to see that if a graph is k-connected then it is k-edge-connected, and if it is k-edge-connected then its minimum degree is at least k. Therefore, $\kappa \leq \kappa^{\mathrm{e}} \leq \delta$ for any graph, and therefore if $\mathcal{X}$ is a finite set in $\mathbb{R}^d$ with more than $k+1$ elements,

$$T_k(\mathcal{X}) \geq T_k^{\mathrm{e}}(\mathcal{X}) \geq M_k(\mathcal{X}). \tag{13.1}$$

Except for Section 13.7, this chapter is mainly concerned with demonstrating the considerable extent to which asymptotic equality holds in (13.1), in the context of geometric random graphs when $d \geq 2$ and the support of the underlying distribution is connected. Thus, in this setting we obtain identical limit theorems for $T_k(\mathcal{X}_n)$ to those already derived for $M_k(\mathcal{X}_n)$. In view of (13.1), all results proved in this section on asymptotic equivalence between $T_k(\mathcal{X}_n)$ and $M_k(\mathcal{X}_n)$ will immediately imply similar results for $T_k^{\mathrm{e}}(\mathcal{X}_n)$, so henceforth we discuss only $T_k(\mathcal{X}_n)$.

There is an alternative formulation for k-connectivity which will be useful to us. Suppose G is a graph with vertex set V. By a *k-separating pair* for G we shall mean a pair of non-empty disjoint sets of vertices $U \subset V$, $W \subset V$ such that (i) the subgraph of G induced by vertex set U is connected, and likewise for W; (ii) no element of U is adjacent to any element of W; and (iii) the number of elements of $V \setminus (U \cup W)$ lying adjacent to $(U \cup W)$ is at most k. If (U, W) is a k-separating pair, then both U and V are are *k-separated sets*, in the sense (given earlier in Section 7.1) of having external vertex boundary consisting of at most k vertices.

Lemma 13.1 *Suppose G is a graph with more than $k+1$ vertices. Then either G is $(k+1)$-connected, or it has k-separating pair, but not both.*

Proof If G is not $(k+1)$-connected, then it is possible to disconnect G by removing at most k vertices. By taking two components of the resulting disconnected graph we obtain a k-separating pair.

Conversely, if a graph G with vertex set V has a k-separating pair (U, W), then we can disconnect G by removing the vertices of $V \setminus (U \cup W)$ adjacent to $(U \cup W)$, so G is not $(k+1)$-connected. □

The case $d = 1$ is special, and is not considered in detail here. For $d = 1$, $T_k(\mathcal{X}_n)$ is the maximum k-spacing amongst the points of $\mathcal{X}_n$, which is discussed in Holst (1980), Barbour *et al.* (1992). Interestingly, for points uniformly distributed on the unit interval $[0,1]$, the limit distribution of $T_k(\mathcal{X}_n)$, suitably scaled and centred Holst (1980, Theorem 1) is the same as that of $2M_k(\mathcal{X}_n)$, scaled and centred in the same way (see Theorem 8.4). In brief, the reason for this goes as follows. For M_k to exceed r there needs to be a point X with fewer than k other points in the interval $(X - r, X + r)$, while for T_k to exceed $2r$, there needs to be a point X with fewer than k other points in $(X, X + 2r)$. For the Poissonized process $\mathcal{P}_n$, by Palm theory (Theorem 1.6) the number of such points X, denoted M in either case, has the same expectation in both cases. For the asymptotic theory one chooses $r = r_n$ so that $E[M]$ tends to a finite limit and obtains the same Poisson limit for M in both cases.

13.2 Strong laws for points in the cube or torus

This section is concerned with strong laws of large numbers for the $(k_n + 1)$-connectivity threshold for a given sequence of integers $(k_n)_{n \geq 1}$, in cases where the support Ω of f is *a product of finite intervals* (e.g. the unit cube). We specify for the duration of this section that $d \geq 2$ and

$$\Omega = [0, \omega_1] \times [0, \omega_2] \times \cdots \times [0, \omega_d], \tag{13.2}$$

with $\omega_j > 0, 1 \leq j \leq d$. We also assume that the norm $\|\cdot\|$ used for the geometric graphs be one of the l_p norms, $1 \leq p \leq \infty$.

We do *not* require f to be uniform on Ω. Recall that $f_0 := \operatorname{ess\,inf}(f|_\Omega)$. For $1 \leq j \leq d$, let ∂_j denote the union of all $(d-j)$-dimensional 'edges' (intersections of j hyperplanes bounding Ω), and let f_j denote the infimum of f over ∂_j. We assume further that $f_0 > 0$ and that the discontinuity set of $f|_\Omega$ contains no element of $\partial\Omega$.

We consider the case where k_n grows like a constant times $\log n$. The constant might be zero, so cases with k_n fixed, and in particular the case with $k_n = 0$ for all n (i.e. the threshold for simple connectivity), are included in the result given. We use again the function $H : [0, \infty) \to \mathbb{R}$, first seen in Section 1.6, defined by $H(a) = 1 - a + a \log a$ for $a > 0$, and $H(0) = 1$.

Theorem 13.2 *Suppose $(k_n)_{n\geq 1}$ is a sequence of non-negative integers satisfying $\lim_{n\to\infty}(k_n/n) = 0$ and $\lim_{n\to\infty}(k_n/\log n) = b \in [0,\infty]$. In the case $b<\infty$, assume also that the sequence $(k_n)_{n\geq 1}$ is nondecreasing, and define $a_0,\ldots,a_{d-1}$ in $[0,1)$ by*

$$\frac{a_j}{H(a_j)} = \frac{bd}{d-j}.$$

If $b=\infty$, then with probability 1,

$$\lim_{n\to\infty} \left(n\theta T_{k_n+1}(\mathcal{X}_n)^d/k_n\right) = \max\{2^j/f_j; 0\leq j\leq d\}, \tag{13.3}$$

whereas if $b<\infty$, then with probability 1,

$$\lim_{n\to\infty}\left(\frac{n\theta T_{k_n+1}(\mathcal{X}_n)^d}{\log n}\right) = \max\left(\max_{0\leq j\leq d-1}\left(\frac{2^j(d-j)}{df_jH(a_j)}\right), \frac{2^d b}{f_d}\right). \tag{13.4}$$

Much of the work in proving Theorem 13.2 has already been done. By (13.1) and Theorem 7.8, we have, with probability 1, that if $b=\infty$ then

$$\liminf_{n\to\infty}\left(\frac{n\theta(T_{k_n+1}(\mathcal{X}_n))^d}{k_n}\right) \geq \max_{0\leq j\leq d}\left(\frac{2^j}{f_j}\right), \tag{13.5}$$

or if $b<\infty$, then

$$\liminf_{n\to\infty}\left(\frac{n\theta(T_{k_n+1}(\mathcal{X}_n))^d}{\log n}\right) \geq \max\left(\max_{0\leq j\leq d-1}\left(\frac{2^j(d-j)}{df_jH(a_j)}\right), \frac{2^d b}{f_d}\right), \tag{13.6}$$

so it remains only to prove an inequality the other way. For each $n>0$, define

$$\rho_n := (k_n/n)^{1/d} \text{ if } b=\infty, \qquad \rho_n := ((\log n)/n)^{1/d} \text{ if } b<\infty.$$

If $b=\infty$, fix t satisfying the inequality

$$\theta t^d > \max_{0\leq j\leq d}\left(2^j/f_j\right), \tag{13.7}$$

or if $b<\infty$, fix t satisfying

$$\theta t^d > \max\left(\max_{0\leq j\leq d-1}\left(\frac{2^j(d-j)}{df_jH(a_j)}\right), \frac{2^d b}{f_d}\right). \tag{13.8}$$

We shall prove that with probability 1, $T_{k_n+1}(\mathcal{X}_n) \leq t\rho_n$ for large enough n. Since t satisfying (13.7) or (13.8) is arbitrary, this, along with (13.5) or (13.6), will suffice to prove Theorem 13.2.

By Lemma 13.1, it suffices to prove non-existence of a k_n-separating pair for $G(\mathcal{X}_n; t\rho_n)$. The next two results establish this. The first of these is a re-statement of Proposition 7.10, which has already been proved. Therefore, to prove Theorem 13.2 it suffices to prove the second result, Proposition 13.4 on non-existence of 'large' k_n-separating pairs.

Proposition 13.3 *Suppose the hypotheses of Theorem* 13.2 *hold. Let* $K > 0$. *Let* $E'_n(K;t)$ *be the event that there exists a* k_n*-separated set* U *for* $G(\mathcal{X}_n; t\rho_n)$ *with* $\mathrm{diam}(U) \leq K\rho_n$. *Then with probability* 1, *events* $E'_n(K;t)$ *occur for only finitely many* n.

Proposition 13.4 *Suppose the hypotheses of Theorem* 13.2 *hold. For* $K > 0$, *let* $H_n(K;t)$ *be the event that there exists a* k_n*-separating pair* (U, W) *for* $G(\mathcal{X}_n; t\rho_n)$ *with* $\mathrm{diam}_\infty(U) > K\rho_n$ *and* $\mathrm{diam}_\infty(W) > K\rho_n$. *Then there exists* $K > 0$ *such that, with probability* 1, *the events* $H_n(K;t)$ *occur for only finitely many* n.

We work towards a proof of Proposition 13.4. With t fixed and satisfying (13.7) if $b = \infty$ or (13.8) if $b < \infty$, pick $\varepsilon_1 \in (0, 1)$ such that

$$\varepsilon_1 < \min(t, 1)/(4d). \tag{13.9}$$

The proof is based on discretization; we wish to divide Ω into cubes of side $\varepsilon_1\rho_n$, but such cubes in general will not fit exactly. Therefore we define 'nearly-cubes' as follows. For $n \geq 1$ and $1 \leq j \leq d$, with the side-length ω_j defined at (13.2), set

$$\delta_{n,j} := \frac{\omega_j}{\lceil \omega_j/(\varepsilon_1\rho_n) \rceil}.$$

Then $\delta_{n,j} \leq \varepsilon_1\rho_n$, but $\delta_{n,j} \sim \varepsilon_1\rho_n$ as $n \to \infty$, and importantly, $\omega_j/\delta_{n,j}$ is an integer. Define the lattice

$$\delta_n\mathbb{Z}^d := \{(\delta_{n,1}z_1, \ldots, \delta_{n,d}z_d) : (z_1, \ldots, z_d) \in \mathbb{Z}^d\}.$$

For $y = (\delta_{n,1}z_1, \ldots, \delta_{n,d}z_d) \in \delta_n\mathbb{Z}^d$, let $z(y) := (z_1, \ldots, z_d) \in \mathbb{Z}^d$, and

$$C_n(y) := \prod_{j=1}^{d} [\delta_{n,j}z_i, (\delta_{n,j} + 1)z_i).$$

The rectangular solid $C_n(y)$ is 'nearly' a cube of side $\varepsilon_1\rho_n$ and has y at one of its corners. These nearly-cubes fit exactly into Ω, in the sense that, for all $y \in \delta_n\mathbb{Z}^d$, either $C_n(y) \subseteq \Omega$ or the interior of $C_n(y)$ is disjoint from Ω. Since $\delta_{n,j} \leq \varepsilon_1\rho_n$ and we use an l_p norm,

$$\mathrm{diam}(C_n(y)) \leq d\varepsilon_1\rho_n. \tag{13.10}$$

Define the finite lattice $\mathcal{L}_n$ by

$$\mathcal{L}_n := \{y \in \delta_n\mathbb{Z}^d : C_n(y) \subseteq \Omega\}. \tag{13.11}$$

The nearly-cubes associated with the elements of $\mathcal{L}_n$ form a partition of Ω (not counting some of the faces of Ω). The idea of the discretization is that instead of the precise configuration $\mathcal{X}_n$, one considers the set of $z \in \mathcal{L}_n$ for which $\mathcal{X}_n(C_n(z)) > 0$, and applies counting arguments to those possibilities for this set which are compatible with the existence of 'large' k_n-separating pairs.

For $U \subseteq \mathcal{X}_n$ and $r > 0$, set

$$U_r := U \oplus B(\mathbf{0}; r) \qquad \left(= \bigcup_{x \in U} B(x; r) \right), \tag{13.12}$$

the r-neighbourhood of U. A non-empty subset U of $\mathcal{X}_n$ is k_n-separated for $G(\mathcal{X}_n; t\rho_n)$, and connected, if and only if $\mathcal{X}_n[U_{t\rho_n} \setminus U] \leq k_n$ and $U_{t\rho_n/2}$ is connected. The key observation is that if U is a k_n-separated set, then a region near the boundary of $U_{t\rho_n/2}$ contains at most k_n points of $\mathcal{X}_n$. We discretize this region into near-cubes of side $\delta_{n,j} \approx \varepsilon_1 \rho_n$, and count the number of possibilities for the discretized region using a Peierls argument.

We shall say a set $\sigma \subseteq \mathcal{L}_n$ is $*$-connected if the corresponding set in the integer lattice, namely $\{z(y) : y \in \sigma\}$, is $*$-connected (see Section 9.2). For integer $i > 0$ let $\mathcal{C}_{n,i}$ denote the collection of $*$-connected sets $\sigma \subseteq \mathcal{L}_n$ of cardinality i. By a Peierls argument (Corollary 9.4), there are constants $\gamma = \gamma(d) > 0$ and $c > 0$, such that for all large enough n, with $\mathrm{card}(\cdot)$ denoting cardinality,

$$\mathrm{card}(\mathcal{C}_{n,i}) \leq c\rho_n^{-d} \exp(\gamma i) \leq n \exp(\gamma i). \tag{13.13}$$

Lemma 13.5 *For all $n \geq 1$, if (U, W) is a k_n-separating pair for $G(\mathcal{X}_n; t\rho_n)$, then there exists $\sigma \in \mathcal{C}_{n,i}$ with $\mathcal{X}_n[\cup_{y \in \sigma} C_n(y)] \leq k_n$, for some i with*

$$i\varepsilon_1\rho_n \geq \min(\mathrm{diam}_\infty(U), \mathrm{diam}_\infty(W)).$$

Proof Suppose (U, W) is a k_n-separating pair for $G(\mathcal{X}_n; t\rho_n)$. The sets $U_{t\rho_n/2} \cap \Omega$ and $W_{t\rho_n/2} \cap \Omega$ are disjoint connected subsets of Ω. So $\Omega \setminus U_{t\rho_n/2}$ has a connected component which contains $W_{t\rho_n/2}$; denote this component W', and let $U' := \Omega \setminus W'$. Then the closures of U' and W' are connected and their union is Ω, so their intersection, a part of the boundary of $U_{t\rho_n/2}$ denoted ∂U, is connected by the unicoherence of Ω; see Section 9.1. Also, $U \subseteq U'$ and $W \subseteq W'$, so *any path in Ω from a point of U to a point of W must pass through ∂U*. We assert that

$$\mathrm{diam}_\infty(\partial U) \geq \min(\mathrm{diam}_\infty(U), \mathrm{diam}_\infty(W)). \tag{13.14}$$

To see this, assume the contrary. Then there would exist a rectilinear cube C of side $b < \min(\mathrm{diam}_\infty(U), \mathrm{diam}_\infty(W))$, such that $\partial U \subseteq C$. By the condition on b there would be points $X \in U$ and $Y \in W$, which were not in C; it would then be possible to get from X to Y by a path avoiding the cube C, and hence avoiding ∂U, a contradiction.

Let DU denote the set of $y \in \mathcal{L}_n$ such that $C_n(y)$ has non-empty intersection with ∂U (see Fig. 13.1). Then DU is $*$-connected. We assert that

$$\mathcal{X}_n \left[\bigcup_{y \in DU} C_n(y) \right] \leq k_n. \tag{13.15}$$

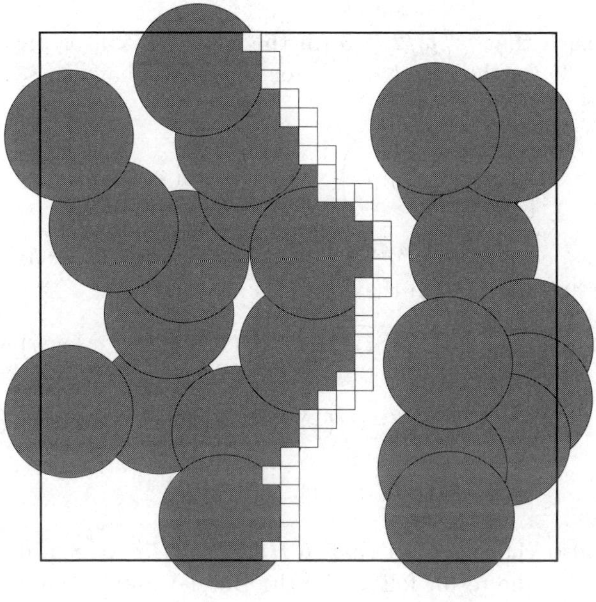

FIG. 13.1. The disks have radius $t\rho_n/2$. The little squares are the 'nearly-cubes' $C_n(y), y \in DU$.

This is because if $y \in DU$ then there exists $x \in C_n(y)$ such that $x \in \partial U$, and therefore $\mathrm{dist}(x, U) = t\rho_n/2$. By (13.9) and (13.10), $C_n(y) \subseteq B(x; t\rho_n/4)$, and therefore by the triangle inequality

$$C_n(y) \subseteq U_{3t\rho_n/4} \setminus U_{t\rho_n/4},$$

and then (13.15) follows because U is a k_n-separated set for $G(\mathcal{X}_n; t\rho_n)$.

Finally, $\varepsilon_1 \rho_n \, \mathrm{card}(DU) \geq \mathrm{diam}_\infty(\partial U)$, and by (13.14), the conclusion of the lemma follows by taking $\sigma = DU$. □

Proof of Proposition 13.4 If $H_n(K; t)$ occurs, there exists a k_n-separating pair (U, W) for $G(\mathcal{X}_n; t\rho_n)$, with $\mathrm{diam}_\infty(U) \geq K\rho_n$ and $\mathrm{diam}_\infty(W) \geq K\rho_n$. By Lemma 13.5, there exists $\sigma \in \mathcal{C}_{n,i}$ with $\mathcal{X}_n[\cup_{y\in\sigma} C_n(y)] \leq k_n$, for some i with $i\varepsilon_1\rho_n \geq K\rho_n$. Hence,

$$P[H_n(K;t)] \leq \sum_{i \geq K/\varepsilon_1} \sum_{\sigma \in \mathcal{C}_{n,i}} P\left[\mathcal{X}_n\left(\bigcup_{y\in\sigma} C_n(y)\right) \leq k_n\right].$$

For n large, if $\sigma \in \mathcal{C}_{n,i}$ then, since the side-lengths $\delta_{n,j}$ of the nearly-cubes comprising σ are asymptotic to $\varepsilon_1\rho_n$,

$$nF\left(\bigcup_{y\in\sigma} C_n(y)\right) \geq n i f_0 (\varepsilon_1 \rho_n)^d / 2 := \mu_{n,i}.$$

Provided i is such that $i\varepsilon_1^d f_0/2 \geq 3$ (in the case $b = \infty$), or provided i is such that $i\varepsilon_1^d f_0/2 \geq 3b$ (in the case $b < \infty$), for large n we have $k_n \leq \mu_{n,i}/2$ so that, by Lemma 1.1,

$$P\left[\mathcal{X}_n\left(\bigcup_{y\in\sigma} C_n(y)\right) \leq k_n\right] \leq \exp(-H(\tfrac{1}{2})\mu_{n,i}).$$

Therefore, provided $K \geq 6\varepsilon_1^{1-d}/f_0$ in the case $b = \infty$ or $K \geq 6b\varepsilon_1^{1-d}/f_0$, $b < \infty$, for large n we have by (13.13) that

$$\begin{aligned} P[H_n(K)] &\leq \sum_{i\geq K/\varepsilon_1} n\exp(\gamma i)\exp\left(-iH(1/2)(\varepsilon_1^d f_0/2)\log n\right) \\ &\leq n \sum_{i\geq K/\varepsilon_1} \exp(-iH(\tfrac{1}{2})(\varepsilon_1^d f_0/4)\log n) \quad (\text{since } \log n \gg \gamma) \\ &\leq cn^{1-KH(1/2)f_0\varepsilon_1^{d-1}/4}. \end{aligned} \tag{13.16}$$

Provided we also choose K so that $KH(1/2)f_0\varepsilon_1^{d-1} > 8$, this expression is summable in n, so the result follows by the Borel–Cantelli lemma. □

Points in the torus. A similar result to Theorem 13.2 holds for the case where the points are distributed in the d-dimensional torus, $d \geq 2$.

Theorem 13.6 *Suppose that* $d \geq 2$ *and the points are distributed on the torus, with* $f_0 > 0$. *Suppose* $(k_n)_{n\geq 1}$ *is a sequence of positive integers with* $k_n/\log n \to b \in [0,\infty]$, *and* $k_n/n \to 0$ *as* $n \to \infty$. *In the case* $b < \infty$, *assume also that the sequence* $(k_n)_{n\geq 1}$ *is non-decreasing, and define* $a \in [0,1)$ *by* $a/H(a) = b$. *Then, if* $b = \infty$,

$$\lim_{n\to\infty}\left(\frac{n\theta T_{k_n}(\mathcal{X}_n)^d}{k_n}\right) = \frac{1}{f_0} \quad \text{a.s.}$$

If $b < \infty$,

$$\lim_{n\to\infty}\left(\frac{n\theta T_{k_n}(\mathcal{X}_n)^d}{\log n}\right) = \frac{1}{f_0 H(a)} \quad \text{a.s.}$$

In other words, when $d \geq 2$ the statement of Theorem 7.1 remains true with M_{k_n} (the largest k_n-nearest-neighbour link) replaced by T_{k_n} (the k_n-connectivity threshold). The argument to prove this is the same as that just given for Theorem 13.2, except for the fact that *the torus is not unicoherent.* However, by Lemma 9.2 it is bicoherent, and hence the set DU described in the proof of Lemma 13.5 is the union of at most two toroidally $*$-connected sets in $\mathcal{L}_n$. If $\mathcal{C}'_{n,i}$ denotes the collection of sets in $\mathcal{L}_n$ with total cardinality i, and with at most two toroidally $*$-connected components, then by Lemma 9.5, we can choose $\gamma > 0$ such that for large n we have

$$\text{card}(\mathcal{C}'_{n,i}) \leq n^2 i\exp(\gamma i), \tag{13.17}$$

and it is not hard to modify the proof of Proposition 13.4 to the torus, using (13.17) instead of (13.13).

13.3 SLLN in smoothly bounded regions

This section contains strong laws of large numbers for the k_n-connectivity threshold, analogous to those in the previous section, for the case where the common density f of the points X_i has connected compact support $\Omega \subset \mathbb{R}^d$ with smooth boundary $\partial\Omega$. More precisely, we assume $\partial\Omega$ is a $(d-1)$-dimensional C^2 submanifold of $\mathbb{R}^d$ (see Section 5.2). As before, we assume that $d \geq 2$, and that $f|_\Omega$ is continuous at x for all $x \in \partial\Omega$. Set $f_0 := \text{ess inf}_\Omega f$, and $f_1 := \text{ess inf}_{\partial\Omega} f$. Unlike the case of points in the cube, we can assume *the norm* $\|\cdot\|$ *used to define our geometric graphs is arbitrary.*

Theorem 13.7 *Suppose $d \geq 2$. Suppose that Ω is bounded and connected in $\mathbb{R}^d$, and $\partial\Omega$ is a $(d-1)$-dimensional C^2 submanifold of $\mathbb{R}^d$. Suppose that $f_0 > 0$, and the discontinuity set of $f|_\Omega$ contains no element of $\partial\Omega$. Suppose $(k_n)_{n\geq 1}$ is a sequence of non-negative integers with* $\lim_{n\to\infty}(k_n/n) = 0$ *and* $\lim_{n\to\infty}(k_n/\log n) = b \in [0,\infty]$. *In the case $b < \infty$, assume also that the sequence $(k_n)_{n\geq 1}$ is nondecreasing, and define numbers a_0 and a_1 in $[0,1)$ by*

$$\frac{a_0}{H(a_0)} = b, \qquad \frac{a_1}{H(a_1)} = \frac{bd}{d-1}. \tag{13.18}$$

Then if $b = \infty$, with probability 1 we have

$$\lim_{n\to\infty}\left(\frac{n\theta T_{k_n+1}(\mathcal{X}_n)^d}{k_n}\right) = \max\left(\frac{1}{f_0}, \frac{2}{f_1}\right),$$

whereas if $b < \infty$, with probability 1 we have

$$\lim_{n\to\infty}\left(\frac{n\theta T_{k_n+1}(\mathcal{X}_n)^d}{\log n}\right) = \max\left(\frac{1}{f_0 H(a_0)}, \frac{2(d-1)}{d f_1 H(a_1)}\right).$$

Let κ_n denote the connectivity of $G(\mathcal{X}_n; r_n)$. Using Theorems 13.2, 13.6, and 13.7, one can obtain a strong law of large numbers for κ_n. The statement of this is the same as the statement of Theorem 7.14 with the minimum degree δ_n replaced by κ_n, and with the extra conditions that $d \geq 2$ and Ω is connected. The proof is the same as that given earlier for Theorem 7.14.

We prove Theorem 13.7 under the extra assumption that the norm $\|\cdot\|$ satisfies

$$\|x\| \leq \|x\|_2 \quad \text{for all } x \in \mathbb{R}^d, \tag{13.19}$$

which involves no loss of generality, since if Theorem 13.7 holds for a given norm $\|\cdot\|$, it also holds for the norm $c\|\cdot\|$ for any strictly positive constant c.

As in the case of the analogous result for points in the cube, Theorem 13.7 is already half proved. By Theorem 7.2, and (13.1), we have at once with probability 1 that if $b = \infty$, then

$$\liminf_{n\to\infty}\left(\frac{n\theta T_{k_n+1}(\mathcal{X}_n)^d}{k_n}\right)\geq \max\left(\frac{1}{f_0},\frac{2}{f_1}\right), \tag{13.20}$$

whereas if $b<\infty$,

$$\liminf_{n\to\infty} n\left(\frac{n\theta T_{k_n+1}(\mathcal{X}_n)^d}{\log n}\right)\geq \max\left(\frac{1}{f_0H(a_0)},\frac{2(d-1)}{df_1H(a_1)}\right). \tag{13.21}$$

As in the preceding section, define

$$\rho_n := (k_n/n)^{1/d} \text{ if } b=\infty, \qquad \rho_n := ((\log n)/n)^{1/d} \text{ if } b<\infty.$$

Fix arbitrary t satisfying (in the case $b=\infty$)

$$\theta t^d > \max\left(\frac{1}{f_0},\frac{2}{f_1}\right), \tag{13.22}$$

or (in the case $b<\infty$)

$$\theta t^d > \max\left(\frac{1}{f_0H(a_0)},\frac{2(d-1)}{df_1H(a_1)}\right). \tag{13.23}$$

In view of (13.20) and (13.21), to prove Theorem 13.7, it suffices to prove that with probability 1, $T_{k_n+1}(\mathcal{X}_n)\leq t\rho_n$ for large enough n. As before, we use the concept of k_n-separating pairs, described in Section 13.1. In view of Proposition 7.4, the following result is sufficient to give us Theorem 13.7.

Proposition 13.8 *Suppose the hypotheses of Theorem 13.7 hold. For $K>0$, let $H_n(K)$ be the event that there exists a k-separating pair (U,W) for $G(\mathcal{X}_n;t\rho_n)$ with $\min(\mathrm{diam}_\infty(U),\mathrm{diam}_\infty(W))>K\rho_n$. Then there exists $K>0$ such that, with probability 1, the events $H_n(K)$ occur for only finitely many n.*

The proof uses discretization and a Peierls argument, as for the corresponding result for points in the cube. In the present case, matters are complicated by the fact that part of the discretized boundary region can lie outside Ω. We shall show that a non-vanishing proportion of the boundary region lies inside Ω, which is sufficient to get the Peierls argument to work.

Let c_1 denote the diameter of the unit cube in the chosen norm, as at (7.14). In this section, we choose ε_2 to satisfy

$$\varepsilon_2 < \min(t,1)/(4\max(c_1,d)). \tag{13.24}$$

Let $\varepsilon_2\rho_n\mathbb{Z}^d$ denote the lattice $\{\varepsilon_2\rho_n z : z\in\mathbb{Z}^d\}$. Also, for $z\in\varepsilon\rho_n\mathbb{Z}^d$ we define the cube

$$C_n(z) := B(\varepsilon_2\rho_n)\oplus\{z\}.$$

We say $\tau\subseteq\varepsilon_2\rho_n\mathbb{Z}^d$ is $*$-connected if $\{(\varepsilon_2\rho_n)^{-1}z : z\in\tau\}$ is a $*$-connected subset of $\mathbb{Z}^d$ (see Section 9.2).

Given $\eta>0$, let $\mathcal{C}_{n,i}(\eta)$ denote the collection of $*$-connected sets $\sigma\subseteq\mathcal{L}_n$ of cardinality i such that at least ηi of the points z of σ satisfy $C_n(z)\subseteq\Omega$. The main step in proving Proposition 13.8 is the following topological lemma, the proof of which is deferred until later on.

Lemma 13.9 *There exist constants $\eta_1 > 0, \eta_2 > 0$, and $n_1 \in \mathbb{N}$, such that for all $n \geq n_1$, if (U, W) is a $k_n$separating pair for $G(\mathcal{X}_n; t\rho_n)$, then there exists $\sigma \in \mathcal{C}_{n,i}(\eta_2)$ with $\mathcal{X}_n[\cup_{z\in\sigma} C_n(z)] \leq k_n$, for some i with*

$$i\varepsilon_2\rho_n \geq \min(\mathrm{diam}_\infty(U), \mathrm{diam}_\infty(V), \eta_1/2).$$

Proof of Proposition 13.8 By a Peierls argument (Corollary 9.4), there are constants $\gamma = \gamma(d) > 0$ and $c > 0$ such that, for all large enough n and all $i \in \mathbb{N}$,

$$\mathrm{card}(\mathcal{C}_{n,i}(\eta_2)) \leq c\rho_n^{-d} \exp(\gamma i) \leq n \exp(\gamma i). \tag{13.25}$$

Choose K so that $K\eta_2 f_0 \varepsilon_2^{d-1} > 2$. If $H_n(K)$ occurs, there exists a k_n-separating pair (U, W) for $G(\mathcal{X}_n; t\rho_n)$, with $\min(\mathrm{diam}_\infty(U), \mathrm{diam}_\infty(W)) \geq K\rho_n$. If also n is large enough so that $K\rho_n \leq \eta_1/2$, and also $n \geq n_1$ with n_1 and η_1, η_2 appearing in Lemma 13.9, then by that result there exists $\sigma \in \mathcal{C}_{n,i}(\eta_2)$ with $\mathcal{X}_n[\cup_{z\in\sigma} C_n(z)] \leq k_n$, for some i with $i\varepsilon_2\rho_n \geq K\rho_n$. Hence,

$$P[H_n(K)] \leq \sum_{i \geq K/\varepsilon_2} \sum_{\sigma \in \mathcal{C}_{n,i}(\eta_2)} P\left[\mathcal{X}_n\left(\bigcup_{y\in\sigma} C_n(y)\right) \leq k_n\right]. \tag{13.26}$$

If $\sigma \in \mathcal{C}_{n,i}(\eta_2)$ then

$$nF\left(\bigcup_{y\in\sigma} C_n(y)\right) \geq ni\eta_2 f_0(\varepsilon_2\rho_n)^d := \mu_{n,i}.$$

Provided i is such that (in the case $b = \infty$) $i\eta_2\varepsilon_2^d f_0 \geq 3$ or (in the case $b < \infty$) $i\eta_2\varepsilon_2^d f_0 \geq 3b$, for large n we have $k_n \leq \mu_{n,i}/2$ so that, by Lemma 1.1,

$$P\left[\mathcal{X}_n\left(\bigcup_{y\in\sigma} C_n(y)\right) \leq k_n\right] \leq \exp(-H(\tfrac{1}{2})\mu_{n,i}).$$

Therefore, provided $K \geq 3\varepsilon_2^{1-d}/(f_0\eta_2)$ in the case $b = \infty$, or $K \geq 3b\varepsilon_2^{1-d}/(f_0\eta_2)$ in the case $b < \infty$, by (13.25), (13.26), and the fact that $n\rho_n^d \geq \log n$, for large n we have

$$\begin{aligned} P[H_n(K)] &\leq \sum_{i \geq K/\varepsilon_2} n \exp\left(\gamma i - iH(\tfrac{1}{2})\eta_2\varepsilon_2^d f_0 \log n\right) \\ &\leq n \sum_{i \geq K/\varepsilon_2} n^{-iH(1/2)\eta_2\varepsilon_2^d f_0/2} \leq cn^{1-KH(1/2)f_0\eta_2\varepsilon_2^{d-1}/2}. \end{aligned}$$

Provided we also choose K so that $KH(1/2)f_0\eta_2\varepsilon_2^{d-1} > 4$, this expression is summable in n, so the result follows by the Borel–Cantelli lemma. □

It remains to prove Lemma 13.9. As in Section 11.3, for $a > 0$ let $\mathcal{A}_a$ denote the class of sets A of the form $A = \cup_{i=1}^k (B(a) \oplus \{az_i\})$, with $\{z_1, \ldots, z_m\} \subseteq \mathbb{Z}^d$, such that A has connected interior. For $A \in \mathcal{A}_a$, let A^o be the interior of A. Let Ω^o be the interior of Ω.

Let the constant δ_1 and the finite collection of pairs (ξ_i, e_i), $1 \leq i \leq \mu$ be given by Proposition 5.10. Define the 'disk' D_i by

$$D_i = \{\xi_i + 4\delta_1 e_i + y : y \cdot e_i = 0, \|y\|_2 < 4\delta_1 e_i\}. \tag{13.27}$$

Then, for each $i \leq \mu$, we assert that

$$D_i \subseteq \Omega \setminus \bigcup_{j=1}^{\mu} D(\xi_j; \delta_1, e_j) := \Omega^{\mathrm{I}}. \tag{13.28}$$

This inclusion holds because any point in the disk D_i lies at an l_2 distance at least $3\delta_1$ from Ω^{c}, since $D^*(\xi_i; 10\delta_1, 0.1, e_i) \subseteq \Omega$, whereas for all $j \leq \mu$, any point in $D(\xi_j; \delta_1, e_j)$ lies an l_2 distance at most $2\delta_1$ from Ω^{c}, since $D^*(\xi_j, \delta_1, 0.1, -e_j) \subseteq \Omega^{\mathrm{c}}$.

By (13.28), the 'interior' set Ω^{I} is non-empty. Pick $x_0 \in \Omega^{\mathrm{I}}$. For integer m, let A_m be the maximal element A of $\mathcal{A}_{2^{-m}}$ (possibly the empty set) such that $x_0 \in A$ and $A \subseteq \Omega^o$. Then, by Lemma 11.12, $A_1 \subseteq A_2 \subseteq A_3 \subseteq \cdots$ and the union of the sets A_m^o is Ω^o.

Since Ω^{I} is a compact set contained in Ω^o, there exists m_1 with $\Omega^{\mathrm{I}} \subseteq A_{m_1}^o$. The set A_{m_1} is a connected finite union of hypercubes with $A_{m_1} \subseteq \Omega^o$. Also, we can (and do) take $m_3 > m_2 > m_1$ such that $A_{m_1} \subseteq A_{m_2}^o$, and $A_{m_2} \subseteq A_{m_3}^o$. We re-label these sets as follows for later reference:

$$\Omega_1 := A_{m_1}, \qquad \Omega_2 := A_{m_2}, \qquad \Omega_3 := A_{m_3}. \tag{13.29}$$

Note that $\Omega_1 \subset \Omega_2 \subset \Omega_3 \subset \Omega$, with the boundaries of these sets all being disjoint. Note also that Ω_1 is non-empty and Ω_3 is a connected finite union of dyadic hypercubes with common side-length 2^{-m_3}; set $\eta_1 := 2^{-m_3}$.

Proof of Lemma 13.9 Suppose (U, W) is a k_n-separating pair for $G(\mathcal{X}_n; t\rho_n)$. First consider the case with $U \cap \Omega_2 \neq \emptyset$ and $W \cap \Omega_2 \neq \emptyset$. The sets $U_{t\rho_n/2}$ and $W_{t\rho_n/2}$ are disjoint connected subsets of $\mathbb{R}^d$ (here we use the notation introduced at (13.12)). So $\mathbb{R}^d \setminus U_{t\rho_n/2}$ has a connected component which contains $W_{t\rho_n/2}$; denote this component W', and let $U' := \mathbb{R}^d \setminus W'$. Then the closures of U' and W' are connected and their union is $\mathbb{R}^d$, so their intersection, a part of the boundary of $U_{t\rho_n/2}$ denoted ∂U, is connected by the unicoherence of $\mathbb{R}^d$ (Lemma 9.1). Also, $U \subseteq U'$ and $W \subseteq W'$, so *any path from a point of U to a point of W must pass through ∂U*. We claim that

$$\mathrm{diam}_\infty(\partial U) \geq \min(\mathrm{diam}_\infty(U), \mathrm{diam}_\infty(W)). \tag{13.30}$$

To see this, assume the contrary. Then there would exist a rectilinear cube C of side $b < \min(\mathrm{diam}_\infty(U), \mathrm{diam}_\infty(W))$ such that $\partial U \subseteq C$. By the condition on b

there would be points $X \in U$ and $Y \in W$ which were not in C; it would then be possible to get from X to Y by a path avoiding the cube C, and hence avoiding ∂U, a contradiction.

Also, $\partial U \cap \Omega_2 \neq \emptyset$, since by assumption we can pick $\tilde{X} \in U \cap \Omega_2$ and $\tilde{Y} \in W \cap \Omega_2$, and take a path in Ω_2 from $\tilde{X}$ to $\tilde{Y}$. Pick $x_1 \in \partial U \cap \Omega_2$, and let $\partial_1 U$ denote the component including x_1 of $\partial U \cap (B(\eta_1) \oplus \{x_1\})$. Since ∂U is connected, if $\partial_1 U \subseteq (B(\eta_3) \oplus \{x_1\})$ for some $\eta_3 < \eta_1$, then $\partial_1 U = \partial U$. Hence, by (13.30),

$$\mathrm{diam}_\infty(\partial_1 U) \geq \min(\mathrm{diam}_\infty(U), \mathrm{diam}_\infty(W), \eta_1/2). \tag{13.31}$$

Let DU denote the set of $z \in \varepsilon_2 \rho_n \mathbb{Z}^d$ such that $C_n(z)$ has non-empty intersection with $\partial_1 U$. Then DU is $*$-connected, and since $\partial_1 U \subset \Omega_3$, provided $c_1 \varepsilon_2 \rho_n \leq \mathrm{dist}(\Omega_3, \partial\Omega)$, we also have $\cup_{z \in DU} C_n(z) \subseteq \Omega$. Also, since $\mathrm{dist}(x, \mathcal{X}_n) = t\rho_n/2$ for each $x \in \partial U$, the condition $c_1 \varepsilon_2 < t/4$ from (13.24) and an argument using the triangle inequality similar to that used for (13.15) gives us

$$\mathcal{X}_n[C_n(z)] \leq k_n, \quad \forall\, z \in DU. \tag{13.32}$$

Finally, $\varepsilon_2 \rho_n \,\mathrm{card}(DU) \geq \mathrm{diam}(\partial_1 U)$, and by (13.31), the conclusion of the lemma follows for this case.

The other more complicated case to be considered is the case where $U \cap \Omega_2$ and $W \cap \Omega_2$ are not both non-empty. Assume, without loss of generality, that $U \cap \Omega_2 = \emptyset$. Let W' be the component of $\mathbb{R}^d \setminus U_{t\rho_n/2}$ which includes Ω_1. Let $U' := \mathbb{R}^d \setminus W'$. Then the closures of U' and W' are connected and their union is $\mathbb{R}^d$, so their intersection, a part of the boundary of $U_{t\rho_n/2}$ denoted ∂U, is connected by unicoherence. Let DU denote the set of $z \in \varepsilon_2 \rho_n \mathbb{Z}^d$ such that $C_n(z)$ has non-empty intersection with ∂U. Then DU is $*$-connected, and since $\mathrm{diam}_\infty(\Omega_1) \geq \eta_1$, by an argument similar to that for (13.30),

$$\varepsilon_2 \rho_n \,\mathrm{card}(DU) \geq \mathrm{diam}_\infty(\partial U) \geq \min(\mathrm{diam}_\infty(U), \eta_1).$$

Also, (13.32) holds for the same reasons as in the previous case. We shall show that the proportion of DU lying inside Ω is bounded away from zero.

For $z \in DU$ with $C_n(z) \cap \Omega^c \neq \emptyset$, we shall define $\phi(z) \in \varepsilon_2 \rho_n \mathbb{Z}^d$ in such a way that $\phi(z) \in DU$, and $C_n(\phi(z)) \subseteq \Omega$. The general idea goes as follows; the reader should refer to Fig. 13.2. Given z (the centre of the higher small square, representing $C_n(z)$, in Fig. 13.2), look for a nearby point X of U (the centre of the more darkly shaded disk), which must be in Ω but near the boundary, and hence in one of the cylinders $D(\xi_i, \delta_1, e_i)$ defined in Proposition 5.10. This cylinder is represented by the large vertical rectangle in Fig. 13.2. Move from X in the direction of e_i (that is, towards the interior of Ω), until the last exit within the cylinder from $U_{t\rho_n/2}$ (there is a last exit because U is assumed to lie entirely near the boundary of Ω). The nearest point of $\varepsilon_2 \rho_n \mathbb{Z}^d$ to this exit point (the centre of the lower small square in Fig. 13.2) is $\phi(z)$.

FIG. 13.2. The horizontal line represents part of the upper boundary of Ω and the shaded region represents $U_{t\rho_n/2}$.

Here is the formal definition of $\phi(z)$, given $z \in DU$ with $C_n(z) \cap \Omega^c \neq \emptyset$. Pick $y = y(z) \in C_n(z) \cap \partial U$. Then pick a point $X = X(z)$ of U with $\|X - y\| = t\rho_n/2$. If there are several possible choices for y or for X, make the choice using the lexicographic ordering on $\mathbb{R}^d$.

By the assumption on U, $X \notin \Omega_2$, so by the definition (13.28) of Ω^{I} and the fact that $\Omega^{\mathrm{I}} \subseteq \Omega_1 \subseteq \Omega_2$, X lies in a cylinder $D(\xi_i; \delta_1, e_i)$ for some $i \leq \mu$; let $i(z)$ be the smallest such i. Take $\lambda_1(z) \in (0, 5\delta_1]$ such that $X + \lambda_1(z)e_{i(z)}$ is in the disk $D_{i(z)}$ (defined at (13.27)). Then by (13.28), $X + \lambda_1(z)e_{i(z)} \in \Omega_I \subseteq U^c_{t\rho_n/2}$. Let

$$\lambda(z) := \max\{\lambda \in (0, \lambda_1(z)] : X(z) + \lambda e_{i(z)} \in U_{t\rho_n/2}\},$$

and let $w(z) = X(z) + \lambda(z)e_{i(z)}$. Let $\phi(z)$ be the point $u \in \varepsilon_2\rho_n\mathbb{Z}^d$ such that $w(z) \in C_n(u)$.

Clearly, $w(z)$ lies on the boundary of $U_{t\rho_n/2}$, and we claim that additionally $w(z)$ is on the boundary of W'. Indeed, $w(z)$ is connected by a path in the complement of $U_{t\rho_n/2}$ to $X + \lambda_1(z)e_{i(z)}$, which lies in Ω^{I} and hence in Ω_1. Hence, $w(z) \in \partial U$ and $\phi(z) \in DU$.

We assert that $C_n(\phi(z)) \subseteq \Omega$. To prove this, let $x \in C_n(\phi(z))$, and set $i = i(z)$. Write $x = X(z) + ae_i + v$, with $v \cdot e_i = 0$. Then we have

$$((t/2) - d\varepsilon_2)\rho_n \leq a \leq 5\delta + d\varepsilon_2\rho_n$$

and $\|v\|_2 \leq d\varepsilon_2\rho_n$, so that by the condition (13.24) on ε_2, $\|v\|_2 \leq a$. Hence, by the definition (5.22) and the property (5.25) of $D^*(x; r, \eta, e)$,

$$x \in D^*(X(z), 3a/2, 0.1, e_i) \subseteq \Omega,$$

which proves the assertion.

The mapping ϕ is many-to-one, but there is a uniform bound on the number of points z which ϕ can map to the same point u, as we now show. Fix $u \in \varepsilon_2\rho_n\mathbb{Z}^d$ and $i \leq \mu$, and suppose $z \in DU$ satisfies $C_n(z) \cap \Omega^c \neq \emptyset$ and $\phi(z) = u$, and $i(z) = i$. Let $X = X(z)$, and observe first that

$$X - t\rho_n e_i \notin \Omega. \tag{13.33}$$

Indeed, if this were not the case then $D^*(X - t\rho_n e_i; 2t\rho_n, 0.1, e_i)$ would be contained in Ω, and hence $B(X; t\rho_n/2)$ would be contained in the interior of Ω (here we use the assumption (13.19)). However, we know from the construction of X that there is a point of the boundary of Ω in $B(X; t\rho_n/2)$, and this contradiction gives us (13.33).

With ψ_i defined in Proposition 5.10, since $X \in \Omega$ and $X - t\rho_n e_i \notin \Omega$, we have $\|\psi_i(w(z)) - X\|_2 \leq t\rho_n$. Also, by the last part of Proposition 5.10,

$$\|\psi_i(w(z)) - \psi_i(u)\| \leq c_2c_1\rho_n,$$

and hence,

$$\begin{aligned}\|z - \psi_i(u)\|_2 &\leq \|z - X(z)\|_2 + \|X(z) - \psi_i(w(z))\|_2 + \|\psi_i(w(z)) - \psi_i(u)\|_2 \\ &\leq (c_1\varepsilon + t/2)\rho_n + t\rho_n + c_2c_1\rho_n.\end{aligned}$$

The number of points $z \in \varepsilon_2\rho_n\mathbb{Z}^d$ satisfying this inequality is bounded by a constant, denoted c_4, independent of n or u; hence, the number of points z mapped by ϕ to u is bounded by $c_4\mu$, where μ is the number of cylinders in Proposition 5.10. Therefore, the proportion of points u of DU satisfying $C_n(u) \subseteq \Omega$ is at least η_2, where we set $\eta_2 := 1/(c_4\mu + 1)$. Thus DU is the required set σ. □

13.4 Convergence in distribution

In this section we assume $f = f_U$, that is, the distribution of points X_i is uniform on the unit cube $C := [-\frac{1}{2}, \frac{1}{2}]^d$, $d \geq 2$. We also assume that the metric on C is given either by the restriction to C of an l_p norm with $1 < p \leq \infty$, or by a toroidal metric based on an arbitrary norm. For this setting, in Chapter 8 we derived convergence in distribution results for the largest k-nearest-neighbour link $M_k(\mathcal{X}_n)$, suitably scaled and centred, with k fixed. We now prove convergence in distribution for the k-connectivity threshold $T_k(\mathcal{X}_n)$; note that here we make the extra assumption that $d \geq 2$. As at (8.2), we set

$$\psi^k_{n,r}(x) := (nF(B(x; r)))^k \exp(-nF(B(x; r)))/k! \tag{13.34}$$

Theorem 13.10 *Let $k \in \mathbb{N} \cup \{0\}$. Suppose $\beta > 0$ and $(r_n = r_n(\beta), n \geq 1)$ is chosen so that*

$$\lim_{n\to\infty} (nr_n^d) = \infty, \qquad \lim_{n\to\infty} n \int_C \psi_{n,r_n}^k(x)\mathrm{d}x = \beta. \tag{13.35}$$

Then

$$\lim_{n\to\infty} P[T_{k+1}(\mathcal{X}_n) \leq r_n] = \lim_{n\to\infty} P[M_{k+1}(\mathcal{X}_n) \leq r_n] = \mathrm{e}^{-\beta}. \tag{13.36}$$

Theorem 13.10 demonstrates an equivalence between the limiting distribution of $T_{k+1}(\mathcal{X}_n)$, suitably transformed, and that of $M_{k+1}(\mathcal{X}_n)$, under the same transformation. The latter limit was given in Theorem 8.4. Later on, in Theorem 13.17, we shall give a stronger form of equivalence between $T_{k+1}(\mathcal{X}_n)$ and $M_{k+1}(\mathcal{X}_n)$; they are actually equal with probability tending to 1.

Let K_n denote the number of components of $G(\mathcal{X}_n; r_n)$. The proof of Theorem 13.10 will also give us the following Poisson limit for K_n.

Theorem 13.11 *Suppose $(r_n)_{n\geq 1}$ is chosen so that (13.35) holds, with $k = 0$. Then $K_n - 1 \xrightarrow{\mathcal{D}} \mathrm{Po}(\beta)$ as $n \to \infty$.*

Specific choices of $(r_n)_{n\geq 1}$ to satisfy (13.35) were described in the proof of Theorem 8.4.

Fix k, let $\beta \in \mathbb{R}$ and take $r_n = r_n(\beta)$ to satisfy (13.35). As we shall see below, Theorems 13.10 and 13.11 follow from the the following two propositions. Recall the definition of a k-separating pair in Section 13.1 and of a k-separated set in Section 7.1.

Proposition 13.12 *Let $E_n(K) = E_n(K;\beta)$ be the event that there exists a k-separated set U for $G(\mathcal{X}_n; r_n)$, with at least two elements and with* $\mathrm{diam}(U) \leq Kr_n$. *Then for all $K > 0$, it is the case that* $\lim_{n\to\infty} P[E_n(K)] = 0$.

Proposition 13.13 *Let $F_n(K) = F_n(K;\beta)$ be the event that there is a k-separating pair (U, W) for $G(\mathcal{X}_n; r_n)$, such that* $\mathrm{diam}(U) > Kr_n$ *and* $\mathrm{diam}(V) > Kr_n$. *Then there exists $K > 0$ such that* $\lim_{n\to\infty} P[F_n(K)] = 0$.

Proof of Theorem 13.10 The second equality in (13.36) comes from Theorem 8.1; we need to prove the first equality. By (13.1), we have $T_{k+1}(\mathcal{X}) \geq M_{k+1}(\mathcal{X})$ for any point set $\mathcal{X}$. Therefore, to prove (13.36), it suffices to prove that

$$\lim_{n\to\infty} P[M_{k+1}(\mathcal{X}_n) \leq r_n < T_{k+1}(\mathcal{X}_n)] = 0. \tag{13.37}$$

Using Proposition 13.13, choose K such that $P[F_n(K)] \to 0$. If $M_{k+1}(\mathcal{X}_n) \leq r_n < T_{k+1}(\mathcal{X}_n)$ then $G(\mathcal{X}_n; r_n)$ is not $(k+1)$-connected but has minimum degree at least $k+1$. By the first of these conclusions, and Lemma 13.1, $G(\mathcal{X}_n; r_n)$ has a k-separating pair (U, V), and by the second conclusion, each of U and V

has at least two elements. Hence, $E_n(K) \cup F_n(K)$ occurs. Therefore, by Boole's inequality,

$$P[M_{k+1}(\mathcal{X}_n) \le r_n < T_{k+1}(\mathcal{X}_n)] \le P[E_n(K)] + P[F_n(K)],$$

which tends to zero by Propositions 13.12 and 13.13, which are proved below. □

Proof of Theorem 13.11 By case $k = 0$ of Propositions 13.12 and 13.13, with probability tending to 1 there is precisely one component of $G(\mathcal{X}_n; r_n)$ of order greater than 1. Combining this fact with the Poisson limit theorem for the number of isolated vertices (Theorem 8.1), we obtain the result. □

It remains to prove Propositions 13.12 and 13.13. As at (13.12), for $A \subset \mathbb{R}^d$, set $A_r := A \oplus B(\mathbf{0}; r)$, the r-neighbourhood of A (in the toroidal case, let A_r be the toroidal r-neighbourhood of A). Set

$$v_r(A) := F(A_r).$$

Throughout this section, we use the notation from (8.19) that for $x \in C$, D_x denotes the set of points in C which are l_1-closer to the centre of C than x is (see Fig. 13.3). Given K, let the region $R_x = R_x(K, n)$ be defined by

$$R_x := D_x \cap B(x; Kr_n).$$

Let $E_n^x(K; m)$ be the event that there is a set U' of m points of $\mathcal{X}_{n-1}$ such that $U' \subseteq R_x$, and such that if we set $U = U' \cup \{x\}$ we have $\mathcal{X}_{n-1}(U_{r_n} \setminus U) \le k$. Let $E_n^x(K) = \cup_{m=1}^{\infty} E_n^x(K; m)$. Proposition 13.12 is proved via the next three lemmas.

Lemma 13.14 *Let $K > 0$. Then with $\psi_{n,r}^k(x)$ defined at (13.34), there is a finite constant c such that*

$$P[E_n(K)] \le c \sup_{x \in C} \left(P[E_n^x(K)] / \psi_{n,r_n}^k(x) \right).$$

Proof Consider $\mathcal{X}_n$ as the union of $\mathcal{X}_{n-1}$ with a single independent uniform point X. Suppose $E_n(K)$ occurs and X is the point of maximal l_1 norm of the points in the set U described in the definition of $E_n(K)$. Then $E_n^X(K)$ occurs. By exchangeability of $X_1, \ldots, X_n$,

$$\begin{aligned} P[E_n(K)] &\le nP[E_n^X(K)] = n \int_C P[E_n^x(K)] \mathrm{d}x \\ &\le \sup_{x \in C} \left(P[E_n^x(K)] / \psi_{n,r_n}^k(x) \right) \times n \int_C \psi_{n,r_n}^k(x) \mathrm{d}x. \end{aligned}$$

Since $n \int_C \psi_{n,r_n}^k(x) \mathrm{d}x$ is bounded by (13.35), the result follows. □

Lemma 13.14 shows that to prove Proposition 13.12, it suffices to prove that for any K, $P[E_n^x(K)]/\psi_{n,r_n}^k(x)$ is small uniformly in x. First we show that this is true for *some* K.

Lemma 13.15 *There exists* $K \in (0,1]$ *such that*

$$\lim_{n\to\infty} \sup_{x\in C} \left(P[E_n^x(K)]/\psi_{n,r_n}^k(x)\right) = 0. \tag{13.38}$$

Proof In this proof we write simply r for r_n. For $0 \le j \le k$ and $m \ge 1$, let $\mu^x(m,j,n)$ be the expected number of subsets U' of $\mathcal{X}_{n-1}$ of cardinality m, contained in R_x, such that if we set $U = U' \cup \{x\}$ we have $\mathcal{X}_{n-1}(U_r \backslash U) = j$. Then

$$\mu^x(m,j,n) = \frac{(n-1)!}{j!m!(n-1-j-m)!} \int_{R_x} \cdots \int_{R_x} \mathrm{d}x_1 \cdots \mathrm{d}x_m v_r(\{x, x_1, \dots, x_m\})^j \times (1 - v_r(\{x, x_1, \dots, x_m\}))^{n-1-m-j}.$$

If $x \in C$ and $R > 1$, then for all $y \in \mathbb{R}^d$ such that $x + Ry \in C$ we have $x + y \in C$ by convexity. Hence,

$$v_{Rs}(\{x\}) \le R^d v_s(\{x\}), \qquad x \in C, R > 1, s > 0. \tag{13.39}$$

Since $\{x, x_1, \dots, x_m\}_r \subseteq \{x\}_{(1+K)r}$ for all $x_1, \dots, x_m \in R_x$, since $v_{(1+K)r}(\{x\}) \le (1+K)^d v_r(\{x\})$ by (13.39), and since $1 - t \le \mathrm{e}^{-t}$ for all t,

$$\mu^x(m,j,n) \le \frac{(1+K)^{dj} v_r(\{x\})^j (n-1)!}{j!m!(n-1-j-m)!} \int_{R_x} \cdots \int_{R_x} \mathrm{d}x_1 \cdots \mathrm{d}x_m \times \exp(-nv_r(\{x, x_1, \dots, x_m\}) + (j+1+m)(1+K)^d \theta r^d).$$

Now,

$$\frac{(n-1)!}{(n-1-j-m)!} = n^{j+m}\left(1 - \frac{1}{n}\right)\left(1 - \frac{2}{n}\right) \cdots \left(1 - \frac{j+m}{n}\right) \le n^{j+m} \exp(-m^2/(2n)).$$

We saw at (8.13) that $nr_n^d = O(\log n)$. Therefore, there are constants c, c' such that for all $j \le k$ and $1 \le m \le n$,

$$(j+1+m)(1+K)^d \theta r_n^d - m^2/(2n) \le cm(c' \log n - m)/n,$$

which is bounded above, uniformly in m and n. So there is a constant c such that

$$\mu^x(m,j,n) \le \frac{c v_r(\{x\})^j n^{j+m}}{j!m!} \int_{R_x} \cdots \int_{R_x} \mathrm{d}x_1 \cdots \mathrm{d}x_m \exp(-nv_r(\{x, x_1, \dots, x_m\})).$$

By symmetry, restricting the above integral to the region in which the maximum $\max_{i \le m} \|x_i - x\|$ is achieved at $i = 1$ reduces it by a factor of m. Also, by

Proposition 5.16 and some easy scaling, there is a constant $\eta_4 > 0$ such that for $x_1, \ldots, x_m$ all in R_x we have

$$v_r(\{x, x_1, \ldots, x_m\}) \geq v_r(\{x\}) + \eta_4 r^{d-1} \max_{i \leq m} \|x_i - x\|.$$

(This statement is also true for points in the torus, by a similar, simpler argument.) Hence, by (13.34), we have

$$\mu^x(m, j, n) \leq \frac{c v_r(\{x\})^j n^{j+m}}{j!(m-1)!} \int_{R_x} \mathrm{d}x_1 \int_{B(x;\|x_1 - x\|)} \cdots \int_{B(x;\|x_1 - x\|)} \mathrm{d}x_2 \cdots \mathrm{d}x_m$$
$$\times \exp(-n v_r(\{x\}) - n\eta_4 r^{d-1} \|x_1 - x\|)$$
$$\leq \frac{c n^m}{(m-1)!} \psi^j_{n,r}(x) \int_{R_x} (\theta \|x_1 - x\|^d)^{m-1} \exp(-n\eta_4 \|x_1 - x\| r^{d-1}) \mathrm{d}x_1.$$

Summing over m and changing variable to $y = (x_1 - x)/r$, we obtain

$$\sum_{m \geq 1} \mu^x(m, j, n) \leq c n \psi^j_{n,r}(x) \int_{\|y\| \leq K} \exp(n\theta r^d \|y\|^d - n\eta_4 r^d \|y\|) r^d \mathrm{d}y.$$

Provided K is chosen sufficiently small, we have $\theta \|y\|^d \leq \eta_4 \|y\|/2$ whenever $\|y\| \leq K$, so that

$$\sum_{m \geq 1} \mu^x(m, j, n) \leq c(nr^d) \psi^j_{n,r}(x) \int_{\mathbb{R}^d} \exp(-n\eta_4 r^d \|y\|/2) \mathrm{d}y$$
$$= c'(nr^d)^{1-d} \psi^j_{n,r}(x).$$

Since $P[E^x_n(K)] \leq \sum_{j=0}^{k} \sum_{m \geq 1} \mu^x(m, j, n)$, and the result (13.38) follows from the condition $\lim_{n \to \infty}(nr^d_n) = \infty$ at (13.35). □

The next lemma extends the range of K for which the conclusion of the previous lemma holds.

Lemma 13.16 *Suppose* $0 < K' < K < \infty$. *Then*

$$\lim_{n \to \infty} \sup_{x \in C} \frac{P[E^x_n(K) \setminus E^x_n(K')]}{\psi^k_{n,r_n}(x)} = 0. \tag{13.40}$$

Proof By Proposition 5.15, we can (and do) choose η_5 such that if $A \subseteq \mathcal{O}_d := [0, \infty)^d$ with $\mathrm{diam}(A) \geq K$ and $x \in A$ with $\|x\|_1 \leq \|y\|_1$ for all $y \in A$, then with $|\cdot|$ denoting Lebesgue measure,

$$|A_1| \geq |A| + |\{x\}_1| + \eta_5. \tag{13.41}$$

Write r for r_n. Define $\varepsilon = \varepsilon(n)$ in such a way that

$$\varepsilon \leq \min(\eta_5/(4d^2(\theta + \eta_5)), 1/(4d), K'/(4d)) \leq 2\varepsilon, \tag{13.42}$$

and such that $(\varepsilon r_n)^{-1}$ is an integer; this is possible for all large enough n.

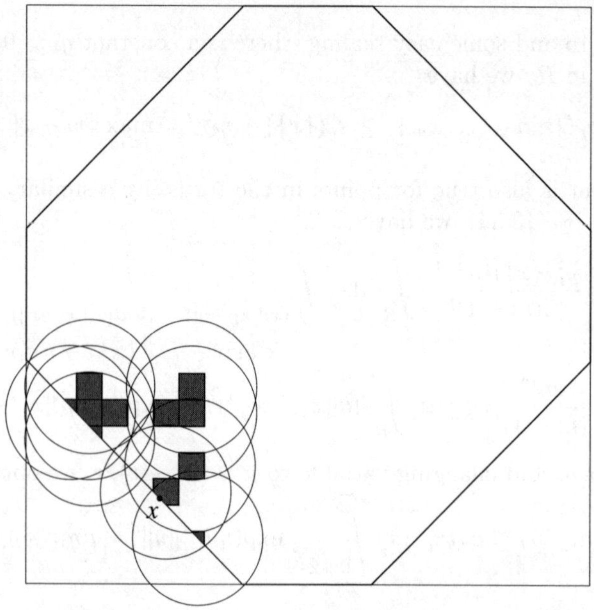

FIG. 13.3. The shaded region is B_σ^x, with $\sigma \in \mathcal{S}(n,x)$ given by the set of centres of shaded or partly shaded squares, with x represented by a point and with D_x bounded by the octagon shown. The union of the disks shown is $\sigma_{(1-d\varepsilon)r}$.

Divide C into little boxes (hypercubes) of side εr_n. Let $\mathcal{L}_n$ be the set of centres of these boxes (a fine lattice of points in C). For each $z \in \mathcal{L}_n$ let B_z be the box centred at z. Let $x \in C$ and let z_x be the $z \in \mathcal{L}_n$ such that x lies in the box B_z.

For $\sigma \subseteq \mathcal{L}_n$, let $B_\sigma^x := \cup_{z \in \sigma}(B_z \cap D_x)$ (see Fig. 13.3). Let $\mathcal{S}(n,x)$ denote the collection of all $\sigma \subseteq \mathcal{L}_n$, such that (i) $z_x \in \sigma$; (ii) σ is contained in $B(x;(K+d\varepsilon)r)$; (iii) σ has diameter at least $(K'-2d\varepsilon)r_n$; and (iv) $B_z \cap D_x \neq \emptyset$ for each $z \in \sigma$. Given $\sigma \in \mathcal{S}(n,x)$, define the event

$$E_n'(\sigma,x) := \{\mathcal{X}_{n-1}(\sigma_{(1-d\varepsilon)r} \setminus B_\sigma^x) \leq k\}.$$

By the triangle inequality, if $y \in B_z$ then $B(y;r) \supseteq B(z;(1-d\varepsilon)r)$. Suppose $E_n^x(K) \setminus E_n^x(K')$ occurs. Then there exists a set U' of points of $\mathcal{X}_n$ such that U' is contained in $\{x\}_{Kr} \cap D_x$, but not in $\{x\}_{K'r}$, and such that if we set $U = U' \cup \{x\}$ we have $\mathcal{X}_{n-1}(U_r \setminus U) \leq k$. Hence, there exists $\sigma \in \mathcal{S}(n,x)$ such that event $E_n'(\sigma,x)$ occurs, namely, the set of centres of boxes containing the points of U. Since $\mathrm{card}(\mathcal{S}(n,x)) \leq 2^{(2(K+d\varepsilon)/\varepsilon)^d}$, uniformly in x and n, it suffices to prove that

$$\lim_{n\to\infty} \sup_{x\in C} \sup_{\sigma\in\mathcal{S}(n,x)} \frac{P[E_n'(\sigma,x)]}{\psi_{n,r}^k(x)} = 0. \qquad (13.43)$$

Setting $v(\sigma, x) := F(\sigma_{(1-d\varepsilon)r} \backslash B_\sigma^x)$, we have

$$\begin{aligned} P[E_n'(\sigma,x)] &= \sum_{m=0}^{k} \binom{n-1}{m} v(\sigma,x)^m (1-v(\sigma,x))^{n-1-m} \\ &\leq \sum_{m=0}^{k} (nv(\sigma,x))^m \exp(-nv(\sigma,x)+m+1)/m!. \end{aligned} \tag{13.44}$$

We require useful upper and lower bounds on $v(\sigma, x)$. For the upper bound, note that the condition $\sigma \subseteq \{x\}_{(K+d\varepsilon)r}$ implies $\sigma_{(1-d\varepsilon)r} \subseteq \{x\}_{(K+1)r}$, so that by (13.39),

$$v(\sigma,x) \leq (1+K)^d v_r(\{x\}).$$

For the lower bound, note that $v(\sigma, x) \geq v_{(1-2d\varepsilon)r}(B_\sigma^x) - F(B_\sigma^x)$. By the definition of $\mathcal{S}(n,x)$, and (13.42), σ has diameter at least $(K'/2)r$. It can be close to at most one of the corners of C. By the definition of η_5 to satisfy (13.41), and some easy scaling, and (13.39),

$$\begin{aligned} v(\sigma,x) &\geq v_{(1-2d\varepsilon)r}(\{x\}) + \eta_5(1-2d\varepsilon)^d r^d \\ &\geq (1-2\varepsilon d)^d (v_r(\{x\}) + \eta_5 r^d) \geq (1-2\varepsilon d^2)(v_r(\{x\}) + \eta_5 r^d) \\ &\geq v_r(\{x\}) + \eta_5 r^d - 2\varepsilon d^2(\theta+\eta_5) r^d \\ &\geq v_r(\{x\}) + (\eta_5/2) r^d \qquad \text{by (13.42).} \end{aligned}$$

Combining these upper and lower bounds at (13.44), we have for some constant c that

$$P[E_n'(\sigma,x)] \leq c \sum_{m=0}^{k} \psi_{n,r}^m(x) \exp(-\eta_5 n r^d/2),$$

and since $nr^d \to \infty$, this gives (13.43) as required. □

Proof of Proposition 13.12 Immediate from Lemmas 13.14 – 13.16. □

Proof of Proposition 13.13 Take $\varepsilon = \varepsilon(n)$ as in the proof of Lemma 13.16. Divide C into cubes of side εr_n, let $\mathcal{L}_n$ denote the set of cube centres, and for $z \in \mathcal{L}_n$ let B_z denote the cube centred at z. For integer $i > 0$, if the points are uniformly distributed on the cube then let $\mathcal{C}_{n,i}$ denote the collection of $*$-connected subsets of $\mathcal{L}_n$, of cardinality i. If the points are uniformly distributed on the torus, then let $\mathcal{C}_{n,i}$ denote the collection of subsets of $\mathcal{L}_n$, of cardinality i, that have at most two toroidally $*$-connected components. By Corollary 9.4 in the case of the cube, or Lemma 9.5 in the case of the torus, there are constants c, γ such that, for all n and i,

$$\mathrm{card}(\mathcal{C}_{n,i}) \leq c r_n^{-2d} i \exp(\gamma i).$$

Suppose $F_n(K)$ occurs, that is, there is a k-separating pair (U, W) for $G(\mathcal{X}_n; r_n)$, with $\mathrm{diam}(U) > K r_n$ and $\mathrm{diam}(V) > K r_n$. Then $U_{r_n/2} \cap C$ and

$V_{r_n/2} \cap C$ are disjoint connected subsets of C. Then by the same argument as the proof of Lemma 13.5, there exists $\sigma \in \mathcal{C}_{n,i}$ satisfying $\mathcal{X}_n[\cup_{y\in\sigma} C_n(y)] \leq k$ and $i\varepsilon r_n \geq K r_n$. Therefore,

$$P[F_n(K)] \leq \sum_{i\geq K/\varepsilon} \sum_{\sigma\in\mathcal{C}_{n,i}} P\left[\mathcal{X}_n\left(\bigcup_{z\in\sigma} B_z\right) \leq k\right]$$

$$\leq \sum_{j=0}^{k} \sum_{i\geq K/\varepsilon} c r_n^{-2d} i \exp(\gamma i) \binom{n}{j} (i\varepsilon^d r_n^d)^j (1 - i\varepsilon^d r_n^d)^{n-j},$$

and since $r_n^d = \Theta((\log n)/n)$ by (8.12) and (8.13), so that $(nr_n^d)^j \leq n$ for large n,

$$P[F_n(K)] \leq c' n^3 \sum_{i\geq K/\varepsilon} i^{k+1} \exp(\gamma i) \exp(-ni\varepsilon^d r_n^d).$$

Take γ' such that $i^{k+1} e^{\gamma i} \leq e^{\gamma' i}$ for all i. Since $nr_n^d = \Theta(\log n)$ and ε is bounded away from zero and infinity, we can choose $\delta > 0$ and $n_2 > 0$ such that for $n \geq n_2$ we have $n\varepsilon^d r_n^d - \log \gamma' \geq \delta \log n$, and hence

$$P[F_n(K)] \leq c' n^3 \sum_{i\geq K/\varepsilon} n^{-i\delta} \leq \frac{c' n^{3-\delta K/\varepsilon}}{1 - n^{-\delta}},$$

which tends to zero, provided we choose K so that $\delta K/\varepsilon > 3$. □

13.5 Further results on points in the cube

We now use Theorem 13.10 to deduce a stronger equivalence between $T_{k+1}(\mathcal{X}_n)$ and $M_{k+1}(\mathcal{X}_n)$, which is analogous to a result in the theory of Erdös–Rényi random graphs (Bollobás 1985, Section VII.2), but has an entirely different proof. We assume throughout this section, as in the previous section, that $d \geq 2$, that $f = f_U$, and that $\|\cdot\|$ is either an l_p norm with $1 < p \leq \infty$, or a toroidal metric based on an arbitrary norm.

Theorem 13.17 *Let $k \in \mathbb{N} \cup \{0\}$. Then*

$$\lim_{n\to\infty} P[T_{k+1}(\mathcal{X}_n) = M_{k+1}(\mathcal{X}_n)] = 1. \tag{13.45}$$

Thus, with high probability for n large, if one starts with isolated points and then adds edges connecting the points of $\mathcal{X}_n$ in order of increasing length, then the resulting graph becomes $(k+1)$-connected at the same instant when it achieves a minimum degree of $k+1$. This is illustrated (for $k = 0$) by the realization shown in Fig. 1.1, where the graph still has an isolated vertex just before sufficiently long edges are added for it to become connected.

In the proof it is convenient to use the specific choice of $r_n(\alpha)$, satisfying (13.35) with $\beta = e^{-\alpha}$, that was identified in the proof of Theorem 8.4. For points in the cube, the specific choice was given by (8.36) for $k+1 < d$, by (8.37) in the case $k+1 > d$, and by (8.39) in the case $k+1 = d$. That is, with $\gamma_1 = \gamma_1(d,k)$ and $\gamma_2 = \gamma_2(d,k)$ defined in Lemma 8.5, in the three respective cases we define $r_n = r_n(\alpha)$ by

$$\frac{n\theta r_n^d}{2^{d-2}} = \frac{2}{d}\log n + (k+3-d-(2/d))\log\left(\frac{2^{d-1}}{\theta d}\log n\right) + \log(\gamma_2) + \alpha, \qquad (13.46)$$

$$\frac{n\theta r_n^d}{2^{d-1}} = \frac{\log n}{d} + (k+2-d-(1/d))\log\left(\frac{2^{d-1}\log n}{d\theta}\right) + \log(\gamma_1) + \alpha, \qquad (13.47)$$

and

$$\begin{aligned}\frac{n\theta r_n^d}{2^{d-1}} = {} & \frac{\log n}{d} + \left(1 - \frac{1}{d}\right)\log\left(\frac{2^{d-1}\log n}{d\theta}\right) \\ & - \log\left(\frac{\gamma_1}{2\gamma_2}\left(\sqrt{4(\gamma_2/\gamma_1^2)\mathrm{e}^{-\alpha}+1} - 1\right)\right),\end{aligned} \qquad (13.48)$$

respectively. For points in the torus, we use the choice of r_n used in the proof of Theorem 8.3, that is,

$$n\theta r_n^d = \log n + k\log(\log n) - \log(k!) + \alpha. \qquad (13.49)$$

For $r_n(\alpha)$ defined in this way, it is then immediate that for any $-\infty < \alpha < \alpha' < \infty$, we have $r_n(\alpha) < r_n(\alpha')$.

Lemma 13.18 *There is a constant c_3 such that for all α, α' with $-\infty < \alpha < \alpha' < \infty$, with $r_n(\alpha)$ as just described,*

$$\limsup_{n\to\infty}\sup_{x\in C} n(F(B(x;r_n(\alpha'))) - F(B(x;r_n(\alpha)))) < c_3(\alpha' - \alpha). \qquad (13.50)$$

Proof By (13.39), for all large enough n and all $x \in C$,

$$\frac{n(F(B(x;r_n(\alpha'))) - F(B(x;r_n(\alpha))))}{\alpha' - \alpha} \leq \frac{n\theta(r_n(\alpha')^d - r_n(\alpha)^d)}{\alpha' - \alpha}, \qquad (13.51)$$

which is clearly uniformly bounded if r_n is defined by (13.46), (13.47), or (13.49). When using (13.48), by the mean value theorem (see, e.g., Hoffman (1975)) the right-hand side of (13.51) is bounded by

$$2^{d-1}\sup_{x\in\mathbb{R}}\frac{\mathrm{d}}{\mathrm{d}x}\left(-\log\left(\frac{\gamma_1}{2\gamma_2}\left(\sqrt{4(\gamma_2/\gamma_1^2)\mathrm{e}^{-x}+1} - 1\right)\right)\right).$$

By an exercise in calculus, this derivative tends to a constant as $x \to \infty$ or $x \to -\infty$, and therefore by continuity it remains uniformly bounded. □

Lemma 13.19 *Let $-\infty < \alpha < \alpha' < \infty$ with $\alpha' - \alpha \leq 1$. Let $H_n(\alpha, \alpha')$ denote the number of points X of $\mathcal{X}_n$ with at most k other points of $\mathcal{X}_n$ in $B(X; r_n(\alpha))$ and at least two points of $\mathcal{X}_n$ in $B(X; r_n(\alpha')) \setminus B(X; r_n(\alpha))$. Then, for all $\alpha < \alpha'$,*

$$\limsup_{n\to\infty} E[H_n(\alpha, \alpha')] \leq c_3^2 (\alpha' - \alpha)^2 \exp(c_3(\alpha' - \alpha)) \mathrm{e}^{-\alpha'}. \tag{13.52}$$

Proof Writing just V_x for $F(B(x; r_n(\alpha)))$ and V'_x for $F(B(x; r_n(\alpha')))$ similarly, we have

$$\begin{aligned} &E[H_n(\alpha, \alpha')] \\ &= n \sum_{j=0}^{k} \sum_{m=2}^{n-1-j} \frac{(n-1)!}{j!m!(n-1-m-j)!} \int_C V_x^j (V'_x - V_x)^m (1 - V'_x)^{n-1-m-j} \mathrm{d}x \\ &\leq \sum_{j=0}^{k} \sum_{m \geq 2} n^{1+j+m} \int_C \mathrm{d}x (V_x^j / j!) \left(\left(\frac{V'_x - V_x}{1 - V'_x} \right)^m / m! \right) \\ &\qquad \times \exp(V'_x(-n + 1 + j)). \end{aligned}$$

By Lemma 13.18 and the fact that $r_n(\alpha') \to 0$, we have

$$\limsup_{n\to\infty} \sup_{x\in C} \left(\frac{n(V'_x - V_x)}{1 - V'_x} \right) \leq c_3(\alpha' - \alpha).$$

By the bound $\mathrm{e}^t - 1 - t \leq t^2 \mathrm{e}^t$ for $t \geq 0$,

$$\limsup_{n\to\infty} \sup_{x\in C} \sum_{m\geq 2} \left(\frac{n(V'_x - V_x)}{1 - V'_x} \right)^m / m! \leq c_3^2 (\alpha' - \alpha)^2 \exp(c_3(\alpha' - \alpha)).$$

Since also $V_x^j \leq (V'_x)^j$, and $r_n(\alpha') \to 0$,

$$\limsup_{n\to\infty} E H_n(\alpha, \alpha') \leq c_3^2 (\alpha' - \alpha)^2 \mathrm{e}^{c_3(\alpha' - \alpha)} \sum_{j=0}^{k} \int_C \frac{n(nV'_x)^j}{j!} \exp(-nV'_x) \mathrm{d}x.$$

By the defining property (13.35) of r_n, the kth term in the sum in the last expression converges to $\mathrm{e}^{-\alpha'}$, and all lower terms ($j < k$) tend to zero because $nr_n^d \to \infty$, so (13.52) follows. □

Proof of Theorem 13.17 We use a 'squeezing argument'. Let $\varepsilon > 0$. Choose $I \in \mathbb{N}$ and $\alpha_1 < \alpha_2 < \cdots < \alpha_I$ such that $\exp(-\alpha_I) < \varepsilon$, such that $\exp(-\mathrm{e}^{-\alpha_1}) < \varepsilon$, such that $\alpha_{i+1} - \alpha_i < 1$ for each i, and such that

$$c_3^2 \exp(c_3) \sum_{i=1}^{I-1} (\alpha_{i+1} - \alpha_i)^2 \exp(-\alpha_1) < \varepsilon. \tag{13.53}$$

For each α let the sequence $(r_n(\alpha), n \geq 1)$ be defined by (13.46), (13.47), or (13.48), for points in the cube according to whether $k + 1 < d, k + 1 > d$, or

$k+1 = d$; let $r_n(\alpha)$ be defined by (13.49) for points on the torus. In each case, $r_n(\alpha)$ is such that $n\int_C \psi^k_{n,r_n(\alpha)}(x)\mathrm{d}x$ converges to $\mathrm{e}^{-\alpha}$. By (13.37), for $i = 1, 2, \ldots, I$,

$$\lim_{n\to\infty} P[M_{k+1}(\mathcal{X}_n) \leq r_n(\alpha_i) < T_{k+1}(\mathcal{X}_n)] = 0. \tag{13.54}$$

It remains to consider the possibility that $M_{k+1}(\mathcal{X}_n)$ and $T_{k+1}(\mathcal{X}_n)$ are distinct, but are squeezed between the same pair α_i, α_{i+1}. Define the event

$$Q_n(i) := \{r_n(\alpha_i) < M_{k+1}(\mathcal{X}_n) < T_{k+1}(\mathcal{X}_n) \leq r_n(\alpha_{i+1})\}.$$

Suppose that $Q_n(i)$ occurs, and also that the inter-point distances of $\mathcal{X}_n$ are distinct. Then there is a unique pair $\{X, Y\} \subseteq \mathcal{X}_n$ with $\|X-Y\| = T_{k+1}(\mathcal{X}_n)$, and it is possible to remove k vertices from $G(\mathcal{X}_n; T_{k+1}(\mathcal{X}_n))$ leaving the remaining graph connected, but disconnected if additionally the edge joining X to Y is removed. Removing the same set of vertices from $G(\mathcal{X}_n; r_n(\alpha_i))$ leaves X and Y in distinct components, and if also events $E_n(K; \mathrm{e}^{-\alpha_i})$ and $F_n(K; \mathrm{e}^{-\alpha_i})$ (defined in Propositions 13.12 and 13.13) fail to occur, then X or Y must have at most k points within distance $r_n(\alpha_i)$. But X has at least $k+2$ points within distance $r_n(\alpha_{i+1})$, as does Y, since its $(k+1)$st nearest neighbour lies within distance $M_{k+1}(\mathcal{X}_n)$, and also by assumption $\|X - Y\| = T_{k+1}(\mathcal{X}_n) \leq r_n(\alpha_{i+1})$. To sum up this discussion, recalling the definition of $H_n(\alpha, \alpha')$ in Lemma 13.19, we have for any $K > 0$ that

$$Q_n(i) \subseteq E_n(K; \mathrm{e}^{-\alpha_i}) \cup F_n(K; \mathrm{e}^{-\alpha_i}) \cup \{H_n(\alpha_i, \alpha_{i+1}) \geq 1\}.$$

By Propositions 13.12 and 13.13, Lemma 13.19, and Markov's inequality,

$$\limsup_{n\to\infty} P[Q_n(i)] \leq c_3^2 \exp(c_3)(\alpha_{i+1} - \alpha_i)^2 \exp(-\alpha_1).$$

Hence, by (13.53),

$$\limsup_{n\to\infty} P\left[\bigcup_{i=1}^{I-1} Q_n(i)\right] < \varepsilon. \tag{13.55}$$

By Theorem 8.1 and the conditions given on α_1 and α_I,

$$\lim_{n\to\infty} P[M_{k+1}(\mathcal{X}_n) \leq r_n(\alpha_1)] = \exp(-\mathrm{e}^{-\alpha_1}) < \varepsilon, \tag{13.56}$$

and

$$\lim_{n\to\infty} P[M_{k+1}(\mathcal{X}_n) > r_n(\alpha_I)] = 1 - \exp(-\mathrm{e}^{-\alpha_I}) \leq \mathrm{e}^{-\alpha_I} < \varepsilon. \tag{13.57}$$

By (13.54)–(13.57), $\limsup_{n\to\infty} P[M_{k+1}(\mathcal{X}_n) < T_{k+1}(\mathcal{X}_n)] < 3\varepsilon$. Since $\varepsilon > 0$ is arbitrary, $P[M_{k+1}(\mathcal{X}_n) < T_{k+1}(\mathcal{X}_n)] \to 0$ as $n \to \infty$, and together with (13.1) this gives us (13.45). □

For the record, we give the convergence in distribution results for $M_k(\mathcal{X}_n)$, for uniformly distributed points in the torus or cube. Let Z be a random variable with the double exponential extreme-value distribution, that is, with $P[Z \leq \alpha] = \exp(-e^{-\alpha})$ for all $\alpha \in \mathbb{R}$.

Corollary 13.20 *Let $\|\cdot\|$ be an arbitrary norm on $\mathbb{R}^d$, $d \geq 2$, and suppose the chosen metric on C (with opposite faces identified) is the toroidal metric* $\mathrm{dist}(x,y) = \min_{z\in\mathbb{Z}^d} \|x+z-y\|$. *Let $k \in \mathbb{N} \cup \{0\}$. Then*

$$n\theta T_{k+1}(\mathcal{X}_n)^d - \log n - k\log(\log n) + \log(k!) \xrightarrow{\mathcal{D}} Z.$$

Proof Immediate from Theorems 13.17 and 8.3. □

Corollary 13.21 *Suppose that $\|\cdot\| = \|\cdot\|_p$ with $1 < p \leq \infty$. Let θ_{d-1} be the Lebesgue measure of the unit radius l_p ball in $\mathbb{R}^{d-1}$. Let $k \in \mathbb{N} \cup \{0\}$. Then if $1 \leq k+1 < d$,*

$$\left(\frac{n\theta T_{k+1}(\mathcal{X}_n)^d}{2^{d-2}} - \frac{2}{d}\log n - (k+3-d-2d^{-1})\log(\log n)\right.$$
$$\left. + \log\left(k!\binom{d}{2}^{-1} 2^{1-k-(2/d)}\,(\theta d)^{3-d-(2/d)}\,d^k\theta_{d-1}^{d-2}\right)\right) \xrightarrow{\mathcal{D}} Z.$$

If $2 \leq d < k+1$, then

$$\left(\frac{n\theta T_{k+1}(\mathcal{X}_n)^d}{2^{d-1}} - \frac{\log n}{d} - (k+2-d-d^{-1})\log(\log n)\right.$$
$$\left. + \log\left(k!2^{2-d-(1/d)}d^{k-1}\,(d\theta)^{2-d-(1/d)}\,\theta_{d-1}^{d-1}\right)\right) \xrightarrow{\mathcal{D}} Z.$$

If $k+1 = d \geq 2$, then if we set $\tau_n := n\theta 2^{1-d}T_{k+1}(\mathcal{X}_n)^d - d^{-1}\log n - (1-d^{-1})\log(\log n)$, we have

$$\tau_n - \log\left(e^{-\tau_n} + (2d\theta)^{1-(1/d)}\theta_{d-1}^{-1}/(d-1)\right)$$
$$+ \log\left((d-2)!(2\theta)^{3-d-(2/d)}\theta_{d-1}^{d-2}d^{1-(2/d)}\right) \xrightarrow{\mathcal{D}} Z.$$

Proof Immediate from Theorems 13.17 and 8.4. □

In the special case with $k = 0$ and $d = 2$, the result simplifies to simply $n\theta T_1(\mathcal{X}_n)^2 - \log n \xrightarrow{\mathcal{D}} Z$.

13.6 Normally distributed points

This section is concerned with the connectivity threshold for points having a multivariate normal distribution. For some of the potential applications such as Rohlf's test for multivariate outliers (Rohlf 1975) it is more relevant to consider cases such as this, in which the distribution of points has unbounded support, rather than the case of uniformly distributed points.

We assume in this section that $d \geq 2$, that $\|\cdot\|$ is the Euclidean (l_2) norm, and that the underlying probability density function of the points X_i is the standard multivariate normal density function

$$\phi(x) := (2\pi)^{-d/2} \exp(-\|x\|^2/2).$$

Let Z be a random variable with the double exponential distribution, that is, with $P[Z \leq \alpha] = \exp(-\mathrm{e}^{-\alpha})$ for all $\alpha \in \mathbb{R}$. The next result says that in the standard normal case, as in the uniform case, the asymptotic distribution of the connectivity threshold is the same as that of the largest nearest-neighbour link, as given by Theorem 8.7. As in that result, we set $\log_2 n := \log(\log n)$ and $\log_3 n := \log(\log_2 n)$.

Theorem 13.22 *Suppose $f = \phi$. Then, as $n \to \infty$,*

$$T_1(\mathcal{X}_n)\sqrt{2\log n} - (d-1)\log_2 n + \left(\frac{d-1}{2}\right)\log_3 n + \log \kappa_d \xrightarrow{\mathcal{D}} Z,$$

where $\kappa_d := 2^{-d/2}(2\pi)^{-1/2}\Gamma(d/2)(d-1)^{(d-1)/2}$.

We also have a Poisson limit theorem for the number of components K_n.

Theorem 13.23 *Suppose $f = \phi$, $\alpha \in \mathbb{R}$, and suppose $(r_n)_{n\geq 1}$ is chosen so that*

$$\lim_{n\to\infty}\left(r_n\sqrt{2\log n} - (d-1)\log_2 n + \left(\frac{d-1}{2}\right)\log_3 n\right) = \alpha. \tag{13.58}$$

Then $K_n - 1 \xrightarrow{\mathcal{D}} \mathrm{Po}(\mathrm{e}^{-\alpha}/\kappa_d)$ as $n \to \infty$.

The main step towards proving these theorems is the following result.

Lemma 13.24 *Let $\alpha \in \mathbb{R}$ and let $(r_n)_{n\geq 1}$ satisfy (13.58). Let D_n be the event that $G(\mathcal{X}_n; r_n)$ has two or more components of order greater than 1. Then $\lim_{n\to\infty} P[D_n] = 0$.*

Proof If D_n occurs, then there is at least one component of $G(\mathcal{X}_n; r_n)$, of order greater than 1, in which the nearest point to $\mathbf{0}$, at X_j say, has $\|X_j\| \geq r_n/2$. This point must also satisfy

$$\mathcal{X}_n((B(X_j; r_n) \cap B(\mathbf{0}; \|X_j\|)) \setminus \{X_j\}) = 0, \quad \mathcal{X}_n(B(X_j; r_n) \setminus \{X_j\}) > 0.$$

Let $\mathcal{E}_n$ be the mean number of points X_j of $\mathcal{X}_{n-1}$ satisfying these conditions; then $P[D_n] \leq \mathcal{E}_n$, and

$$\mathcal{E}_n = n \int_{\|x\|\geq r_n/2} P[\{\mathcal{X}_{n-1}(B(x; r_n) \cap B(\mathbf{0}; \|x\|)) = 0\} \cap \{\mathcal{X}_{n-1}(B(x; r_n)) > 0\}]\, \phi(x) \mathrm{d}x. \tag{13.59}$$

For $r_n/2 \leq \|x\| \leq 1$, the ball of radius $r_n/4$, centred at $(\|x\| - r_n/4)\|x\|^{-1}x$, is contained in the set $B(x; r_n) \cap B(\mathbf{0}; \|x\|)$. Therefore, there exists $c > 0$ such that

$$P[\mathcal{X}_{n-1}(B(x; r_n) \cap B(\mathbf{0}; \|x\|)) = 0] \leq \exp(-ncr_n^d), \qquad r_n \leq \|x\| \leq 1.$$

Thus, if we set γ_n to be the contribution to $\mathcal{E}_n$ from x with $\|x\| \leq 1$, that is,

$$\gamma_n := n \int_{r_n/2 \leq \|x\| \leq 1} P[\{\mathcal{X}_{n-1}(B(x; r_n) \cap B(\mathbf{0}; \|x\|)) = 0\} \cap \{\mathcal{X}_{n-1}(B(x; r_n)) > 0\}]\, \phi(x) \mathrm{d}x,$$

then $\gamma_n \leq n \exp(-ncr_n^d)$ which converges to zero.

Now consider $\|x\| \geq 1$. As at (8.43) (and as illustrated in Fig. 8.1), let $B_\delta(x; r) := \{y \in B(x; r) : \|x\|^{-1} x \cdot y \leq \|x\| - (1 - \delta) r\}$, where $x \cdot y$ is the Euclidean inner product. Also, as at (8.44), set $I(x; r) := F(B(x; r))$ and set $I_\delta(x; r) := F(B_\delta(x; r))$. We can (and do) pick $\delta \in (0, 1)$ such that

$$B_\delta(x; r) \subseteq B(x; r) \cap B(\mathbf{0}; \|x\|) \qquad \text{if } \|x\| \geq 1, r \leq \tfrac{1}{2}.$$

Also for $\|x\| \geq 1$ and $0 < r \leq \frac{1}{2}$,

$$\phi(y) \leq (2\pi)^{-d/2} \exp(-(\|x\| - (1 - \delta) r)^2 / 2), \qquad y \in B(x; r) \setminus B_\delta(x; r),$$

and hence, setting $c = \theta(2\pi)^{-d/2}$ here, we have

$$\begin{aligned} &P[\mathcal{X}_{n-1}(B(x; r_n)) > 0 | \mathcal{X}_{n-1}(B(x; r_n) \cap B(\mathbf{0}; \|x\|)) = 0] \\ &\leq \frac{nF(B(x; r_n) \setminus B(\mathbf{0}; \|x\|))}{1 - F(B(x; r_n) \cap B(\mathbf{0}; \|x\|))} \leq 2n(I(x; r_n) - I_\delta(x; r_n)) \\ &\leq 2cnr_n^d \exp((1 - \delta) r \|x\| - \|x\|^2 / 2), \end{aligned}$$

so that

$$\begin{aligned} &P[\{\mathcal{X}_{n-1}(B(x; r_n) \cap B(\mathbf{0}; \|x\|)) = 0\} \cap \{\mathcal{X}_{n-1}(B(x; r_n)) > 0\}] \\ &\leq \exp(-nI_\delta(x; r_n)) v_n(\|x\|) \qquad \text{if } \|x\| \geq 1, \end{aligned} \tag{13.60}$$

with

$$v_n(\|x\|) := 3 \min\left(1, ncr_n^d \exp((1 - \delta) r_n \|x\| - \|x\|^2 / 2)\right).$$

As in Section 8.3, set $a_n := \log n + ((d/2) - 1) \log_2 n - \log(\Gamma(d/2))$, and set $\rho_n(t) := (2(t + a_n))^{1/2}$. By an argument similar to the one leading up to (8.54), with g_n defined at (8.51)

$$\begin{aligned} \mathcal{E}_n &\leq \gamma_n + n \int_{\|x\| \geq 1} \exp(-nI_\delta(x; r_n)) v_n(\|x\|) \phi(x) \mathrm{d}x \\ &\leq \gamma_n + \int_{-a_n}^{\infty} \exp(-nI_\delta(\rho_n(t); r_n)) v_n(\rho_n(t)) g_n(t) \mathrm{d}t, \end{aligned} \tag{13.61}$$

The integrand converges pointwise to zero since by (8.57), (8.58) and (8.60),

$$v_n(\rho_n(t)) \leq c' \left(\frac{(\log_2 n)^d}{(\log n)^{d/2}}\right) \mathrm{e}^{-t}(\log n)^{1-d/2}((\log n)^{d-1}(\log_2 n)^{(1-d)/2})^{(1-\delta)}$$
$$= c'\mathrm{e}^{-t}(\log n)^{-\delta(d-1)}(\log_2 n)^{d+(1-\delta)(1-d)/2},$$

which converges to zero, while the other factor remains bounded by (8.63). Also $v_n(t) \leq 3$, so the integrand in (13.61) is uniformly bounded by $g_n(t)\exp(-nI_\delta(\rho_n(t); r_n))$; thus, by the same domination argument as in the proof of Proposition 8.10, the integral in (13.61) converges to zero, hence so does $\mathcal{E}_n$ and so does $P[D_n]$. □

Proof of Theorem 13.23 Immediate from Theorem 8.13 and Lemma 13.24. □

Proof of Theorem 13.22 This is deduced from Theorem 13.23 in the same way that Theorem 8.7 is deduced from Theorem 8.13. □

13.7 The component count in the thermodynamic limit

Given any graph G, let $K(G)$ denote the number of components of G. Recall that K_n (respectively, K'_n) denotes the total number of components in a binomial (respectively, Poisson) sample, that is, $K_n = K(G(\mathcal{X}_n; r_n))$ and $K'_n := K(G(\mathcal{P}_n; r_n))$. The following law of large numbers holds for K_n.

Theorem 13.25 *Suppose $nr_n^d \to \rho \in (0, \infty)$ as $n \to \infty$. Then, as $n \to \infty$,*

$$n^{-1}K_n \xrightarrow{L^2} \int_{\mathbb{R}^d} \left(\sum_{k=1}^{\infty} k^{-1}p_k(\rho f(x))\right) f(x)\mathrm{d}x. \tag{13.62}$$

Theorem 13.25 holds for any choice of the density f. The intuition behind the result is as follows. Since K_n is a sum of contributions from each vertex, where the contribution of a vertex is the reciprocal of the order of the component containing that vertex. After re-scaling, the point process $\mathcal{X}_n$ in the vicinity of a Lebesgue point x resembles a homogeneous Poisson process $\mathcal{H}_{\rho f(x)}$, so the probability that the contribution of a vertex at x to K_n is approximately $p_k(\rho f(x))$, and hence n^{-1} converges to the right-hand side of (13.62).

We do not prove Theorem 13.25 here, but refer the reader to Penrose and Yukich (2003) for a proof. The method of that paper can also be applied to obtain a similar result for K'_n.

The main subject of this section is a central limit theorem associated with the above law of large numbers, which holds under fairly mild conditions on f.

Theorem 13.26 *Suppose that $d \geq 1$, f has bounded support and f is Riemann integrable. Suppose $nr_n^d \to \rho \in (0, \infty)$. Then there exists $\tau > 0$ such that, as $n \to \infty$, we have $n^{-1}\mathrm{Var}(K_n) \to \tau^2$ and*

$$n^{-1/2}(K_n - EK_n) \xrightarrow{\mathcal{D}} \mathcal{N}(0, \tau^2).$$

Before proving this, we give a Poissonized version of the result.

Theorem 13.27 *Suppose that $d \geq 1$, f has bounded support and f is Riemann integrable. Suppose $nr_n^d \to \rho \in (0,\infty)$, as $n \to \infty$. Then there exists $\sigma > 0$ such that, as $n \to \infty$, we have $n^{-1}\mathrm{Var}(K'_n) \to \sigma^2$ and*

$$n^{-1/2}(K'_n - EK'_n) \xrightarrow{\mathcal{D}} \mathcal{N}(0,\sigma^2).$$

The proof of Theorem 13.27 yields a formula for σ^2; it is given by the right-hand side of (13.71) below. The proof is related to that of Theorem 10.22 (central limit theorem for the order of the largest component), but in some ways the present quantity of interest (number of components) is easier to deal with; the increment in the number of components due to the insertion of an extra point is uniformly bounded, and also the 'stability' property that we shall describe below is stronger than that which holds for the order of the largest component. These technical advantages are reflected in the fact that here, unlike in Theorem 10.22, we consider a general class of underlying probability density functions, and not just the case $f = f_{\mathrm{U}}$.

The proof of Theorem 13.27 uses the following notion of 'stability'. For $x \in \mathbb{R}^d$, let $C(x;r)$ be the (rectilinear) cube of side r centred at x, that is, set $C(x;r) := B(r) \oplus \{x\}$, where $B(r)$ is the cube of side r centred at the origin as at (9.11). For $x \in \mathbb{R}^d$, $r, s \in (0,\infty)$, with $s - r > 4\,\mathrm{diam}(B(\mathbf{0};r))$, we shall say that a finite set $\mathcal{X} \subset C(x;s)$ is (x,r,s)*-stable* if at most a single component of the graph $G(\mathcal{X} \setminus C(x;r);r)$ has a vertex set that comes within distance r both of $C(x;r)$ and of $\mathbb{R}^d \setminus C(x;s)$, that is, at most one component approaches near to both the inner and the outer boundary of the annulus $C(x;s) \setminus C(x;r)$.

The significance of this notion of stability is that it means that any 'local' change to $\mathcal{X}$ made by changing the point configuration in $C(x;r)$ has only a 'local' effect on the number of components. To be more precise, suppose $\mathcal{X}$ is (x,r,s)-stable, let $\mathcal{Y} \subset \mathbb{R}^d \setminus C(x;s)$ and $\mathcal{W} \subset C(x;r)$ be arbitrary finite sets. Let $\mathcal{X}'$ be the set obtained by replacing the points of $\mathcal{X}$ in $C(x;r)$ with the point set $\mathcal{W}$, that is, set $\mathcal{X}' := (\mathcal{X} \setminus C(x;r)) \cup \mathcal{W}$. We assert that

$$K(G(\mathcal{X} \cup \mathcal{Y};r)) - K(G(\mathcal{X}' \cup \mathcal{Y};r)) = K(G(\mathcal{X};r)) - K(G(\mathcal{X}';r)).$$

To see this, it suffices to consider the case where $\mathcal{W}$ is the empty set. In this case, $\mathcal{X}'$ is contained entirely in the annulus $C(x;s) \setminus C(x;r)$. The effect of adding points of $\mathcal{X} \cap C(x;r)$ is to (possibly) create some new components and to (possibly) join together previously distinct components of $G(\mathcal{X}';r)$. Any two such components that could possibly get joined together in this way must reach the r-neighbourhood of $C(x;r)$, and so by the stability assumption, at most one such component reaches the r-neighbourhood of $\mathbb{R}^d \setminus C(x;s)$. Therefore, any pair of distinct components of $G(\mathcal{X}';r)$ that get joined together by the addition of points of $\mathcal{X} \cap C(x;r)$ remain distinct when we add the points of $\mathcal{Y}$, justifying the assertion above.

As in Section 9.6, let $\mathcal{H}_\lambda$ be a homogeneous Poisson process of intensity λ on $\mathbb{R}^d$, and for $s > 0$, let $\mathcal{H}_{\lambda,s}$ be its restriction to $B(s)$.

Lemma 13.28 *For $\lambda \geq 0$ and $s > 1 + 4\,\mathrm{diam}(B(\mathbf{0};1))$, let $\zeta_s(\lambda)$ be the probability that $\mathcal{H}_{\lambda,s}$ is not $(\mathbf{0},1,s)$-stable. Then, for any $\lambda_0 \in (0,\infty)$,*

$$\lim_{s\to\infty} \sup_{\lambda\in[0,\lambda_0]} \zeta_s(\lambda) = 0.$$

Proof First consider the case of fixed λ. Let E_s be the event $\mathcal{H}_{\lambda,s}$ is not $(\mathbf{0},1,s)$ stable, that is, the event that there exist two (or more) disjoint components of $G(\mathcal{H}_{\lambda,s} \setminus C(\mathbf{0};1);1)$, both of them containing elements in the 1-neighbourhoods both of $C(\mathbf{0};1)$ and of $\mathbb{R}^d \setminus C(\mathbf{0};s)$. Then E_s is a decreasing event in s, and by uniqueness of the infinite component (Theorem 9.19), $P[E_s] \to 0$ as $s \to \infty$, that is,

$$\lim_{s\to\infty} \zeta_s(\lambda) = 0. \tag{13.63}$$

Moreover, $\zeta_s(\lambda)$ is decreasing in s, for each λ, and for each s the function $\zeta_s(\lambda)$ is continuous in λ, as can be seen using the superposition theorem (Theorem 9.14).

A compactness argument using the above properties of $\zeta_s(\lambda)$ (Dini's theorem; see, e.g., Hoffman (1975)) shows that, for each λ_0, the convergence in (13.63) is uniform on the interval $[0,\lambda_0]$. □

To prove Theorem 13.27, we shall also need the following non-probabilistic result. Given a finite set $\mathcal{X} \subset \mathbb{R}^d$, set $K(\mathcal{X}) := K(G(\mathcal{X};1))$. Let C denote the unit cube $C(\mathbf{0};1)$.

Lemma 13.29 *There exists a constant $c < \infty$, depending only on the dimension and the choice of norm, such that for all finite $\mathcal{X} \subset C$, $\mathcal{Y} \subset \mathbb{R}^d \setminus C$, we have $|K(\mathcal{X} \cup \mathcal{Y}) - K(\mathcal{Y})| < c$.*

Proof For all finite $\mathcal{X} \subset C$, $\mathcal{Y} \subset \mathbb{R}^d \setminus C$, an upper bound for $K(\mathcal{X}\cup\mathcal{Y}) - K(\mathcal{Y})$ is given by the number of components of $K(\mathcal{X})$. This is bounded by the maximum number of disjoint balls of radius $\frac{1}{2}$ whose centres can be packed into C.

On the other hand, $K(\mathcal{Y}) - K(\mathcal{X} \cup \mathcal{Y})$ is bounded above by the number of components of $G(\mathcal{Y};1)$ which approach within a distance 1 of C, since the only way in which adding points in C can reduce the number of components is by connecting together such components. Thus, $K(\mathcal{Y}) - K(\mathcal{X} \cup \mathcal{Y})$ is bounded by the maximum number of disjoint balls of radius $\frac{1}{2}$ whose centres can be packed into the 1-neighbourhood of C. □

Let $\mathcal{H}'_\lambda$ be an independent copy of $\mathcal{H}_\lambda$, and for $s > 1$ set

$$\Delta_s(\lambda) := K(\mathcal{H}_{\lambda,s}) - K((\mathcal{H}_{\lambda,s} \setminus C) \cup (\mathcal{H}'_\lambda \cap C)).$$

Let A be the set of $x \in \mathbb{R}^d$ which precede or equal the point $(\frac{1}{2}, \frac{1}{2}, \ldots, \frac{1}{2})$ in the lexicographic ordering. Let $\mathcal{F}_A$ be the σ-field generated by the positions of the points of $\mathcal{H}_\lambda$ in A (cf. the proof of Theorem 9.19). Set $\tilde{D}_s(\lambda) := E[\Delta_s(\lambda)|\mathcal{F}_A]$ and $h_s(\lambda) := E[\tilde{D}_s(\lambda)^2]$.

Lemma 13.30 *The function $h_s(\lambda)$ is a (Lipschitz) continuous function of λ. Also, $h_s(\lambda)$ tends to a limit $h_\infty(\lambda)$ as $s \to \infty$.*

Proof Given λ, λ' we can couple the Poisson process $\mathcal{H}_\lambda$ to $\mathcal{H}_{\lambda'}$ and couple $\mathcal{H}'_\lambda$ to $\mathcal{H}'_{\lambda'}$ using the superposition theorem (Theorem 9.14), and use the uniform boundedness of $\tilde{D}_s(\lambda)$, $\tilde{D}_s(\lambda')$, $\Delta_s(\lambda)$, and $\Delta_s(\lambda')$ (Lemma 13.29), along with the conditional Jensen inequality, to obtain

$$\begin{aligned} E[\tilde{D}_s(\lambda)^2] - E[\tilde{D}_s(\lambda')^2] &\leq cE[\,|\tilde{D}_s(\lambda) - \tilde{D}_s(\lambda')|\,] \\ &= cE(\,|E[\Delta_s(\lambda) - \Delta_s(\lambda')|\mathcal{F}]\,|\,) \\ &\leq cE[\,|\Delta_s(\lambda) - \Delta_s(\lambda')|\,] \\ &\leq c'P[\Delta_s(\lambda) \neq \Delta_s(\lambda')] \leq c''|\lambda - \lambda'|. \end{aligned}$$

This shows that $h_s(\cdot)$ is Lipschitz continuous.

By definition, the variables $\Delta_s(\lambda), s \geq 1$, are coupled together. With this coupling, $\Delta_s(\lambda)$ tends to a limit $\Delta_\infty(\lambda)$ as $s \to \infty$, almost surely. In fact, we have $\Delta_s(\lambda) = \Delta_\infty(\lambda)$ once s is so large that for any two of the finitely many points of $\mathcal{H}_\lambda \setminus C$ lying inside the 1-neighbourhood of C, such that there is a path in $G(\mathcal{H}_\lambda \setminus C; 1)$ connecting these two points, the shortest such path is contained in $C(\mathbf{0}; s)$.

By Lemma 13.29, the quantity $\|\Delta_s(\lambda)\|_\infty$ remains uniformly bounded as $s \to \infty$. Therefore, by the conditional dominated convergence theorem (see, e.g., Williams (1991)), $\tilde{D}_s(\lambda) \to \tilde{D}_\infty(\lambda) := E[\Delta_\infty(\lambda)|\mathcal{F}]$, almost surely, and thus $h_s(\lambda) \to h_\infty(\lambda) := E[\tilde{D}_\infty(\lambda)^2]$ as $s \to \infty$. □

Proof of Theorem 13.27 Let $\mathcal{P}$ be a homogeneous Poisson process of unit intensity in $\mathbb{R}^d \times [0, \infty)$. Without loss of generality, assume $\mathcal{P}_n$ is the image, under projection onto $\mathbb{R}^d$, of the restriction of $\mathcal{P}$ to points lying under the graph of $nf(\cdot)$, that is, to points (x, t) with $t \leq nf(x)$. This is a Poisson process in $\mathbb{R}^d$ with intensity function $nf(\cdot)$, by the mapping theorem (Kingman 1993). The purpose of this construction is for coupling to certain homogeneous Poisson processes, as will become apparent later on. Let $\mathcal{P}'$ be an independent copy of $\mathcal{P}$, and let $\mathcal{P}'_n$ be the image, under projection onto $\mathbb{R}^d$, of the restriction of $\mathcal{P}$ to points lying under the graph of $nf(\cdot)$.

Given n, divide $\mathbb{R}^d$ into cubes of side r_n. Label those cubes which intersect the support of f, in the lexicographic ordering, as $C_1, C_2, \ldots, C_{k_n}$, with centres denoted $x_1, \ldots, x_{k_n}$ respectively. Let $\mathcal{F}_0$ be the trivial σ-field, and for $1 \leq i \leq k_n$ let $\mathcal{F}_i$ be the σ-field generated by the positions of all points of $\mathcal{P}$ in the union of regions $C_j \times [0, \infty)$, $1 \leq j \leq i$. Then

$$K'_n - EK'_n = \sum_{i=1}^{k_n} D_i,$$

where we set $D_i := E[K'_n|\mathcal{F}_i] - E[K'_n|\mathcal{F}_{i-1}]$. Set

$$F_i := K'_n - K(G((\mathcal{P}_n \setminus C_i) \cup (\mathcal{P}'_n \cap C_i); r_n)).$$

In other words, $-F_i$ is the increment in the number of components if one replaces the points of $\mathcal{P}_n$ lying in C_i by an independent Poisson process on C_i with intensity function $nf(\cdot)$. Then

$$D_i = E[F_i|\mathcal{F}_i].$$

By orthogonality of martingale differences, $\mathrm{Var}(K'_n) = E\sum_{i=1}^{k_n} D_i^2$. By this fact, along with the central limit theorem for martingale differences (Theorem 2.10), it suffices to prove that

$$\sup_{n \geq 1} E\left[\max_{1 \leq i \leq k_n} \left(n^{-1/2}|D_i|\right)^2\right] < \infty, \tag{13.64}$$

$$n^{-1/2} \max_{1 \leq i \leq k_n} |D_i| \xrightarrow{P} 0, \tag{13.65}$$

and

$$n^{-1} \sum_{i=1}^{k_n} D_i^2 \xrightarrow{L^1} \sigma^2 \in (0, \infty). \tag{13.66}$$

The first two of these conditions are not hard to check. Indeed, we have

$$n^{-1} E[\max_{i \leq k_n} D_i^2] \leq n^{-1} \sum_{i=1}^{k_n} E[D_i^2],$$

and by Lemma 13.29, the variables D_i are uniformly bounded by a constant depending only on the dimension and norm. Since $k_n = O(n)$, (13.64) follows.

For the second condition (13.65), use Boole's and Markov's inequalities to obtain

$$P\left[\max_{1 \leq i \leq k_n} |D_i| \leq n^{1/2}\varepsilon\right] \leq \sum_{i=1}^{k_n} \frac{E[D_i^4]}{n^2 \varepsilon^4},$$

which tends to zero since the variables D_i are uniformly bounded and $k_n = O(n)$.

It remains to prove (13.66). Let $R > 0$ be an odd integer. Set $C_{i,R} := C(x_i; Rr_n)$. Set

$$F_{i,R} := K(G(\mathcal{P}_n \cap C_{i,R}; r_n)) - K(G((\mathcal{P}_n \cap C_{i,R} \setminus C_i) \cup (\mathcal{P}'_n \cap C_i); r_n)),$$

so that $-F_{i,R}$ is the increment in the number of components if one replaces the points of $\mathcal{P}_n \cap C_{i,R}$ lying in C_i by an independent Poisson process of intensity $nf(\cdot)$ on C_i. Let $D_{i,R} := E[F_{i,R}|\mathcal{F}_i]$. Then $D_{i,R}$ is determined by the points of $\mathcal{P}_n$ and $\mathcal{P}'_n$ in $C_{i,R}$ so is independent of $D_{j,R}$ for $\|x_i - x_j\|_\infty > 2Rr_n$.

We now use the $(d+1)$-dimensional Poisson process $\mathcal{P}$ to construct a 'homogenized' approximation $\tilde{D}_{i,R}$ to $D_{i,R}$. Let $\mathcal{Q}_{i,R}$ be the image, under projection onto $\mathbb{R}^d$, of the restriction of $\mathcal{P}$ to $C_{i,R} \times [0, nf(x_i)]$, and let $\mathcal{Q}'_{i,R}$ be defined similarly using $\mathcal{P}'$ instead of $\mathcal{P}$. Then $\mathcal{Q}_{i,R}$ is a homogeneous Poisson process on the cube $C_{i,R}$, of intensity $nf(x_i)$, and is coupled to the non-homogeneous Poisson processes $\mathcal{P}_n \cap C_{i,R}$ in such a way that 'most' points in $C_{i,R}$ are common to both of these Poisson processes.

Define the variable

$$\tilde{F}_{i,R} := K(G(\mathcal{Q}_{i,R}; r_n)) - K(G((\mathcal{Q}_{i,R} \setminus C_i) \cup (\mathcal{Q}'_{i,R} \cap C_i); r_n)).$$

Set $\tilde{D}_{i,R} := E[\tilde{F}_{i,R}|\mathcal{F}_i]$. By some easy scaling, $\tilde{D}_{i,R}$ has the same distribution as $\tilde{D}_R(nr_n^d f(x_i))$.

By the coupling, $\tilde{F}_{i,R}$ differs from $F_{i,R}$ only if $\mathcal{Q}_{i,R} \neq \mathcal{P}_n \cap C_{i,R}$ or $\mathcal{Q}'_{i,R} \neq \mathcal{P}'_n \cap C_{i,R}$. Hence

$$P[\tilde{F}_{i,R} \neq F_{i,R}] \leq 2E[\mathrm{card}(\mathcal{Q}_{i,R} \Delta (\mathcal{P}_n \cap C_{i,R}))] \leq 2 \int_{C_{i,R}} |nf(x) - nf(x_i)| \mathrm{d}x.$$

Also, by the conditional Jensen inequality and the fact that the variables $F_{i,R}$, $\tilde{F}_{i,R}$, $D_{i,R}$ and $\tilde{D}_{i,R}$ are uniformly bounded because of Lemma 13.29,

$$\begin{aligned} E[\,|D_{i,R}^2 - \tilde{D}_{i,R}^2|\,] &= E[\,|(D_{i,R} + \tilde{D}_{i,R})(D_{i,R} - \tilde{D}_{i,R})|\,] \\ &\leq cE[\,|D_{i,R} - \tilde{D}_{i,R}|\,] = cE[\,|E[F_{i,R} - \tilde{F}_{i,R}|\mathcal{F}_i]\,|\,] \\ &\leq cE[\,|F_{i,R} - \tilde{F}_{i,R}|\,] \leq c'n \int_{C_{i,R}} |f(x) - f(x_i)| \mathrm{d}x. \end{aligned} \tag{13.67}$$

By the Riemann-integrability of f and the fact that no point of $\mathbb{R}^d$ lies in more than $(R+1)^d$ of the cubes $C_{i,R}$, we find that

$$n^{-1} \sum_{i=1}^{k_n} E[|D_{i,R}^2 - \tilde{D}_{i,R}^2|] \leq c'' \sum_{i=1}^{k_n} \int_{C_{i,R}} |f(x) - f(x_i)| \mathrm{d}x \to 0, \tag{13.68}$$

as $n \to \infty$.

Since $h_R(\cdot)$ is continuous by Lemma 13.30, the function $h_R \circ f$ is Riemann-integrable. We have

$$n^{-1} E\left[\sum_{i=1}^{k_n} \tilde{D}_{i,R}^2\right] = n^{-1} \sum_{i=1}^{k_n} h_R(nr_n^d f(x_i)),$$

which converges to $\rho^{-1} \int_{\mathbb{R}^d} h_R(\rho f(x)) \mathrm{d}x$. Combined with (13.68) this gives us

$$E\left[n^{-1} \sum_{i=1}^{k_n} D_{i,R}^2\right] \to \rho^{-1} \int_{\mathbb{R}^d} h_R(\rho f(x)) \mathrm{d}x. \tag{13.69}$$

Next consider the variance. By Lemma 13.29, $\|D_{i,R}\|_\infty$ is uniformly bounded by a constant. Since $\mathrm{Cov}(D_{i,R}, D_{j,R}) = 0$ unless $\|x_i - x_j\|_\infty \leq Rr_n$, it follows that

$$n^{-2}\mathrm{Var}\sum_{i=1}^{k_n} D_{i,R}^2 \leq cn^{-2}k_n(R+1)^2.$$

This tends to zero, so

$$n^{-1}\sum_{i=1}^{k_n} D_{i,R}^2 \xrightarrow{L^2} \rho^{-1}\int_{\mathbb{R}^d} h_R(\rho f(x))\mathrm{d}x. \tag{13.70}$$

Next we take the limit $R \to \infty$. By Lemma 13.30 and dominated convergence, as $R \to \infty$,

$$\rho^{-1}\int_{\mathbb{R}^d} h_R(\rho f(x))\mathrm{d}x \to \rho^{-1}\int_{\mathbb{R}^d} h_\infty(\rho f(x))\mathrm{d}x. \tag{13.71}$$

To complete the proof, it suffices to show that

$$\lim_{R\to\infty}\left(\limsup_{n\to\infty} n^{-1}\sum_{i=1}^{k_n} E[|D_i^2 - D_{i,R}^2|]\right) = 0. \tag{13.72}$$

Given R, let $A_{i,R}$ be the event that $\mathcal{P}_n \cap C_{i,R}$ is not (x_i, r_n, Rr_n)-stable. The probability of this event is bounded by $P[\mathcal{P}_n \cap C_{i,R} \neq \mathcal{Q}_{i,R}] + P[\tilde{A}_{i,R}]$, where $\tilde{A}_{i,R}$ is the event that $\mathcal{Q}_{i,R}$ is (x_i, r_n, R)-stable. Given $\varepsilon > 0$, by Lemma 13.28 and a scaling argument, we can choose R_0 such that for all $R > R_0$, $P[\tilde{A}_{i,R}] < \varepsilon$, for all i. Since also, by the coupling, $P[\mathcal{P}_n \cap C_{i,R} \neq \mathcal{Q}_{i,R}] \leq \int_{C_{i,R}} n|f(x) - f(x_i)|\mathrm{d}x$, and since F_i, D_i, $F_{i,R}$, and $D_{i,R}$ are uniformly bounded by a constant, by an argument similar to (13.67) we have

$$\begin{aligned} n^{-1}\sum_{i=1}^{k_n} E[|D_i^2 - D_{i,R}^2|] &\leq cn^{-1}\sum_{i=1}^{k_n} P[F_i \neq F_{i,R}] \leq cn^{-1}\sum_{i=1}^{k_n} P[A_{i,R}] \\ &\leq cn^{-1}\sum_{i=1}^{k_n}\left(\varepsilon + \int_{C_{i,R}} n|f(x) - f(x_i)|\mathrm{d}x\right) \\ &= cn^{-1}k_n\varepsilon + c\sum_{i=1}^{k_n}\int_{C_{i,R}} |f(x) - f(x_i)|\mathrm{d}x. \end{aligned}$$

The first term on the right-hand side is bounded by a constant times ε because f is assumed to have bounded support, and the second term tends to zero by the Riemann-integrability assumption, as at (13.68). Since ε is arbitrary, this gives us (13.72). Combined with (13.71) this gives us (13.66) with σ^2 given by the right-hand side of (13.71). The strict positivity of σ will be verified in the course of the next proof. □

Proof of Theorem 13.26 Let $H(\mathcal{X}) := K(\mathcal{X})$, the number of components of $G(\mathcal{X};1)$. Then for $\lambda > 0$, by uniqueness of the infinite component of $G(\mathcal{H}_\lambda;1)$

(Theorem 9.19) and an argument similar to the discussion of stability just before Lemma 13.28, the functional $H(\cdot)$ is strongly stabilizing on $\mathcal{H}_\lambda$, in the sense of Definition 2.15, with limiting add one cost $\Delta(\mathcal{H}_\lambda)$ given by minus the number of distinct components of $G(\mathcal{H}_\lambda; 1)$ which include a vertex in $B(\mathbf{0}; 1)$ (or by $+1$ if there are no such components). Hence, $\Delta(\mathcal{H}_\lambda)$ has a non-degenerate distribution, for all $\lambda > 0$.

Also the change in $H(\mathcal{X})$ induced by inserting another point into $\mathcal{X}$ is uniformly bounded by a constant. Therefore by Theorem 13.27, together with the de-Poissonization result at Theorem 2.16, we get the result, including the strict inequality $\tau > 0$, which implies that also $\sigma > 0$. □

13.8 Notes and open problems

Notes *Section 13.2.* Appel and Russo (2002) proved Theorem 13.2 in the case of uniformly distributed points on the unit cube, with $k = 0$, using the l_∞ norm. All other cases of this result are new.

Section 13.3. In the special case $k = 0$, Theorem 13.7 was proved in Penrose (1999*b*). Statistical motivation for this result comes from Tabakis (1996) and is described in Penrose (1999*b*).

Sections 13.4 and 13.5. Theorems 13.10 and 13.17 are from Penrose (1999*c*). The explicit limit law in Corollary 13.20 is stated but not fully proved in Penrose (1999*c*), while the one in Corollary 13.21 is new in terms of the generality given here, although the special case with $k = 0, d = 2$ dates back to Penrose (1997). Gupta and Kumar (1998) consider the case with $k = 0, d = 2$ and points that are uniformly distributed in a *disk*; they show that for any sequence $(a_n)_{n\geq 1}$, $P[n\theta T_1(\mathcal{X}_n)^2 - \log n > a_n]$ tends to zero if and only if $a_n \to \infty$, which can be viewed as weaker version of the anticipated extension (see below) of Corollary 13.21 to points in the disk.

Section 13.6. Theorem 13.22 is from Penrose (1998). Theorem 13.23 is an extension that goes beyond Penrose (1998). Recently, Hsing and Rootzén (2002) have extended Theorem 13.22 to a general class of two-dimensional distributions having densities with a logarithm satisfying certain regularlity conditions including a form of regular variation. In particular, elliptically contoured densities such as the correlated bivariate normal are included in their result.

Rohlf (1975) proposed a test based on the connectivity threshold to look for outliers in multivariate normal data. See Simonoff (1991), Hadi and Simonoff (1993), Caroni and Prescott (1995) for more recent discussions. The use of this test has been hindered by a lack of knowledge about the distribution of the test statistic M_n; a gamma distribution with unknown parameters was suggested by Rohlf on heuristic grounds. Caroni and Prescott (1995) found in a simulation study that the gamma assumption was 'too inaccurate in the tail of the distribution'. As we have seen in Sections 13.5 and 13.6, at least in the case of uniformly or normally distributed points the asymptotic distribution of the connectivity threshold, suitably transformed, is actually the double exponential distribution. This suggests that it might be worth reassessing Rohlf's test using this distribution. However, further simulations by Caroni and Prescott (2002) suggest that the convergence in distribution given here is very slow, especially for normally distributed points.

Section 13.7. Theorems 13.27 and 13.26 are new. Its proof uses ideas in Lee (1999), where central limit theorems are proved for minimal spanning trees on non-uniformly distributed points, following similar results for uniformly distributed points in Kesten and Lee (1996), and Lee (1997). The method of proof of Theorem 13.27 is applicable elsewhere, providing, for example, an alternative approach to Theorem 3.11.

Open problems *Sections 13.4 and 13.5.* In the case $d = 2$, we know from Corollary 13.21 that $n\theta T_1(\mathcal{X}_n)^2 - \log n$ is asymptotically double exponential, for uniformly distributed points on the unit square. It seems likely that the same is true for uniformly distributed points on any two-dimensional domain with unit area and with a smooth or polygonal boundary. The result of Gupta and Kumar (1998) for uniformly distributed points in a disk is consistent with this conjecture.

An extension to Theorem 13.17 would be to show a similar equivalence between $M_{k_n}(\mathcal{X}_n)$ and $T_{k_n}(\mathcal{X}_n)$ for a sequence of integers $(k_n)_{n\geq 1}$ with k_n growing (slowly) as a function of n.

Other results similar to Theorem 13.17 which are known to be true for Erdös–Rényi random graphs but are not known for geometric graphs include the following: Asymptotic equivalence between the threshold for Hamiltonian paths and the threshold for the degree to be at least 2 (see Bollobás (1985, Theorem VIII.11)), and asymptotic equivalence between the threshold for existence of a bipartite matching and the threshold minimum degree at least 1 in a bipartite geometric random graph (see Bollobás (1985, Theorem VII.11)); if true at all, this latter equivalence will not hold except for $d \geq 3$; see Shor and Yukich (1991), which shows that for $d \geq 3$, with probability 1, the threshold for a matching is of the same order of magnitude, in probability as the threshold for the minimum degree to be at least 1.

Section 13.6. An extension of Theorem 13.22 would be to consider density functions other than ϕ. See Hsing and Rootzén (2002) for recent progress in this direction.

Section 13.7. It may be possible by an extension of the methods used here to extend Theorems 13.26 and 13.27 to cases where f has unbounded support.

REFERENCES

Alexander, K. S. (1991). Finite clusters in high density continuous percolation: compression and sphericality. *Probability Theory and Related Fields* **97**, 35–63.

Alexander, K. S., Chayes, J. T., and Chayes, L. (1990). The Wulff construction and asymptotics of the finite cluster distribution for two-dimensional Bernoulli percolation. *Communications in Mathematical Physics* **131**, 1–50.

Alon, N., Spencer, J. H., and Erdös, P. (1992). *The Probabilistic Method.* Wiley-Interscience, New York.

Ambartzumian, R. V. (1990). *Factorization Calculus and Geometric Probability.* Cambridge University Press, Cambridge.

Appel, M. J. B. and Russo, R. P. (1997*a*). The maximum vertex degree of a graph on uniform points in $[0,1]^d$. *Advances in Applied Probability* **29**, 567–581.

Appel, M. J. B. and Russo, R. P. (1997*b*). The minimum vertex degree of a graph on uniform points in $[0,1]^d$. *Advances in Applied Probability* **29**, 582–594.

Appel, M. J. B. and Russo, R. P. (2002). The connectivity of a graph on uniform points in $[0,1]^d$. *Statistics and Probability Letters* **60**, 351-357.

Appel, M. J. B., Najim, C. A., and Russo, R. P. (2002). Limit laws for the diameter of a random point set. *Advances in Applied Probability* **34**, 1–10.

Arratia, R., Goldstein, L., and Gordon, L. (1989). Two moments suffice for Poisson approximations: the Chen–Stein method. *The Annals of Probability* **17**, 9–25.

Auer, P. and Hornik, K. (1994). On the number of points of a homogeneous Poisson process. *Journal of Multivariate Analysis* **48**, 115–156.

Auer, P., Hornik, K., and Révész, P. (1991). Some limit theorems for homogeneous Poisson processes. *Statistics and Probability Letters* **12**, 91–96.

Avram, F. and Bertsimas, D. (1993). On central limit theorems in geometrical probability. *The Annals of Applied Probability* **3**, 1033–1046.

Baillo, A. and Cuevas, A. (2001). On the estimation of a star-shaped set. *Advances in Applied Probability* **33**, 717–726.

Baldi, P. and Rinott, Y. (1989). On normal approximations of distributions in terms of dependency graphs. *The Annals of Probability* **17**, 1646–1650.

Barbour, A. D. and Eagleson, G. K. (1984). Poisson convergence for dissociated statistics. *Journal of the Royal Statistical Society B* **46**, 397–402.

Barbour, A. D., Holst, L. and Janson, S. (1992). *Poisson Approximation.* Clarendon Press, Oxford.

Barraez, D., Boucheron, S., and Fernandez de la Vega, W. (2000). On the fluctuations of the giant component. *Combinatorics, Probability and Computing*

9, 287–304.

Beardwood, J., Halton, J., and Hammersley, J. M. (1959). The shortest path through many points. *Proceedings of the Cambridge Philosophical Society* **55**, 299–327.

Berry, J. W. and Goldberg, M. K. (1999). Path optimization for graph partitioning problems. *Discrete Applied Mathematics* **90**, 27–50.

Bhatt, S. N. and Leighton, F. T. (1984). A framework for solving VLSI graph layout problems. *Journal of Computer and System Sciences* **28**, 300–343.

Bhattacharya, R. N. and Ghosh, J. K. (1992). A class of U-statistics and asymptotic normality of the number of k-clusters. *Journal of Multivariate Analysis* **43**, 300–330.

Bickel, P. J. and Breiman, L. (1983). Sums of functions of nearest neighbour distances, moment bounds, limit theorems and a goodness of fit test. *The Annals of Probability* **11**, 185–214.

Billingsley, P. (1968). *Convergence of Probability Measures.* Wiley, New York.

Billingsley, P. (1979). *Probability and Measure.* Wiley, New York.

Bingham, N. H., Goldie, C. M., and Teugels, J. L. (1987). *Regular Variation.* Encyclopedia of Mathematics, vol. 27, Cambridge University Press, Cambridge.

Bock, H. H. (1996*a*). Probabilistic models in cluster analysis. *Computational Statistics and Data Analysis* **23**, 5–28.

Bock, H. H. (1996*b*). Probability models and hypotheses testing in partitioning cluster analysis. In *Clustering and Classification* (eds P. Arabie, L. J. Hubert, and G. De Soete). World Scientific, River Edge, NJ, pp. 377–453.

Bollobás, B. (1979). *Graph Theory: An Introductory Course.* Springer, New York.

Bollobás, B. (1985). *Random Graphs.* Academic Press, London.

Bollobás, B. and Leader, I. (1991). Edge-isoperimetric inequalities in the grid. *Combinatorica* **11**, 299–314.

Borgs, C., Chayes, J. T., Kesten, H., and Spencer, J. (2001). The birth of the infinite cluster: finite-size scaling in percolation. *Communications in Mathematical Physics* **224**, 153–204.

Brito, M. R., Cháves, E. L., Quiroz, A. J., and Yukich, J. E. (1997). Connectivity of the mutual k-nearest-neighbor graph in clustering and outlier detection. *Statistics and Probability Letters* **35**, 33–42.

Bui, T., Chaudhuri, S., Leighton, T., and Sipser, M. (1987). Graph bisection algorithms with good average case behavior. *Combinatorica* **7**, 171–191.

Burago, Yu. D. and Zalgaller, V. A. (1988). *Geometric Inequalities.* Springer, Berlin (Russian original 1980).

Byers, S. and Raftery, A. E. (1998). Nearest-neighbor clutter removal for estimating features in spatial point processes. *Journal of the American Statistical Association* **93**, 577–584.

Caroni, C. and Prescott, P. (1995). On Rohlf's method for the detection of outliers in multivariate data. *Journal of Multivariate Analysis* **52**, 295–307.

Caroni, C. and Prescott, P. (2002). Inapplicability of asymptotic results on the minimal spanning tree in statistical testing. *Journal of Multivariate Analysis* **83**, 487–492.

Cerf, R. (2000). *Large deviations for three dimensional supercritical percolation.* Astérisque, vol. 267, Société Mathématique de France.

Chalker, T. K., Godbole, A. P., Hitczenko, P., Radcliff, J., and Ruehr, O. G. (1999). On the size of a random sphere of influence graph. *Advances in Applied Probability* **31**, 596-609.

Clark, B. N., Colbourn, C. J., and Johnson, D. S. (1990). Unit disk graphs. *Discrete Mathematics* **86**, 165–177.

Cressie, N. (1991). *Statistics for Spatial Data.* Wiley, New York.

Deheuvels, P., Einmahl, J. H. J., Mason, D. M., and Ruymgaart, F. H. (1988). The almost sure behavior of maximal and minimal multivariate k_n-spacings. *Journal of Multivariate Analysis* **24**, 155–176.

Dette, H. and Henze, N. (1989). The limit distribution of the largest nearest-neighbour link in the unit d-cube. *Journal of Applied Probability* **26**, 67–80.

Dette, H. and Henze, N. (1990). Some peculiar boundary phenomena for extremes of rth nearest neighbor links. *Statistics and Probability Letters* **10**, 381–390.

Deuschel, J.-D. and Pisztora, A. (1996). Surface order large deviations for high-density percolation. *Probability Theory and Related Fields* **104**, 467–482.

Díaz, J., Penrose, M. D., Petit, J., and Serna, M. (2000). Convergence theorems for some layout problems on random lattice and random geometric graphs. *Combinatorics, Probability and Computing* **9**, 489–511.

Díaz, J., Penrose, M. D., Petit, J., and Serna, M. (2001*a*). Approximating layout problems on random geometric graphs. *Journal of Algorithms* **39**, 78–116.

Díaz, J., Petit, J., Serna, M., and Trevisan, L. (2001*b*). Approximating layout problems on random sparse graphs. *Discrete Mathematics* **235**, 245–253.

Diekmann, R., Monien, B., and Preis, R. (1995). Using helpful sets to improve graph bisections. In *Interconnection Networks and Mapping and Scheduling Parallel Computations* (eds D. F. Hsu, A. L. Rosenberg, and D. Sotteau). American Mathematical Society, Providence, RI. DIMACS series in discrete mathematics and theoretical computer science, vol. 21, pp. 57–73.

Dugundji, J. (1966). *Topology.* Allyn and Bacon, Boston.

Durrett, R. (1991). *Probability: Theory and Examples.* Wadsworth and Brooks/Cole, Pacific Grove.

Eilenberg, S. (1936). Sur les espaces multicohérents I. *Fundamenta Mathematicae* **27**, 153–190.

Feller, W. (1968). *An Introduction to Probability Theory and its Applications, Volume I* (3rd edn). Wiley, New York.

Feller, W. (1971). *An Introduction to Probability Theory and its Applications, Volume II* (2nd edn). Wiley, New York.

Friedman, J. H. and Rafsky, L. C. (1979). Multivariate generalizations of the Wolfowitz and Smirnov two-sample tests. *The Annals of Statistics* **7**, 697–717.

Garey, M. R. and Johnson, D. S. (1979). *Computers and Intractability: A Guide to the Theory of NP-Completeness.* W. H. Freeman, San Francisco.

Gibbs, N. E., Poole, W. E., Jr., and Stockmeyer, P. K. (1976). An algorithm for reducing the bandwidth and profile of a sparse matrix. *SIAM Journal on Numerical Analysis* **13**, 236–250.

Gilbert, E. N. (1961). Random plane networks. *Journal of the Society for Industrial Applied Mathematics* **9**, 533–553.

Glaz, J. and Balakrishnan, N. (eds) (1999). *Scan Statistics and Applications.* Birkhäuser, Boston.

Glaz, J., Naus, J., and Wallenstein, S. (2001). *Scan Statistics.* Springer, New York.

Godehardt, E. (1990). *Graphs as Structural Models* (2nd edn). Wieweg, Braunschweig.

Godehardt, E. and Jaworski, J. (1996). On the connectivity of a random interval graph. *Random Structures and Algorithms* **9**, 137–161.

Godehardt, E., Jaworski, J., and Godehardt, D. (1998). The application of random coincidence graphs for testing the homogeneity of data. In *Classification, Data Analysis, and Data Highways: Proceedings of the 21st Annual Conference of the Gesellschaft für Klassifikation e.V., University of Potsdam, 12–14 March 1997* (eds. I. Balderjahn, R. Mathar, and M. Schader). Springer, Berlin, pp. 35–45.

Gower, J. C. and Ross, G. J. S. (1969). Minimum spanning trees and single linkage cluster analysis. *Applied Statistics* **18**, 54–65.

Grimmett, G. (1999). *Percolation* (2nd edn). Springer, Berlin.

Grimmett, G. R. and Stirzaker, D. R. (2001). *Probability and Stochastic Processes* (3rd edn). Oxford University Press, Oxford.

Gupta, P. and Kumar, P. R. (1998). Critical power for asymptotic connectivity in wireless networks. In *Stochastic Analysis, Control, Optimization and Applications: A Volume in Honor of W. H. Fleming* (eds W. M. McEneany, G. Yin, and Q.Zhang). Birkhäuser, Boston, pp. 547–566.

Hadi, A. S. and Simonoff, J. S. (1993). Procedures in the identification of multiple outliers in linear models. *Journal of the American Statistical Society* **88**, 1264–1272.

Hadwiger, H. (1957). *Vorlesungen über Inhalt, Oberfläche und Isoperimetrie.* Grundlehren, Band 093, Springer, Berlin.

Hafner, R. (1972). The asymptotic distribution of random clumps. *Computing* **10**, 335–351.

Hale, W. K. (1980). Frequency assignment: theory and applications. *Proceedings of the IEEE* **68**, 1497–1514.

Hales, T. C. (2000). Cannonballs and honeycombs. *Notices of the American Mathematical Society* **47**, 440–449.

Hall, P. (1986). On powerful distributional tests based on sample spacings. *Journal of Multivariate Analysis* **19**, 201–224.

Hall, P. (1988). *Introduction to the Theory of Coverage Processes.* Wiley, New York.

Harris, B. and Godehardt, E. (1998). Probability models and limit theorems for random interval graphs with applications to cluster analysis. In *Classification, Data Analysis and Data Highways* (eds I. Balderjahn, R. Mathar, and M. Sachader). Springer, Berlin, pp. 54–61.

Hartigan, J. A. (1975). *Clustering Algorithms.* Wiley, New York.

Hartigan, J. A. (1981). Consistency of single linkage for high-density clusters. *Journal of the American Statistical Association* **76**, 388–394.

Hartigan, J. A. and Mohanty, S. (1992). The RUNT test for multimodality. *Journal of Classification* **9**, 63–70.

Henze, N. (1982). The limit distribution of the maxima of 'weighted' rth nearest-neighbour distances. *Journal of Applied Probability* **19**, 344–354.

Henze, N. (1983). Ein Asymptotisher Satz über den maximalen Minimalabstand von unabhängigen Zufallsvektoren mit Anwendung auf einen Anpassungstest im $\mathbb{R}^p$ und auf der Kugel. *Metrika* **30**, 245–259.

Henze, N. (1987). On the fraction of random points with specified nearest-neighbour interrelations and degree of attraction. *Advances in Applied Probability* **19**, 873–895.

Henze, N. and Klein, T. (1996). The limit distribution of the largest interpoint distance from a symmetric Kotz sample. *Journal of Multivariate Analysis* **57**, 228–239.

Hoffman, K. (1975). *Analysis in Euclidean Space.* Prentice-Hall, Englewood Cliffs, NJ.

Holst, L. (1980). On multiple covering of a circle with random arcs. *Journal of Applied Probability* **16**, 284–290.

Hsing, T. and Rootzén, H. (2002). Extremes on trees. Preprint, Texas A&M University and Chalmers University of Technology. `http://www.math.chalmers.se/~rootzen/`

Huang, K. (1987). *Statistical Mechanics* (2nd edn). Wiley, New York.

Illanes Mejia, A. (1985). Multicoherence and products. *Topology Proceedings* **10**, 83–94.

Jammalamadaka, S. R. and Janson, S. (1986). Limit theorems for a triangular scheme of U-statistics with applications to inter-point distances. *The Annals of Probability* **14**, 1347–1358.

Janson, S., Łuczak, T. and Rucinski, A. (2000). *Random Graphs.* Wiley, New York.

Jardine, N. and Sibson, R. (1971). *Mathematical Taxonomy.* Wiley, London.

Johnson, D. S., Aragon, C. R., McGeoch, L. A., and Schevon, C. (1989). Optimization by simulated annealing: an experimental evaluation; part I, graph partitioning. *Operations Research* **37**, 865–892.

Karlin, S. and Taylor, H. M. (1975). *A First Course in Stochastic Processes* (2nd edn). Academic Press, New York.

Karp, R. M. (1976). The probabilistic analysis of some combinatorial search algorithms. *Algorithms and Complexity: New Directions and Recent Results* (ed. J. F. Traub). Academic Press, New York, pp. 1–19.

Karp, R. M. (1977). Probabilistic analysis of partitioning algorithms for the traveling-salesman problem in the plane. *Mathematics of Operations Research* **2**, 209–224.

Karp, R. M. (1993). Mapping the genome: some combinatorial problems arising in molecular biology. *Proceedings of the Twenty-fifth Annual ACM Symposium on the Theory of Computing, San Diego, 16–18 May 1993.* ACM Press, New York, pp. 278–285.

Kesten, H. (1982). *Percolation Theory for Mathematicians.* Birkhäuser, Boston.

Kesten, H. and Lee, S. (1996). The central limit theorem for weighted minimal spanning trees on random points. *The Annals of Applied Probability* **6**, 495–527.

Kingman, J. F. C. (1993). *Poisson Processes.* Oxford University Press, Oxford.

Lang, K. and Rao, S. (1993). Finding near-optimal cuts: an empirical evaluation. In *Proceedings of the Fourth Annual ACM-SIAM Symposium on Discrete Algorithms, Austin, TX, 1993.* Association for Computing Machinery, New York; Society for Industrial and Applied Mathematics, Philadelphia, pp. 212–221.

L'Écuyer, P., Cordeau, J.-F., and Simard, R. (2000). Close-point spatial tests and their applications to random number generators. *Operations Research* **48**, 308–317.

Ledoux, M. (1996). Isoperimetry and Gaussian analysis. *Lectures on Probability Theory and Statistics: École d'été de Probabilités de Saint-Flour XXIV – 1994*: R. Dobrushin, P. Groeneboom, M. Ledoux (ed. P. Bernard). Springer, Berlin, pp. 165–294.

Lee, A. J. (1990). *U-Statistics: Theory and Practice.* Dekker, New York.

Lee, S. (1997). The central limit theorem for Euclidean minimal spanning trees I. *The Annals of Applied Probability* **7**, 996–1020.

Lee, S. (1999). The central limit theorem for Euclidean minimal spanning trees II. *Advances in Applied Probability* **31**, 969–984.

Leese, R. and Hurley, S. (eds) (2002). *Methods and Algorithms for Radio Channel Assignment.* Oxford University Press, Oxford.

Leighton, F. T. (1992). *Introduction to Parallel Algorithms and Architectures.* Morgan Kaufman, San Mateo, CA.

van Lieshout, M. N. M. (2000). *Markov Point Processes and their Applications.* Imperial College Press, London.

Ling, R. F. (1973). A probability theory of cluster analysis. *Journal of the American Statistical Association* **68**, 159–164.

McDiarmid, C. (2003). Random channel assignment in the plane. *Random Structures and Algorithms* **22**, 187–212.

McDiarmid, C. and Reed, B. (1999). Colouring proximity graphs in the plane. *Discrete Mathematics* **199**, 123–127.

McKee, T. A. and McMorris, F. R. (1999). *Topics in Intersection Graph Theory.* Society for Industrial and Applied Mathematics, Philadelphia.

McLeish, D. L. (1974). Dependent central limit theorems and invariance principles. *The Annals of Probability* **2**, 620–628.

Månsson, M. (1999). On Poisson approximation for continuous multiple scan statistics in two dimensions. *Scan Statistics and Applications* (eds J. Glaz and N. Balakrishnan). Birkhäuser, Boston, pp. 225–247.

Mardia, K. V., Kent, J. T., and Bibby, J. M. (1979). *Multivariate Analysis.* Academic Press, London.

Meester, R. and Roy, R. (1996). *Continuum Percolation.* Cambridge University Press, Cambridge.

Mitchison, G. and Durbin, R. (1986). Optimal numberings of an $n \times n$ array. *SIAM Journal on Algebraic and Discrete Methods* **7**, 571–582.

Molchanov, I. (1997). *Statistics of the Boolean Model for Practitioners and Mathematicians.* Wiley, Chichester.

Monien, B. and Sudborough, H. (1990). Embedding one interconnection network in another. *Computational Graph Theory* (eds G. Tinholfer, E. Mayr, H. Noltemeier, and M. M. Sysko). Computing Supplementum, vol. 7, Springer, Berlin, pp. 257–282.

Oesterlé, J. (2000). Densité maximale des empilements de sphères en dimension 3 (d'après Thomas C. Hales et Samuel P. Ferguson). *Astérisque* **266**, 405–413.

Pach, J. and Agarwal, P. K. (1995). *Combinatorial Geometry.* Wiley, New York.

Peierls, R. (1936). On Ising's model of ferrromagnetism. *Proceedings of the Cambridge Philosophical Society* **36**, 477–481.

Penrose, M. D. (1991). On a continuum percolation model. *Advances in Applied Probability* **23**, 536–556.

Penrose, M. D. (1995). Single linkage clustering and continuum percolation. *Journal of Multivariate Analysis* **53**, 94–109.

Penrose, M. D. (1996). Continuum percolation and Euclidean minimal spanning trees in high dimensions. *The Annals of Applied Probability* **6**, 528–544.

Penrose, M. D. (1997). The longest edge of the random minimal spanning tree. *The Annals of Applied Probability* **7**, 340–361.

Penrose, M. D. (1998). Extremes for the minimal spanning tree on normally distributed points. *Advances in Applied Probability* **30**, 628–639.

Penrose, M. D. (1999*a*). A strong law for the largest nearest-neighbour link between random points. *Journal of the London Mathematical Society. Second Series* **60**, 951–960.

Penrose, M. D. (1999*b*). A strong law for the longest edge of the minimal spanning tree. *The Annals of Probability* **27**, 246–260.

Penrose, M. D. (1999*c*). On k-connectivity for a geometric random graph. *Random Structures and Algorithms* **15**, 145–164.

Penrose, M. D. (2000*a*). Central limit theorems for k-nearest neighbour distances. *Stochastic Processes and their Applications* **85**, 295–320.

Penrose, M. D. (2000*b*) Vertex ordering and partitioning problems for random spatial graphs. *The Annals of Applied Probability* **10**, 517–538.

Penrose, M. D. (2001) A central limit theorem with applications to percolation, epidemics, and Boolean models. *The Annals of Probability* **29**, 1515–1546.

Penrose, M. D. (2002). Focusing of the scan statistic and geometric clique number. *Advances in Applied Probability* **34**, 739–753.

Penrose, M. D. and Pisztora, A. (1996). Large deviations for discrete and continuous percolation. *Advances in Applied Probability* **28**, 29–52.

Penrose, M. D. and Yukich, J. E. (2001). Central limit theorems for some graphs in computational geometry. *The Annals of Applied Probability* **11,** 1005–1041.

Penrose, M. D. and Yukich, J. E. (2003). Weak laws of large numbers in geometric probability. *The Annals of Applied Probability* **13**, 277–303.

Petit, J. (2001). *Layout Problems.* Unpublished D.Phil thesis, Departament de Llenguantges i Sistemes Informàtics, Univeritat Polytècnica de Catalunya. `http://www.lsi.upc.es/~jpetit/`.

Quintanilla, J., Torquato, S., and Ziff, R. M. (2000). Efficient measurement of the percolation threshold for fully penetrable discs. *Journal of Physics A. Mathematical and General* **33**, L399–L407.

Rintoul, M. D. and Torquato, S. (1997). Precise determination of the critical threshold and exponents in a three-dimensional continuum percolation model. *Journal of Physics A. Mathematical and General* **30**, L585–L592.

Rogers, C. A. (1951). The closest packing of convex two-dimensional domains. *Acta Mathematica* **86**, 309–321.

Rogers, C. A. (1964). *Packing and Covering.* Cambridge University Press, Cambridge.

Rohlf, F. J. (1975). Generalization of the gap test for the detection of multivariate outliers. *Biometrics* **31**, 93–101.

Roy, R. and Sarkar, A. (1992). On some questions of Hartigan in cluster analysis: an application of BK-inequality for continuum percolation. Unpublished manuscript, Indian Statistical Institute, New Delhi.

Rudin, W. (1987). *Real and Complex Analysis* (3rd edn). McGraw-Hill, New York.

Saad, Y. (1996). *Iterative Methods for Sparse Linear Systems.* PWS Publishing Company, Boston.

Sangionvanni-Vincentelli, A. (1987). Automatic layout of integrated circuits. In *Design Systems for VLSI Circuits; Logic Synthesis and Silicon compilation*, NATO Advanced Study Institute (eds G. De Micheli, A. Sangionvanni-Vincentelli, and P. Antognetti). M. Nijhoff, Dodrecht/Boston, pp. 113–195.

Santalo, L. A. (1976). *Integral Geometry and Geometric Probability.* Addison-Wesley, Reading, MA.

Shiryayev, A. N. (1984). *Probability.* Springer, New York.

Shor, P. W. and Yukich, J. E. (1991). Minimax grid matching and empirical measures. *The Annals of Probability* **19**, 1338–1348.

Shorack, G. R. and Wellner, J. A. (1986). *Empirical Processes with Applications to Statistics.* Wiley, New York.

Silverman, B. W. (1986). *Density Estimation for Statistics and Data Analysis.* Chapman and Hall, London.

Silverman, B. W. and Brown, T. (1978). Short distances, flat triangles and Poisson limits. *Journal of Applied Probability*, **15**, 815–825.

Simonoff, J. S. (1991). General approaches to stepwise identification of unusual values in data analysis. In *Directions in Robust Statistics and Diagnostics: Part II* (eds W. Stahel and S. Weisberg). Springer, New York, pp. 223–242.

Sneath, P.H.A. and Sokal, R. R. (1973). *Numerical Taxonomy: The Principles and Practice of Numerical Classification.* W. H. Freeman, San Francisco.

Solomon, H. (1967). Random packing density. *Proceedings of the Fifth Berkeley Symposium on Probability and Statistics* **3**, 119–134.

Stauffer, D. and Aharony, A. (1994). *Introduction to Percolation Theory* (2nd edn). Taylor and Francis, London.

Steele, J. M. (1997). *Probability Theory and Combinatorial Optimization.* Society for Industrial and Applied Mathematics, Philadelphia.

Steele, J. M. and Tierney, L. (1986). Boundary domination and the distribution of the largest nearest-neighbor link in higher dimensions. *Journal of Applied Probability* **23**, 524–528.

Stein, C. (1986). *Approximate Computation of Expectations.* Institute of Mathematical Statistics, Hayward, CA.

Stoyan, D., Kendall, W. S., and Mecke, J. (1995). *Stochastic Geometry and its Applications* (2nd edn). Wiley, Chichester.

Tabakis, E. (1996). On the longest edge of the minimal spanning tree. *From Data to Knowledge* (eds W. Gaul and D. Pfeifer). Springer, Berlin, pp. 222–230.

Tanemura, H. (1993). Behavior of the supercritical phase of a continuum percolation model in $\mathbb{R}^d$. *Journal of Applied Probability* **30**, 382–396.

Tanemura, H. (1996). Critical behavior for a continuum percolation model. In *Probability Theory and Mathematical Statistics: Proceedings of the Seventh Japan–Russia Symposium, Tokyo 1995* (eds S. Watanabe, M. Fukushima, Yu. V. Prohorov, and A. N. Shiryayev). World Scientific, River Edge NJ, pp. 485–495.

Torquato, S. (2002). *Random Heterogeneous Materials: Microstructure and Macroscopic Properties.* Springer, Berlin.

Turner, J. S. (1986). On the probable performance of heuristics for bandwidth minimization. *SIAM Journal on Computing* **15**, 561–580.

Weber, N. C. (1983). Central limit theorems for a class of symmetric statistics. *Mathematical Proceedings of the Cambridge Philosophical Society* **94**, 307–313.

Wells, M. T., Jammalamadaka, S. R., and Tiwari, R. C. (1993). Large sample theory of spacings statistics for tests of fit for the composite hypothesis. *Journal of the Royal Statistical Society B* **55**, 189–203.

Whitt, W. (1980). Some useful functions for functional limit theorems. *Mathematics of Operations Research* **5**, 67–85.

Williams, D. (1991). *Probability with Martingales.* Cambridge University Press, Cambridge.

Yang, K. J. (1995). *On the Number of Subgraphs of a Random Graph in* $[0,1]^d$. Unpublished D.Phil. thesis, Department of Statistics and Actuarial Science, University of Iowa.

Yukich, J. E. (1998). *Probability Theory of Classical Euclidean Optimization Problems.* Lecture Notes in Mathematics, vol. 1675, Springer, Berlin.

INDEX

add one cost, 42
adjacent, 13
almost surely, a.s., 14
ancestor, 249
Azuma's inequality, 33, 36, 78

bandwidth, 8, 260
Bernoulli process, 180
Bernoulli random variable, 16
bicoherent, 177
Bieberbach inequality, 102
bifurcation, 249
binomial random variable, 16
bisection, 8, 260
Boole's inequality, 14
Boolean model, 21
Borel–Cantelli lemma, 14
brain cortex, 8, 261
Brunn–Minkowski inequality, 102, 136

Cauchy–Schwarz inequality, 14
central limit theorem
  for Γ-component count, 65, 68
  for Γ-subgraph count, 60
  for giant component, 225, 252
  for martingale difference arrays, 34
  for subgraph count, 65
  for total component count, 309
chaining, 6
Chebyshev's inequality, 14
Chernoff bound, 16, 17
chromatic number, 109, 130
classification, v, 4
clique number, 109, 126, 134
cluster analysis, 4
cluster at the origin, 180
communications networks, v, 3, 281
comparable sequence of boxes, 226
complete convergence, 15
complete graph, 109
complete-linkage cluster, 7
component, 14, 47
compound Poisson approximation, 55
connected, A-connected, $*$-connected, r-connected, 178
connectivity of a graph, 282
connectivity regime, 10
connectivity threshold, 6, 281, 282
constant approximation algorithm, 269
continuum percolation threshold, 188
convergence in distribution, 15
convergence in distribution, 10
  for (k-)connectivity threshold, 296, 306, 307
  for largest (k-)nearest-neighbour link, 160, 161, 167
Cox process, 43
Cramér–Wold device, 15
critical value, 188
crossing, k-crossing, 200

Delaunay graph, 21
dense limiting regime, 9
dependency graph, 22
descendant, 249
diameter, 13
disk graph, 1
DNA sequence reconstruction, 261
dominated convergence theorem, 15
double exponential distribution, 160, 167, 306, 307, 316
down-set, 103, 182

Erdös–Rényi random graph, 2, 8, 19, 55, 73, 134, 194, 216, 302, 317
edge, 13
equivalence of norms, 12
ergodic theorem, 187
Euclidean norm, 12
exponential decay, 12
  for binomial distribution, 16
  for continuum percolation, 195
  for lattice percolation, 181
  for Poisson distribution, 17

Fatou's lemma, 15
feasible graph, 47
focusing, 110, 134
fractional consistency, 240
frequency assignment, 109

Gaussian process, 74
geometric graph, 1, 13
goodness-of-fit, 4
graph, 13

Hamiltonian path, 317
heierarchical clustering, 6

heuristic, 7
homogeneous Poisson process, 19

independence number, 131
independent paths, 13
induced subgraph, 47
integration by parts formula, 14
interval graph, 1
isodiametric inequality, 102
isomorphic graphs, 13
isoperimetric inequality, 103, 182

Jensen's inequality, 15

k-connectivity threshold, 282
k-edge-connected, 282
k-nearest-neighbour distance, 74
k-separated, 140

l_p norm, 12
largest k-nearest-neighbour link, 11, 136
lattice packing density, 130
law of large numbers, 10
  for Γ-component count, 69, 72
  for Γ-subgraph count, 70, 71
  for k-connectivity threshold, 284
  for chromatic number, 130, 131
  for clique number, 118, 127, 128
  for largest k-nearest-neighbour link, 137, 145
  for largest component, 199, 205, 232, 240
  for maximum degree, 118, 125
  for minimum degree, 152
  for ordering problems, 262, 275
  for smallest k-nearest-neighbour link, 121
  for vertex count of given degree, 76
layout on a graph, 259
Lebesgue density theorem, 16, 50, 52, 57, 95
Lebesgue point, 16, 49, 51
left-most point, 48
locally finite, 13

Markov's inequality, 14
martingale, 15, 33
matching, bipartite, 317
maximum degree, 109
Menger's theorem, 14
metric diameter, 205
minimal spanning tree (MST), 6, 281
minimum cut, 260
minimum linear arrangement, 8, 259
minimum sum cut, 260
minimum vertex separation, 260
Minkowski addition $\oplus$, 102

monotone increasing property, 9
multimodal, 247
multivariate normal, 16

nearest-neighbour graph, 21, 46
norm, 12
normal random variable, 16
nowhere constant, 248
NP-complete problems, 7
number of vertices of fixed degree, 55
numerical analysis, 8, 260

open cluster, open r-cluster, 180
order of a graph, 13
ordering on a graph, 259
outlier, 4, 281, 306, 316

packing density, 130
packing number, 97, 147
Palm theory, Palm point process, 19
parallel computing, 8, 260
Peierls argument, 178
percolation, 9
  continuum, 188
  lattice, 180
percolation probability, 188
Poisson approximation theorem, 22
  for subgraph count, 52
  for total number of components, 296
  for vertex count of given degree, 113, 156
  multivariate, 25
    for Γ-component count, 55
    for Γ-subgraph count, 54, 55
Poisson process, 19
Poisson random variable, 16
population cluster, 240
profile of a matrix, 261
projection layout, 269
proximity graph, 1

random d-vector, 15
random connection model, 21
range, 134, 176
regular height, 247
RUNT, 247

scaling theorem, 190
scan statistic, 4, 109, 134
simulation, 7, 261
single-linkage cluster, 5, 240
Skorohod space, Skorohod topology, 91, 94
Slutsky's theorem, 15
smallest k-nearest-neighbour link, 109
sparse limiting regime, 9
splitting, 249

stability number, 131
stabilization, 42, 46, 226
Stein(–Chen) method, 22, 23, 27
strong k-linkage cluster, 6
sub-exponential decay, 12
  for lattice percolation, 181
  for continuum percolation, 210
  for largest component, 220
subadditivity, 135, 280
subconnective regime, 10
subcritical Bernoulli process, 180
subcritical thermodynamic limit, 9
submanifold, 97
subsequence trick, 123
superconnectivity regime, 10
supercritical Bernoulli process, 180
supercritical thermodynamic limit, 9
superposition theorem, 189
support, 96

taxonomy, 4
thermodynamic limit, 9
thinning theorem, 189
threshold distance, 9
total variation distance, 15
trifurcation, 250

U-statistics, 60, 73
unicoherent, 177
uniformly integrable, 15
unimodal, 247

vertex, 13
very large scale integration (VLSI), 260

weak convergence, 10
white noise, 78